# Real World Drug Discovery

# Real World Drug Discovery
# A Chemist's Guide to Biotech and Pharmaceutical Research

Robert M. Rydzewski

**ELSEVIER**

Amsterdam • Boston • Heidelberg • London • New York • Oxford
Paris • San Diego • San Francisco • Singapore • Sydney • Tokyo

Elsevier
The Boulevard, Langford Lane, Kidlington, Oxford OX5 1GB, UK
Radarweg 29, PO Box 211, 1000 AE Amsterdam, The Netherlands

First edition 2008

**British Library Cataloguing in Publication Data**
A catalogue record for this book is available from the British Library

**Library of Congress Cataloging-in-Publication Data**
A catalog record for this book is available from the Library of Congress

ISBN: 978-0-08-046617-0

For information on all Elsevier publications
visit our web site at books.elsevier.com

Printed and bound in Slovenia
08 09 10 11 12   10 9 8 7 6 5 4 3 2 1

To Jan and Melanie
for their patience and unwavering support

# Contents

# Preface

*What's it like being in industry? What do you do all day? Do you get projects assigned? What do you look for in a job candidate?*

—Questions posed to an industrial researcher by members of an academic group[1]

Drug discovery research in both pharma and biotech has always been a magnet that attracts keen scientific minds. Every year eager science graduates and postdocs enter this great game and bring with them new talents and abilities. Those hired as medicinal chemists will know the latest synthesis techniques and be capable of making just about anything. But most new researchers will be surprised to find just how many other disciplines and types of expertise are involved. Knowing how to synthesize complex compounds is an important part of the job for a medicinal chemist and certainly provides the entry ticket. But there're more things involved in inventing new drugs than compound synthesis. Many other scientific disciplines—biochemistry, molecular biology, pharmacology, etc.—are also important in finding new drugs. These scientific disciplines are "like the fingers of a hand: of the same origin, but no longer in contact."[2] Coordinating them all so that they work together is vital to new drug discovery. And even non-scientific disciplines like law and business can have a major impact on what the researcher does.

One estimate is that it can take "up to 15 years of experience . . . to learn the skills and acquire the intuition to be successful as an independent medicinal chemist in the pharmaceutical industry."[3] This long time-frame, which roughly corresponds to the length of time it takes to get a new drug these days, isn't surprising considering all that's involved. By one analysis, requirements include "an understanding of the biology that relates to the target disease, an understanding of the pharmacological tests . . . sufficient knowledge of the factors that influence ADME [absorption, distribution, metabolism, excretion] characteristics of chemicals in vivo" as well as "an understanding of clinical medicine that pertains to the target disease; knowledge of the regulatory requirements for related drugs; a current knowledge of competitive therapies, both in market and under development by competitors; a thorough knowledge of the literature . . . familiarity with many newer technologies . . . an entrepreneurial attitude in behaving as innovator and inventor" and, not least of all, the interpersonal skills required "to interact effectively with colleagues from other disciplines to achieve project goals."[4]

The new researcher is unlikely to have studied many of these in school. The whole drug discovery process can appear to be so complex and intricate that at first he feels like an actor who's been cast in a play at the last minute on opening night and handed a script for the first time just as the curtain goes up. Those around him have been through it before and understand the nuances of how each line contributes to the play's progression while, until he's been through it himself, he's not even sure whether he's in Act II.

Having been involved in a number of such productions over the years (and pelted with just a few metaphorical tomatoes) I've always felt the need for a practical guidebook that tells the new drug discovery researcher what to expect in industry today. Of course, articles, reviews, treatises, whole volumes, even encyclopedias exist for many of the disciplines and subdisciplines the researcher will interact with. These can certainly be helpful but require more discipline to peruse and are meant to impart a much more extensive and specialized knowledge than the average researcher is likely to be interested in obtaining for each of the many fields involved. They represent a series of intellectual mountain-climbing expeditions when what's usually wanted is a brisk walk through the hills.

Even in his own field, the challenges that the researcher faces will be many. The same problems crop up over and over again in many different projects. Medicinal chemists have actually built up a rich lore over the decades about problems commonly encountered with issues like cell permeability, solubility, toxicity, etc. and how they can sometimes be addressed. Unfortunately, this seems to be largely an oral tradition handed down to the new researcher only over the course of many years and through many fellow scientists and other sources. It was hoped that putting a lot of these tips and tricks down in writing in a single place would speed up the learning process.

The purpose of this book, then, is threefold: (1) To present scientists with a basic overview of how modern industrial drug discovery works, (2) To introduce them to the many disciplines involved, why they're important and how they impact the job of the medicinal chemist, and (3) To provide some practical insights into common problems in drug discovery and ways that they can sometimes (no guarantees here!) be overcome.

All in all this is a tall order. Writing a single-author volume that covers so many different areas of expertise involves many problems, not least of which is the size of ego it would take to assume oneself an expert at them all. Jared Diamond, the author of *Guns, Germs, and Steel*, which brings together a theory based on subjects as diverse as the history of metallurgy, comparative linguistics, and plant genetics, points out that this kind of "diversity of disciplines poses problems for the would-be author" of such a book. "The author must possess a range of expertise," he writes, "so that relevant advances can be synthesized. These requirements seem at first to demand a multi-author work. Yet that approach would be doomed from the outset, because the essence of the problem is to develop a unified synthesis. That consideration dictates single authorship, despite the difficulties that it poses. Inevitably, that single author will have to sweat copiously in order to assimilate material from many disciplines, and will require guidance from many colleagues."[5]

I leave it to the readers to decide whether adequate perspiration has been shed in the writing of this book. As for guidance from colleagues, I've been consistently surprised and impressed by the gracious response to my inquiries by many experts sorely pressed for time and in some cases not knowing me from Adam. Their names and the nature of their assistance can be found in the Acknowledgments section.

Several other subjects deserve mention. This book is written as a "funnel," going from the broadest aspects of industrial drug discovery to the particulars of a real-world job therein. It begins with a background to the history, business aspects, and possible future of the industry in Chapters 1 and 2, then narrows down to practical aspects of how projects are conducted in industry in Chapters 3 and 4. Different project types are presented in Chapter 5, and the progression from hit-finding strategies through to compound optimization for in vivo experiments, the core of the medicinal chemist's day-to-day job, is covered sequentially in Chapters 6 through 10. Finally, Chapter 11 presents a perspective on drug discovery research as a career with an emphasis on the practical aspects of finding, keeping, and excelling at a job therein. A way of familiarizing the new researcher to current bestselling prescription drugs, their structures, targets, and sales figures is shown in Appendix. As these will change annually, updates along with additional material will be made available at this book's website, www.realworddrugdiscovery.com.

Hopefully all of this makes for a consistent story, but there's no reason why it needs to be read in that order and can't be "cherry-picked" for the subject of interest. This book is ultimately meant as an aid to you, the scientist, so feel free to go through it as you will, by chapter, section, subject, or drug name. One note on these: A given drug might have dozens of names, which can make for great confusion. In many books a drug will be referred to solely by its non-proprietary name. This is consistent and logical, but it provides no hint to readers trying to familiarize themselves with real-world drugs that, for example, atorvastatin is the same drug as Lipitor, a name they may hear more often. In this book non-proprietary names are always used, but for drugs familiar to many patients and/or researchers the brand name by which this author knows them are usually listed first, followed by the non-proprietary name in parentheses, as in "Lipitor (atorvastatin)".

Drug discovery research is obviously not exclusively the province of men, but unfortunately the English language makes it difficult and awkward to write many sentences without referring to the researcher's gender. Most authors (and not only those of drug discovery books!) avoid the use of he/his and she/her whenever possible and when not possible use the former. As slightly more than half the world's population is the latter, this makes no sense to me: logic would dictate use of the feminine pronouns as defaults. In this book I've tried to alternate between the genders, at least whenever I've remembered to. If this raises hackles somehow I make no excuse.

Finally, this book necessarily touches on a number of subjects that invite controversy, especially in the first two chapters. The cost of developing a new drug, possible effects of price control, the increasing outsourcing of labor, academia in the wake of the Bayh–Dole Act, and other subjects are seen in very different ways by different parties. Quite different world-views can sometimes be found in biotech and big pharma as well when it comes to their respective roles in drug discovery. In all such areas I've attempted to present as objective and unbiased a view as possible and have literally spent hours scrutinizing phrases and adjectives to try and achieve such a balance. This being the real world, though, if a roughly equal number of readers object to my perceived bias in one direction as the other, I'll feel that this has been achieved.

But in the end all such issues are irrelevant. What I truly hope to accomplish with this book has nothing to do with opinion or controversy. Instead, the measure of its success or failure will be the extent to which it helps the drug discovery scientist to understand the complex and challenging scientific and business environment in which she plays a role and to maximize her contributions to the making of new medicines, our common goal.

# ☐ Notes

1. Dr Frank Woolard relating questions posed to him by members of a U.C. Berkeley group during his sabbatical in academia. From Wilkinson, S.L. On sabbatical: A refreshing pause. *Chem. Eng. News*, April 20, 2001, pp. 22–25.

2. Janssen, P.A. Drug research. *Rev. Méd. Brux.* **1980**, 1, 643–645.

3. Greenlee, W.J., Desai, M.C. The role of medicinal chemists in drug discovery. *Curr. Opin. Drug Discovery Dev.* **2005**, 8, 419–420.

4. Lombardino, J.G., Lowe III, J.A. The role of the medicinal chemist in drug discovery—then and now. *Nat. Rev. Drug Discovery* **2004**, *3*, 853–862. This excellent review of what's involved in the job is highly recommended reading for anyone thinking of going into industrial drug discovery research.

5. Diamond, J. *Guns, Germs, and Steel: The Fates of Human Societies* (New York, W.W. Norton & Co., 1997).

# Acknowledgements

This book owes much to the many people who generously took time out of their busy schedules to provide comments, quotes, feedback, and suggestions that have helped make it a better one. In particular the author would thank the interviewees, all experts in their fields, whose insights provided both substance and "spice" for a number of sections. These were, in chronological order:

Dr Matthew S. Bogyo (Stanford University)
Dr Brian Shoichet (UCSF)
Dr Hans Maag (Roche Palo Alto)
Dr Mark Murcko (Vertex Pharmaceuticals)
Dr Guy Bemis (Vertex Pharmaceuticals)
Dr Michael C. Venuti (BioSeek)
Dr Cynthia Robbins-Roth (BioVenture Consultants)
Dr David Brown (Alchemy Biomedical Consulting)
Dr Albert I. Wertheimer (Temple University)
Dr Sarvajit Chakravarty (Medivation)
Mr David Whitman (Pharmadyn), and
Mr Wayne Montgomery (Exelixis).

Although, owing to page limitations, their comments have been "sliced and diced" herein, more complete versions of their interviews, which are well worth reading should soon be available at this book's website, www.realworlddrugdiscovery.com.

Those who read through chapters and provided feedback deserve special thanks too. This list includes Drs Barry Bunin, James M. Clark, and Hans Maag, Mr Tom M. Moran, Mr Steven Rydzewski, and Drs Jeffrey R. Spencer, David Szymkowski, Julien P. Verheyden, and Walter S. Woltosz.

Thanks are due as well to Drs Gordon Amidon, Greg Berger, Joseph DiMasi, Miklos Feher, Michael Green, Victor Hruby, Chuck Johnson, Tong Lin, James T. Palmer, Daniel H. Rich, Camille Wermuth, and Wendy Young. Special thanks are due to Dr Jan Rydzewski, whose technical advice, sound judgment, and common sense I've come to rely on time after time. Finally, the help and support of the editor, Dr Adrian Shell, as well as of Mr Derek Coleman at Elsevier are gratefully acknowledged.

It goes without saying that outside of quotes attributed to individual interviewees and others, any opinions expressed herein are entirely the author's own independent views and are not necessarily endorsed by any other person or any organization.

# About the Author

Robert M. Rydzewski conducted his first chemistry experiments at the age of 7 and has never lost his love of science since. A Chicagoan by birth, he received his B.S. and M.S. degrees in organic chemistry from DePaul University. His industrial experience in drug discovery began at G.D. Searle & Co. in 1981 and continued at Syntex Corporation, Gensia Pharmaceuticals, and Celera Genomics. In these positions he came to see the importance of cross-disciplinary learning in career growth. Starting with chemical synthesis, his interests and responsibilities expanded over the years to include authorship, supervision, inventorship, and project leadership. His publications and expertise extend to nucleosides as well as small molecule inhibitors of proteolytic enzymes, in particular cysteinyl cathepsins and proteasomes. He's had both first-hand involvement in facilitating external research with academic and government groups and the privilege of contributing to a number of exciting biotech-pharma collaborations, one of which has resulted in a current Phase III clinical candidate. Mr Rydzewski lives in the San Francisco Bay Area, and he is an independent consultant, family man, and collector of state of the art electronics from the 1920s.

# Chapter 1

# The Drug Discovery Business to Date

## 1.1 Introduction

Drug discovery research as carried out today represents the co-evolution of a number of disciplines, some scientific and some not. Although chemistry and biology lie at the heart of the process, they're certainly not the only important issues involved. The ways in which modern research is done, the organizations that do it, and the rules that govern it are the result of dynamic changes in science, business, and law that have been going on for well over a century. The goal has always been the same: finding new medicines to cure diseases and alleviate suffering. But everything else has changed over the years, including the diseases themselves, our understanding of their mechanisms and points of therapeutic intervention, the technologies available for use, the corporate structures of organizations doing such research, and the legal, economic, and regulatory constraints under which they operate.

How all of these came about might seem abstract and remote to today's focused researcher. After all, it's not likely that a research job candidate will be quizzed on the history of chemistry or asked to explain what Paragraph IV certification is. He or she is rightly going to concentrate on making a good research presentation and sounding knowledgeable about current drug discovery. So why should anyone care about how the industry got to be the way it is or how hard it might be to raise venture capital or what the implications of the Orphan Drug Act are?

Beyond the fact that good scientists—especially, these days, good drug discovery scientists—are by nature curious, there's another answer. The successful research job candidate will soon find that he has a defined role to play within the organization. Although the scientific knowledge and skills brought to that role are crucial for doing a good job, alone they're not enough: not enough to explain why a small biotech company might be trying to develop new drugs it can't afford to bring to the market; not enough to explain why a pharma company's executives might use the strange word "omics" in a derogatory sense or why a project to develop a novel drug may be competing with one that uses an old drug originally developed for a totally different indication; not enough to allow a researcher to decide whether to worry about his job disappearing or his being replaced by another researcher overseas and, if so, what additional skills might be needed to make that less likely to ever happen.

Of course, he can still do his job without knowing any of these things. His hiring and evaluations won't immediately depend on it. But in the end, understanding the broader issues involved in drug discovery will not only answer such questions but will also add value to the contributions he's able to make to the search for new medicines. To get that kind of big picture of what's going on, we need to go back in time a bit (history) and look a little farther afield (economics). The remainder of this chapter, along with the listed references, is dedicated to providing a starting point for this kind of exploration.

Since the primary purpose of this book is to help readers in their exploration of the industrial drug discovery landscape, those only interested in the scientific aspects of the business can always skip to Chapter 3 and refer back to the first two chapters when issues like repurposing, chiral switching, or pharmacogenomics arise.

## 1.2  The Past

### 1.2.1  Pharma Roots

Nowadays it's hard to see the connection between *INDs* (*Investigational New Drugs*, drug candidates being tested in the clinic) and mauve Victorian gowns, or between modern drug research facilities brimming with computer-driven robotics and the gaslit parlors of yester-years. But look hard enough and you'll find a continuum both of capital and of intellect that links today's high-tech new medicines with a lowly starting point, coal.

Before the dawn of electric lighting the gaslamp held sway, and the "gas" that was burned in these wasn't the oilfield-derived hydrocarbon we think of today, which only became available later, but was instead a product of the large-scale pyrolysis of coal. The process tended to be optimized for its valuable components, "town gas" (also known as "coal gas" or "illuminating gas") and coke, but another product was always obtained, coal tar.[1] Coke was used to fuel blast furnaces, and illuminating gas became all the rage, but the darksome, foul-smelling coal tar was initially greeted with the same enthusiasm that an organic chemist feels for the black sludges formed from reactions gone wrong. Although a few very limited uses had been found for it, for the most part it represented a disposal problem which was solved quite simply in those pre-EPA days by dumping it into the nearest river or stream.

About a century and a half ago, young William Perkin, a student of August Wilhelm von Hoffmann, set about trying to make quinine—a compound with known anti-malarial properties that could then only be obtained through isolation—by chemical synthesis starting with organic bases derived from coal tar. As part of these efforts, he tried oxidizing aniline and toluidines. You can see from Figure 1.1 that had anyone understood the complexity of quinine's structure at the time its synthesis would never have been attempted back then. In Perkin's day, when even the structure of benzene was not yet understood, turning aniline or its kin into quinine was, from our modern point of view, about as feasible as turning lead into gold. All the same, a lot of gold came out of his efforts: instead of quinine, his serendipitous experiments and artist's eye allowed him to quickly discover the first popular synthetic dye, mauve,[2] thus—shortsighted managers take note—failing in his original goal while founding an entire industry centered about organic chemistry that greatly increased the value of, and research into, coal tar.[3]

Dyes, in turn, became the basis for industrial chemical manufacturing concerns like Bayer in Germany, where Dr Felix Hoffman synthesized the analgesic and antipyretic compound Aspirin in 1897. Many dyes were found to be useful in tissue histology, where selective staining was often observed. This differential effect, observable under an ordinary micro-scope, led Dr Paul Ehrlich at the University of Strasbourg to propose the existence of different "chemoreceptors" in cells which might be exploited to cure diseases, thus laying

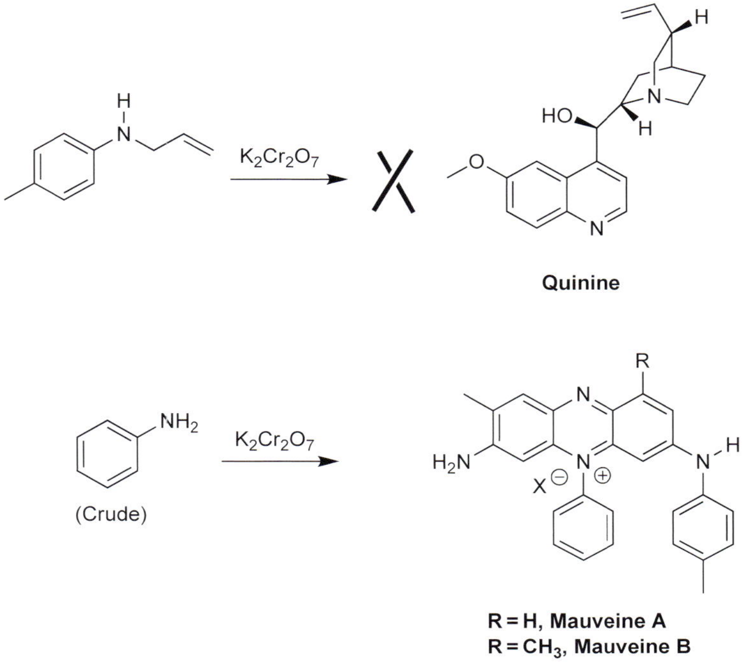

**Figure 1.1** ▶ Results of W.H. Perkin's 1856 experiments. The crude aniline used contained toluene and toluidines. The structures of mauveines were determined more than 130 years later by Meth-Cohn and Smith.[2]

the theoretical groundwork for all modern chemotherapeutic agents.[4] The concept of this kind of "magic bullet" would seem to be confirmed by Ehrlich's own arsenical drug, the anti-syphilitic Salvarsan (arsphenamine),[5] and later by the azo dye and sulfanilamide pro-drug, Prontosil (Figure 1.2). The chemical industry, initially developed to exploit the commercial potential of mauve and other dyes, found itself uniquely positioned to produce these new and profitable medicines on the necessary scale.

Of course, the path from coal tar to medicines, important though it was, was only one of several roots of modern pharma. The others that would come together in time included analytical chemistry (to reproducibly extract, isolate, and quantitate natural products like quinine, morphine, and salicylic acid obtained from plants known since ancient times to have therapeutic potential), advances in structural and synthetic organic chemistry (to allow the preparation of these and other substances), pharmacology, and animal physiology (to provide a theoretical basis for understanding the actions of new medicines as well as a practical way of testing them in animal models of human diseases). Even a brief chronicle of the many achievements that made this possible is well beyond the scope of this book. Those interested in more details might start with the listed references[6–7] for pharmaceutical industry history, other references[8–9] for an introduction to the classic source of drugs and natural products, and another one[10] for insights into the histories of the major classes of therapeutic drugs now in use.

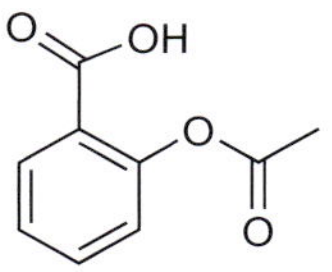

**Aspirin**

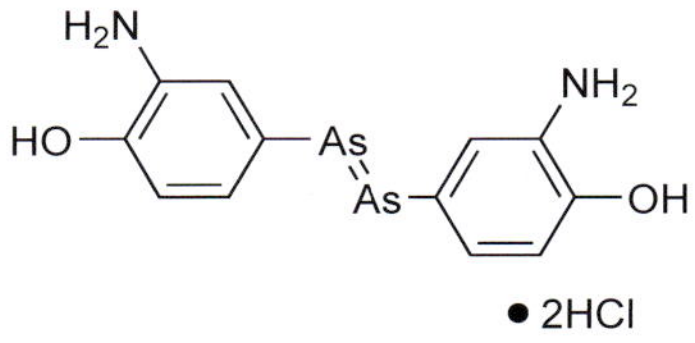

**Salvarsan**

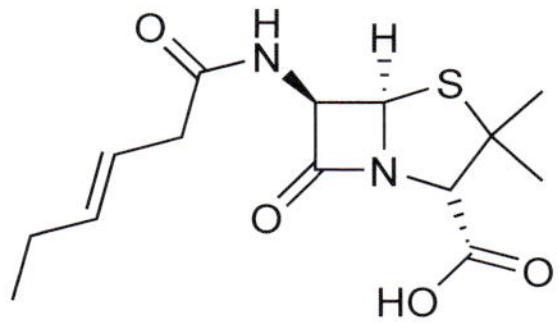

**Penicillin F**

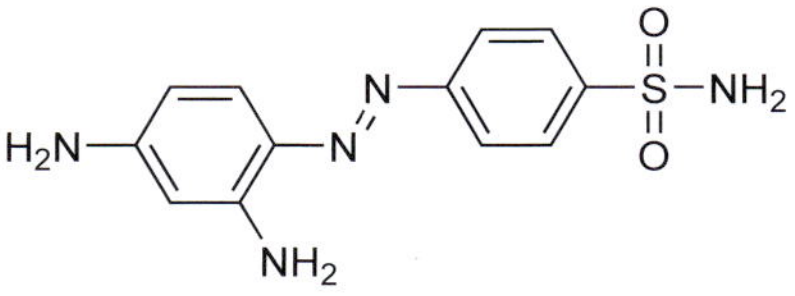

**Prontosil**

**Figure 1.2** ▶   Some early, successful drugs: Aspirin, Salvarsan (arsphenamine), Penicillin F, and Prontosil.

By the second half of the twentieth century, industrial drug discovery had come to be seen as a godsend, providing "miracle cures" like penicillin for bacterial infectious diseases that were previously incurable and often fatal. With these new treatments, the classic scourges of typhoid, cholera, pneumonia, tuberculosis, and many others were at last tamed in the industrialized world. Even diseases for which there were no known cures, like polio and smallpox, could be all but eliminated by large-scale vaccination programs. The average life span in developed countries increased as infectious diseases became a smaller and smaller part of overall mortality. Effective new antibiotics were brought to market faster than resistant bacterial strains could emerge as problems. In those heady days could anyone be faulted for daring to speculate that someday a cure for *every* disease might exist? Figure 1.3 shows some examples of the major progress made in various therapeutic areas, which was particularly rapid back then. A career in the drug industry— back when the word "drug" did not itself connote illegal narcotics—was popularly viewed as a wonderful achievement combining scientific acumen with humanitarian dedication. It seemed for a time that there were no limits to what the pharmaceutical industry could do and that mankind had found a powerful new set of benefactors including names like Merck and Lilly.

With expectations running so far in advance of anything humanly possible, it's easy to see what came next. The success of chemotherapy in treating bacterial and parasitic diseases and the increasing life span that went along with it paradoxically left the drug discovery industry facing a set of diseases much more difficult to treat: viruses, cancer, cardiovascular diseases, etc. "Magic bullets" tended to bounce off of targets that weren't living foreign organisms. Both determining the appropriate points of intervention and measuring the success of the approach became difficult when the responsible parties couldn't be seen with a microscope. Much more detailed biological work and pathway mapping, often difficult to do with the biochemical tools of the day, became necessary for the resulting programs.

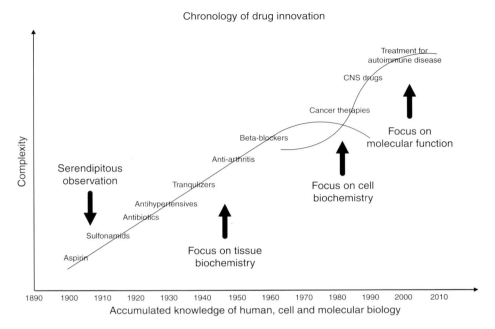

Chronology of drug innovation

**Figure 1.3 ▶** Some major therapeutic innovations by decade. (Reprinted with permission from "Biopharmaceutical Industry Contributions to State and US Economics." Available at www.milkeninstitute.org/pdf/biopharma_report.pdf, Milken Institute.)

Useful and profitable new drugs could still be obtained, but increasingly they were for indications such as contraception and CNS diseases, where chronic dosing over years and even decades would lead to increasing concerns about long-term toxicity. A new and devastating form of toxicity, teratogenicity, was encountered with thalidomide, discussed below, around 1960. The much-touted and highly financed "war on cancer," perhaps in retrospect every bit as ambitious as Perkin's attempt to make quinine from coal tar, was begun in the 1970s, but did not result in quick victory and has still not been won. Deadly new diseases began to emerge for which cures couldn't always be found, like Ebola and HIV, shaking the public's confidence in the omnipotence of the pharmaceutical industry. Old diseases like tuberculosis and staph infections came back with a vengeance in drug-resistant form. Concerns about the safety of vaccines, unethical clinical trial practices, and especially the high cost of prescription drugs came to the fore. The recession-proof profitability of the pharmaceutical industry had made it into the darling of Wall Street, and pharma, in line with all other industries, began to focus increasingly on the interests of stockholders, which certainly did not include expensive long-term research failures or drugs for impoverished third-world countries. Increased productivity was demanded from both outside as well as within the industry, with much of the burden ultimately falling upon the shoulders of the research scientist.

By the close of the twentieth century the golden age of drug discovery research had become a fading memory, to be passed on by research old-timers to fresh new faces in the industry who would need to face uncertainties and challenges unknown to their peers of a few decades before. A silver lining to these clouds, however, would come in the form of opportunities afforded by scientific breakthroughs inconceivable to the generations of Perkin and Ehrlich. The most important of these was the new recombinant DNA technology, which would give rise to an industry separate from, but inevitably allied with, big pharma: *biotechnology*.

### 1.2.2  Biotech is Born

The birth of biotechnology, of course, wasn't without an important gestation period. Many important contributions to the field of molecular biology had been made since the elucidation of the structure of DNA by Watson and Crick. Some early automated methods to put together as well as to take apart and analyze DNA and proteins needed to be in place, all of which involved major efforts and creative work. But the real cornerstone for the modern biotechnology industry was laid by Dr Herbert Boyer of UCSF and Dr Stanley Cohen of Stanford University. Their method gave scientists a practical way of producing desired proteins in cell culture by introducing the corresponding coding sequences into their DNA. Proteins, of course, had been used therapeutically at least since the discovery of insulin, but prior to Cohen's and Boyer's method in the 1970s, production of these large biomolecules was limited to isolation from tissues (like that of insulin from porcine pancreas) or chemical peptide synthesis, which could sometimes be used to make very short peptides but was useless for longer sequences.[11] Having a more practical method of protein production not only enabled the preparation of peptide therapeutics but also facilitated the production of drug targets and the discovery of new small molecules acting upon them. To screen compounds for their effects on a protein, you need the protein, and having to exhaust a slaughterhouse of its entire supply of a particular animal organ to get an animal version of it might be enough to discourage you from starting such a project in the first place.

Recombinant DNA technology would eventually become so omnipresent in discovery research that people would rarely think about it anymore, like Manhattanites going about their jobs oblivious to the technology that made the skyscrapers they work in possible. The license revenues from Boyer's and Cohen's patents would bring over a quarter billion dollars to Stanford. In 1976 the first company to exploit this technology commercially, Genentech, founded by Dr Boyer along with venture capitalist Robert Swanson, would produce revolutionary new medicines, save lives, enrich shareholders, and would eventually come to be seen as a sort of milestone in the history of modern therapeutics.

Another important contribution to what biotechnology could do was the discovery by Kohler and Milstein in the mid 1970s that normally short-lived B-cells, which didn't proliferate in cell culture, could be fused with a cell line that did, to form *hybridomas* capable of producing *monoclonal antibodies* (*mAbs*), end products of a wonderful kind of in vivo combinatorial library, that could bind with exquisite selectivity to a desired target.[12] Although early results using murine mAbs as clinical drugs turned out not to be particularly encouraging, the evolution of this technology has allowed for the production of chimeric and humanized antibodies which have been much more successful so that mAbs like Rituxan (rituximab) and Humira (adalimumab) now constitute the majority of recombinant proteins in the clinic.[13] Between recombinant DNA technology and the promise of antibody therapeutics, young biotechnology had quite an exciting package to sell.

But for the most part, the pharmaceutical industry wasn't buying it. A couple of companies, including Eli Lilly, which was intimately familiar with the limitations of protein isolation, seemed to "get it," buying recombinant human insulin from Genentech outright and making a serious effort to incorporate biotechnology into their programs. By and large, though, the industry's response was an exceedingly cool one. Rather than embrace the new technology, the consensus action was to wait and see, a response that turns out to be, statistically speaking, the appropriate one for most highly touted, brand-new technologies. But it was a major mistake for this one. The failure of pharma to aggressively colonize this new territory allowed for university professors, ex-big pharma scientists, and enthusiastic new graduates to move in and stake their claims. The biotechnology business was born, and drug discovery would never be the same.

According to Nobel laureate Dr David Baltimore, the pharmaceutical industry was "asleep at the wheel, unable to understand the profound opportunities provided by molecular biology and, therefore, unable to take advantage of them. In fact, the pharmaceutical companies were initially blind to the biotechnology revolution because drug companies were so based on making small-molecule drugs by traditional chemical means."[14]

In fairness to the pharmaceutical industry, there were a couple of mitigating factors. Although vaccines (for long a low-profit, high-liability business, but one recently showing signs of rebirth[15]) and peptide therapeutics like insulin existed long before the birth of biotechnology, they constituted a minor part of pharmaceutical sales at that point. Most drug discovery was directed at small molecules and the resulting culture was oriented more toward chemists than biologists. So when biotechnology enthusiasts spoke to pharma managers, they spoke, metaphorically, if not in a foreign language, at least with a heavy accent.

And the kind of paradigm shift that biotechnology represented was bound to be greeted with skepticism by the then-current practitioners of drug discovery. Imagine an established house painter being told that a new kind of electric wall panel that could be programmed to display any desired color had been invented and was going to make his profession, at best, an antiquarian curiosity. His first response would *not* be to go out and invest in the new products. Likewise pharmaceutical researchers at the time tended to react with denial. How could such drugs be made and purified economically? How much of a market would there be for drugs that required IV injection? How could these drugs make it to market when these new start-up companies had no experience in drug development, and nobody knew what FDA might require? These questions were all eventually answered, but it was a long time and many biopharmaceutical product launches later that the real importance of biotechnology became widely acknowledged in pharma.

In time, the acceptance of biotechnology became, for the most part, enthusiastic. The fact of the matter is that, as we'll see, new drug discovery is just so challenging that every available tool, especially one as useful as recombinant DNA technology, needs to be at the disposal of the researcher. And far from signaling the demise of the medicinal chemist, after about a decade of focusing only on proteins, biotechnology companies began to move into small-molecule research as well, thereby providing a new venue for the chemist to show her abilities, a new industry and new companies where she might work. At that point some "biotech" companies could be found that focused on small molecules from the start and did little, if any, recombinant DNA research. How they got the moniker at all is a bit mysterious, but back in the 1980s it became attached to any new start-up company that tried to make human therapeutic agents, regardless of type. "Biotech" was almost taken as a synonym for "entrepreneurial." Meanwhile, pharma companies, which had traditionally been directed toward small molecules, slowly began to incorporate recombinant proteins and monoclonal antibodies into their repertoire.[16]

Dr Cynthia Robbins-Roth, founder of BioVenture Consultants and author and founding Editor-in-Chief of BioVenture Publications, defines how the word "biotech" is used today. She says, "realistically, a biotech company is a relatively young, entrepreneurial, relatively small company that is using the tools of biotechnology to discover and develop novel products. They might be therapeutics, they might be diagnostics, they might be industrial—it doesn't matter. They might be proteins but they're probably going to be small molecules."

So if a focus on biologics versus small molecules can't always distinguish biotech from pharma other things can. Size, age, market capitalization, and sales of existing drugs were all greater for the latter, youth (both corporate and individual) and risk-taking for the former. Many scientists left big pharma in a move to what they saw as the greener fields of biotech, with stock options often being a major driving force. The story of the scientist—always someone else, alas!—who got rich on his stock options and retired by 40 was told over and over at candidate interviews. In addition, the "corporate culture" of biotech was remarkably different from that of pharma. In a company with 40 employees, in theory at least, one might have 100 times as much input as he would have in a company of 4000. It was possible for a scientist to work at a big pharma company for years and never even shake hands with the CEO, but in biotech the two of them might well be on a first name basis. Catering services flourished as biotech companies became notorious for feeding their employees well, either for free or at heavily subsidized prices, to reward and encourage their contributions. The Friday afternoon "TG" or "Ho-Ho" encouraged informal companywide communications and engendered team spirit, while the dire warnings of corporate attorneys about serving alcoholic beverages at company functions could still fall on deaf ears. Many scientists found the biotech environment stimulating and positive, but not all of them understood the price that needed to be paid.

That price could be measured in "burn rate," the dollars per year it took to keep these companies alive when so many more years remained before a profitable new drug could possibly be had. This necessitated extremely high productivity, not to mention luck, in biotech on both the individual and the corporate level. Start-up companies were funded primarily through venture capital, IPOs, and secondary stock offerings. But Wall Street's clock was ticking, and when the alarm would go off, biotech companies that looked unlikely to have drugs anytime soon would find themselves starved of capital, ultimately making them every bit as beholden to investors as big pharma, if not more so. And unlike big pharma, where a clinical failure could be counterbalanced by success with one of the many other irons in the fire, in biotech a single negative Phase II or Phase III result could, and did, wipe out most of the company's market capitalization in a single day.[17] Furthermore, funding often wasn't sufficient to cover the costs of development and clinical trials that would be the most expensive, and most critical, part of bringing new drugs to market.

Few biotechs were able to reach the critical mass of successful science, business acumen, investor enthusiasm, and subsequent market capitalization required to bring enough drugs to the clinic to have a real chance for success. As Dr Hans Maag, Vice President of Medicinal Chemistry at Roche Palo Alto, points out, "What hasn't changed is the success rate. You have to be lucky to get one compound out of the ten you put into the clinic into the market. And somehow that hasn't really changed." And so of the myriad biotechs formed by 1990s, only a few, most notably Amgen and Genentech, would succeed in becoming profitable and relatively autonomous organizations.

For most biotechs, deeper pockets with a longer-term view were needed—in two words: *big pharma*. The era of the biotech/big pharma research collaboration (and occasional acquisition) began and continues unabated today. This unique union, made possible as pharma in time began to realize the possibilities inherent in biotechnology and the industry that had grown up about it, proved capable of keeping most biotechs, if not flush with cash, at least out of bankruptcy. At the same time it proved so successful in providing *New Molecular Entities* (*NME*s, defined as a new chemical or biological therapeutic agent not previously used in man) to big pharma that licensed and acquired compounds, mostly from biotech, constitute about half the candidates in current clinical pipelines,[18] and it's been projected that by 2010 about 40% of pharmaceutical product sales will derive from drugs obtained in this way.[19]

### 1.2.3 The Genomics Revolution

In 1977, when biotech was still in its infancy, two teams, one headed by Dr Allan Maxam and Dr Walter Gilbert at Harvard and another led by Dr Frederick Sanger at the UK Medical Research Council (MRC), independently developed methods for sequencing DNA.[20] In time, advances in robotics, electrophoresis, software, and informatics and the invention of *polymerase chain reaction (PCR)*, which allowed researchers to produce copious amounts of DNA to study, would turn the laborious manual job of reading DNA sequences into an automated and incredibly fast process. As Figure 1.4 demonstrates, over time the cost of sequencing would decrease while the amount of known sequence information increased in an exponential manner. But in the 1980s, as biotech blossomed, gene sequencing wasn't there just yet.

The first complete *genome*—a new word to describe the entire DNA sequence of an organism—of a free-living organism, *Haemophilus influenza*, with its 1.8 million base pairs, was published in 1995.[21] But 10 years before that, a meeting was convened by Chancellor Robert Sinsheimer on the beautiful redwood-shaded campus of the University of California at Santa Cruz to discuss what seemed to many to be an outlandish idea considering the technology of the time: the possibility of sequencing the human genome, all 3 *billion* base pairs of it. By 1988 efforts were underway both in Europe and in America to put together a publicly funded, large-scale effort to do just that.[22] The age of genomics had begun.

In the United States, the effort initially involved the Department of Energy (DOE) and, shortly thereafter, the NIH. Increased funding was made possible by constant reminders of the importance of the work, which would provide insights into the fundamental code of life itself, and the implicit promise of medical advances to follow. Scientists with public visibility spoke of its importance. Nobel laureate James Watson, then head of the NIH part of the effort, said, "it's essentially immoral not to get it done as fast as possible."[23] Objections to the

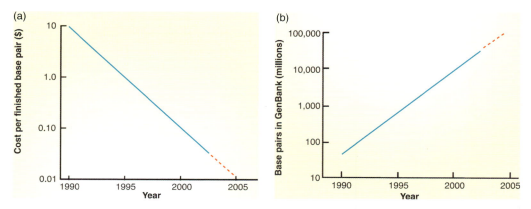

**Figure 1.4** ▶  (a) Decrease in sequencing costs, 1990–2005. (b) Increase in DNA sequence information in GenBank 1990–2005. (Reprinted with permission from Collins, F.S., et al. The human genome project: Lessons from large-scale biology. *Science* **2003**, *300*, 286–290, AAAS.)

scale of project funding, the possible diversion of funds away from other projects, and the idea that the eventual completion of the sequence would immediately lead to new cures were always present, but were drowned out by the chorus of high-tech enthusiasm. If the promise of recombinant DNA technology had not been fully appreciated back in the 1970s, the same mistake was not going to be made for genomics in the 1990s.

In boldness, scope, and excitement, the Human Genome Project (HGP) rivaled NASA's efforts to put a man on the moon in the 1960s. In keeping with the times, of course, it was an international effort, with the participation of groups like the Sanger Centre in the United Kingdom, which would go on to sequence a full one-third of the genome. And although the effort was initially viewed as completely unilateral, with the HGP in 1993 announcing plans to sequence the entire human genome by 2005, an unanticipated element of competition was introduced in May of 1998 when scientist and entrepreneur J. Craig Venter announced that with support from instrument manufacturer Perkin Elmer he was founding a new company, Celera Genomics, which would complete the project privately in 2001, ahead of the government effort. His suggestion to the HGP? "You can do mouse."[24]

The best way to understand the resulting shock is to imagine NASA suddenly finding itself in competition not with the Soviets, but instead with a group of American investors who proposed to land a man on the moon sooner and at no public expense—although they would charge people to watch the landing, auction off moon rocks, plant a corporate flag, and perhaps try to levy a small fee on anyone caught looking at it in the future! Wall Street's applause could be heard in the rustle of billions of dollars pouring into Celera and other private companies like Human Genome Sciences (HGS) and Millennium Pharmaceuticals that were associated with the new genomics business. Some discerned the hisses and boos of the publicly funded HGP in efforts to discredit the method Celera used, whole genome shotgun sequencing (WGS),[25] which it later adopted as its own standard, as well as claims that Celera's sequence was largely based on the necessarily publicly available HGP data.[26–27]

Conflicts and controversies that arose, of course, only served to heighten the general public's interest in this "race" and fan the flames of what became known as "genohype," uninformed and wild expectations about what this new technology could do.[28] And expectations were high even among scientists. According to Dr Mark Murcko, Chief Technology Officer at Vertex Pharmaceuticals, "if you go back and read the sort of comments that people were making at the time, there were a couple of different worldviews."

"Some people basically said, 'Well, we have to own all the genes because it's a gold rush!' It was the gold rush mentality: you have to stake your claim to all the genes because whoever owns the genes will own the pharmaceutical industry in the future. The reality is that own the target' is a complex concept and in reality may enable you to achieve a small royalty position. So if you own all the genes that people care about you could theoretically get a royalty on revenues from drugs that hit those targets. But you wouldn't dominate or 'shut down' the pharmaceutical industry. It's just a cost of doing business for them that they pay their small royalty to whoever holds the patent on that gene."

This gene gold rush has to date resulted in the patenting of about 20% of all human genes, with the top US patent assignee, Incyte, having rights to 2000 of them, mostly for use in microarrays.[29] Obtaining such patents is, of course, much more difficult than simply forwarding *expressed sequence tags* (*ESTs*) to the US Patent and Trademark Office (PTO)—gene function as well as potential utility must be established.[30] Even so, a recent study found that many such patent claims are problematic and thus might be challenged in court.[31] But overall, gene patenting has proven to be a real and modestly profitable offshoot of human genome sequencing, if not exactly the fabled Mother Lode.

The second major expectation about what completion of the human genome sequencing might provide has proven less tractable. Dr Murcko explains: "The other worldview was 'biology will make sense to us now.' Some people lost sight of the distinction between having a parts list and knowing what to do with it. There's that great quote from Eric Lander, one of the major players in the human genome project, and clearly he knows what he's talking about. He said, 'It's just a parts list. And if you have a parts list for a 747 that doesn't mean you can fly the plane. It doesn't mean you can build the plane. It doesn't even mean you know what the parts do. It just means you have a list of all the parts.' So the hype there was to lose the distinction between having the parts list and knowing what the parts do and how they interact with each other. In fact, to take that further, Lander also said that 'understanding the genome is the work of the coming century.' Not decade, but *century*. I think he's right."

In retrospect, having a parts list does have its advantages. Dr Murcko provided some examples: "We work in kinases quite a bit and we now know that there are roughly 500 kinases and we have all those gene sequences. When a paper pops up in the literature saying 'Kinase X plays an important role in the proliferation of such-and-such a tumor type' we can very quickly assess the similarity between kinase X and the other 500. So we can quickly answer questions like, 'Well, do we think we're likely to own molecules that inhibit kinase X?' In that sense having all the protein sequences is a useful thing. It also helps you to design reagents like RNAi. So we and everybody else can use this for target validation studies. Having the genome is useful in that sense."

But at the height of the "gene race," expectations were much higher. The confidence that novel drug discovery targets would quickly emerge, prospects for potentially lucrative gene patents, and the vision of the "personal genome" with a service to provide information on health risks and disease susceptibilities to 100 million subscribers were, in fact, major drivers behind the sequencing of the human genome.[32] At one point, several pharmaceutical and biotech companies were sufficiently convinced of the time-sensitive value of the emerging sequence to pay $25 million each to Celera for a license which included a "90-day sneak peek" at the data before it was made public.[33]

Asked what he thought the initial expectations for genomic data were, Dr Michael C. Venuti, former head of research at Arris (later Axys) Pharmaceuticals, and now CEO of BioSeek, put it wryly. "I think everyone in biotech and pharma hoped that we would take out our secret decoder ring and there'd be all kinds of things in there that would be obvious if we turned on our computers at the right time of day," he said. "I think everyone eventually came to the realization that it's what everyone did with the information that matters, and it's still what anyone does with the information that matters."

But by the time the "draft sequence" of the human genome[34–35] was declared complete and a truce between Celera and HGP was publicly celebrated at a Presidential press conference on June 26, 2000, few knew what to do with the information, and investors were beginning to sense that. Their expectations had been brought down somewhat by another Presidential press conference 3 months earlier that sounded like a call to end or restrict gene patents. Optimistic predictions on what genomics could do might still be found, for example, in one report that concluded that the "brave new genomics world" would increase R&D productivity, shave $300 million off the cost of each drug and bring it to market 2 years sooner.[36] But not every analysis of this new technology came to a similar conclusion as can be seen by the title of another report, "The Fruits of Genomics: Drug Pipelines Face Indigestion until the New Biology Ripens".[37] This predicted that "genomics will lead to the inclusion of more 'unprecedented' drug candidates in research and development pipelines. A significant portion of these candidates may progress into 'proof of concept' or Phase II testing, despite better screening technologies, before they fail. Accumulated spending to this stage will be significant, and companies will not be able to compensate for it with higher success rates in Phase III developments. Furthermore . . . new technologies are taking longer to deliver than originally expected."

In the investment community, where results from the next quarter are frequently considered long-term, taking longer than expected to deliver is not met with forgiveness. And the concern about the new technology producing a plethora of novel targets that, as we'll see in Chapter 5, historically are less likely to turn into drugs than already-known targets unfortunately proved to be "right on the money." Furthermore, even winnowing down the list of new genes to those likely to provide new drug targets was not easy: no codon specifying "drug target" was found. Only the old slow process known as *target validation* (TV), which began with correlative data and educated guesses, proceeded through in vitro and in vivo experiments, and finally ended in clinical results—good or bad—could tell you this. TV quickly became the rate-limiting step as more and more gene products piled up in the "To Do" bin.

Beginning in 2001 it became obvious to investors that true value of genomics lay not in owning the genome or running some sort of subscription informatics service, but in the old-fashioned, profitable pursuit of new prescription drugs, and, further, that genomic knowledge alone would not immediately enable this. Expectations for the rapid discovery and exploitation of new, genomic-derived targets had been dashed. Within a few years, articles with questions for titles—always a bad sign—like "Molecular genetics: the Emperor's clothes of drug discovery?",[38] "Genomics: success or failure to deliver drug targets?",[39] and "Is Genomics Advancing Drug Discovery?"[40] began to appear with regular frequency in scientific journals. What was needed was the next step, an understanding of what the various genes' products and classic targets of drugs, proteins, do in health and disease, a blueprint showing how the genomics parts list was assembled. A new name was given to this field: *proteomics.*

The cutting edge of science shifted back from DNA to peptides, and market capital followed. Plummeting stock prices for genomics companies such as Millennium and Celera, which saw about four-fifth of their market capitalization disappear in just over 1 year, and the biotech "nuclear winter" of 2002, when *initial public offerings* (*IPO*s, which raise funds for new companies) were few and funding was scarce, convinced executives that a new model was needed. Genomics companies became, like biotech and pharma, drug discovery and development companies with, perhaps, more of an emphasis on proteomics, some gene patents, and a large collection of DNA sequencers to distinguish themselves from the others. Management was reshuffled in the name of bringing in more drug development expertise. Even Craig Venter himself wasn't immune, leaving the company he founded and going on to start a non-profit research institute instead.

Sequencing of the 20,000–25,000 (not 100,000+ as predicted earlier) genes that make up the human genome was essentially completed in 2004.[41] By that time the real-world drug discovery potential of genomics, which should not be judged by its failure to live up to earlier "genohype", had largely become the province of existing pharma, biotech, and related diagnostic companies. Some of the companies founded upon genomics still exist but, despite their best efforts, have not yet managed to establish a major, independent presence in the drug discovery business to the extent that biotech companies have. Whether they will in time remains to be seen.

So together pharma, the largest, most experienced, and most profitable industry with roots going back to the nineteenth century chemical industry, and biotech, child of recombinant DNA and antibody technology which has already borne fruit in protein and antibody therapeutics have, with a certain amount of interdependence and some incorporation of genomic technology, evolved to become the major players in industrial drug discovery today.

## 1.3  Current Economics—Problems

In the real world, a kind of economic Darwinism usually ensures that companies that don't make a profit don't exist for long. In our society, the humanitarian aims of curing diseases and improving the quality and length of life need to exist in a world of economic realities. To understand current industry strategies and future directions, you need to pursue the famous Watergate dictum and "follow the money!" And the truth of the matter is that new drug discovery is expensive as well as risky. In no other industry does it cost so much or take so long to bring a new product to market. None are as heavily regulated, as uncertain, or susceptible to such unforeseeable, disastrous failures. Few other industries are as poorly understood or viewed as negatively by the public, largely because of the everyday economic reality of prescription drug prices.[42–43]

### 1.3.1   Cost of Drug Development

So how much does it cost to bring a new drug to market? Amazingly, this is one of the most controversial questions in the industry. The figure that's most frequently quoted, $802 million, stems from a study headed by Dr Joseph DiMasi at the Tufts Center for Study of Drug Development (CSDD) that was published in 2003.[44] Here 68 NMEs developed between 1983 and 1994, which had not been funded or tested at government expense and which had originated in-house, were selected and industry expense figures were used to calculate the average pre-tax cost, including the cost of projects that failed to deliver a drug. This turned out to be $403 million. However, this kind of calculation assumes that if those millions had not been spent on drug development, they would have sat in a safe or under a mattress somewhere, earning zero interest. In reality, such lazy money does not exist in the business world. So to be more realistic, one needs to figure in the loss of resulting income, which business people know as the *opportunity cost of capital*. Adding this in, based on the 10.5% return thought to be attainable in those years, gave Dr DiMasi's oft-quoted figure of $802 million (2000$) per drug.

Public Citizen, a consumer advocacy group, soon challenged this figure,[45] denying that opportunity cost was a legitimate expense and pointing out that DiMasi's numbers, being pre-tax, did not include generous R&D tax credits which would reduce the out-of-pocket amount substantially. A series of rebuttals followed and, of course, no consensus was ever reached.[46] The admitted limitations of the DiMasi figure are (1) that it wasn't representative of new approved drugs since most aren't NMEs[47] (think of controlled release versions of older drugs, etc.), (2) that it excluded drugs with government sponsorship (quite a few drugs, like AZT[48] did have this), (3) that it excluded in-licensed drugs (which many are these days), (4) that it doesn't consider tax savings, and (5) that it really represents the cost of developing drugs 10–20 years ago, not today.

Estimates of drug development costs done in recent years have ranged from Public Citizen's pre-tax figure of $341 million (2000$) to Bain & Co.'s estimate of $1.7 *billion* (2002$), which includes an average launch cost of a quarter billion dollars.[49] In a very crude, and totally unjustifiable, apples & oranges sort of way one can do a meta-analysis, averaging the pre-tax figures from 11 such studies done by Tufts CSDD, government agencies, the pharmaceutical industry, consulting groups, financial concerns, and Public Citizen since 1987, which have been collected in one of the references,[49] and arrive at a figure of $885 million. In the end, in the absence of any consensus, the $802 million figure from the DiMasi study seems as good a place to start as any in considering the pre-tax cost of bringing a self-originated NME to market. But what the "average cost of bringing a new drug to market" really is today when non-NMEs, in-licensed compounds, current expenses, and after-tax costs are taken into account—except for the fact, to which everyone agrees, that it's a very large number—remains uncertain.

Public debate about the high cost of prescription drugs has led a defensive pharmaceutical industry to emphasize the high cost of drug discovery research so that this figure is repeatedly cited and argued over in debates that usually lead to further debates on whether new-drug research or new-drug advertising costs more. But in the real world, the dollar-and-cents amounts that pharma companies get for their products are generally determined by what government agencies or, increasingly in the United States, managed care organizations are willing to pay, and R&D cost arguments are unlikely to sway those who have their hands on the purse strings.

All of this underscore a fundamental problem: The drug discovery industry increasingly seems to be caught between the hammer of private enterprise with its requirement for return on investment and the anvil of public health, which demands unrestricted access to necessary medicines. "Society posits for the industry inconsistent standards of behavior," observes one commentator. "On some occasions, lawmakers and the general public seem to expect pharmaceutical firms to behave as if they were community-owned, nonprofit entities. At the same time, the firms' owners... always expect the firms to use their market power and political muscle to maximize their owners' wealth. Caught between these inconsistent standards of behavior is an industry that naturally will never get it quite right."[50] And as long as that remains the case, debates about R&D costs are unlikely to end.

Less controversy surrounds the breakdown of where existing R&D funds go. Figure 1.5 shows how much of the R&D money spent by the members of the Pharmaceutical Manufacturers of America (PhRMA) in 2004 went to the various parts of the new-drug process. The early stage researcher may be surprised to find that his part, in itself representing just a fraction of the "Prehuman/Preclinical" category, which includes everything from target identification to IND submission, accounts for less than 26% of the spending, while the majority of the money is consumed in clinical testing where expenses of thousands of dollars per patient enrolled are the norm. According to one analysis, which refers to the *New Drug*

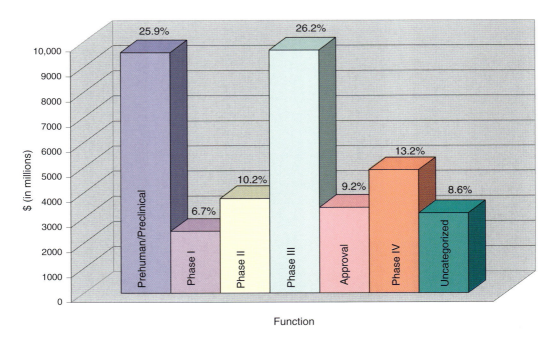

**Figure 1.5 ▶** R&D spending by function for PhRMA member companies in 2004. Source: Pharmaceutical Industry Profile 2006, PhRMA.

*Application* (*NDA*) process by which successful clinical candidates are granted FDA approval for marketing, "the average number of clinical trials per NDA has more than doubled in the last 30 years, and the average number of patients participating in clinical trials has increased two and one half times during the same period."[51] So it should come as no surprise that many of pharma's efforts are focused on reducing the costs and increasing the success rates associated with clinical testing.

Overall, industry investment in R&D is undeniably huge. According to the PhRMA 2006 survey, total biopharmaceutical R&D expenditures for 2005, including that by members and non-members, was $51.3 billion, which amounts to about $170 per year for each person in America. Investors, consumers, and other interested parties are bound to keep a sharp eye on bottom-line results and productivity metrics.

### 1.3.2   The Productivity Gap

Although the absolute figure may be open to debate, no one seriously doubts that drug discovery and development costs are rising. Figure 1.6 shows this trend, again using data from Tufts studies. The increasing cost of bringing a new drug to market far exceeds the inflation rate. It's been blamed on any number of things like increasing clinical trial expenses owing to the need for much larger trials than was formerly the case, more regular demands for Phase IV (postmarket) clinical studies, increasingly extensive preclinical *safety assessment* (toxicology), and even, according to some, the cost of rapidly adopting new technologies like

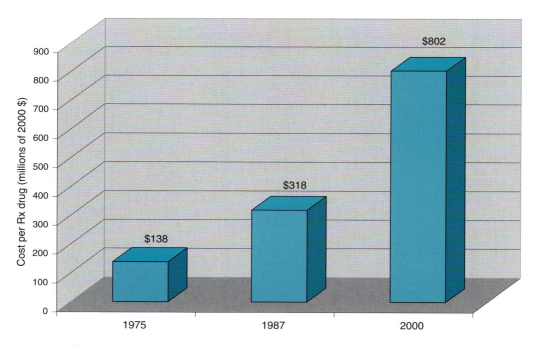

**Figure 1.6** ▶   The increasing cost of new drug development. (Source: DiMasi, J.A., et al. The price of innovation: new estimates of drug development costs. *J. Health Econ.* **2003**, *22*, 151–185.)

combinatorial chemistry, genomics, and proteomics that were designed to make the whole process more efficient and therefore *less* costly!

A recent US Government Accountability Office (GAO) report[52] found that pharma R&D spending grew by 147% between 1993 and 2004 while the overall number of NDAs submitted to the FDA increased only 38% and, worse still, the number of NDAs submitted for the presumably more innovative NMEs increased by only 7% in that time. Some of the factors mentioned above like larger clinical trials and the adoption of expensive, unproven new technologies were cited as being partly responsible for this. Others such as a focus on more complex and therefore more difficult to treat diseases, a concentration on "*me-too*" *drugs* (those similar to existing drugs both in structure and effect), which are considered more likely to make money, and the deleterious effects of mergers and acquisitions on productivity were also implicated. As you might imagine, the reasons behind increasing R&D costs for a similar number of NDAs are hard to sort out, but the trend is clear.

Worse still, it's actually uncommon for a new drug to even earn back the R&D expenses it took to get it to that point. Obviously, all new drugs are not created equal and the amounts of revenue they bring in, or don't bring in, vary tremendously. The best recent estimate is that only about 30% of marketed drugs succeed in recouping or earn revenues exceeding the average R&D cost, not to mention other expenses, making the majority of successfully launched new drugs money-losing propositions.[53] The burden for supporting the whole expensive infrastructure of drug discovery R&D, then, is borne by a small number of new drugs that must therefore be highly profitable.

In a worrisome development, at the same time that drug development prices have increased to the point where relatively few drugs turn out to be profitable, fewer new drugs are being approved overall. Figure 1.7 shows the number of NMEs approved by FDA for each year between 1993 and 2005. The average number of approvals was 28.5 per year over that time period, but you can see that starting in 2000, for every year except one (2004) approval numbers were below that average, a terrifying fact to an industry spending $51 billion per year on R&D. If the reasons for increasing R&D costs are many, complex, and subject to debate, the reasons for declining new drug approval rates are absolutely labyrinthine. Not only the quality of current clinical drugs, but many other factors, including a regulatory environment shaped by crises, lawsuits, and unfavorable public opinion may be involved.

Dr David Brown, Managing Director of Alchemy Biomedical Consulting, Ltd., co-inventor of Viagra (sildenafil), and former Global Head of Discovery Research at Roche, when asked whether the downturn in NDA approvals is a concern for pharmaceutical profitability in the long term replied, "Yes, both in numbers and quality."

"The drug approvals have basically dropped year by year and they're running at less than half the peak they were at in the mid 1990s," he notes. "Just as shocking as the fall in NDAs is the fact that the number of truly innovative drugs reaching NDA has fallen also. Long-term, sales and profits do depend on innovation so we need to see an upswing not only in number of NDAs but also the quality, i.e. more real breakthrough drugs."

Whatever the cause or causes, the very real trends toward increasing expenses, decreasing number of drug approvals, and fewer innovative new drugs have combined to create a lot of angst among stockholders, industry executives, and finally, via the "trickle down" effect, research scientists themselves. Concerns about the resulting "productivity gap" and ways to address it have been the subject of journal articles,[54–55] countless research conferences, and

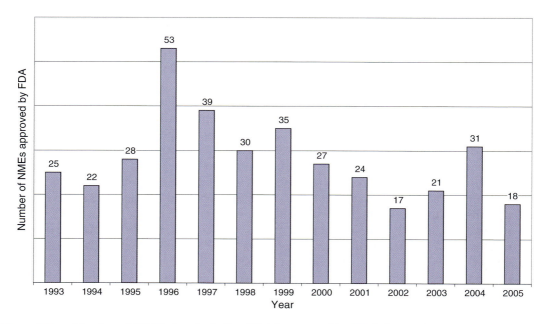

**Figure 1.7** ▶  Number of NMEs (not including BLAs) approved by FDA CDER by year. (Source: FDA.)

boardroom meetings worldwide, providing a boon to analysts and lecturers. Not only have NDA approvals been decreasing, but until 2004, the number of *Investigational New Drug (IND)* submissions in the United States was down from the peak levels of 1998 (see Figure 1.8). And, unfortunately, the "productivity gap" isn't the only problem biotech and pharma have had to face in recent years.

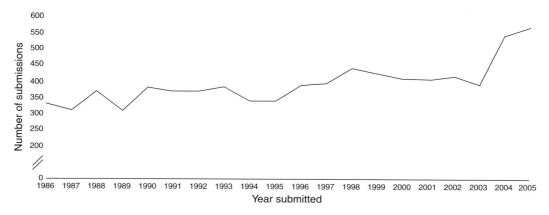

**Figure 1.8** ▶  Commercial IND submissions to FDA, 1986–2005. (From United States Government Accountability Office Report to Congressional Requesters, "New Drug Development: Science, Business, Regulatory, and Intellectual Property Issues Cited as Hampering Drug Development Efforts," GAO-07-49, November 2006.)

## 1.3.3 Market Withdrawals

Between 1995 and 2005, 16 prescription drugs were withdrawn from the US market. Several of them eventually returned to market with warnings and limited access. For example, Tysabri (natalizumab), an anti-integrin antibody for multiple sclerosis, came full circle, being approved, withdrawn, and then, at the bequest of physicians and patient advocacy groups who felt that its benefits outweighed the risks for many patients, returned to the market with a *black-box warning*, FDA's mechanism for alerting physicians and patients to serious safety concerns (see Figure 1.9).

Such market resurrections have happened before, in even more dramatic circumstances. In the late 1950s the drug thalidomide (Figure 1.10), which fortunately hadn't been approved for sale in the United States but was sold in countries like Canada, Japan, and Germany, was taken by pregnant women as a sedative and anti-emetic and led to about 10,000 cases of severe birth defects worldwide. "Children were born with missing (amelia) or abnormal (phocomelia) legs, arms, feet and hands; spinal cord defects; cleft lip or palate; absent or abnormal external ears; heart, kidney and genital abnormalities; and abnormal formation of the digestive system. It is estimated that 40% of thalidomide victims died within a year of birth."[56] The compound seems to have immunomodulatory, anti-inflammatory, and anti-angiogenic properties, the molecular bases for which are still not totally understood. Current thinking is that its anti-angiogenic properties that prevent proper blood vessel sprouting in the developing fetus are responsible for the birth defects.

---

WARNING

TYSABRI® increases the risk of progressive multifocal leukoencephalopathy (PML), an opportunistic viral infection of the brain that usually leads to death or severe disability. Although the cases of PML were limited to patients with recent or concomitant exposure to immunomodulators or immunosuppressants, there were too few cases to rule out the possibility that PML may occur with TYSABRI® monotherapy.

- Because of the risk of PML, TYSABRI® is available only through a special restricted distribution program called the TOUCH™ Prescribing Program. Under the TOUCH™ Prescribing Program, only prescribers, infusion centers, and pharmacies associated with infusion centers registered with the program are able to prescribe, distribute, or infuse the product. In addition, TYSABRI® must be administered only to patients who are enrolled in and meet all the conditions of the TOUCH™ Prescribing Program **(see WARNINGS, Progressive Multifocal Leukoencephalopathy; and WARNINGS, Prescribing, Distribution, and Administration Program for TYSABRI®).**

- Healthcare professionals should monitor patients on TYSABRI® for any new sign or symptom that may be suggestive of PML. TYSABRI® dosing should be withheld immediately at the first sign or symptom suggestive of PML. For diagnosis, an evaluation that includes a gadolinium-enhanced magnetic resonance imaging (MRI) scan of the brain and, when indicated, cerebrospinal fluid analysis for JC viral DNA are recommended **(see CONTRAINDICATIONS and WARNINGS, Progressive Multifocal Leukoencephalopathy).**

**Figure 1.9 ▶** A "black box" warning from the FDA label for Tysabri (natalizumab), approved June 5, 2006. For illustrative purposes only. (Source: FDA.)

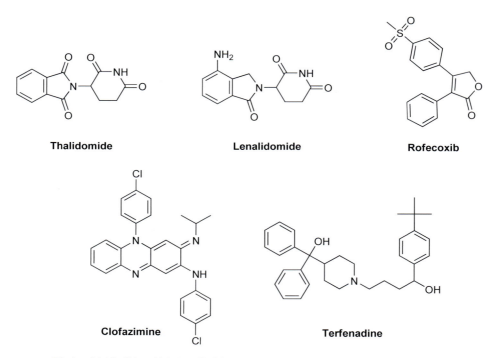

**Figure 1.10** ▶  Thalomid (thalidomide), Revlimid (lenalidomide), Vioxx (rofecoxib), Lamprene (clofazimine), and Seldane (terfenadine).

Despite thalidomide's horrendous teratogenicity and repute that led to an overhaul in drug regulations worldwide, including the United States, it was later found to be useful for the treatment of erythema nodosum leprosum (ENL), a complication of Hansen's disease (leprosy). Although another drug, Lamprene (clofazimine), that targets not only ENL but also the disease's causative bacteria, *Mycobacterium leprae*, eventually came along, the continued availability of thalidomide resulted in the discovery of its usefulness in yet another indication, multiple myeloma, a type of cancer. Today the drug is sold as the anti-cancer drug Thalomid. Its label includes a four-page black-box warning requiring that extreme measures be taken to ensure that the drug can't be used during pregnancy. Thalidomide has even given rise to a structural analog, Revlimid (lenalidomide; Figure 1.10), also suspected of having teratogenic potential, which is approved for the treatment of myelodysplastic syndromes (MDS, once called "preleukemia"), with a similarly rigorous black-box warning. Although the thalidomide story points to some limited possibilities for the revival of very toxic drugs, such toxicity is obviously not a property anyone would want to encounter in the first place.

These days, increasingly large and numerous lawsuits have gone hand in hand with increasing press coverage to make names like Fen-Phen (felfluramine plus phentermine) and Vioxx (rofecoxib) household names. Howard Solomon's question, "Don't drugs do something else other than cause side effects?",[57] is one that might need serious consideration nowadays. In most cases, withdrawals have followed toxicity noted only after tens or hundreds of thousands of patients had taken the drug. Although great strides have been taken in toxicity prediction, as we'll see later, even the best available methods can't always predict this kind of clinical toxicity in advance. Finding toxicity during clinical trials is bad enough for everyone concerned, but finding it after approval can be disastrous.

Consider the case of Vioxx (rofecoxib), Merck's selective COX2 inhibitor, an analgesic agent whose lack of COX1 inhibition promised, and delivered, a lower risk of GI bleeding than pain relievers like aspirin and naproxen. Vioxx was approved in 1999 and then voluntarily withdrawn from the market 5 years later after an analysis of a large postapproval study meant to expand its market instead indicated an increased risk of adverse cardiovascular events in those taking the drug. By then, Vioxx had been marketed via *Direct to Consumer* (*DTC*) ads, and the success of this approach led to many more patients taking the drug than would otherwise have been the case. As a result Merck faces liability suits of staggering magnitude; a clamor has arisen for new and improved mechanisms for postmarketing drug surveillance; and pressure has mounted to ban or limit DTC ads.

In the wake of Vioxx and other drug withdrawals, regulatory scrutiny by FDA and other agencies can, of course, only be expected to increase. The agency, like the pharmaceutical industry itself, seems to be caught between a rock and a hard place. On the one hand, its failure to approve drugs quickly enough brought intense criticism in the AIDS crisis of the 1980s and was a major factor behind the *Prescription Drug User Fee Act* (*PDUFA*). Here money paid by drug sponsors enables FDA to hire more people to reach regulatory decisions faster, thereby getting useful drugs to patients more quickly.

But on the other hand, drugs going on to produce toxicity and patient deaths after approval are likely to be viewed by some as evidence that Federal regulators aren't being strict enough, approving potentially dangerous drugs from their now-paying pharma sponsors. Didn't they approve the drug too quickly, perhaps without requesting more safety studies? The suspicion will be there whether or not any amount of thoroughness, deliberation, or additional studies could have uncovered the problem prior to approval. This puts the agency in a particularly uncomfortable, if familiar, position. FDA Deputy Commissioner Janet Woodcock puts it succinctly: "Whatever action we take, someone's going to be unhappy. That's why it takes a special kind of person to work here—a masochist."[58]

A recent report by the Institute of Medicine[59] has faulted the agency for its handling of drug safety issues, and pressure on Congress to pass legislation addressing this is growing. As the last thing FDA wants in light of all this is the appearance of being "soft on safety," longer review times and more *approvable letters* (see Chapter 3) requesting additional studies prior to marketing approval can be anticipated.

### 1.3.4 Generic Competition

In the United States, the biggest check on pharmaceutical profits comes from competition with generic drugs. According to the Generic Pharmaceutical Association (GPhA) generic drugs now fill 56% of all the prescriptions dispensed in the United States. This translates into about 13% of all prescription dollars spent owing to the reduced costs of these medicines, which is, after all, their *raison d'etre*. On average, generic drugs are said to cost about 60% less than their branded counterparts, saving consumers at least $15 in copayments per prescription. According to Jody Fisher, Vice President of Verispan, drugs sold by the generics company Teva filled more prescriptions in the first half of 2006 than drugs sold by Pfizer.[60]

In 2005 alone, sales of generic drugs grew at a rate of almost 21% according to IMS Health, and the trend toward generics isn't expected to abate anytime soon.

This major move into prescription pharmaceuticals was enabled by the *Hatch–Waxman Act* of 1984, also known as the Drug Price Competition and Patent Term Restoration Act. This legislation enables generic drug manufacturers to receive FDA approvals for *Abbreviated New Drug Applications* (*ANDAs*) without having to go through the full range of clinical studies to establish safety and efficacy that was required of the original drug sponsor. Assuming that the usual requirements for proper regulatory filing, *Good Manufacturing Practices* (*GMP*), etc. have been met, approval can be granted upon the demonstration of only one thing: *bioequivalence*. Studies of the bioequivalence of generics can begin, thanks to the law's so-called *safe-harbor* provision, prior to the patent expiration date for the brand name drug so that the generic equivalent might literally be available at pharmacies the day after the patent expires.

Furthermore, the lucrative nature of the generics market, which Hatch–Waxman Act enabled, has encouraged generics manufacturers to take advantage of what's known as *Paragraph IV certification*, which amounts to challenging allegedly invalid pharma patents for drugs so that they needn't wait for existing patents to expire to sell the generic equivalents. According to Howard Solomon, CEO of Forest Labs, this "rewards a generic company which successfully challenges a patent on a branded product. The law was well intentioned—to encourage challenges to some of the frivolous patents that branded companies have sometimes used to extend their exclusivity. But it has resulted in challenges to every patent in sight regardless of the merits and imposed another vast legal expense in the system."[61]

The Paragraph IV process, like the Act, is complex and controversial, involving a 30-month delay on the ANDA approval while patent litigation is going on, but in the end a successful challenge by the generic manufacturer is rewarded with a 180-day exclusivity period, during which time no other generic can compete, with the important exception of *authorized generics* from the innovator company that originate from their own generics division or are licensed out to a third party. A true 180-day exclusivity period with no competition from authorized generics is far and away the most profitable period for the generic manufacturer as they typically will charge nearly as much as the brand name drug during this time, a practice known as *shadow pricing*.[62] After this period ends and when many other generics are competing, drug costs can drop to less than 20% of the original price, which is the effect the consumer appreciates and the legislation intended.

The effect of availability of generics on brand name sales is predictably disastrous: 3 months after the expiration of Syntex's naproxen patent, sales were said to be down 75%. Few argue that brand name drugs have any advantage over generics that would justify using the more expensive product, and in a cost-constrained environment, it's increasingly unusual for generics not to be used whenever possible.

Hatch–Waxman does contain several concessions to the pharmaceutical industry. NMEs are granted a 5-year exclusivity period before generic competition can begin, even in the absence of any patent. The act also lengthens relevant patent life to make up for the potential market time lost to FDA review, assuming that patent has not been declared invalid through a Paragraph IV-inspired legal challenge. Third, the original legislation permitted the stringing together of a number of 30-month delays on generic entry mentioned above, permitting pharma companies to greatly extend their product's "life cycles" and enraging consumer advocates, until a new law in 2003 limited them to one 30-month stay per generic application.[63]

Loss of income due to generic competition has been a grim reality for the pharmaceutical industry and will continue in coming years as more *blockbuster drugs*, those with annual sales of more than $1 billion (in the United States), go off patent. "In truth," said one report in 2003, "many companies are living on borrowed time until their blockbuster patents run out."[64] Many blockbuster drugs that "go generic" lose patent coverage not through the "natural causes" of patent old age, but instead experience death by Paragraph IV filing. As Figure 1.11 shows, losses due to scheduled patent expiries are expected to be somewhat lower through at least 2010, but will still represent major blows to pharma. Cutting Edge estimates that drugs worth $80 billion in annual sales will have already gone off patent by 2008.

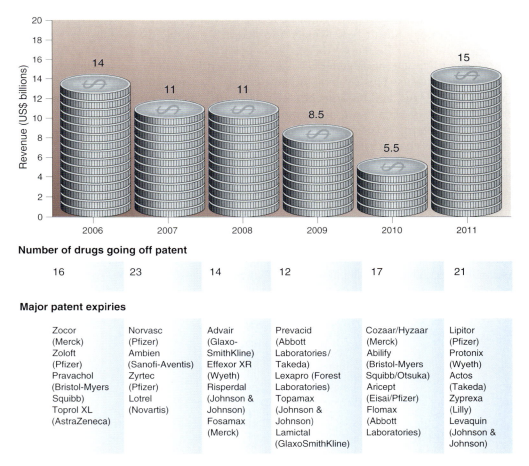

**Figure 1.11** ▶ Major drug patent expiries expected by year. (Reprinted by permission from Frantz, S. Pharma companies becoming more aggressive towards generics firms. *Nat. Rev. Drug. Discov.* **2006**, *5*, 619–620, copyright 2006, Macmillan Publishers Ltd.)

Pharma has adopted a number of different strategies to fight back. The prevailing philosophy seems to be, "if you can't beat 'em, join 'em!" Rather than quickly losing most of the market share for the drugs they pioneered, pharma companies can either come out with their own authorized generics through subsidiaries they own (such as Pfizer's Greenstone Limited), or license another company to do it for them. They then have the advantages of not needing ANDA approval (as they've already received an NDA) and of being able to sell it ahead of other generic copies, cutting into their lucrative exclusivity period, using available manufacturing facilities, and at least retaining more market share than if they hadn't thrown their hat in the ring. The strategy is a popular one with pharma: Novartis, for example, is said to currently be the world's number one generic drug manufacturer through its Sandoz division.[65]

A second aggressive strategy is for the pharma company to directly negotiate with health insurance providers to make their brand name drugs available to them cheaper than the generics, as exemplified by a recent Merck move to place Zocor (simvastatin) in the cheapest tier of a three-tier formulary through price reductions and rebates.[66] Although this kind of price slashing strategy hasn't been widely adopted, it's possible that it may see more use in coming years.

A third way to "out-generic generics" is to switch a drug from prescription status to *OTC* (*over the counter*) status. This is obviously not going to be possible for all but the safest and most widely used drugs, and requires potentially lengthy and expensive submission for FDA approval, but it has been used, for example, for the antihistamine Claritin (loratidine, Figure 1.12). The hope is, of course, that increased sales volume will compensate for the price reductions involved, or at least keep *some* profits flowing. And interestingly, even this strategy doesn't necessarily please consumer advocates because of the resulting shift from prescription drugs which are usually covered by insurers to OTC products which often aren't.

Finally, the most controversial method involves negotiating with generics companies. Since the legal challenges involved in Paragraph IV certifications are expensive and risky for both companies to pursue, they may well find a negotiated settlement preferable. These might involve cash (which would make it a so-called *reverse payment*) and an agreement to allow the generic company to market the drug at a specified date. These kinds of agreements can make business sense to both parties as well as their investors, but to government agencies

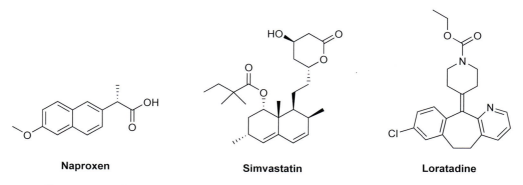

**Naproxen**                    **Simvastatin**                    **Loratadine**

**Figure 1.12 ▶**    Naprosyn (naproxen), Zocor (simvastatin), and Claritin (Loratidine).

and beleaguered consumers they appear suspiciously anticompetitive, a form of price fixing. Investigations into such agreements as potential violations of antitrust laws have been launched both in the United States and in Europe, and it's quite possible that in future new legislation will restrict such practices, which have only further tarnished the public's view of pharma.

One area where generic penetration has consistently been thwarted is that of biopharmaceuticals: recombinant DNA and monoclonal antibody therapeutics. Hatch–Waxman applies to drugs approved by FDA's *Center for Drug Evaluation and Research* (*CDER*) via the NDA process, but most biopharmaceuticals, instead, have been approved by FDA's *Center for Biologics Evaluation and Research* (*CBER*) via the *Biologics License Application* (*BLA*) process. Rather than being an oversight or reflection of the few biopharmaceutical agents approved at the time the Hatch–Waxman Act was written, the difference in treatment under Hatch–Waxman amounts to a tacit acknowledgement of the difficulties involved in producing a true "generic biologic." Most biologics, to date, have enjoyed de facto perpetual market exclusivity as a result.

The physiological effects of a small molecule don't depend on the process used to make it so long as impurity levels are kept sufficiently low, which is easier to do with something made from chemicals instead of cells, and its crystalline properties are consistent and identical. It's easy to run all the analyses necessary (NMR, LCMS, DSC, X-ray methods, etc.) to show whether two batches are equivalent or not. If they are, one has no reason to suspect they might act differently in vivo. But biologics, produced through recombinant DNA technology, can vary in properties like the exact glycoslyation state, aggregation, and denaturation that potentially have major effects on their potency and even safety. Immunogenicity can be problematic for such drugs, and this in turn can depend on things like the cell type used to produce them and subsequent differences in trace impurities. Worse still, these differences are difficult to analyze and characterize, the consensus being that whether or not a follow-on biologic agent is really "equivalent" to the original can only be determined by extensive clinical trials in patients, which are an anathema to the generics industry, of course, but par for the course for biotech and pharma.

For this reason the term *biosimilar* or *follow-on biologic* is preferred to "generic biologic", and in the United States, FDA is yet to develop requirements or a specific mechanism for the approval of such agents. Recently, however, it approved Omnitrope (somatotropin, human growth hormone), made by Novartis's generics division, Sandoz. This is a follow-on to Genentech's hGH, which was originally marketed after NDA, not BLA, approval. The agency took pains to stress that this approval did not set a new precedence or establish a new mechanism for further ones. In Europe, the *European Medicines Agency* (*EMEA*) approved Omnitrope a few months earlier and has begun compiling guidelines for biosimilar approvals for specific agents like hGH and EPO. Hence it is likely that biosimilars will appear first overseas until political pressure inevitably forces FDA to release its set of guidelines.

Global annual sales of biologic therapeutics were $33 billion in 2005[67] and are expected to exceed $100 billion by 2010.[68] The average cost of a 1-day supply of a biologic drug is now about $45, while a 1-day supply of small molecule drugs costs about $1.66.[60] More than $10 billion per year in global sales of biologic agents are already off patent,[69] making this a lucrative area for potential generic competition and a natural source of contention between current and aspiring biosimilar manufacturers and groups.

As the demand mounts to bring down the cost of prescription drugs, it's a good bet that guidelines and procedures for bringing biosimilars to market will be found in coming years, but it's unlikely that development costs or resulting drug prices for such agents will ever be as low as they are for small-molecule generics.

## 1.4 Current Economics—Solutions

### 1.4.1 Pharma Profits and Market Expansion

All in all, recent years have not been particularly good ones for the drug discovery industry. Consider the factors that have changed from the mid 1990s to more recent times, some of which are listed in Box 1.1.

If one were to consider only these trends, one might expect to see an industry with its profits in decline, perhaps even struggling to survive. Yet this isn't the case. A glance at Figure 1.13 reveals that, instead, pharmaceutical sales have actually been *increasing*. Even the casual observer would see some sort of disconnect at work here.

I asked Dr Michael Venuti how the pharmaceutical industry has been able to keep up profits in the face of fewer new products. "Well, there's market penetration to increase the size of the market," he explained. "One thing the pharmaceutical industry has done effectively is it's created markets. Twenty years ago there were no such markets as erectile dysfunction or restless leg syndrome. So they've created markets where there is a treatable condition."

He cited statins as an example of the potential for increased market penetration. "The estimates are that statins only get prescribed for 10 or 15% of the people who might see a benefit from them. If you keep that penetration you can see that with that same compound or its follow-up molecules, you can have profitability without having to spend a whole lot more money."

"What's happened is that pharma companies have become much more effective marketing machines," says Dr David Brown, "so they're actually making more money per drug." He also referred to the example of statins. "You've got Lipitor running at $11 billion or

---

**Box 1.1   Major Pharma Trends 1995–2003**

▶ Cost to bring a drug to market has risen

▶ Number of new drugs in clinical trials has fallen

▶ Number of NDA approvals for NMEs has dropped

▶ Expensive, high-profile withdrawals have occurred

▶ Generic competition has increased

▶ Fewer start-up companies were launched

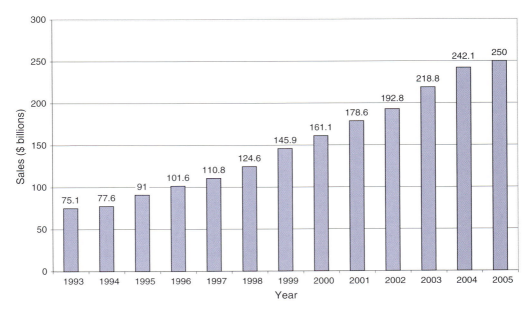

**Figure 1.13** ▶ Total sales for PhRMA member companies by year, 1993–2005. (Source: Pharmaceutical Industry Profile 2006, PhRMA.)

$12 billion, which was almost inconceivable a few years ago. So overall, profits have kept moving in a positive direction. However, the growth rates of the companies have slowed. They're single digit. They were classically running at 10–25% cumulative annual growth in sales and profits in the past, with an average of probably 12–15%. But hardly any companies are achieving double-digit growth anymore. They're in the 5–10% region. And the companies that are achieving the double-digit growth are smaller ones. So yes, they're keeping profits moving, but it's more and more dependent on fewer and fewer compounds. It's basically marketing."

Linking pharmaceutical sales and profits with the number of new drug approvals at any given time involves a couple of false assumptions. The first is that all new drugs bring in equal amounts of money. As we've seen, they don't. The second, and even worse, assumption is that the size of the market for existing products will always be the same and thus can't be significantly expanded. Pharma strategy has so far proven otherwise. But market expansion, like a run of good luck, can't go on forever, or companies would have ended their expensive investments in R&D long ago. In the long run market saturation, generic competition, and other factors will limit sales growth. Long-term, as Dr Brown points out, profits do depend on innovation.

For the present, though, the overall profit situation might be summarized as follows. Once again caught between hammer and anvil, pharma profit margins, which in 2005 averaged 15.7%, making it the fifth most profitable industry in the United States,[70] anger consumer advocates,[71] whereas recent growth rates of about 7% have proven to be far below the investors' comfort level. But fortunately a common ground exists: the need for new drugs to treat existing diseases like cancer and Alzheimer's disease as well as emerging ones. Fortunately, on this both sides agree, and so drug discovery research goes on.

### 1.4.2 Mergers and Acquisitions

But the economics behind it are turbulent, and this is reflected in ways that affect the researcher: how many jobs there are, how well they pay, what kind of companies they work for, what diseases and which projects they work on, etc. The scientist entering the field today can expect to see many such changes over the course of a career. One such trend that's impacted researchers in recent years has been the move toward mergers and acquisitions.

An example of the effects of pharma mergers can be seen in Figure 1.14, which outlines part of the corporate genealogy of Aventis up to 2000. Its later merger with Sanofi, itself the result of a number of mergers, formed Sanofi–Aventis, now the world's third largest pharmaceutical company. Such industry consolidations have significantly reduced the number and increased the size of the remaining pharma companies over the last several decades. The rationale for merging big pharma companies will typically be to promote efficiency by combining complementary product pipelines while eliminating redundancy, which shareholders will approve of, but the employees made redundant (most of whom always seem to work at the smaller of the two companies) won't. Mergers involve disruptions, program realignments, relocations, expenses, priority shifts, and a potential clash of "corporate cultures," and whether or not these really pay off in the end is uncertain.

Many view pharma–pharma "megamergers" as, at best, a temporary fix. "Large scale M&A (merger and acquisition) has regularly failed to deliver long-term benefits, with much

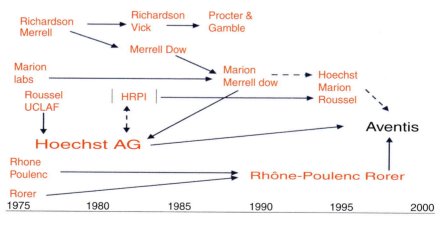

**Figure 1.14** ▸ Aventis corporate realignments, 1975–2000. (Reproduced with permission from Steele P: Patent Insight. Where did that come from? *Curr. Drug Discov.* 2001 *1* (June), 38–41, copyright 2001, The Thomson Corporation.)

consolidation driven by the need to meet short-term growth," concludes one analysis.[72] Another observes that "bosses may have hurt the discovery process with an orgy of dealmaking that has turned Big Pharma into Enormous Pharma," in the process taking out some costs but adding in more bureaucracy[73] and perhaps making it even more difficult for the resulting company to sustain a given rate of growth in the long run.

If mergers & acquisitions have retired many of the familiar names from the past like G.D. Searle and Upjohn as the ranks of pharma thin and the survivors become bigger, new and unfamiliar names of smaller companies like Gilead, Alnylam, and Infinity have sprung up to take their place in the lexicon of drug discovery. These are products of a kind of biotech "big bang" that occurred while pharma was being pruned back. Although the availability of funding for such companies has varied over the years, more than 1400 of them now exist in the United States.[74]

As we'll see in the next chapter, big pharma acquisition of small companies, mostly biotechs, has become extremely popular and shows no sign of abating. All told, more than a thousand life-science mergers & acquisitions (including medical device companies) took place worldwide in 2006 according to Dealogic, which tracks such activity. One of the drivers behind this has been the *American Jobs Creation Act* of 2004, which gave a special tax deduction to US companies for plowing profits earned abroad back into the US economy.[75] The resulting cash windfall has added fuel to the M&A fire. A bigger factor, though, is pharma's thirst for new clinical candidates, described below.

### 1.4.3  Biotech Clinical Candidates to Pharma

As we've seen, biotech companies have become a significant source of NMEs in clinical trials, helping to fill pharma pipelines in return for funding and a potential percentage of sales or sometimes outright acquisitions by pharma. Like pharma mergers, funding by acquisition has become more popular in recent years. Dr Venuti explains that big pharma "pays premium prices for things that have been proven to work. In fact, in many cases now they're compelled by tax law to buy the whole company to get the major asset. If they don't, they have to roll the profits & losses of the smaller company up into the big one because they own the major asset through license. So they wait until they see the pivotal data—and it's actually an arithmetic exercise—and they're willing to spend, for example, $200 million to get a compound that works instead of spending that $200 million on 20 early stage projects that might fail."[76]

Pharma's increasing thirst for clinical stage biotech compounds obtained by licensing or acquisition has been a major trend in recent years. As a risk-minimization strategy it seems to work. Industry studies have consistently shown that in-licensed[77] compounds are about twice as likely to gain marketing approval as compounds originating in-house.[78] Part of this probably reflects the higher approval success rate for biologics, which in-licensing is more likely to supply, than for small molecules, as Figure 1.15 shows.[79] Part of it may also reflect potentially closer or more objective scrutiny of in-licensed drugs versus in-house candidates, which have executives and corporate momentum behind them and are therefore hard for a company to drop. Whatever the reason, in-licensing has been

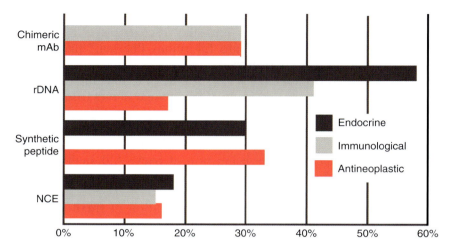

**Figure 1.15 ▶** Comparative approval success rates of US biopharmaceutical products and drugs by therapeutic category. (Reproduced with permission from *Outlook 2005*, Tufts Center for the Study of Drug Development, 2005:4, Tufts Center for the Study of Drug Development.)

so successful that even the 1400+ biotech companies can't keep up with the demand for late-stage clinical candidates.

"If you go back 5 or 7 years," explains Dr David Brown, "all the big pharma companies were trying to in-license Phase III or Phase II compounds, which are close to market. And of course they'd still prefer to do that, overall. But there are a limited number of compounds in late-stage clinical development and buying them is very expensive. So companies search now more often for Phase IIa or Phase IIb compounds. Everybody's competing for those. There's now even a trend growing to in-license Phase I and lead-optimization compounds and much competition there."

"The interesting thing," he says, "is they're now paying probably as much for Phase I compounds as they were for late Phase II compounds only 6 or 7 years ago. This is good because it's actually now supporting the biotech industry that makes these compounds they really can't afford to take into Phase II and would rather sell at Phase I. So in-licensing has become a very, very competitive game for most companies."

Increased competition even for Phase I clinical candidates is now pushing pharma licensing back to preclinical programs, a wonderful sign for early-stage companies that were able to survive the funding drought of a few years ago. According to Dr Carl Spana, CEO of Palatin Technologies, "One of the changes I see occurring is the shift for larger pharma to early-stage programs: the willingness to access external science and projects when it's seen as a strategic fit."[80]

Obtaining clinical candidates through in-licensing and acquisition has become so competitive, in fact, that in 2005 the median acquisition deal cost pharma companies $170 million, almost triple the 2004 price of $57 million.[18] To the biotech industry, strapped for cash in the wake of the "nuclear winter" of 2002, this is a welcome relief, even if it means having to sell an innovative small company lock, stock, and barrel.

## 1.4.4  Academic Contributions

Aside from biotech companies, another and increasingly important contributor to the flow of new drugs exists—academia. The traditional model of "pure" or "basic" research for the public good freely disseminated through publications had been around just about forever. It used to be axiomatic that professors didn't get rich. Research funding coming in the form of grants from government agencies like the NIH and NSF was their primary financial concern. Potential ties ("ensnarements," some thought) to industry did exist in the form of support for postdocs, jobs for ex-group members, industry sponsored achievement awards, and potential consulting agreements, but by and large any professor who dwelt too much on the commercial potentials of his discoveries would probably be viewed with puzzlement or disdain by his colleagues. A research publication, not a patent—which if pursued in those days would probably be granted to the federal funding agency and not the university anyway—was the true "coin of the realm." Success, in the form of tenure and prestige, depended on it.

In this environment, little incentive existed for the rapid incorporation of academic discoveries into the commercial world of drug discovery. In 1980 the federal government held about 28,000 patents, most of them arising from discoveries made by agency-funded academics, but fewer than 5% of them were commercially licensed.[81] One reason for this was that the law required that all such licenses be non-exclusive. If Company A took out a license to practice such an invention, its competitor, Company B could just come along and do the same. So there was less incentive to license a potentially useful invention and lock out the competition. Other disincentives existed as well.

So after much debate, in 1980, to facilitate the transfer of inventions that had been achieved using public funds to private enterprise, thereby benefiting patients and taxpayers, Congress passed the *Bayh–Dole Act*. The primary driving force behind the frequently aggressive licensing, or *technology transfer* that it would come to entail, would be income for the universities, which could now hold title to the inventions themselves and exclusively license them out if desired, as well as the academic inventors themselves who would also be allowed a share of the bounty. So in an ironic twist of fate, professors became more "incentivized" than industrial drug discovery researchers who, as we'll see, generally sign away all remunerative rights to their discoveries to their employers and thus don't stand to directly profit from patents. Bayh–Dole did contain some safeguards such as a requirement that the licenses go to US companies if at all possible and a provision for the federal government to take control of intellectual property if deemed necessary under certain conditions, the so-called *march-in rights* which are yet to be exercised.

In many ways this single piece of legislation shook the established academic research model to its foundations. According to one analysis, "rather than behaving as selfless benefactors of US industry, universities and other owners of patents on government-sponsored research discoveries were quick to see their intellectual property rights as an opportunity to capture a share of the profits that flow from downstream product development."[82] The amount of that flow exceeded the wettest expectations. Licensing income from the Boyer & Cohen patents previously mentioned and a taxol process patent by Dr Robert Holton at Florida State University[83] proved to be true bonanzas, each bringing in fees of hundreds of millions of dollars. University technology transfer offices proliferated as the possibilities began to be grasped.[84] The commercial potential of research that universities were doing didn't seem quite so crass when the new buildings and institutes that it funded could be seen going up. Even the resulting publicity, which focused on the cutting-edge technologies and cures for diseases that technology transfer enabled, and not on a few professors making considerable amounts of money through publicly funded research, didn't seem so bad.[85]

The financial value of Bayh–Dole to the academic community has proven to be incredible. In 2004 alone, American academic and non-profit institutions gained $1.39 billion in licensing revenue and applied for over 10,000 patents.[86] The University of California, for example, currently ranks seventh on the list of top pharma/biotech patent assignees right between Novartis and Boehringer-Ingelheim.[87] In the wake of Bayh–Dole the boundaries between public research and private enterprise have become harder to spot; academia can be expected to contribute an increasing number of clinical candidates to the industry; and academic research has consequently seen a major shift toward *translational research*, which is discussed in the next chapter.

### 1.4.5  Global Outsourcing

Efforts to contain the rising cost and improve the efficiency of drug discovery have increasingly focused on the use of *outsourcing*. Strictly speaking, this word only refers to the contracting out of some corporate task without specifying where. It could be to a company across the street or across an ocean. *Global outsourcing* or *offshoring* are better terms to describe the latter.

Biotech and pharma companies outsource a number of different functions. In the discovery phase, the syntheses of building blocks, intermediates, and especially libraries are frequently outsourced, often to India or Eastern Europe. Offshoring of chemistry is frequent, that of biology less so. But *ADME (Absorption, Distribution, Metabolism, and Excretion)* studies and animal models for preclinical development are now sometimes done in China, for example, at Shanghai's WuXi Pharmatech, whose customers currently include nearly all of big pharma and big biotech.[88]

Pilot plant synthesis and the large-scale manufacturing of marketed drugs are also frequently offshored. Singapore has recently become a major center of *active pharmaceutical ingredient (API)* manufacture, with companies like GlaxoSmithKline (GSK) and Merck having major facilities there.[89] Singapore's government has committed billions of dollars over the next few years and provides generous tax incentives to make it a major research center for biotechnology as well, much of it happening at its impressive Biopolis research complex.[90] Also downstream, many advantages underlie the offshoring of clinical trials. Getting enough patients enrolled quickly is a perennial problem in US clinical trials, but countries like India and China have huge populations of patients available, many of them being "treatment naive," meaning previously untreated, which is often a plus in drug studies. India is also known for the quality of its clinical data handling and informatics. Running clinical studies in China, with its rapidly expanding market for pharmaceuticals, has become especially popular, as pharma companies can envision tremendous future sales-growth in the ever-expanding Chinese market.

Though its use from discovery through clinical trials and on to manufacture is frequent, keep in mind that global outsourcing needn't be confined to just R&D. At many companies parts of non-core functions such as information technology (IT), payroll, and HR have been offshored for some time now. It's so pervasive a force that, as the distinguished chemistry professor, Dr Madeleine Joulliè recently noted, "while in 2001 companies worried about how to justify a non-US provider, in 2006 they worry about justifying a US provider."[91]

To outline how global outsourcing takes place, three basic models can be considered.[92] *Vendor-based* or *contract outsourcing*, such as offshoring the synthesis of a library of compounds, is the simplest and the most frequently encountered in early-stage drug discovery. These are relatively low-cost, low-commitment, and are frequently used on almost an *ad hoc* basis even though it's often argued that an overall long-term outsourcing strategy is really necessary for success. *Partnership* is the next step, where intellectual contributions are made cooperatively by both parties.

The biggest investment, requiring the most confidence, is the *captive R&D center* model. Here "they still keep it in-house, but they'll have a house over in Bangalore or Mumbai," in the words of Dr Albert Wertheimer, Director of the Center for Pharmaceutical Health Services Research at Temple University in Philadelphia. In this model, the pharma company establishes its own research center in the other country, a good example of this being Roche's R&D facility in Shanghai.

According to one report, the overall outsourcing of drug discovery services, which of course doesn't include downstream items like manufacturing, amounted to $4.1 billion in 2005 and is expected to reach $7.2 billion by 2009.[93] Not surprisingly, the cost of labor has so far been the main driving force behind global outsourcing. It may surprise scientists new to industry to learn that in the United States, a *full-time employee* (*FTE*) in pharma or biotech will currently cost her company about $300,000 per year. Unfortunately this isn't all salary, but includes substantial amounts for benefits and overhead. But the corresponding price in India is somewhere between $55,000 and $85,000.[94] The cost–benefit calculation gets a bit more complex, though.

"There is a cost that is associated with managing this," says Dr Sarvajit Chakravarty, Vice President of Medicinal Chemistry at Medivation, Inc. and former VP of Chemistry at Sai Advantium Pharma, Ltd of Pune, India and Mountain View, California, who's familiar with both sides of the outsourcing equation. "So internally when you look at that, your $85,000 FTE is now beginning to cost you more like $100,000 or $120,000, depending on the number of people you have internally managing this."

"The economics are simple," explains Dr Michael Venuti, who as CEO of Discovery Partners International directly observed the migration of almost all of the US-based combi-chem outsource business to India and China in 2004–2005. "Doing the work in the US has a certain cost to it, and right now it's being done for at least 30–50% less overseas. You can't ask a chemist in the US to ask his landlord to give him Shanghai rent. You can't go to a car dealer and say, 'I want a car for what it costs in Shanghai.' And you can't ask for the grocery store to sell you food at Shanghai prices either. So it's all about cost-of-living issues and the associated cost-of-doing-business issues."

It should come as no surprise that in recent years biotech and pharma have been quick to embrace global outsourcing. For example, Bristol-Myers Squibb (BMS) says it now outsources about 90% of its chemical scaffold synthesis.[95] Right now this is an extreme example: other companies and other areas of drug discovery have been affected to less of an extent. But that extent keeps increasing.

Several other advantages, besides cost, lie behind outsourcing as well. Outsourced contract work in general, whether onshore or off, can be terminated more easily than in-house efforts, which would involve reassignments or layoffs. In a sense it's like hiring some skilled temps, but paying by the task rather than by the hour. The resulting flexibility may encourage companies to contract-out projects or parts of projects they might otherwise not be willing to take on, the ones falling into the "it would be nice to try that but we don't have the resources" category. With labor costs abroad running at less than half the US price, that can amount to quite a few things.

To underscore the advantages of such an approach, Dr Chakravarty draws up a fictional scenario where a small company has $5 million per year to invest in research. "At $250,000 that's about 20 chemists in the US. You know, if you're really serious about taking a molecule through and getting something that's going into the clinic, you probably need about 10 chemists per program. Now, imagine a situation where you can take that $5 million and you say, 'alright, instead of 20 chemists internally I'm going to do a 5/5 split. We're going to have five people internally and five people externally.' Suddenly, you've basically added one project."

"You can start to do programs that you couldn't have done through internal growth," he says. "That is where you add value."

Aside from the cost advantage, an oft-touted benefit of offshoring is the ability to dip into the pool of untapped talent that lies outside of America and Europe. Countries like China and India not only have larger populations than the United States, but have also placed more of an emphasis on education in science, and particularly in chemistry, than the United States has since the days of Sputnik. The number of chemistry degrees granted each year in these two countries far exceeds the US total. Many talented Asian scientists contribute to the drug discovery industry by coming to work in America, but more are available back at home, still within the reach of global outsourcing. And why shouldn't a lot of drug discovery work be done offshore when, for example, most papers in the *Journal of Medicinal Chemistry* now originate abroad?[96] Figure 1.16 illustrates the same trend for some other high-impact journals as well.

Expectations as to how big a role global outsourcing will eventually play in industrial drug discovery vary. Few expect to see the day when the biotech boutiques of South San Francisco, for example, stand empty and lifeless, victims of lower offshore labor costs. But few will argue with the fact that in the future, and maybe even today, every piece of drug discovery work done in America or Europe, where labor costs are relatively high, will need to have a reason why it shouldn't be sent offshore. To understand what these reasons might be we need to look at some issues and constraints involved in global outsourcing.

Management and control issues can be problematic in long-distance corporate relationships. As usual, the devil is in the details. Communication difficulties can arise. In-person meetings require days of travel and jetlag. Phone and teleconferences can be inconvenient and awkward owing to the time differences involved. E-mails and fax have some advantages, but can't be relied upon as primary means of communications. Ensuring that both parties are "on the same page" ends up being much harder than it is when they're both in the same country. Although India has the advantage of widespread English language use, that isn't true for some of the other countries involved in global outsourcing, and as few Americans have the necessary multilingual abilities, language difficulties can arise.

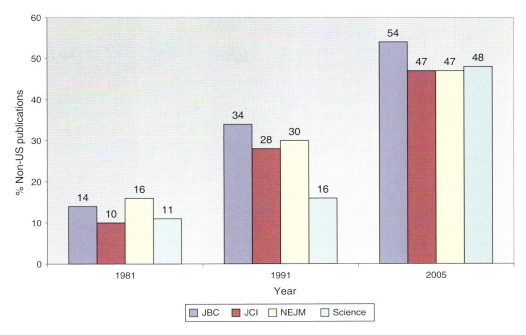

**Figure 1.16** ▶ Percentage of annual publications in the *Journal of Biological Chemistry* (JBC), *Journal of Clinical Investigation* (JCI), *New England Journal of Medicine* (NEJM), and *Science* where corresponding authors are based outside of the United States. (Data from Olefsky, J.M. The US's changing competitiveness in the biomedical sciences. *J. Clin. Invest.* **2007**, *117*, 270–276.)

Managers experienced in global outsourcing find that a lot depends on the person in charge of managing the project at the offshore site. "That is a very key hire," says Dr Chakravarty. "You're essentially hiring chemists that are good managers. They're hard to find in the US, let alone in India. But the answer is not just to put *somebody* in that position, because the moment you do that you compromise where you can go in these relationships."

So, a surprising shortage of qualified labor, particularly the all important client–partner *liaisons*, can sometimes be a problem as recently reported in Shanghai: "Probably the biggest challenge R&D managers face in Shanghai is to secure qualified staff. R&D centers have nearly dried up the pool of qualified scientists who can manage projects and lead teams. Scientists with English-language capabilities are particularly in high demand."[97] Although Asian universities graduate many more scientists than American colleges, the number of candidates with the necessary language skills, experience, proven leadership abilities, and, these days, the required cross-disciplinary understanding is still limited. US and European researchers looking to maximize their long-term employability should particularly take note.

Limitations on the worldwide talent pool aren't particularly surprising, according to Dr Mark Murcko: "Personally, I think chemical outsourcing is going to run into a wall and fundamentally change, because there aren't enough good chemists in the world. Yes, there are large populations in India and China, but if you talk to any of these little companies there they'll all say, 'Well, we're going to hire a hundred chemists next year.' How conceivable would it be for all the major pharmas in the United States to simultaneously hire that many good chemists? It would be impossible."

Even with the right person as *liaison*, an improvement in productivity over what could have been gotten in-house is by no means assured. Maintaining quality can get to be an issue. As that quality becomes more critical the risks become higher. For this reason many managers are understandably reluctant to offshore clinical trials despite all of the advantages. "Misreading, losing, or misentering one piece of data could result in a multibillion-dollar lawsuit," says Russ Bantham, Senior Vice President of PhRMA.[98] With the stakes this high, depending on a global partner takes a lot of trust, and that kind of trust can only be born of long experience with that partner.

The degree of *intellectual property* (*IP*, in this case patent) protection conferred by American patent law has traditionally not been matched in India and China, and right now the anxiety that this causes among US executives is arguably the biggest barrier to the more widespread use of global outsourcing. Until recently, India recognized only process patents, not the stronger and more easily enforced composition-of-matter patents that are all-important to biotech and pharma companies. As a result, it developed into a center of *reverse engineering* of APIs, which under Western laws would be considered willful infringement of patents. This isn't the case anymore, but the memory lingers.

Both India and China have recently tightened up their IP-protection laws. Membership of both nations in the World Trade Organization (WTO) has made them subject to the "Agreement of Trade-Related Aspects of Intellectual Property Rights" (TRIPS), an international standard. The jury is still out on just how broadly and how thoroughly those laws will be enforced. For the present, no one can say. Alternative methods for IP protection which may seem strange by American standards, like limiting the amount of information available to individual researchers, restricting access to different parts of a lab, and even withholding the name of the pharma client from outsourced scientists, have been used.

Some believe that concerns about IP loss are exaggerated. "Novel IP around something that is truly in discovery has never, ever been lost in India," says Dr Chakravarty. "That's the bottom line. I haven't been able to find an example of it." He likens it to a situation the US and European scientists frequently face if they work at a small company involved in a partnered project but split their time with another in-house project. In this situation, information obtained from the partnered project such as synthetic tricks and design improvements will typically need to be excluded from others, even though both involve the same company and maybe even the same scientist. In that kind of case, "Do you ever ask the question, 'How am I keeping the IP separate?' ", asks Dr Chakravarty. "We have an inherent culture that says, 'this is here and that is there.' You learn that from day one in the US pharmaceutical industry. If you build an organization in India where that is imbibed in the culture, over time this issue is going to go away. It's as much about ethics as it is about strategy and tactics. If you make your employees feel like they're part of the organization, see that they're involved, give them a work environment that is supportive, I think it will, over time, become a non-issue."

But other nation-specific risk factors in offshoring exist as well. "If operations are ever disrupted by workforce disputes or animal-rights activists, that would be in India: if by government interference, that would more likely be in China. The infrastructure is also far more reliable in China; India still suffers from interrupted power supplies, antiquated ports and inadequate highways in many regions."[99]

Finally, a frequently expressed concern about offshoring is the negative reaction it's bound to engender among US employees, especially those who see it as a threat to their jobs. New problems in employee morale and more negative publicity are not things that the industry would prefer to deal with. As companies dip their feet in the pool of global outsourcing potential they come to understand these factors. Will a close-up view of the downside be enough to check the growth of offshoring?

"I think it will temper the trend," says Dr Albert Wertheimer, "but they're not going to stop it or reverse it as long as it's less expensive and you can get equivalent quality work done abroad."

The biotech and pharmaceutical industries seem to be on the steep, upward part of the curve when it comes to global outsourcing. Experience from other industries further along, however, shows that at some point down the road expansion slows. In IT, for example, where outsourcing is an older, more established fact of life, there are recent indications of a new movement toward, of all things, insourcing.[100] At least one recent survey, by the analyst group DiamondCluster seems to indicate that the enthusiasm with offshoring, along with expectations for its impact on corporate bottom lines, may be waning.[101] And like almost everything else, outsourcing isn't as cheap as it used to be. Andreas Tschirky, Roche's head of R&D in China, estimates that labor costs are increasing there at 7–8% per year.[99]

"There's a limit," says Dr Venuti. "China and India are going to have an advantage only as long as the price makes up for the inconvenience and the time and the risk involved in putting a project offshore, in terms of the IP leaking out, compounds being lost in the mail, etc. At certain prices, that risk is OK. At higher prices, that risk tolerance will no longer be there, and the business would come back here. Those pricing pressures are going to come from the marketplace."

For now, at least, global outsourcing of drug discovery is most often used in the limited context of what are considered "lower value" parts of the process such as library synthesis and intermediate scaleup, but not as a replacement for the traditional "medchem" functions, perhaps because cross-disciplinary learning and experience with this most important part of drug discovery is not yet widespread overseas. "This is not a replacement for the creativity of medicinal chemists and organic chemists in the West," observes Dr Chakravarty. "That creativity is basically here in the US. You have to accept the fact that the vast majority of medicinal chemistry knowledge in the world today lies either in the US or in Europe. It's never going to go overseas because it's really closely coupled to novel biology, and that novel biology is not going overseas for our lifetimes."

Does global outsourcing result in the loss of American research jobs? One recent study by Duke University and Booz Allen concludes that "the offshoring of high skill-content work does not result in job losses in the originating country; rather, the overall job pool is increasing."[102] A surprising net *gain* of jobs was seen on projects utilizing R&D outsourcing. But the study was not directed specifically to pharmaceutical or biotech outsourcing and so no definition of what "high skill-content" might mean in that context was provided. Those familiar with the effects of offshoring on early-stage pharma and biotech research to date might be forgiven for being less sanguine. These days combinatorial libraries, once the province of high-tech America, are mostly made overseas with the resulting loss of many American jobs. Effects farther along in the drug discovery process have been much smaller so

---

**Box 1.2    Advantages and Disadvantages of Offshoring in Discovery Stage**

*Advantages of Offshoring*

▶ Lower costs mean more work for less money

▶ Adds resources without infrastructure expansion

▶ Lets in-house scientists concentrate on core jobs

▶ Dips into new talent pools

*Disadvantages of Offshoring*

▶ Need to factor in internal management costs

▶ Logistics and communications difficulties

▶ Relies on having the right partner and a good *liaison*

▶ Potential for intellectual property loss

▶ Country-specific problems (infrastructure, etc.)

▶ Unfavorable reaction at home

---

far. For today's medicinal chemist, the biggest observable effect of offshoring might be that libraries and intermediates are cheaper and therefore more readily available for his project. What its effect will be on the next generation of medicinal chemists is anybody's guess.

Some advantages as well as disadvantages of global outsourcing pertaining to drug discovery industry are summarized in Box 1.2. So, all things considered, what can global outsourcing really do for drug discovery? "Your outsource chemistry is never going to solve all the problems for you. It's not a panacea," Dr Chakravarty says. "It's another tool in your tool chest. You've got to use it appropriately and not all programs will be amenable to outsourcing. Not all chemistries will be amenable to outsourcing. In the end it's going to be the time-driver that's going to be the big advantage, not the cost-driver. If you can now push three programs instead of one, you're probably going to increase your overall chances of success and reduce the times to IND. It's certainly going to be cheaper to do it this way than to build out an organization in the US or Europe. However, the cost arbitrage is not going to be the ultimate driver for why something like this will succeed.

"Remember, you can make all the scaffolds and you can make all the libraries, but in the end it's the blood and guts in the trenches that make the molecule go into the clinic. You're still going to have to figure that out. What outsource chemistry does for you is it allows you to do three things instead of one."

Factors impossible to predict will affect future outsource utilization. For example, should a sustained drop in biotech and pharma profitability occur, the pressure to put this particular shoe on whether it fits or not will increase. The situation will obviously vary from company to company and project to project as well. Although the extent to which it's used may vary, global outsourcing will be a fact of life in drug discovery for the foreseeable future.

## 1.4.6  Blockbusters and Orphan Drugs

Owing to the tremendous costs associated with new drug development in recent years, profitability has come to depend on the success of a small number of drugs, the blockbusters. For the most part blockbusters follow one of two basic strategies: high volume or premium pricing, the classic yin and yang of sales approaches. The former is exemplified by Pfizer's small molecule Lipitor (atorvastatin), currently the most widely prescribed drug in existence, which now retails for somewhere around $1200 per year in the United States. Genzyme's Cerezyme (imiglucerase), a biological agent sold to about 5000 people with Gaucher's disease and priced around $200,000 per year, is an example of the latter. Many believe that one of the major reasons for big pharma's original coolness to biotechnology mentioned earlier was that they just didn't believe they could ever charge that much for *any* drug. In fairness to Genzyme, which has perennially taken heat for this price tag, the company does supply the drug for free to a number of uninsured patients. But the dichotomy between widely prescribed and less expensive small molecules and relatively pricey biologics catering to much smaller patient groups where no other good treatment is available is a pervasive theme that characterizes many profitable drugs today.

Cerezyme, along with many other expensive biologics that serve relatively small patient populations, is an example of the success of the *Orphan Drug Act* of 1982. Under this law, FDA can grant orphan drug status to new or existing drugs for indications that affect fewer than 200,000 people in the United States. The drug does not need to be a new one, only new to the indication. Orphan drug status confers a number of benefits, including a guarantee of 7 years of market exclusivity. The Act was intended to promote the development of medicines for relatively rare but serious illnesses that otherwise wouldn't be developed because the market was so small that companies would find them unprofitable. At the time, though, legislators probably didn't have orphan drugs costing hundreds of thousands of dollars per year, or blockbuster orphan drugs, in mind.

Whatever the original intentions and regardless of how it's been applied to date, the Orphan Drug Act has been responsible for catalyzing the development of 287 marketed drugs as of July 2006, with a number of blockbusters among them. (The impact of this Act, as well as of some other acts discussed in this chapter, on pharmaceutical industry are listed in Table 1.1.) Most orphan drugs have been biologics, the mainstay of biotech, like Amgen's

**TABLE 1.1 ▶  Some Federal Laws Having a Major Impact on Pharmaceutical Economics**

| Bill | Year | Major effect |
|------|------|--------------|
| *Bayh-Dole Act* | 1980 | Enable university tech transfer |
| *Orphan Drug Act* | 1982 | Incentivize the development of drugs for diseases with $< 200\,\text{K}$ US patients |
| *Hatch-Waxman Act* | 1984 | Increase generic competition |
| *Prescription Drug User Fee Act (PDUFA)* | 1992 | Shorten FDA review times |
| *Medicare Prescription Drug Improvement & Modernization Act (MMA)* | 2003 | Initially, increase Rx drug sales |
| *American Jobs Creation Act* | 2004 | Help fuel pharm investments, including M&As |

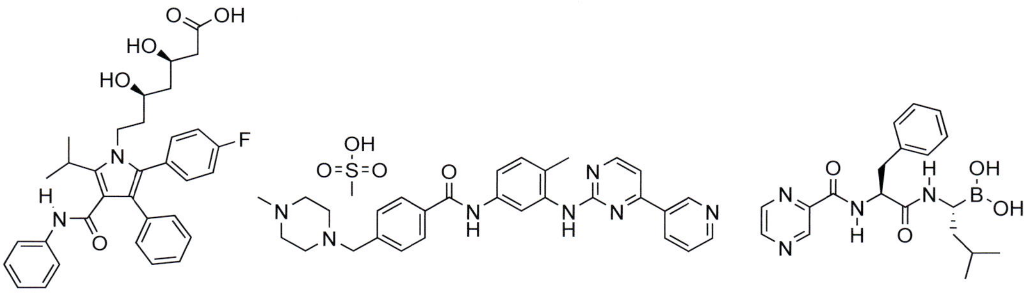

**Atorvastatin**　　　　　　　**Imatinib**　　　　　　　**Bortezomib**

**Figure 1.17** ▶ Blockbuster drug Lipitor (atorvastatin), and orphan drugs Gleevec (imatinib) and Velcade (bortezomib).

Epogen (epoetin alpha, for anemia), but small molecules such as Gleevec (imatinib, a kinase inhibitor for chronic myelogenous leukemia; Figure 1.17) and Velcade (bortezomib, a proteasome inhibitor for multiple myeloma; Figure 1.17) also figure among them. The Orphan Drug Act has proven a boon to biotech, where it's currently applied to almost 50% of all drugs produced,[103] while less benefit has accrued to pharma. Only 12% of orphan drug approvals went to the current top 10 pharma firms,[104] which is probably a reflection of their traditional focus on larger markets. In the past, pharma has concentrated on small molecules for big markets while biotech has mostly done just the opposite. But the gap between their approaches grows narrower every day.

## 1.4.7　Repurposing

The rising cost of bringing an NME to market has of course led companies to try and maximize the value of existing drugs. As we've seen, one way of doing that is through market expansion either by the further penetration of existing markets that methods like DTC ads can provide or through postmarket clinical trials aimed at expanding either the approved indication or the patient base. Fighting generic competition also qualifies as a value maximization strategy.

But there's another strategy that pharma companies are increasingly using, called *repurposing* or *repositioning*.[105] As scientists we like to think that our understanding of drugs is more than sufficient to predict which clinical indications they're likely to work for, but the uncomfortable truth of the matter is that unexpected pharmacological activity in the clinic is sometimes encountered as a kind of positive side effect so that new and better indications can be found. For example, almost 60 years ago Dramamine (dimenhydrate; Figure 1.18), which had been developed by G.D. Searle as an antihistamine, was being investigated at Johns Hopkins for the treatment of hives (urticaria). Unexpectedly, in one patient suffering from both hives and life-long motion sickness, it was found to be effective against both. This observation soon led to a trial aboard a troop transport ship sent from New York to Germany, dubbed "Operation Seasick", where the drug proved its effectiveness.[106] Since then millions of people have taken the drug for motion sickness. This and other cases serve to emphasize the importance of good clinical observation, in the absence of which such an easily missed piece of data might have been overlooked and a new and better indication lost.

A much more recent example showing that such serendipity can still be a factor involved a PDE5 inhibitor being developed by Pfizer. The drug, originally known as UK-92,480 and developed for angina, entered clinical trials in 1991. An unanticipated side effect of the drug, however, appeared in the form of penile erection, eventually leading to the development of the drug, Viagra (sildenafil; Figure 1.18), as a treatment for erectile dysfunction (ED).[107] Sildenafil has also found a use, as Revatio, in treating pulmonary arterial hypertension (PAH) in what might be considered a case of double repurposing.

In these examples, repurposing occurred at the original clinical-candidate stage, but it can also happen for drugs already marketed for different indications. Dozens of repurposed drugs are today either marketed or in development for new indications. Such compounds, having gone through the development process, with many of them already known to be safe in patients, are at that point much cheaper and quicker to bring to market than novel therapeutics which would need to start from "ground zero." And if repositioning can happen essentially by accident, as in the two cases above, what about the strategy of systematically testing known drugs for unexpected activity in a different indication of interest? This too has proven to be a popular strategy with companies like Melior Discovery and Ore Pharmaceuticals (formerly Gene Logic) doing exactly that. Taking a page from their institution's own book, researchers at Johns Hopkins put together a library of approved drugs and recently found, for example, that the histamine receptor antagonist Hismanal (astemizole, interestingly, one of the drugs withdrawn from the US market; Figure 1.18) has activity against the malaria parasite, *Plasmodium falciparum*.[108]

Potential problems exist in trying to develop and market any drug to which a company doesn't have thorough intellectual property rights. This issue didn't confront Searle or Pfizer, which held their own patents on dimenhydrinate and sildenafil, respectively. But testing a library of known drugs for activity is likely to turn up drugs that are either patented by another party or for which the original *composition-of-matter* patents have expired. The latter

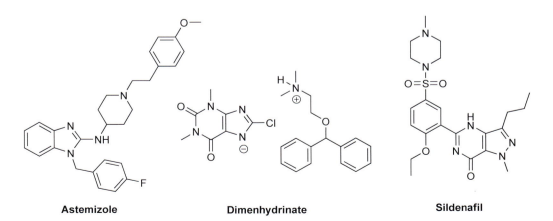

**Astemizole**                **Dimenhydrinate**                **Sildenafil**

**Figure 1.18 ►**   Hismanal (astemizole), Viagra (sildenafil), and Dramamine (dimenhydrinate).

case may allow for workarounds such as obtaining a *method-of-use* patent in the United States, but the former is more difficult, calling for a licensing agreement with a competitor which will probably not be willing to grant such a license but will be interested in knowing what other indication their proprietary drug might expand into. So repurposing is not without its drawbacks.

Also keep in mind that leads obtained by screening old drugs for new indications can also be used as just that—*leads*. Subsequent optimization might result in new, patentable compounds even if these would need to begin the whole process from *lead optimization (LO)* to the clinic again. This strategy, although useful, wouldn't be considered repurposing, as it lacks the speed-to-market that repurposing can bring, instead, calling for a new long-term program.

### 1.4.8   Chiral Switching

A strategy known as *chiral switching* involves the development, approval, and marketing of a single stereoisomer of an existing racemic drug.[109] Before improvements in asymmetric synthesis and chiral purification made stereopure compounds the rule rather than the exception, most drugs having a stereocenter were introduced as racemic mixtures. For some racemic drugs, isolating a single enantiomer would have made little sense because racemization would occur in vivo anyway. Others were developed as racemates because at the time the difficulty involved in obtaining pure enantiomers was felt to outweigh possible advantages. Physiologically, of course, differences in efficacy and toxicity between the eutomer and distomer could be expected, and some advantage to one stereoisomer or the other (or sometimes even both) could frequently be found.

Entrepreneurial companies like Sepracor jumped in to exploit these differences, for example, in developing Xopenex (levalbuterol; Figure 1.19), the *R*-isomer of the asthma drug albuterol. In this case chiral purity was a definite advantage, as the corresponding *S*-isomer actually antagonizes the beneficial effects of the *R*. Big pharma companies came to see chiral switching as a tool for maintaining market share. In this scenario, just before one of its racemic drugs would face generic competition, a drug's manufacturer, perhaps via a licensing agreement with a company like Sepracor, would introduce a new and improved chirally pure version. Racemic generics could still follow, but the newer drug would have some exploitable benefit so that market share could be maintained. The most famous example of this strategy at work is AstraZeneca's introduction of Nexium (esomeprazole; Figure 1.19), the magnesium salt of the *S*-enantiomer

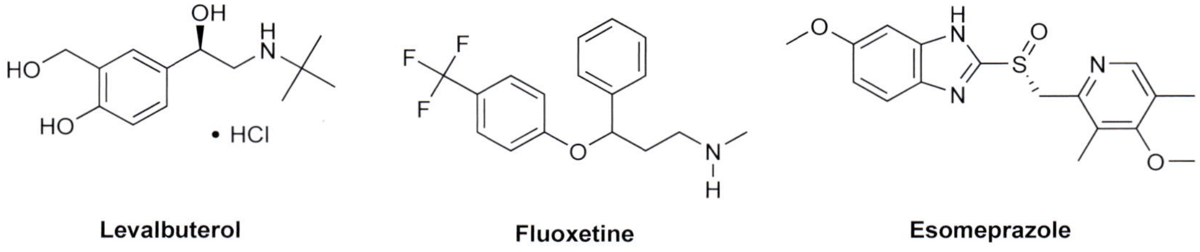

| Levalbuterol | Fluoxetine | Esomeprazole |

**Figure 1.19** ▶ Xopenex (levalbuterol), Prozac (fluoxetine), and Nexium (esomeprazole, actually a magnesium salt of the compound shown).

of its racemic proton pump inhibitor (PPI) Prilosec (omeprazole).[110] Nexium, which demonstrated slightly higher clinical efficacy than its predecessor,[111] has been a hugely successful blockbuster drug for the treatment of gastroesophageal reflux disease (GERD) and has become the poster child for chiral switching.

Many other chiral switch compounds are in the clinic as of this writing. Like any other value maximization approach, however, chiral switching has some potential problems. The legal issues involved can be complex. Patent offices can find it passing strange that companies should wish to patent a particular racemic mixture and then each of its components, claiming novelty, utility, and non-obviousness in each case. Also, the fact that a racemic drug previously passed through clinical trials without safety problems does not ensure that subsequent testing of one of its components will do the same, as Sepracor and Eli Lilly recently found in developing the *R*-enantiomer of Prozac (fluoxetine, Figure 1.19) for depression when a cardiovascular effect, prolonged QT interval, was encountered at high doses. Third, Figure 1.20 shows that chiral switching for *product lifecycle management (PLM)*[112] may well have a limited window of opportunity as the number of new racemic drugs that could be used continues to decrease and more and more drugs are introduced as single enantiomers to begin with. And of course one can't do a chiral switch on achiral compounds, which many known drugs are.

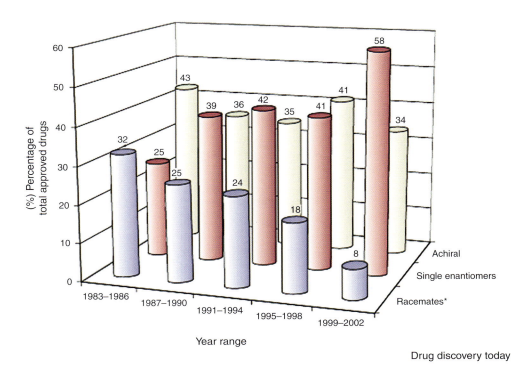

**Figure 1.20** ▶  Distribution of worldwide-approved drugs according to chirality character in 4-year ranges. (Reprinted with permission from Caner, H. et al. Trends in the development of chiral drugs. *Drug Discov. Today* **2004**, *9*,105–110, copyright 2004, Elsevier.)

### 1.4.9　Combination Therapeutics

Another way to maximize the value of known drugs involves combining them.[113] Many diseases, like HIV and Type II diabetes, typically require concomitant treatment by more than one drug, or API. This can be very inconvenient for the patient and almost ensures that there will be more problems with *patient compliance* or *patient adherence* (taking the medicines on schedule). Convenience and compliance are increased by combining them into a single pill, and the resulting fixed combination is considered a new drug with new market life. A recent example of this is GlaxoSmithKline's recently approved anti-diabetic drug Avandamet, which combines two different APIs, Avandia (rosiglitazone, Figure 1.21) and metformin. This type of combination therapeutic, where each component has been used separately to treat the same indication, has been referred to as a *congruous combination drug*.[114]

In at least one case, congruous combination has even been used to overcome the limitations of a company's new development compound. In evaluating the new HIV protease inhibitor that would eventually be called Lopinavir, researchers at Abbott found its plasma half-life to be disappointingly short owing to metabolism by CYP3A4, an enzyme involved in the in vivo breakdown of many drugs. An already-marketed Abbott HIV protease inhibitor, Norvir (ritonavir), was known to be an inhibitor of this metabolic enzyme. This is generally not considered a desirable property for a drug, as we'll see later. But in this case, ritonavir's CYP inhibition allowed researchers to combine the two protease inhibitors in a fixed ratio,

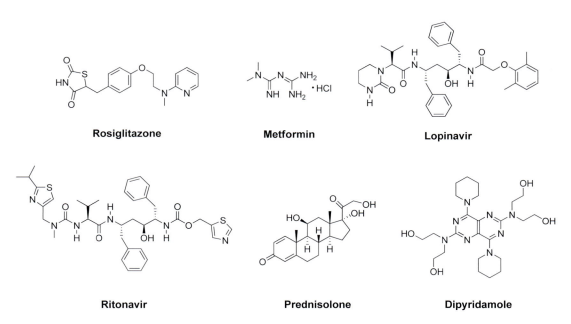

**Figure 1.21 ▶** Avandia (rosiglitazone), Glucophage (metformin), Lopinavir, Norvir (ritonavir), Prednisolone, and Dipyridamole.

using a small amount of ritonavir to improve lopinavir's PK properties, resulting in the recently approved combination drug, Kaletra.[115]

A different type of approach, called *syncretic combination*, can also be used. A company called CombinatoRx, for example, conducts large-scale screening of combinations of known drugs that affect different pathways in disease models and has found combinations of drugs that seem to work synergistically, such as prednisolone (an anti-inflammatory steroid; Figure 1.21) plus dipyridamole (a coronary vasodilator) for immuno-inflammatory diseases. These sorts of combinations can be very unobvious, both scientifically and legally. They also can take advantage of known drugs that are long off-patent and have good safety profiles, at least individually. Keep in mind, however, that toxicity as well as efficacy can sometimes be synergistic, so the need to establish the safety and efficacy of the combination through extensive clinical trials still exists. Both syncretic and congruous combinations, the latter being more frequently encountered, are referred to as *multicomponent therapies.*

It's often taken for granted by those not involved in formulation that compounding any combination pill should be "no big deal," but that's just not the case. Differences in solubility, problems with achieving uniformity, and the physical or chemical effects of pressing two compounds together can create problems that aren't necessarily easy to solve. In late 2006, for example, Merck announced that it would delay its approval submission for MK-524B, a 3-in-1 combination pill, because of an unspecified formulation problem. Formulation, one of the oldest, most mysterious, least acclaimed, and least-funded aspects of drug development, can often turn out to be the most critical, particularly for combination drugs.

### 1.4.10 Reformulation

In recent years, extensive use has been made of new formulations for old drugs, a good example being Alza's Concerta (methylphenidate, extended release; Figure 1.22). This drug, used to treat Attention Deficit Hyperactivity Disorder (ADHD) in children and adolescents, uses sophisticated osmotic pressure delivery to maintain drug concentrations, with once a day (qd) dosing, at levels comparable with those observed with thrice daily

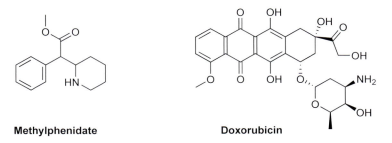

**Methylphenidate**                **Doxorubicin**

**Figure 1.22 ▶**   Ritalin (methylphenidate) and Adriamycin (doxorubicin).

dosing (tid) of standard methylphenidate. This frees young patients from having to take a pill during the school day. The pegylation of biologics is another method for decreasing the necessary dosing frequency for drugs. Interferon alpha-2a (Roferon A), useful in the treatment of hepatitis C (HCV), needs to be injected three times per week, but its pegylated cousin Pegasys (peginterferon alpha-2a) can be administered just once a week instead.

Although not necessarily affecting the frequency of administration, changing a drug's mode of administration can be quite a plus too, particularly when the drug, like insulin, normally requires injection. Exubera, an inhalable form of insulin developed by Nektar and Pfizer, was approved in 2006. Although the product has proven to be a poster child for disappointing sales, other inhalable forms of insulin continue to be explored in hopes of offering diabetic patients a medicine that's easier to take.[116] Some types of reformulation can result in reductions in toxicity. Doxil is a liposomally encapsulated form of the cancer drug doxorubicin, which has reduced cardiotoxicity, alopecia (hair loss), and other side effects relative to the standard drug.[117]

To see the extent to which the drug industry has adopted the methods mentioned above, we need only look at statistics for recently approved drugs. In 2005, of the 78 NDAs that were approved by FDA's CDER, 18 (23%) were NMEs, 11 (14%) were new combinations, and 37 (47%) consisted of new formulations.[118] Such numbers are often cited by critics of the industry to bolster their case that innovation is absent and a "me-too" spirit prevails. Before taking on such a familiar pharma-bashing attitude, though, it's best to consider the kinds of real advantages many of these non-NME products offer to patients.[119]

## 1.5  Summary

As we've seen, the economics of industrial drug discovery has been under increasing pressure in recent years from rising costs associated with drug development, fewer NDA approvals, increasing generic competition, high-profile drug withdrawals, and patent expiries. Blockbuster drugs, for the most part widely prescribed small molecules or

---

**Box 1.3   Some Methods for Maximizing the Value of Existing Drugs**

- ▶ Market expansion
- ▶ Fighting generic competition
- ▶ Reformulation
- ▶ Multicomponent therapies
- ▶ Chiral switching
- ▶ Repurposing

expensive biologics for smaller patient populations, have become the pillars of profitability. Biotechnology has proven to be an important part of the equation both in terms of biologic products and tools for drug discovery research. The inflated initial expectations for genomics have not been met, but few doubt that this new science has an important role to play in future.

Through it all, the pharmaceutical industry has remained profitable, and some recent signs may even point to a possible turnaround in productivity (see Figure 1.8).[120] But pharma's growth in profits has slowed, and other industries have nudged it out of its lead in this area.[70] A two-pronged approach for grappling with current economic problems has been taken: a focus on improving the success rate for new clinical candidates as well as efforts to maximize the value of existing drugs. The latter, largely a business strategy, is summarized in Box 1.3. The former, a matter of devising and implementing the best possible methods to discover and develop successful new drugs, will be the main concern of the reader and the rest of this book.

# Notes

1. Interestingly, a therapeutic agent that, despite its drawbacks, is still sometimes used in the treatment of psoriasis today.

2. The actual structure of what Perkin produced was only determined in relatively recent times. See Meth-Cohn, O., Smith, M. What did W.H. Perkin actually make when he oxidized aniline to obtain mauveine? *J. Chem. Soc. Perkin I* **1994**, 5–7.

3. An account of Sir William Perkin, whose name formerly graced two Royal Society journals, and his most notable discovery can be found in Garfield, S. *Mauve: How One Man Invented a Color That Changed the World* (New York, W.W. Norton & Co., 2001).

4. Drews, J. Drug discovery: A historical perspective. *Science* **2000**, *287*, 1960–1964.

5. A rather toxic drug with low efficacy, requiring IV injections over 18 months, but for all that, a welcomed cure far preferable to things like mercury compounds and malaria parasites (!) which had previously been tried.

6. Drews, J. *In Quest of Tomorrow's Medicines* (New York, Springer-Verlag, 1999).

7. Anon. *The Pharmaceutical Century: Ten Decades of Drug Discovery* (American Chemical Society, 2000).

8. Brown, D. "Sources of New Drugs", Vol. 1, pp. 321–353 in *Comprehensive Medicinal Chemistry* Triggle, D.J., Taylor, J.B., Eds. (London, Elsevier, 2006).

9. Mann, J. *Murder, Magic, and Medicine* (Oxford, Oxford University Press, 1992).

10. See Scriabine, A. "Discovery and Development of Major Drugs Currently in Use", Chapter 2 in *Pharmaceutical Innovation*, Landau, R., Achilladelis, B., Scriabine, A., Eds. (Philadelphia, Chemical Heritage Press, 1999).

11. For example, Fuzeon (Enfuvirtide), a 36-mer, requires no fewer than 106 chemical steps to make. See Matthews, T., et al. Enfuvirtide: The First therapy to inhibit the entry of HIV-1 into host CD4 lymphocytes. *Nat. Rev. Drug Discov.* **2004**, *3*, 215–225.

12. Kohler, G., Milstein, C. Continuous cultures of fused cells secreting antibody of predefined specificity. *Nature* **1975**, *256*, 495–497.

13. Reichert, J.M., et al. Monoclonal antibody successes in the clinic *Nat. Biotechnol.* **2005**, *23*, 1073–1078. For an excellent review of current antibody therapeutics, also see Carter, P.J. Potent antibody therapeutics by design. *Nat. Rev. Immunol.* **2006**, *6*, 343–357.

14. Quoted in "Science for Life: A Conversation with Nobel Laureate David Baltimore" by Barbara J. Culliton, found at http://content.healthaffairs.org/cgi/content/full/hlthaff.25.w235v1/DC1.

15. Merck's Gardasil, recently approved by FDA for the prevention of two types of human papillomavirus (HPV) that can lead to cervical cancer, is expected to be quite profitable, and money and some liability protection have become available in recent years in light of potential bioterrorist threats like anthrax. For more on the future profitability of vaccines, see Ginsberg, T. Making vaccines worth it. *Philadelphia Enquirer*, September 24, 2006.

16. For an early review of biotechnology and its potential effects on drug discovery, see Venuti, M.C. "Impact of Biotechnology on Drug Discovery", Chapter 31 in *Annual Reports in Medicinal Chemistry Volume 25*, Bristol, J.A. Ed. (San Diego, Academic Press, 1990).

17. For example, in one 30-day period in the late summer of 2006 no fewer than four small companies, Icagen, Corcept, SGX, and Adolor, lost 73%, 56%, 41%, and 45% of their stock values, respectively, in one single day because of bad news on late-stage clinical trials. In contrast, Pfizer stock lost 10.6% of its value (still about $19 billion!) after the company announced that it was discontinuing clinical trials on its most promising candidate, torcetrapib, for safety reasons.

18. Data from Bain & Co. quoted in Herskovits, B. Pharma-biotech M&A prices nearly triple. *Pharm. Exec.*, August 16, 2006. Available at http://www.pharmexec.com/pharmexec/article/articleDetail.jsp?id=365544.

19. "Falling Innovation Levels Fuel Pharma/Biotech Collaborations", May 10, 2005, available at www.drug-researcher.com/news/printNewsBis.asp?id=67611.

20. For a timeline of genomics, see Roberts, L., et al. A history of the human genome project. *Science* **2001**, *291*, 1195.

21. Fleischmann, R., et al. Whole-genome random sequencing and assembly of *Haemophilus influenzae* Rd. *Science* **1995**, *269*, 496–512.

22. National Research Council, Mapping and Sequencing the Human Genome (Washington, DC, National Academy Press, 1988).

23. Quoted in Angier, N. Great 15-Year Project to Decipher Genes Stirs Opposition. *New York Times*, June 5, 1990.

24. A fascinating account of the resulting race between HGP and Celera to sequence the human genome is available. See Shreeve, J. *The Genome War* (New York, Ballantine Books, 2004).

25. First proposed by James Weber and Eugene Myers before Celera existed and later eagerly adopted by the company. See Weber, J.L., Myers, E.W. Human whole-genome shotgun sequencing. *Genome Res.* **1997**, *7*, 401–409.

26. Waterston, R.H., et al. More on the sequencing of the human genome. *Proc. Natl. Acad. Sci. U.S.A.* **2003**, *100*, 3022-3024.

27. Adams, M.D., et al. The independence of our genome assemblies. *Proc. Natl. Acad. Sci. U.S.A.* **2003**, *100*, 3025–3026.

28. Caulfield, T. Biotechnology and the popular press: Hype and the selling of science. *Trends Biotech.* **2004**, *22*, 337–339.

29. Jensen, K., Murray, F. Intellectual property landscape of the human genome. *Science* **2005**, *310*, 239–240.

30. Crease, D.J., Schlich, G.W. Is there a future for "speculative" gene patents in Europe? *Nat. Rev. Drug Discov.* **2003**, *2*, 407–410.

31. Paradise, J., et al. Patents on human genes: An analysis of scope and claims. *Science* **2005**, *307*, 1566–1567.

32. Interestingly, the personalized genome services originally envisioned are just now beginning to become a reality. See Anon. Within spitting distance. *The Economist*, November 22, 2007.

33. Shreeve, J. *The Genome War* (New York, Ballantine Books, 2004), Chapter 12.

34. Lander, E.S., et al. Initial sequencing and analysis of the human genome. *Nature* **2001**, *409*, 860–921.

35. Venter, J.C., et al. The sequence of the human genome. *Science* **2001**, *291*, 1304–1351.

36. Tollman, P., et al. A Revolution in R&D: How Genomics and Genetics are Transforming the Biopharmaceutical Industry, November, 2001. Available at www.bcg.com/publications/files/eng_genomicsgenetics_rep_11_01.pdf.

37. Anon. *The Fruits of Genomics: Drug Pipelines Face Indigestion until the New Biology Ripens* (Lehman Brothers, McKinsey, & Co., 2001).

38. Higgs, G. Molecular genetics: The Emperor's clothes of drug discovery? *Drug Discov. Today* **2004**, *9*, 727–729.

39. Betz, U.A.K., et al. Genomics: Success or failure to deliver drug targets? *Curr. Opin. Chem. Biol.* **2005**, *9*, 387–391.

40. Buehler, L.K. Is genomics advancing drug discovery? *Pharm. Discov.* **2005**, 26–28.

41. International Human Genome Sequencing Consortium, Finishing the euchromatic sequence of the human genome. *Nature* **2004**, *431*, 931–945.

42. A Kaiser Family Foundation *Health Poll Reports Survey* conducted February 3–6, 2005 showed that a total of 50% of respondents had an overall unfavorable view of pharmaceutical companies while only 44% were favorably disposed. Only oil and tobacco companies were viewed more unfavorably among the industries included in the survey. In addition, 70% of respondents felt that pharmaceutical companies were most interested in making profits, putting profits ahead of people, and 81% felt that prescription drug prices are "not usually justified because companies charge more than necessary." Data available at www.kff.org/healthreport/feb_2005/index.cfm.

43. Honestly, it's difficult to think of any other industry where paying $11 for lunch and giving away free pens and sticky pads would be thought a scandal worthy of major press coverage. See Saul, S. "Drug Makers Pay for Lunch as They Pitch" *New York Times*, July 28, 2006.

44. DiMasi, J.A., et al. The price of innovation: New estimates of drug development costs. *J. Health Econ.* **2003**, *22*, 151–185.

45. "Tufts Drug Study Sample is Skewed: True Figure of R&D Costs Likely is 75 Percent Lower", December 4, 2001. Available at www.citizen.org/pressroom/release.cfm?ID=954.

46. See also "The cost of new drug and development" by Michael Dickson and Jean Paul Gagnon. *Discov. Med.* **2004**, *4*, 172–179, and Hileman, B. Many doubt the $800 million pharmaceutical price tag. *Chem. Eng. News*, June 19, 2006, p. 50.

47. In 2004, for example, only 31 of 113 FDA new drug approvals (27%) were for NMEs. See www.fda.gov/cder/rdmt/pstable.htm.

48. See the letter to the editor, Credit government scientists with developing Anti-AIDS drug, by Mitsuya, H., et al. *New York Times*, September 28, 1989.

49. *Parexel's Bio/Pharmaceutical R&D Statistical Sourcebook 2006/2007* (Waltham, MA, Parexel Intl., 2006).

50. Reinhardt, U.E. Perspectives on the pharmaceutical industry. *Health Aff.* **2001**, *20*, 136–149.

51. Paddison, C., et al. Outsourcing Beyond the Comfort Zone. *Pharmaceutical Technology Outsourcing Resources 2005*, p. 24. Available at http://www.pharmtech.com/pharmtech/article/articleDetail.jsp?id=174014.

52. United States Government Accountability Office Report to Congressional Requesters, New Drug Development: Science, Business, Regulatory, and Intellectual Property Issues Cited as Hampering Drug Development Efforts, GAO-07-49, November, 2006. Available at www.gao.gov/new.items/d0749.pdf.

53. Grabowski, H., et al. Returns on research and development for 1990s new drug introductions. *Pharmacoecon.* **2002**, *20*, suppl. 3, 11–29.

54. Carney, S. How can we avoid the productivity gap? *Drug Discov. Today* **2005**, *10*, 1011–1013.

55. Brown, D., Superti-Furga, G. Rediscovering the sweet spot in drug discovery. *Drug Discov. Today* **2003**, *8*, 1067–1077.

56. Pannikar, V. The return of thalidomide: New uses and renewed concerns. *WHO Pharm. Newsletter* **2003**, No. 2, 11–12.

57. Forest Laboratories 2006 Annual Report, p. 13. Available at http://library.corporate-ir.net/library/83/831/83198/items/203929/Stockholders-Letter_AR06.r3.pdf.

58. Quoted in Simons, J. FDA damned if it does, damned if it doesn't. *Fortune*, November 9, 2007. Available at http://money.cnn.com/2007/11/08/magazines/fortune/simons_fda.fortune/index.htm.

59. The Future of Drug Safety, Institute of Medicine, September 2006. Available at http://newton.nap.edu/execsumm_pdf/11750.

60. Quoted in Mayer, R. Generics gain ground. *MedAd News*, November 2006, p. 1.

61. See Forest Laboratories 2006 Annual Report. Forest was in the middle of successfully fending off a challenge to one of its patents when this was written.

62. Estimates vary from about 60% to 95% of the brand name drug cost for the first generic entry.

63. See "President Takes Action to Lower Prescription Drug Prices by Improving Access to Generic Drugs", available at www.whitehouse.gov/news/releases/2002/10/20021021-4.html.

64. Gilbert, J., et al. Rebuilding big pharma's business model. *In Vivo* **2003**, *21*, 73–82.

65. Christensen, C. Big pharma's prognosis. *Forbes*, August 1, 2006. Available at www.forbes.com/2006/07/31/leadership-innovation-pharmaceutical-cx_cc_0801pharma.html.

66. Frantz, S. Pharma companies becoming more aggressive towards generics firms. *Nat. Rev. Drug Discov.* **2006**, *5*, 619–620.

67. Source: IMS Health. See www.imshealth.com.

68. Datamonitor estimate quoted in Belsey, M.J., et al. Biosimilars: Initial excitement gives way to reality. *Nat. Rev. Drug Discov.* **2006**, *5*, 535–536.

69. Moran, N. After recommendation, first biosimilar approval imminent. *BioWorld Today*, February 1, 2006.

70. Source: Fortune Magazine, April 2006. Available at http://money.cnn.com/magazines/fortune/fortune500/performers/industries/return_on_revenues/index.html.

71. For a polemic view of the pharmaceutical industry by the former Editor-in-Chief of the *New England Journal of Medicine*, see Angell, M. *The Truth About the Drug Companies: How They Deceive Us and What to Do About It* (New York, Random House, 2004).

72. Pavlou, A.K., Belsey, M.J. BioPharma licensing and M&A trends. *Nat. Rev. Drug Discov.* **2005**, *4*, 273–274.

73. Anon. Billion dollar pills. *The Economist*, January 25, 2007. Available at www.economist.com/business/displaystory.cfm?story_id=8585891.

74. Source: *Guide to Biotechnology 2005–2006* published by the Biotech Industry Organization (BIO) and available at www.bio.org/speeches/pubs/er/BiotechGuide.pdf.

75. See the whitepaper "American Jobs Creation Act of 2004", PriceWaterhouseCooper, available at www.pwc.com/images/gx/eng/about/ind/chemicals/ajca-section-199-whitepaper-ind-manu.pdf.

76. Recent changes in accounting rules by the International Accounting Standards Board and Financial Accounting Board may affect this equation. See Anon. Changes on the horizon. *R&D Directions* **2006**, *12*, 12.

77. Not all in-licensed compounds, of course, originate in biotech, but the majority do.

78. See Lou, K., de Rond, M. The "not invented here" myth. *Nat. Rev. Drug Discov.* **2006**, *5*, 451–452.

79. Anon. *Outlook 2005*, Tufts Center for the Study of Drug Discovery. Available at www.csdd.tufts.edu/InfoServices/OutlookPDFs/Outlook2005.pdf.

80. Quoted in Carroll, J. A hot market for licensing deals. *FiereceBiotech*, February 9, 2007. Available at www.fiercebiotech.com/story/feature-palatin-technologies-finds-a-hot-market-for-licensing-deals/2007-02-09.

81. "The Bayh-Dole Act. A Guide to the Law and Implementing Regulations", Council on Governmental Relations, October 1999. Available at www.cogr.edu/docs/Bayh_Dole.pdf.

82. Eisenberg, R.E. The shifting functional balance of patents and drug regulation. *Health Aff.* **2001**, *19*, 119–135.

83. See Stephenson, F. "A Tale of Taxol", available at www.research.fsu.edu/research/fall2002/taxol.html.

84. For an interesting insight into the nuts-and-bolts of university tech-transfer licenses, see Leute, K. Stanford's licensing and equity practices with biotechnology companies. *J. Commer. Biotech.* **2005**, *11*, 318–324.

85. To be fair, these scientists have proven to be enormously generous, funneling large amounts of royalty income back into their institutions to further research. For one such example see Silverman, E. The trouble with tech transfer. *The Scientist* **2007**, *21*, 40. Available at www.the-scientist.com/article/home/39379/.

86. See Bayhing for blood or Doling out cash? *The Economist*, December 20, 2005. Available at www.economist.com/science/PrinterFriendly.cfm?story_id=5327661.

87. Steele, P. "International Patenting of Pharmaceuticals and Biotechnology, 2004–2006: Top Innovators" reprinted in Paraxel's Sourcebook, Reference 49.

88. See www.pharmatechs.com/about.htm.

89. Tremblay, J.-F. An unlikely center for pharmaceuticals. *Chem. Eng. News*, February 27, 2006, pp. 12–15.

90. Arnaud, C.H. A magnet for talent. *Chem. Eng. News*, February 27, 2006, pp. 10–14.

91. Joullié, M.M. Outsourcing: Blessing or curse? *Chem. Eng. News*, September 25, 2006, pp. 116–117.

92. Goodall, S., et al. The promise of the East: India and China as R&D options. *Nat. Biotechnol.* **2006**, *24*, 1061–1064.

93. "Outsourcing in Drug Discovery, 2nd Edition", Kalorama Information. Available at http://www.kaloramainformation.com/pub/1099949.html.

94. Frantz, S. Chemistry outsourcing going global. *Nat. Rev. Drug Discov.* **2006**, *5*, 362–363.

95. According to Dr Arvind Mathur, External Resource Leader, Drug Discovery, for BMS, as quoted in "BMS says outsourcing is 'a strategic function asset' ", Chu, W.L. DrugResearcher.com, September 18, 2006. Available at http://drugresearcher.com/news/ng.asp?id=70641-bms-outsourcing-scaffolds.

96. Portoghese, P.S. The state of the journal at its 50th publication volume. *J. Med. Chem.* **2007**, *50*, 1.

97. Tremblay, J.-F. R&D takes off in Shanghai. *Chem. Eng. News*, August 21, 2006, pp. 15–22.

98. Quoted in Paddison, Reference 51, p. 30.

99. This quotation is from Goodall, S., et al., reference 92. In the opinion of the author of this volume (RR), inadequate highways and interrupted power supplies can be found in other places as well, such as California.

100. Barnes, K. IT Outsourcing Bucking Pharma Industry Trends, September 20, 2006. Available at www.drugresearcher.com/news/printNewsBis.asp?id=70479.

101. See O'Connell, B. Have business, will travel: Outsourcing in the global commerce age. *BioPharm Int.*, 18–19, October 1, 2005. Available at www.biopharminternational.com/biopharm/article/articleDetail.jsp?id=187653.

102. Couto, V., et al. The Globalization of White-Collar Work, 2006. Available at www.boozallen.com/media/file/Globalization_White_Collar_Work_v2.pdf.

103. Anand, G. Federal law gives monopoly for seven years, fueling surge in biotech profits. *Wall St. J.*, November 15, 2005, p. A1.

104. Source: FDA website. www.fda.gov/orphan/designat/list.htm. Note that approvals originally granted to companies later merged with the current top 10 pharma companies, such as orphan drug approvals to Pharmacia & Upjohn, now part of Pfizer, were not counted in this percentage.

105. Ashburn, T.T., Thor, K.B. Drug repositioning: Identifying and developing new uses for existing therapeutics. *Nat. Rev. Drug Discov.* **2004**, *3*, 673–683.

106. Gay, L.N., Carliner, P.E. The prevention and treatment of motion sickness I. Seasickness. *Science* **1949**, *109*, 359.

107. Ghofrani, H.A., et al. Sildenafil: From angina to erectile dysfunction to pulmonary hypertension and beyond. *Nat. Rev. Drug Discov.* **2006**, *5*, 689–702.

108. Chong, C.R., et al. A clinical drug library screen identifies astemizole as an anti-malarial agent. *Nat. Chem. Biol.* **2006**, *2*, 415–416.

109. Agranat, I., et al. Putting chirality to work: The strategy of chiral switches. *Nat. Rev. Drug Discov.* **2002**, *1*, 753–768.

110. Olbe, L., et al. A proton-pump inhibitor expedition: The case histories of omeprazole and esomeprazole. *Nat. Rev. Drug Discov.* **2003**, *2*, 132–139.

111. Rohss, K., et al. Esomeprazole 40 mg provides more effective acid control than standard doses of all other proton pump inhibitors. *Gastroenterology* **2001**, *120*, A419.

112. For a brief review of this and other strategies, see Ho, J. Extending the product lifeline. *Pharm. Exec.* **2003**, *7*, 70–76.

113. For some real world advantages of combination drugs, see Wertheimer, A.I., Morrison, A. Combination drugs: Innovation in pharmacotherapy. *Pharm. Ther.* **2002**, *27*, 44–49.

114. Keith, C.T., et al. Multicomponent therapeutics for networked systems. *Nat. Rev. Drug Discov.* **2005**, *4*, 1–8.

115. Qazi, N.A., et al. Lopinavir/ritonavir (ABT-378/r). *Expert Opin. Pharmacother.* **2002**, *3*, 315–327.

116. Costello, D. Small rival takes deep breath. *Los Angeles Times*, October 30, 2007.

117. O'Brien, M.E.R., et al. Reduced cardiotoxicity and comparable efficacy in a phase III trial of pegylated liposomal doxorubicin HCl (CAELYX/Doxil) versus conventional doxorubicin for first-line treatment of metastatic breast cancer. *Ann. Oncol.* **2004**, *15*, 440–449.

118. Source: FDA website, www.fda.gov/cder.

119. For some examples, see Wertheimer, A.I., et al. Drug delivery systems improve pharmaceutical profile and facilitate medication adherence. *Adv. Ther.* **2005**, *22*, 559–577.

120. "New drugs entering clinical testing in top 10 firms jumped 52% in 2003–2005", *Tufts Impact Report*, Vol. 8, No. 3, May/June 2006.

**Chapter 2**

# The Drug Discovery Business to Come

*"Prediction is very difficult, especially about the future."*—attributed most often to Niels Bohr, but sometimes to Mark Twain, Albert Einstein, or Yogi Berra.

## 2.1 Introduction

Anyone who could really predict where drug discovery will be in 5 or 10 years would be worth billions. It's obvious from the previous chapter that just trying to understand where the field stands today with its many scientific, legal, and business factors is challenging enough. Every one of those factors is constantly changing. New discoveries are made, a few are retracted; new theories are propounded, discounted, then brought back again; clinical trials succeed or fail (or sometimes do both for the same drug in the same indication); investor strategies change; regulatory agencies are alternately urged to speed up or slow down drug approvals; new technologies excite, then disappoint expectations; laws benefiting or hurting the industry are passed; new pandemics are feared; new patient advocacy groups and philanthropic foundations become factors; court decisions suddenly change the whole picture . . . With so many variables and this much complexity involved, no supercomputer or pedigreed panel of analysts, much less this humble author, would be up to the task of correctly predicting where the industry will be in 2020.

In place of a definitive prediction, this chapter offers something of more limited, but still real, value: a glimpse into recent macroscopic trends for different types of organizations involved in drug discovery, the kinds of drugs being sought, and how some new technologies are being used in their pursuit. Whether these will be decisive factors in coming years will always be open to debate. Talk to two experts and you're likely to get two very different answers. Sometimes, as in the section that immediately follows, diametrically opposed movements seem to be going on simultaneously. In such cases, both directions are presented and the reader is free to decide, in effect, which, if either, will win out. The hope is that presenting a realistic view of recent trends in new drug discovery, while not exactly a crystal ball into the future, will at least familiarize the researcher with many of today's important issues as well as provide a starting point for his own explorations into tomorrow's.

## 2.2 New Models for Pharma

### 2.2.1 R&D Minus R

Aside from pharmaceuticals, there's another industry where success depends on the introduction of new products which are complex, risky, and expensive to produce, and where profits are largely derived from a few blockbusters. This is the motion picture industry.

Once upon a time a few Hollywood studios owned everything related to the movies, from scripts to cameras to (essentially) actors to the theaters their films were shown in. A few studio moguls had near-complete control over what movies most of the world saw through this process of *vertical integration*, which refers to ownership of every aspect (production, distribution, etc.) of a product from conception to retail sale.

The "studio system" of movie production had its heyday in the 1930s and 1940s, but federal court decisions and a new medium, television, eventually caused its demise. Outside of the legacy of movies it resulted in, little evidence of it can be seen in the industry today. No handful of studios controls the market anymore. Some of the grand old studio names still live on, but today's films are the products of partnerships between studios, of which there are now more, and the contributions of many independent contractors. "In the movie industry, people don't work for a single company, but join projects at points in which their particular skills are required—then move on to another project. The end result provides distributed returns, and fixed costs are minimized."[1]

In the same way, the landscape of drug discovery research today differs quite a bit from what it was a few decades ago, when *FIPCOs* (*Fully Integrated Pharmaceutical Companies*) ruled the world of prescription drugs. Think of an environment without biotech companies, revenue-draining generic competition, and major pricing pressures, where university research labs shied away from "commercial research" and a new first-in-class drug might remain without competition for years, and you'll begin to glimpse the world of big pharma before the 1980s. Big pharma, not biotech or academia or anyone else, made drugs. "It used to be that pharma companies were the only game in town," says Michael Rosenberg, CEO of Health Decisions. "They did all the discovery, all the development, all the marketing. Now, new smaller players can compete pretty well against big companies."[2] Although many new drugs and the research that makes them possible still do originate in-house in big pharma, the number of products brought in from other sources is increasing, leading some to wonder whether big pharma has lost its touch.

"It isn't a problem," says Dr David Brown of Alchemy Biomedical Consulting. "In any company there are a few hundred or a few thousand scientists, but there are hundreds of thousands of scientists around the world. So you've only got a very small amount of global research going on at any one company. There's always going to be more innovation outside than there is internally. You've just got to access that innovation.

"The question is: where does it settle? What percentage of sales come from outside? Well, we've got the figures. We know most of big pharma are settling in the 40–60% range. Companies tend to be grown on the basis of one big product or one class of products, and then it's almost impossible to repeat that trick, and you have to access things from outside. I don't see that as a problem. It's actually very positive."

As far as most people are concerned, big pharma exists to manufacture and sell prescription drugs. Coming up with the drugs in the first place is less strongly connected with the industry. Disconcertingly, polls show that in the public's mind new drug research is largely the job of universities and government labs. Because of this belief, much of the public doesn't really understand the difference between "innovator companies" and generic manufacturers, aside from what their products cost. It may also explain part of the reluctance to buy into the "expensive R&D" argument for high drug prices. But, as any pharmaceutical executive can tell you, the focus on prescription drug manufacture and sales by big pharma is real enough and is reflected in their employment numbers.

"The fact is," continues Dr Brown, "that pharma companies are really manufacturing and marketing companies. Most of the people in big pharma work in manufacturing and marketing. This hadn't occurred to me until GlaxoWellcome sent me up to business school. I remember being in a course with 20 or 30 other people. I was the only research person in the room. I thought that was a bit odd until I realized that, pro rata, that's about right. These companies are manufacturing and marketing companies. So it's reasonable that they're going to access products from outside."

One view is that the trend toward the increasing in-licensing of clinical candidates from biotech and other sources is part of an evolutionary process. The goal here would be an overall increase in efficiency through the specialization of biotech and academia in the early stages of drug research while pharma concentrates on the later stages, beginning with drug development and ending at pharmacy counters and hospital dispensaries worldwide. A graphical representation of what this type of future might look like has been outlined by Dr Jürgen Drews, former President of Global R&D at Hoffmann-LaRoche and a thought-leader in the field, as one possible future scenario. This is depicted in Figure 2.1. Here

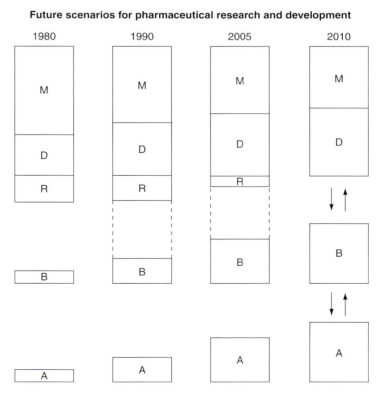

**Future scenarios for pharmaceutical research and development**

**Figure 2.1 ▶** Possible future model of pharmaceutical R&D. M = Marketing, D = Development, R = Research, B = Biotechnology industry, and A = Academic centers. (Reproduced with permission from Drews, J. *In Quest of Tomorrow's Medicines* (New York, Springer-Verlag, 1999), copyright 1999, Springer Science and Business Media.)

discovery research has become altogether separated from pharma, which essentially in-licenses everything. "Research directed toward the invention or discovery of new substances will be found more and more in the small biotechnology firms . . . The biotechnology industry is increasingly taking up what the pharmaceutical industry is neglecting, namely, carrying out future-directed research, thereby creating the conditions for the discovery of new classes of medicines that attack more directly the root causes of disease."[3]

The importance of biotech's contributions to drug discovery is almost universally acknowledged, but not everyone thinks that big pharma's role will ever be confined to development and marketing. "No, I don't agree with that one bit," says Dr Brown. "Imagine the scenario where a pharmaceutical company has no research. If you haven't got R&D people who understand R&D, you're going to make serious mistakes in what you in-license. They're going to be *big* disasters."

*Due diligence*, in this case the process of thoroughly investigating a potential in-license candidate before making a decision, would be hard to exercise in the absence of research facilities and the experienced research personnel to drive the process. It would be a little like buying a car solely based on a nice picture and what the salesman told you about it. To do it right, you need to get an unbiased view of how it compares with other products, then kick the tires, and give it a test drive. Although it's possible that a kind of buyer's guide could be brought in through business development or in-licensing advisory services or that some minimal research department might be maintained to evaluate potential candidates, it just seems a difficult way to go and one predicated on the assumption that licensing in the products of someone else's research will always be more cost-effective than doing one's own. Merck, for example, "is in disagreement with the speculation that big pharma will become 'virtual' entities performing little or no research" and finds that "internal strengths are essential to provide the required expertise that is needed to assess the suitability of a potential partner."[4] So separating the "R" from the "D" isn't likely to work unless the two can be put back together again, and in the right order.

Although many doubt that pharma will see a wholesale trend toward the divestiture of in-house research in coming years, at least one example of pharma R&D without the "R" already exists. In February of 2006, Procter & Gamble Pharmaceuticals announced it was dropping all of its in-house discovery phase efforts in favor of in-licensing, leaving it to others to supply their development candidates.[5] This was an extension of a strategy they had developed called "C&D" for "connect and develop". They had observed that "important innovation was increasingly being done at small and midsize entrepreneurial companies." They found that "even individuals were eager to license and sell their intellectual property. University and government labs had become more interested in forming industry partnerships, and they were hungry for ways to monetize their research." With the ready market availability of such IP, they predict that "connect and develop will become the dominant innovation model in the twenty-first century."[6]

### 2.2.2  D Plus R

Although the general trend toward increasingly outsourcing what was formerly in-house research is there for all to see, a number of cases of the opposite philosophy, adding in-house research where it previously didn't exist, is also occasionally in evidence. Such cases point to a

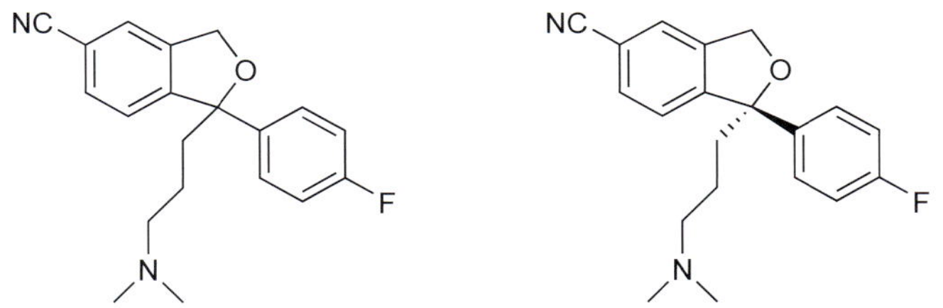

**Figure 2.2** ▶ Left: Celexa (citalopram). Right: Lexapro (escitalopram).

smaller, but real, corporate movement in the opposite direction to the one noted above. A company called Forest Labs, which does about $3 billion in annual sales and is obviously on the smaller side of big pharma, has grown rapidly in the last few years. Forest has so far done little early-stage research in-house, but has managed to be quite successful through in-licensing to supply their product pipeline. To date, this approach has done remarkably well, with their biggest product now being Lexapro (escitalopram; Figure 2.2), a chiral switch version of their previous racemic CNS drug, Celexa (citalopram). But how long the company intends to operate along these lines is uncertain: Its recent annual report states that "we think it is time for Forest to become involved in the earliest stages of drug development," although "for the time being" early-stage (drug discovery) activities will be done through collabora-tors.[7] Increasingly intense competition for available clinical candidates might be one factor behind this decision.

In a second interesting example of potential "R" addition, some generics companies, which one thinks of as being profitable precisely because they don't rely upon their own risky and expensive R&D efforts, are now experimenting with in-house drug discovery programs. Ranbaxy Labs, a very large generics company, has evolved its own *NCE* (*New Chemical Entity*) pipeline and states that it channels about 6% of its sales into R&D.[8–9] Another Indian generics company, Dr. Reddy's Labs, reported having seven NCEs, four of them in clinical development, in 2005.[10] Such efforts raise the possibility that in future, companies like Ranbaxy and Dr. Reddy's Labs might become independent players in the US new drug market.

Dr Albert Wertheimer of Temple University believes that the economics involved make it likely that this will happen. "Ranbaxy started out by doing joint ventures with Lilly and doing subcontracting with Lilly, and now they're on their own more and more," he says. "If, hypothetically, they discover a product in India, what they would have done before is license it to their partner, Lilly, and Lilly would give them, let's just arbitrarily say, 7% royalties on it.

"Well, why should they be satisfied with 7%? Now that they've had some experience in the joint venture, they have some staff in America, and they can now get a drug through the US FDA themselves, they can sell it and make 100% of what comes through. Those companies now all have US facilities. They have deep pockets, the money to invest that a small, local generic company might not have otherwise. So that will be their foot in the door."

To what extent big generics manufacturers will really be willing to forsake the low risks and near-guaranteed returns involved in producing and selling drugs someone else has brought to market for the expense and uncertainty of developing their own remains to be seen. For example, the giant Indian generics company Sun Pharma recently "de-merged" its

R&D efforts as a separate corporation to cut down on research costs and protect profit margins. But overall, as they now have pools of money to invest, lower labor costs, and contract research, with collaborative, regulatory, and manufacturing experience to boot, it's likely that there will be some movement in this direction by Indian generics companies.

What seems certain is that the monopoly on all phases of drug discovery and development which was formerly held by big pharma, like the monopoly Hollywood studios once held on movies, is over. New sources of clinical candidates as well as independent organizations capable of doing every stage of the R&D process now exist and are increasingly being utilized by pharma itself. In-house research remains an important part of the equation, and as real-world drug development and sales requires deep pockets, experience, and large candidate portfolios to spread out the risk, big pharma is not likely to be supplanted as the major producer of new drugs anytime soon. But in place of the previous monolithic model, today's and tomorrow's research scenarios are significantly more diversified, with ideas, work, and ultimately candidates flowing from a variety of new sources, both foreign and domestic.

### 2.2.3  Smaller is Better

Another issue which has drawn close attention lately is the effect of an organization's size on productivity and innovation. Drug discovery research is a dynamic process that calls for constant, informal contact between the many disciplines involved as well as an environment where imagination and experimental tinkering have room to roam. Rigid structures and inflexibility are anathema to the creative mind. In the words of Dr Drews, "Hierarchy, bureaucracy, unnecessary diversion on account of the requirements of other areas of the enterprise—these are all counterproductive influences on a research organization. Nonhierarchical organization, informality, openness, and a congruence between scientific capability and personal integrity on the one hand and responsibility on the other are typical indicators of successful research organizations"[11]

As the size of an organization increases, however, these organizational qualities becomes harder and harder to achieve. Flat reporting schemes which may work well with a few hundred employees become untenable at an organization with 10,000. Decisions call for more buy-in from senior management, which is more and more remote from scientists in their labs. Decisions come from the top down while too little information manages to flow upward from the rank-and-file.

"Size does not correlate with superior performance," notes one report. "Among the top 20 pharma companies, the largest firms perform no better than the smaller companies."[12] "My prejudice," remarks biotech industry consultant and writer Dr Cynthia Robbins-Roth, "is that every large organization, whether it's pharmaceutical or otherwise, has a very difficult time making decisions and going into new territory. And everything takes longer. It just does."

As with a Mack truck as opposed to a sports car, the advantages of size bring with them a concomitant loss in speed and maneuverability. These observations have been noted in biotech companies, which are of course generally smaller and sportier (and carry less freight), for a long time now. But in recent years, in the face of the "productivity gap" and investor unrest, big pharma itself has bought into this argument and is making an effort to address it.

"I think that most drugs will come out of smaller organizations," says Dr Mark Murcko. "I agree with the sentiment that smaller, more focused organizations are more likely to produce good development candidates. But the question then is, 'What do you mean by smaller organization?' It could be biotechs or retooled big pharmas."

One model of this kind of retooling involves the breaking down of a single, large, centralized research organization into smaller, more focused, semi-autonomous business units. The poster child for this approach is probably GlaxoSmithKline with its "Centres for Excellence in Drug Development" (CEDD). Currently GSK has seven of these units in addition to a Centre for Excellence in External Drug Development (CEEDD), which focuses on external collaborations and licensing. The focus areas for the CEDDs include biopharmaceuticals, cardiovascular and urogenital diseases, and others.

GSK notes that "One of the historical contradictions in the pharmaceutical industry has been the need to leverage the advantages of a large organization without losing the creative spirit of the research environment." While some areas like HTS can benefit from large scale, "other areas flourish to their best advantage if the structural units remain small: the units can respond quickly to the changing environment; the opportunity for scientists to interact is optimized; and the need for return on investment is focused through the fostering of an entrepreneurial, accountable culture." The resulting centers are "focused on specific disease areas and designed to be nimble and entrepreneurial with the range of skills and resources required to drive mid-stage development projects from lead optimization through to their key decision-point, proof of concept."[13]

The biggest of all big pharma companies, Pfizer, recently announced a corporate reorganization along similar lines. It aims to "create smaller, more focused and entrepreneurial business units that will enhance innovation and draw on the advantages of our scale and resources" by dividing up into four separately accountable business units organized around specific therapeutic areas.[14] This is a big move and one not very easy for such a massive organization to pull off. But it's one that's likely to be repeated by other pharma companies in coming years as a possible antidote to low productivity and problems of scale.

Along a separate line, one could argue that another way of maintaining an entrepreneurial spirit in pharma is to avoid crushing and digesting smaller, more entrepreneurial companies in the aftermath of an acquisition. Allowing acquired biotechs a certain amount of autonomy instead might do wonders. For example Roche, which for years now has been the majority owner of Genentech, has seen its relatively hands-off approach rewarded handsomely in terms of both innovation and profits. Unfortunately, many pharma companies still don't follow such an enlightened policy, preferring to acquire small companies for their products or clinical candidates and then either attempt a Borg-like assimilation or simply discard the rest. Either way destroys something potentially more valuable than the products themselves, namely, the research culture that produced them. "If a small company is allowed to retain its autonomy, it benefits from more intangible human factors, such as the fact that people work harder and more effectively when they feel they are in control of their own destiny."[15] But instead, big pharma often seems to pluck the ripe fruits of biotech companies, then chop down the trees for firewood.

For both pharma and biotech "grow or die" always seemed to be the rule. But how to grow innovation and subsequent profits without growing an enormous infrastructure and its inevitable inefficiencies is the problem. Increasing willingness to use contractors and collaborators for parts of the R&D process and corporate reorganizations designed to provide smaller, more productive business units are probably part of the answer.

### 2.2.4  Specialty Drugs

Recent years have seen a shift in the kinds of prescription drugs that companies are developing. A number of factors have gone into making drugs prescribed by primary care physicians for large patient groups less desirable than those designed for the *specialty pharmaceutical* market. The latter have been defined as "high-cost injectable, infused, oral, or inhaled drugs that generally require close supervision and monitoring of the patient's drug therapy."[16] They're often (e.g. Tysabri), but not always (e.g. Cellcept, a small-molecule immunosuppressant), biologics and tend to be prescribed by medical specialists like oncologists or rheumatologists. They're characteristically directed at diseases having smaller patient populations too.

In these post-Vioxx days, drugs meant to be widely prescribed require huge, expensive clinical trials as a hedge against potential toxicity to a small subset of patients. Widespread usage means that a drug must be incredibly safe, particularly if other drugs are available as alternatives and the indication it treats isn't immediately life-threatening. As we've seen, regulatory agencies are likely to adopt a conservative stance and eschew making decisions too rapidly under these conditions. So the development and approval of widely prescribed drugs tend to be expensive and slow.

Specialty drugs are a different matter. Restricting the size of the potential market doesn't seem like a good thing to do from a profit perspective, but higher prices associated with specialty drugs can be used to compensate for this. Although one might wince at a drug priced in the tens or even hundreds of thousands of dollars per year, without the financial incentive this provides, such drugs would simply never be developed and such treatment options wouldn't exist for patients. Specialty drugs often bring with them the additional benefit of orphan drug designation (if the indication affects fewer than 200,000 people in the United States) with the resulting tax advantages, priority review, and market exclusivity. Recent growth in specialty drugs sales has far outstripped that of prescription drugs in general. For 2007, IMS Health has predicted 10–11% growth in specialty drug sales, about twice the growth rate expected for the global pharmaceutical market.[17]

Something that's obvious to pharma executives, if not necessarily to research scientists, is that a focus on specialty drugs means that fewer sales reps (detailers) are needed, as the number of specialists prescribing such drugs is bound to be much smaller than the number of primary care physicians prescribing more widely used medicines. And last but not least, public perception of an industry that makes new drugs for cancer, crippling rheumatoid arthritis, or serious inherited disorders, is likely to be a lot more favorable than that of an industry engaged in producing *life style drugs*, even though the resulting medicines may be pricey. At this point, the industry can use all the public goodwill it can get.

This being the case, the shift toward drugs prescribed by specialists in oncology, rheumatology, and several other areas, "where reimbursement is still comparatively generous, regulations still reasonable, and lawyers morally challenged to wage aggressive war,"[18] is likely to continue for the foreseeable future.

### 2.2.5  Pricing Pressures and Price Controls

Prices are always too high: ask any buyer. Prices are always too low: ask any seller. Pharmaceuticals aren't exempt from this dynamic that makes the economic world go round. But given the importance of medicines—which, after all, are directly tied to at least two of the unalienable rights to "life, liberty, and the pursuit of happiness"—it's really no wonder that drug prices should be the subject of keen and passionate debate.

According to the Kaiser Family Foundation, a non-profit foundation that researches health care issues, prescription drug spending in the United States totaled $188.5 billion in 2004, more than 4½ times the total spent in 1990, and it's estimated that it will reach $446.2 billion in 2015.[19] Although such spending represented only about 10% of national health care spending in 2004, it was until recently increasing at double digit rates, faster than any other segment as Figure 2.3 illustrates.

Figures on prescription drug spending, however, can be misleading. Increases in dollars spent reflect, in part, the increasing use of medications. According to the Kaiser report, the number of retail prescriptions per capita increased from 7.9 in 1994 to 12.3 in 2005. This in turn can be attributed to factors like growth in US population, especially the high growth in the population segment above 65 years of age, which tends to use many more drugs than the general population, and the use of medications to treat conditions patients might not have been willing or able to treat with prescription drugs before.[20] So a real economic analysis of what these numbers mean is a difficult and complex undertaking.

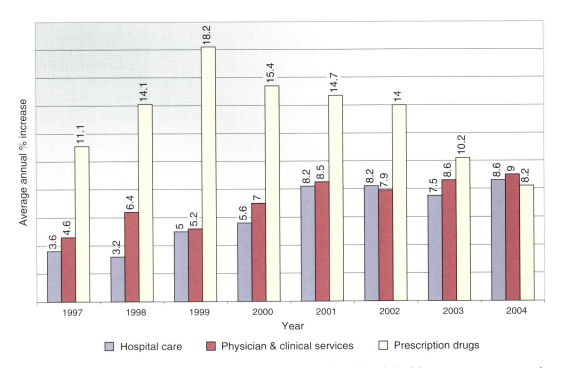

**Figure 2.3** ▶  Annual percentage change over previous year in selected health care expense categories. Source: HHS Centers for Medicare & Medicaid Services (CMS).

But one thing at least is simple: the average American is now paying more, possibly much more, for prescription drugs than she used to, and that has not sat well with the public. Consider the following, gleaned from recent surveys:

▶ 83% of those surveyed consider brand name prescription drugs "unreasonably priced."[21]

▶ A full 50% of Americans surveyed agreed that "Drug/insurance companies making too much money" is one of the single biggest factors in rising health care costs, while only 23% believed that the aging population is responsible, and a mere 12% attributed it to more people getting better medical care.[22]

▶ Nearly two-thirds of Americans (65%) believe there should be more government regulation of drug prices, and 46% of those believe it's necessary *even if it leads to less new drug R&D*.[23]

Rising health insurance costs, higher deductibles, and prescription copayment creep have managed to drown out industry arguments, for example, that for every extra dollar spent on prescription drugs, $3.65 in hospital care expenses are saved,[24] or that such high prices are necessary to produce new drugs. Facing an uphill battle in the court of public opinion, public relations and legislative lobbying campaigns have ensued.

But to see how things might change in future, we need a glimpse at where they are today. Most people are aware that outside of the United States, most countries with large markets for prescription drugs like Canada, Japan, Australia, and EU member nations already employ some form of price control. And we've already seen that although as researchers we may be sympathetic to the argument that drug prices need to be high enough to provide an incentive to produce the next generation of drugs, this argument tends not to sway those governments' regulatory agencies. Their chief concern is really what benefit a particular drug will bring to patients at what cost compared with existing therapeutics, if there are any.

Dr Albert Wertheimer, the Founding Director of the Center for Pharmaceutical Health Services Research at Temple University, a former Director of Outcomes Research at Merck, and an expert in the field of pharmacoeconomics, says, "Price is a function of what the company can do, given the laws and regulations in that country. The same new drug in Australia comes out at about 20% of the US price. It really has much more to do in Western Europe with the company negotiating with the government over the wholesale price of the drug. The ministry can say, 'Well, we're not going to pay more than 40 cents and that's it.' "

A system often used to determine what the price should be in these situations is called *reference pricing*, which is what's done in Germany, for example. Dr Wertheimer explains: "So the first beta blocker goes on sale for, say, 75 cents a tablet. And if you have another one and you go to the ministry and say, 'we want to sell it for 90 cents,' they might say, 'Well, we already have one for 75 cents a tablet, so why should we grant you one for 90 cents?'

"But if you come to them with a pharmacoeconomics study or a cost–benefit calculation and show that yours is a once-a-day dosage form and the others are three-times-a-day, and people are more compliant, there'll be better blood pressure control and fewer strokes, saving them tons of money, they'll say, 'Alright, 90 cents it is.' "

Reference pricing isn't the only scheme used to control drug prices. In the United Kingdom, for example, profits of pharmaceutical companies—not drug prices *per se*—are capped by a voluntary agreement between the Department of Health and the Association of the British Pharmaceutical Industry (ABPI) so that "the UK system allows companies to price new products, but within the overall constraints of a company's PPRS (Pharmaceutical Price Regulation Scheme) profit cap."[25] So there's no single, standard method for price controls, and more than one regulatory mechanism can be used within a given country as well.

One would think that in the United States the situation would be very different; no such government agencies or schemes currently exist and drug prices should be market-based. "Well that's true," says Dr Wertheimer, "but to a smaller extent every single day. It would be misleading to say that we have a *laissez-faire* system that isn't controlled in any way. Let's say tomorrow GSK says 'We want $2 a tablet for this drug.' And they take this to the managed care organizations and *they* say, 'We're not giving you $2. If you want that on our formulary you'll have to beat the price of the drugs that are there now, and we don't pay more than $1.40 for any of them.' So there is that sort of negotiation that does take place."

Even in the United States, it seems, there are now major factors working against drug price increases: not only price negotiations with managed care organizations, *pharmacy benefits managers* (*PBMs*), and government agencies like the Veterans Administration, but the proliferation of generics brought about by Hatch–Waxman, a related phenomenon called *class cannibalization* whereby the first drug in a therapeutic class to go generic takes away much of the business of all the others as formularies redirect patients toward them (e.g. generic simvastatin now biting into sales of Lipitor and other statins),[26] pressures on end users resulting from *tiered formularies* which tie more expensive drugs to higher copayments, and, last but not least, the fear of provoking draconian price control laws. The cumulative effect of these factors has probably had a lot do with the downward curve in drug price increases seen in the last few years and illustrated in Figure 2.3. Signs that the industry is getting the message can be seen from time to time. Recently Genentech voluntarily decided to cap the price for its cancer drug Avastin (bevacizumab, a VEGF receptor antibody) at $55,000 per year for patients making less than $75,000 per year, and Amgen priced its new cancer drug Vectibix (panitumumab, an EGFR antibody) at 20% less than another EGF receptor antibody, Erbitux (cetuximab), in moves that were immediately unpopular with investors.[27]

Some of the most concentrated efforts to mandate lower drug prices have come not from the federal government but instead from individual states. These have seen rising drug prices place increasing demands on their budgets via programs like Medicaid that the federal government mandates but requires states to pay for. Legislation designed to coerce drug price reduction for low-income residents has been introduced and often passed, literally from Maine to California, in recent years.[28] The 2006 California Prescription Drug Initiative, for example, gives pharma companies until 2010 to voluntarily lower the costs of brand-name drugs up to 40% and generic drugs up to 60% for uninsured low-income Californians. If such major reductions haven't happened by then, the state could take their products off its preferred drug list for Medi-Cal, which currently spends $4 billion per year on drugs to elderly and impoverished patients.[29]

Opinions about how much more legislation the industry should expect in the future, of course, are divided. When asked whether he believed that the US government will step in and take on a role in drug pricing similar to that of foreign governments, Dr Wertheimer replied, "I don't see that happening for a long time, at least not directly. It's at odds with the free enterprise *ethos*. But that doesn't mean the government won't continue to be an aggressive buyer, through the VA, for example. They can do that more strongly. But I don't see them dictating market price for the private sector."

If opinions diverge on whether or not drug price control is coming to America, it isn't surprising that a similar chasm exists on what effect it might have on the industry if it should arrive. One school of thought holds that the adoption of a scheme like reference pricing, which benchmarks a new drug against others in its class, might actually serve as an incentive for innovation. "One likely consequence will be a shift in research toward diseases not currently treated by multiple drug therapies. The reason is simple. These drugs would be literally incomparable... Drugs that are not first or second to market will have to demonstrate clear superior efficacy for a targeted set of patients—or get out."[30]

Sidney Taurel, past president of PhRMA and former CEO of Eli Lilly & Co., presents a very different prediction: "I believe pharmaceutical companies would focus first on how to survive and later on how to succeed in the new environment. The first thing they would likely do is to focus much more attention and resources on maximizing sales of existing products. That means much greater marketing expenditures to try to gain market share, and the only place that money can come from is the research budget." He goes on to predict that price controls would cause "the collapse of true innovation in biomedicine."[31]

In 1993 and 1994 the Clinton Administration looked into health care reform, took a critical look at the pharmaceutical industry, and it was widely suspected that drug price controls were on the table. Although in the long run no new legislation would come out of this effort, even this had a major effect on investments in pharma and biotech. A detailed economic analysis has associated a 52% decline in market-adjusted pharmaceutical stock prices with it,[32] while biotech stocks plunged 40%.[33] Interestingly, it was found that during this time pharma largely held back on price increases,[34] so "political pressure can restrain prices even absent direct regulation,"[32] and there was little difference in the decline between innovator pharma and generics stocks. The latter probably indicates that investors didn't discriminate between R&D-based and non-R&D-based organizations: they punished both. And finally, during this short period, although R&D spending continued to increase, its growth rate suffered a sharp decline.[31]

It seems that any perception that the US government might regulate drug prices could set off a stampede of investors fleeing the sector. Industry response, at least initially, would probably involve a concentration of effort and capital on sales of existing products and those already in the pipeline. More extensive offshoring of R&D to cut costs would be a given. Mergers and acquisitions to consolidate revenues and market share would probably increase. Diversification of pharma companies into other sectors, which has happened before,[35] might once again become fashionable. At least for a while, drug discovery R&D would likely suffer cutbacks and hard times. "If you reduce the reward for an activity," says Dr Joseph DiMasi, "then you will get less of it."[36]

Although there's no crystal ball that can see into the future of drug price regulation, some current trends have enough momentum behind them to last for decades. One of these is drug pricing pressure. Whether explicit in the form of government regulation or implicit in the growth of tiered formularies and unfavorable public opinion, this is a trend that's here to stay. It brings with it a sea change in what a marketed drug needs to achieve. Up until now, the key attributes for a clinical drug to be successful have been *safety* and *efficacy*. As difficult as these have proven to achieve, alone they're no longer enough. A third property, *economy*, now increasingly needs to be demonstrated.[37]

In the face of pricing pressures, "outcome-based medicine will predominate, increasingly requiring product innovators to demonstrate clear pharmacoeconomic benefits."[38] Cost–benefit analyses and comparisons with other therapeutics are now *de rigueur* for new drugs and will only increase in importance in future.[39]

## 2.3  New Models for Academia and Biotech

### 2.3.1  Translational Research

As mentioned in the previous chapter, the Bayh–Dole Act of 1980, which allowed academic institutions and researchers funded by government agencies to hold title to their inventions and license them out as desired, represented a turning point in the direction of academic research, which inevitably became more product oriented. It soon became apparent that the most useful and most lucrative avenues of research were those that helped provide commercial products, new drugs, and that the closer an invention was to one of these, the more value it had. A new paradigm was needed to wring the full value out of earlier-stage academic inventions, which were often related to basic biochemistry and molecular biology, and not drug candidates *per se*.

"The trend that I've noticed in the last year," explains Dr Michael Venuti, "is that more and more, universities aren't waiting for somebody to come in and simply license their patents. What they're doing is proactively moving a biological finding into chemistry somehow. In fact, there are a few universities that are willing to pay for that chemistry so that they actually have an asset that's licensable as a drug candidate rather than as a biological assay. It's a different way to leverage the biology, instead of reading *PNAS* and setting up screens. This is a big change, a very big change.

"They're now looking to leverage the foundational biology that comes out of big universities into a chemistry situation where there's an actual molecule that will make money. They see, for example, Emory University making hundreds of millions of dollars on some of the HIV drugs from Pharmasset. And those are molecules that went into the clinic. They're not assays. So a lot of that innovation may actually come out of collaborations, joint ventures, and risk sharing deals with universities."

The ultimate goal for university biomedical research has become the "bench to bedside" translation of basic scientific findings into medical therapies, a process called *translational research*. No one's likely to argue with the fact that, ultimately, basic biology does result in new medicines and is the true foundation for today's as well as tomorrow's therapeutic armamentarium. Current discussions on translational research instead revolve around ways that the parties involved should be organized and funded to streamline the process.

A look at how the process has typically worked in the past provides a starting point. Keep in mind that this only represents one model. In the real world not all drugs are in-licensed, pharma often does its own basic biology research (Novartis's Institutes for Biomedical Research being one example), and plenty of different schemes exist.

The translation of ideas into new drugs requires, among other things, good science, hard work, a fair amount of luck, and enormous amounts of money. Since prospective patient groups can't be billed in advance for the basic biology research that might or might not give rise to potential future treatments, funding typically begins with government agencies, principally the NIH, which does its own research through an *intramural* program and has its own tech transfer office. Mostly, however, NIH allocates grant funding, primarily to universities, as part of its *extramural* focus. Inventions resulting from this government-funded extramural research would then (under Bayh–Dole) belong to their academic inventors and the institutions they represented. Their technology transfer offices could aggressively license this intellectual property out, either exclusively or non-exclusively, to existing biotech or pharma companies. In return the universities could get licensing fees, milestone payments, and potentially even royalties on sales of drugs, should any eventually result from the use of the inventions. But that, of course, is never a simple matter.

Since big pharma traditionally has been most interested in downstream IP—read *real clinical candidates*—but academic IP traditionally related to earlier-stage discoveries like potential targets, models, technologies, assays, etc., some sort of intermediate party would have to do the "translation." Often academic out-licensing to biotech would accomplish this, but sometimes, academic patents would engender new companies founded by the faculty member/inventor involved, who would also be a co-owner of that IP. These would be set up with the explicit aim of doing that translation, bringing their inventions closer to the clinic. This gives rise to a whole new set of challenges, not the least of which are the gathering up of the various disciplines needed to make a go of it—not only sciences but even law and business—and the transition from a focus on long-term, publishable, basic science to one based on shorter-term, investor-supported, proprietary applied research. Other researchers in the field, which the founder might still view as competitors but investors would like to welcome as board members, can cause potential conflicts, and *founderitis* in the form of discord between businessmen/investors and the company's originator, can set in.[40]

Big pharma would traditionally become interested only relatively late in the game, preferably when a resulting candidate had reached Phase II or Phase III trials. At that point the costs associated with the trials would be burdensome for small companies, and relatively few would have the resources or the willingness to gamble required to see it through all the way to approval. Instead, big pharma would tend to step in and buy either the candidate or the whole company.

In this modular model four different organizations, the NIH, a university, a biotech or small pharmaceutical company (itself probably funded by venture capital or stock offerings), and a big pharma company, are required to provide the variety of disciplines, expertise, and funding that it takes to go from bench to bedside. Dr Henry Grabowski, Professor of Economics at Duke University puts it this way: "Government-supported research gets you to the 20-yard line, biotech companies get you to the 50-yard line and [the big pharmaceutical companies] take you the rest of the way to the goal line."[41] Splitting up the process that way may allow each organization to concentrate on what it does best (and hopefully not fumble or be forced to punt!), but at the same time it has several drawbacks.

Lines of communication between parties won't necessarily be open. Contributions can be thrown "over the wall" from one organization to the next. But the new drug process requires the combined interplay of all disciplines up and down the line. "It is difficult, if not downright impossible, to successfully develop a drug by solving problems individually in isolation, because each technical choice (the target you pursue, the molecule you develop, the formulation, the design of the clinical trial, the choice of the target patient population, and the choice of manufacturing process) has implications for the others. Arriving at a solution requires that different kinds of scientists repeatedly exchange huge amounts of information. In other words, they must work together in a highly integrated fashion."[42]

The "over the wall" model implies fragmented development, loss of input, loss of control, and even loss of information on development issues and status. It can be particularly frustrating for the academician who originated the idea or the researchers who made the compound to be kept "out of the loop" later on. But most importantly as pointed out earlier, this system, which uses tech transfer licensing of something less than a clinical candidate, brings far less money to its inventor and the university than a potential drug would, because of the years of work required to develop it into one, the odds that it never *will* result in one, and the fact that yet another party is involved, trying to maximize its own income. Both the organizational and the economic issues could be addressed by carrying out translational research completely in academia or other non-profit institutions at least to the point of clinical testing.

Most of the disciplines required to develop a drug are well represented within academic halls, although personnel and funding for the more basic sciences like molecular biology and organic synthesis far outstrip resources for more applied ones such as formulations. So in theory, there's no reason why most things from target identification to clinical trials can't be done there. But a couple of major problems exists. Funding is one of them. Over the years the NIH, as we've seen, has funded basic, not applied, research, which explains why few academic groups focusing on formulations can be found. Since much of translational research consists of applied research, lack of government funding in years past has been a problem. Furthermore, in academia "research" is rarely interpreted to mean the continuous running of nuts-and-bolts assays for things like solubility or cell permeability, nor does it include the more "industrial" parts of the process like API manufacture. Parts of the process might therefore need to be contracted out, and that means money, or perhaps equity. A company called Bridge Pharmaceuticals, for example, a spin off of SRI that outsources contract work to Asia, has recently been willing to trade work for an equity stake in the project.[43] Still, at least most of preclinical developement could be done in academia.

But the bigger problem is a cultural one. If walls exist between academia and biotech or biotech and big pharma, much higher and thicker ones can sometimes be found between departments and even research groups at a given university. Compared to the way things are done in industry, university research usually takes place within deep and narrow research groups and they operate, for the most part, independently of one another. Connections between them tend to be too few. So although every individual problem encountered in developing a given drug might be solved there, and solved brilliantly, by the various groups, they aren't set up to work on the same project, much less to work together. The broad, interdisciplinary integration that's so vital to drug discovery and development has been the exception in academia, not the rule.

In light of the benefits of more successful translational research to universities as well as patients, steps are being taken to address these difficulties. The recent *NIH Roadmap for Medical Research*[44] includes grant funding meant to encourage the multidisciplinary training of researchers through Clinical and Translational Science Awards (CTSA) and a pilot program called Rapid Access to Interventional Development (NIH-RAID) which makes resources like compound scaleup and formulation development available to researchers who otherwise would lack these necessary components of translational research.

Non-profit foundations have shown an interest in doing their part to enable translational research as well. Notably the Howard Hughes Medical Institute has funded both physician-scientists engaged in translational research[45] and graduate programs designed to introduce Ph.D. students to the world of clinical medicine.[46] Organizations like the Leukemia and Lymphoma Society have established grant funding to encourage translational research as well.[47] One of the largest non-profit research organizations, the Scripps Research Institute, is building a Translational Research Institute at its new Florida facility, to be headed by Dr William Roush, former Chairman of the chemistry department at the University of Michigan, Ann Arbor.[48]

And academia itself has made an effort to break down the walls between disciplines and groups that stand in the way of translational research. One example is the Broad Institute, an interdisciplinary collaboration between Harvard and MIT made possible by donor funding, which includes in its faculty diverse and luminary scientists such as chemical biologist Stuart Schreiber and geneticist Eric Lander of HGP fame.[49] Another organization operating under the aegis of Harvard is the Laboratory for Drug Discovery in Neurodegeneration (LDDN), a "not-for-profit biotech" which seeks to identify and develop new lead compounds into promising CNS therapeutic candidates.[50]

To carry this concept further, the suggestion was recently made that National Medicinal Chemistry Resource Centers should be established, presumably with public funding, to translate academic hits into saleable drug candidates.[51] Translational research, which adds the many challenges of drug discovery and development to those academia already faces in basic science, is anything but easy.[52] A lot of funding, effort, and experimentation probably need to happen before "bench to bedside" becomes a major force in new therapeutics. But the driving force is there: the potential for more rapid, efficient, and profitable new drug development utilizing large pools of postdoctoral talent unavailable to industry except at a higher price. The advantages are just too great for it not to happen in time, one way or another.

## 2.3.2  The Standard Biotech Model

To calculate a future direction a current starting point needs to be defined. So far, the model by which basic scientific discoveries are translated into marketed drugs via entrepreneurial organizations owes much to the example of Genentech. As a new type of company and a for-profit enterprise able to attract many millions of investor dollars long before achieving profitability, collaborating with big pharma, and then later going on to develop its own successful drugs, the Genentech paradigm was permanently etched into the collective psyches of Wall Street investors, entrepreneurs, and scientists alike. It seemed to offer hope not only of revolutionary new therapeutics but an equally revolutionary business model as well, one that almost every other biotechnology or small pharma company formed since then has tried, without success except in a very few cases, to emulate.

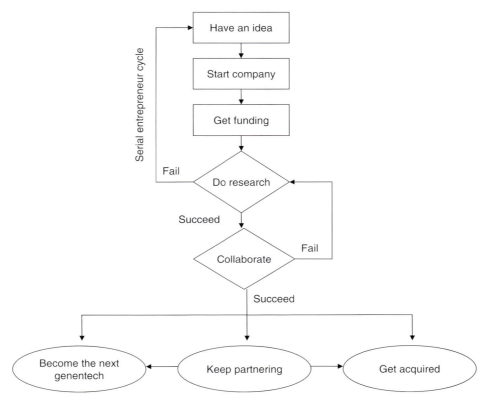

**Figure 2.4** ▶ Simplified model of drug discovery startup development. "Succeed" and "Fail" first refer to the ability to attract a corporate partner, then later to the production of a promising drug candidate in clinical trials.

Figure 2.4 shows a rough outline of how companies intent on discovering new drugs have often started out and developed in the past. An idea sets the whole process in motion and becomes the seed from which the enterprise springs. So *seed funding* becomes the next order of business for the inventor, who now, having presented a valid business plan and in seeking to develop and commercialize the idea, meets the definition of entrepreneur. Formation of an initial team—nobody goes it alone—and successful acquisition of *angel*, venture capital (VC), government grant, or foundation funding brings initial access to the lab space, equipment, and the people required to test the feasibility of the approach. Multiple, potentially difficult, and increasingly large rounds of financing are normally required to keep the process going. These in turn depend on demonstrating that good progress is being made on the path to *proof of concept* (*POC*), scientific evidence that the idea could result in an effective therapeutic agent.

Should the money run out before preclinical POC experiments can be performed or if the concept can't be demonstrated in those experiments, both of which happen frequently, the company fails, lays off employees, attempts to license out whatever IP it may have at that point, and sells off its equipment. Anyone who's gone to an auction of equipment previously used by a defunct biotech or seen vacant labs with their scattered lab books waiting for the next experiments to be entered knows what a sad event that is.

Every once in a while, though, things go right. Management teams succeed in keeping a new company afloat through those difficult early stages, progress is made, milestones are met, and compounds are produced that give promising positive results in animal models. Sufficient funding might even be found for early clinical trials that would greatly enhance the value of the new drug. At this point spirits are high, expectations are higher, PowerPoint slides proliferate, presentations are rehearsed, and corporate partners are courted or come courting on their own.

Just about every small company with positive results of any kind has traditionally wanted to be a major, independent supplier of new drugs ("the next Genentech"), but few expect to go it alone anymore. One company, Genzyme, did just that, growing to become a profitable biotech company and supplier of new drugs and diagnostics without a single big pharma collaborator, but it remains very much the exception to the rule.[53] More typically, the huge costs and many different kinds of expertise required to get a drug to market mean that, at least initially, a big pharma partner is needed in order to get there. The plan would generally be to use the milestone payments, royalties, and other income from these types of collaborations to build up in-house resources, increase the value of the company's stock (having "gone IPO" already), and, as quickly as possible, to change corporate focus from external collaborations to internal development programs thus achieving the desired vertical integration. After all, it worked for Genentech!

But it hasn't worked for almost anyone else. Success in this "California model" of biotech has turned out to be elusive to many hundreds of companies. The transition from research collaborator to independent drug seller has proven more difficult than anyone could imagine, even with the best business and scientific acumen behind it. A very few companies were ever able to make this model work, and that was some time ago. As Martha Amram, Founder and Managing Partner of Growth Options Insights puts it, "In the California model, you take great work from UC Berkeley, form a company, get venture capital, do an IPO, and grow up to be Genentech. That model is dead."[54]

A much larger number of entrepreneurial companies have found that a continuing stream of collaborations with big pharma in itself might provide a sustainable model for solvency and growth, if on a more modest scale. Such an arrangement amounts to the outsourcing of a big pharma company's R&D project, usually at the clinical candidate stage, as we've seen earlier, but also sometimes at the discovery stage, especially if the smaller company features an interesting technology, platform, or area of expertise (e.g. antisense oligonucleotides for Isis) rather than one particular drug. Figure 2.5 shows the growth in the number of such big pharma collaborations there's been in recent years, while Figure 2.6 demonstrates biotech's increasing dependence on them for its funding.

Outright acquisition by a big pharma company has always been a third possible endpoint. Investors, of course, would probably react warmly to this, as it means that a premium to the current stock price would be offered for the purchase, allowing many of them to make money on their investment. Record amounts of money are now flowing into such acquisitions, a recent example being Merck's acquisition of the RNAi company Sirna Therapeutics for $1.1 billion, a premium of just over 100%.

Biotech employees will normally have mixed feelings about being acquired, as it does provide a measure of validation and, importantly, valuation for the company's IP. And it could, theoretically, still leave some autonomy to the company and solve its financial

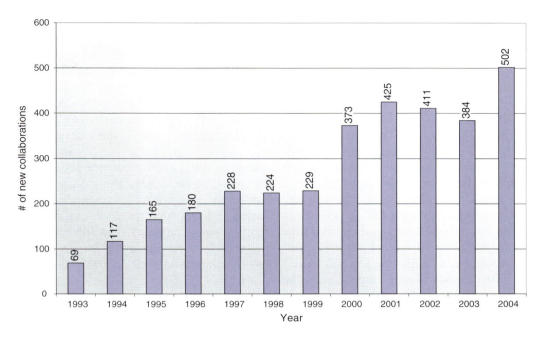

**Figure 2.5 ▶**  Number of new biotech–pharma collaborations by year. (Source: Burrill and Company.)

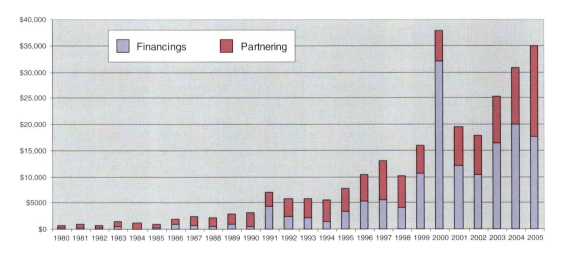

**Figure 2.6 ▶**  Biotech capital raised (in $ million) 1980–2005. Note the financing spike during the "genohype" year of 2000. (Source: Burrill and Company. Used with kind permission of Burrill & Co.)

problems, allowing it to move faster using newfound resources. But that hasn't always been how it's worked out. The acquisition might take place solely to bring in a specific clinical drug and everything and everyone else might prove to be expendable. As a result, biotech acquisitions tend to raise at least as much fear as hope among employees. Other than the founders (perhaps), early employees, and investors, who stand to gain the most financially, few dream of their company being acquired.

These, then, are the three different possible outcomes for a successful start-up biotech or small pharma company. Admittedly, this simplified schema can capture neither the complexities nor the dynamics involved. "Thing is, most companies are spinning, bouncing, three steps forward, two steps backward, chaos-in-motion sort of things," notes David Whitman, Chief Operating Officer of Pharmadyn. Each one will find its own unique path to "success or a reasonable alternative," but in the end they're likely to share one of these three common fates, if they stay in business long enough.

The California model by which a new company might grow to "become the next Genentech" has been the most sought after, but least frequently achieved. These days the idea is pretty much a non-starter. A second route, long-term existence as a "research boutique" collaborating on a succession of projects partnered with big pharma is viewed as less of a success, but is much more achievable and many small companies owe their continuing existence to it. Acquisition by big pharma (or, possibly these days, big bio) represents a kind of success in validating the company's IP and bringing in some return on investment, but it often means a loss of independence and jobs.

### 2.3.3  "Is it a project or a company?"

Taking a step back, two big questions need to be asked about this model. The first concerns the process: "Does it really represent the best way to turn innovative ideas into drugs?" The second addresses the endpoints: "Are these really the best outcomes for new companies to achieve?"

To deal with the first point first, one thing that might be viewed as a fundamental flaw in the usual start-up process is the assumption that a single idea, however brilliant, can engender its own corporate entity. "Is it a project or a company?" asks Dr Robbins-Roth. "Just because you have some success with research on campus doesn't mean that you form a company. That could be the absolute worst thing in the world to do, because the probability is very high that you might raise the first little bit of money and you'll never get money ever again and it will die. It might not be because of science. It could be because right now we're seeing that mechanisms don't really exist to support that kind of funding. The market is saying, 'We don't see enough of a reason to invest in this anymore: we have plenty of those.' "

Even for a funded idea that plays out brilliantly, unless the breakthrough it enables is of deep and broad value, like the recombinant DNA technology of Genentech, it's likely to result in a "one-trick pony". If that particular trick is spectacular enough and results in, say, a fantastic new treatment for non-small cell lung cancer, there's room for plenty of success, and no one will argue with the paradigm that produced it. But keep in mind that everything depends on one's definition of "success." Even coming up with a successful new drug or two doesn't necessarily another Genentech make. Agouron, for example, ended up merging with Warner-Lambert (now part of Pfizer) in the aftermath of its successful new HIV protease inhibitor Viracept (nelfinavir; Figure 2.7). Pharma's hunger for promising drugs and its willingness to pay extremely large amounts of money for them and the companies that produce them is a major reason why few new Genentechs or Genzymes are likely to emerge.

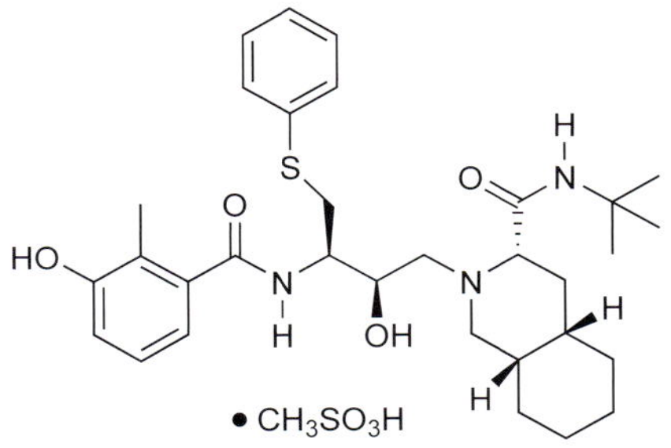

**Figure 2.7 ▶** Viracept (nelfinavir mesylate).

Not surprisingly, most collaborative projects, like most projects in general, don't end up producing new approved drugs. So in this more common case, the real viability of a company based on a single project portfolio (and probably hastily trying to add in a second and a third with too few scientists) will be open to question. So alternatives like translational research done in academia or the out-licensing of IP to an existing company seem like better choices in many cases. These avoid the terrible inefficiencies involved in building up a new company to develop a single idea, then knocking it down or watching it crumble when the funding or the science doesn't pan out.

Looked at from a more global perspective, perhaps the brakes have been put on the proliferation of start-ups for a reason. "We needed lots of companies in the early days because there was no existing corporate infrastructure willing to take a chance on developing products from our new science," says Dr Robbins-Roth, referring to biotechnology. "That certainly is not the case today—for most new technologies, there are plenty of corporate homes for in-licensing and development."[40] These days it isn't so much a question of whether or not there is a home for innovative projects, but, instead, a question of which home is best.

The risks involved in new start-up companies might be minimized by having a larger assortment of new projects *ála* big pharma, but this isn't usually possible for companies forced to operate with very limited funding and resources. There are occasional exceptions, though. Recently, in a reversal of the usual flow of clinical candidates, a new company in South San Francisco called Synosis Therapeutics in-licensed five CNS compounds from Roche. The company plans to use diagnostic imaging techniques and small clinical trials to establish proof of concept for members of this portfolio and decide which ones to take further. Five reasonable clinical candidates isn't a lot for big pharma, but it's often an unattainable number for smaller companies: it's simply too hard for them to grow that many in their own backyards. The CEO of Synosis, Dr Ian Massey, former head of research and preclinical development at Roche Palo Alto, hopes to build up a larger portfolio by in-licensing additional compounds and thereby enjoy the luxury of developing the best-looking ones while pruning away the rest.[55]

The reverse flow of clinical candidates from big pharma to start-up companies is actually not that unusual: many pharma companies are now willing to part with drugs they don't wish to develop for various, oftentimes business, reasons. Others besides Synosis have been born this way, for example the Swiss company Speedel which obtained clinical POC for the first-in-class renin inhibitor Tekturna (aliskiren, Figure 2.8), which was recently approved.[56]

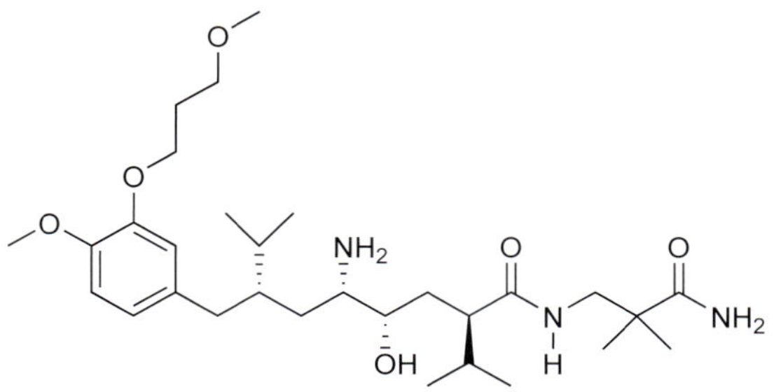

**Figure 2.8** ▶  Tekturna (aliskiren).

Right now there almost seems to be a boom in pharma executives starting up spinout companies with compounds their former employers weren't interested in developing. Often the contracts involved give the pharma company the opportunity to re-acquire rights to their compounds for a considerable fee if they look promising down the road, which of course Tekturna did. This makes their originators less reluctant to part with them. But keep in mind that although the new companies involved may, like Speedel, eventually develop their own discovery programs, it isn't a given that anything but development will ever be done there, which makes them fundamentally different from most biotech/small pharma start-ups.

### 2.3.4  Leaner, Meaner Start-ups

If portfolio diversification is one problem start-up companies need to face, another is the need to build expensive infrastructure to investigate each new idea, something that investors particularly dislike. "When I look at the maybe 30 or more companies I've been involved with at various stages," says Olav Bergheim, Founder and Managing Director of Fjord Ventures, "I've realized that about 80 percent of the money is spent on things that don't create value. That money is being used to build corner offices, computers, coffee pots—all the overhead. On the other hand, maybe only 20 percent is used for things that help foster breakthroughs."[57]

One newer model aims to minimize the infrastructure costs incurred. Early-stage research typically is directed at proof of concept, and this drives the need for equipment and the facilities to prove that concept with. Not even the brightest minds can predict whether it will work out or not. This being the case, wouldn't it be better to rent or lease the vehicle needed to get there rather than buy it outright and take a huge loss if forced to sell? If every new project doesn't warrant its own company, does every new company need to start from a brand new lab and an open VWR catalog?

What would be preferable is a facility where early proof-of-concept experiments can be carried out with minimal commitments on infrastructure, taking advantage of shared resources like lab equipment, facilities, and analytical services. In fact, such *incubators* do exist today. Organizations like Synergenics, whose Chairman is Dr William Rutter, co-founder of Chiron and the San Jose Biocenter, provide start-up companies with not only facilities and shared capital equipment, but expertise in a variety of science (e.g. bioinformatics) as well as business issues. Another advantage they provide in bringing together a variety of start-ups is an atmosphere that may make the cross-pollination of ideas possible.

Especially close contacts with experienced drug developers would be provided by industry-sponsored incubators such as the one the biotech company Biogen-Idec recently announced, called BI3. They propose to give those with exceptional enough ideas facilities and money as well as science *and* business expertise in return for a pre-negotiated option for product rights. In what almost sounds like every scientist's dream, everything from animal facilities to NMR support and DNA sequencing would be part of the offer.[58] No shortage of applicants is anticipated.

Universities are also beginning to provide facilities for start-ups. At the UCSF's Mission Bay campus entrepreneurs affiliated with the University can rent space in the "QB3 Garage," so-called as a reminder of the garage experiments that spawned Hewlett-Packard.[59] Leasing small amounts of hood and office space located on the campus of a major research university on a short-term basis comes with a lot of advantages, in a sense easing the transition from concept to company.[60]

Another example of quick access to infrastructure is provided by the non-profit Molecular Medicine Research Institute in Sunnyvale, California, founded by Dr Edward P. Amento, a veteran of both academia and the biotech industry.[61] The Institute conducts its own basic research through grant funding and through the support of its affiliate program, which provides turnkey access to offices, labs, equipment, and services for start-up companies seeking to do early studies without having to give up any IP.[62] So far, about 30 companies have been born there.

So by a variety of mechanisms and through various private, government, foundation, and university sponsorships, the requirement for infrastructure buildup and the consequent energy of activation for start-ups to reach proof-of-concept is being partly met. If start-up funding remains tight at times, these mechanisms represent at least one way to make the process more efficient.

## 2.3.5  Biotech Alternatives

The second question about the biotech model represented in Figure 2.4 is whether its three eventual endpoints are really the best places for newly successful companies to be. The best case seems to be impossible to reach, the next best has been described as "selling off their children for a chance to stay hungry,"[63] and the third fate involves the possible loss of independence. It isn't too surprising that many in the industry are looking for another way, a new and better place for entrepreneurial prowess in the drug discovery and development process. After all, the kind of people who found start-up companies have a risk tolerance level more akin to that of wildcat oilmen than accountants or insurance agents. Eking out a modest but reliable existence isn't really what they have in mind. More likely, they subscribe to architect Daniel Burnham's philosophy: "Make no little plans. They have no magic to stir men's blood. Think big."

But in this case, is bigger really better? If it *were* possible for start-up companies to grow into major new forces in marketed drugs, would that be the most efficient route to innovation? We've seen in a previous section that a number of big pharma companies are experimenting with dividing themselves up into smaller business units to increase efficiency. With the success and subsequent expansion of companies like Amgen, Genentech, Genzyme, and Gilead, potential problems of size now confront biotechs as well as big pharma. Unlimited expansion isn't always the best item on biotech's menu.

As a biotech consultant and writer, this is an issue Dr Cynthia Robbins-Roth has thought about a lot. "The thing about small companies," she says, "is that they have all kind of gotten used to the idea that they're all going to follow some model. They think they need to get huge, to become Amgen or Genentech, with 10,000 employees to be successful. I think that is not the goal.

"Our job is to get innovative, important products that can make a difference to patients into the marketplace at a price the market can handle. And that means we have to do our business in a cost-effective way. So what does that mean? It means that one of the things to contemplate is that we should stop thinking that the only business model to be defined as a success is to turn into Genentech. Success has to have a very different kind of definition, and part of that definition is to look more like a relatively nimble R&D institute that may incorporate, as time goes by, certain elements of downstream capabilities, but instead of building everything in-house, will do it through a global network of low-cost, high-quality providers."

Dr Robbins-Roth suggests that a new model for biotech may be in order, one that doesn't depend on big pharma to develop and market new drugs. Smaller companies, she says, should "wake up to the fact that big pharma may not be their best partners. Because if your return to your shareholders is in the form of royalties based on net profit are you going to want to work with the most expensive partner out there? I'm thinking no. Right now people say, 'But, oh my God, who would you partner with if you don't partner with them?' " She suggests that an alternative does exist in light of years of mergers and downsizing that have freed up thousands of experienced development and marketing people. "A whole bunch of those people know a whole lot about sales and marketing and downstream development issues. And they're forming small, entrepreneurial companies that are completely happy to interact in the context of an outsource network of collaborators."

*Contract research organizations* (*CRO*s) thus provide important resources at multiple stages of the drug discovery and development process and hold the potential to be a bigger part of it still. A word about the definition of CRO is in order first, though. "Contract research" can refer to an organization involved in anything from screening to clinical trials, all of which can be outsourced, but the term is frequently used to specifically mean an organization under contract for clinical trials. This makes it a potentially confusing term, so it's important to note the context in which it's used.

The use of CROs for a given task obviates the need to build infrastructure, thus keeping down costs and soothing early-stage investors' concerns. By using CROs a small company can "hit the ground running," making use of capabilities that are present today, not 6 months or a year from now when the organizational structure, personnel, and facilities required are in place and the kinks are just beginning to be worked out. Getting reliable data faster, and often cheaper, is the major strength of this approach. If the data obtained are positive, they bring momentum and probably attract more investment capital. If they're negative, the sooner it's known the better. The decision to modify or scrap a project or a company can then be made earlier on, saving futile expense and effort.

Clinical CROs are extensively employed even by big pharma itself, but right now small companies don't use them much for late-stage trials on unpartnered, in-house projects for the simple reason that few such projects exist. Pharma is now willing to pay so much for promising clinical candidates and the companies that produced them that those producers rarely get that far without a partner. If early trials are successful, big pharma will be waiting in the wings with very nice offers to collaborate or acquire, thus replenishing a biotech company's depleted coffers. Going beyond that without a partner means greater expense and greater risk, though potentially greater reward.

Federal funding is sometimes an option for a small company taking a candidate, unpartnered, into clinical trials. One example of this involves a proteasome inhibitor originally called PS-341, which was being developed by a small company called ProScript, later bought by Millennium Pharmaceuticals. The drug is now sold as Velcade (bortezomib, Figure 1.17), a treatment for multiple myeloma. ProScript worked with the National Cancer Institute (NCI) on some of the preclinical and formulations issues of the compound, and NCI, along with other funding sources, then became involved in the compound's Phase I trials under the sponsorship of its Cancer Therapy Evaluation Program (CTEP).[64]

A newer method of financing clinical trials for companies that may choose to go further into clinical development *sans* big pharma is called *collaborative development financing*. In an example of this type of deal, in early 2006 Isis Pharmaceuticals and Symphony Capital entered into an agreement whereby $75 million from the latter became available to the former for the development of three drug candidates. B. Lynne Parshall, Executive VP and CFO of Isis noted that "The transaction represents an attractive alternative to partnering these very promising compounds. In contrast to working with a pharmaceutical company, we retain control of development of these drugs and exclusive rights to them. The opportunity to reacquire all rights to these drugs in the future without residual royalty obligations, after using the funding to increase their value, is very attractive."[65] In addition to the money, access to expertise from Symphony's collaborators, a company called RRD International, was part of the deal.

Isis was essentially trading to Symphony rights to the compounds and a chance to buy into the company via stock warrants in return for the money to move Isis further along in their development. Should the data look promising, it could reacquire those rights at a price that would guarantee Symphony about a 30% annual interest rate.[66] This they in fact did in late 2007, partnering out two of the now clinically advanced candidates to Ortho-McNeil and retaining full ownership of a third.[67] Collaborative development financing thus represents a new way to move potential new drugs further downstream into clinical trials. This explains why Isis, Exelixis, Alexza, and other companies have negotiated such agreements recently.

But even if it could arrange financing and run all of the necessary trials independent of big pharma, a small company using CROs to take a drug through to approval *sans* partner would still need to take on the challenge of *selling* its product. After all, even though the availability of cheap printers, pdf files, and the World Wide Web theoretically means that authors don't need publishers for *their* own products anymore, how many have gone that route, and how successful have they been? Companies could decide to go it alone or use contract marketing organizations, particularly with specialty drugs that, again, don't require a "Mongol horde" sales approach, vast teams of sales reps that seem to go along with sales of more widely used medications.[68] Many of Genzyme's products fit into such a strategy. And Agouron independently sold Viracept (Figure 2.7) quite successfully at one time. But such cases have so far been the exceptions. Sales and marketing, like everything else, can be outsourced, but usually in the current paradigm big pharma becomes involved by the time the drug is marketed, if not long before. Most of the time for small companies the question isn't *whether* they should partner their candidates out, but only when to do so and for how much. As long as valuations for clinical candidates remain sky high, this seems unlikely to change, but should they drop for any reason in future, or if more examples of the success of this marketing approach appear, the independent route to drug sales via extensive CRO outsourcing should become more viable and more frequently used.

If the increasing use of "on demand" outsourcing and CROs can speed things up and make the overall drug discovery process more efficient by minimizing infrastructure and start-up time, then why not just outsource everything from concept to pharmacy counter? This used to be impossible because pharma research was always done in-house and so few, if any, outsource providers existed for many necessary parts of the drug discovery and development process. That's no longer the case; absolutely everything can now be outsourced.[69] This has given rise to the concept of the *virtual company.*

A number of such organizations currently exist. RegeneRx of Bethesda, Maryland is one of them. With a total of eight employees, the company has progressed the 43-amino acid peptide, thymosin $\beta4$ (T$\beta4$), under license from NIH, into a series of Phase I and Phase II trials, the earliest ones being for wound healing, through outsourcing to CROs and by supplying the compound to interested academic investigators via *Material Transfer Agreements* (*MTAs*). The results of the latter can be seen in a publication by researchers at the University of Texas Southwestern Medical Center where T$\beta4$ was shown to hold the potential for promoting cardiac repair after ischemic injury, a finding that opens up a large and lucrative market should the compound prove to be effective in post-MI situations.[70]

The minimization of infrastructure costs attained by setting up a virtual company is catalyzing the formation of an increasing number of them, such as San Diego-based Anaborex and Baltimore's Ruxton Pharmaceuticals to name just two. Typically virtual companies focus on early-stage drug candidates, usually obtained through in-licensing, thereby bypassing the need for most early-stage research as well as the scientists to do it. This deliberate strategy of starting later on in development is very much in line with the shorter time-to-profitability sought throughout the industry in recent years. So although virtual companies can carry out early-stage concept-to-clinic work via CROs, many of them are essentially development organizations.

In addition to using the outsourcing business model, RegenRx decided early on to do things differently from the standard model, according to the President and CEO of RegeneRx, J.J. Finkelstein. "This was not going to be a project where we come in, raise $50 million and then just jump right ahead into clinical development. It was going to be a series of small steps, with each milestone being identified and achievable within a specific period of time. We believed that as we achieved those milestones, our market valuation would increase. Accordingly, we could raise the next tranche of capital at a higher valuation and with less dilution, kind of step our way up into greater valuation in a deliberate and controlled manner. We have been successful in doing that … We went from a market cap of a few million dollars to one of about $120 million now."[71]

Admittedly the virtual company model has its own unique drawbacks and limitations and isn't meant to be universally applicable. "You have to manage outside companies more closely than you probably would internal staff," Mr Finkelstein says. "Just because they're experts does not mean everything will happen exactly the way you want it to. If your contractors aren't working out, you have to get rid of them and find someone else. There's a lot of oversight."[72]

Once again, breaking the drug discovery and development process into different pieces and sending each one of them out requires an unusual talent for coordinating and eventually re-integrating the results while maintaining the flow of important information throughout. There are other caveats too. One thing outside-contractors can't necessarily supply is the creativity that's so vital to early-stage research projects. Nor can they read their client's mind, so precise, timely communications about expectations and experimental details become particularly critical. And there's the chance—probably a very small chance, but one that looms large in the minds of project managers—that the kind of critical IP generated at an early-stage discovery company, by far the most valuable thing it has, could be compromised by extensive outsourcing to a variety of contractors and collaborators, either foreign or domestic, despite the *confidentiality agreements* (*CDA*s) and MTAs involved.

Successful and innovative business models are rare in any industry, so many eyes will be trained on companies like RegenRx to see how well they do in coming years. Should they succeed, the virtual company model is likely to proliferate. Independently of that, though, business strategies involving multiple rounds of smaller funding triggered by incremental milestones and the increased use of outsourcing via CROs to reduce expensive infrastructure buildup are likely to become the norm for start-ups and are much in evidence already.

Just as the landscape of big pharma has changed significantly in the last 20 years as the industry was forced to adapt to new realities, so too biotech and its place in the discovery of new medicines has undergone a sea change. The transition hasn't been easy for either of them nor for those who depend on them for their livelihoods. Not all the problems have been solved, and uncertainties and inefficiencies remain.

Recently an insightful analysis of the biotech industry by Harvard business professor Gary P. Pisano has appeared.[42,73] From the perspective of someone familiar with other innovative industries as well, he points out some unique difficulties that the biotech sector faces. One of these is that in biotech, "the most valuable IP is often not a specific molecule but data, understanding, and insights relating to how that molecule behaves, what it can do, what its potential problems are, and how it might be developed."[42] Such knowledge resides in the minds of experienced researchers and not necessarily in filings at the US Patent and Trademark Office (PTO), which is rather a scary thought from a patent attorney's point of view.

Although it increases the value of the knowledgeable scientist to the drug discovery industry, this creates problems for biotech, according to Dr Pisano. So many types of expertise and disciplines are involved in bringing a new drug to market that to bring together the number of such walking IP repositories sufficient to achieve critical mass has been an almost impossible task for a small company, any one of which could only get a few. "The result was hundreds of islands of specialized expertise," he says.[42] Ways to bridge these islands, presumably through collaborations with other biotechs, big pharma, universities, and even contractors, just haven't proven to be enough, at least partly because of reluctance to truly share such unprotected IP with other parties who are often seen as temporary allies, not committed co-developers.

One of the solutions to this problem, Dr Pisano suggests, amounts to a call to bring back the Genentech model, at least as that company exists today. In such a "quasi-public corporation," big-pharma ownership of the majority of a biotech company's stock assures financial stability and a long-term outlook while a hands-off approach allows creativity and

---

**Box 2.1   Some Predicted Industry Trends for the Next 5–10 years**

Increases in:

- ▶ Biotech in-licensing and acquisition by big pharma

- ▶ Use of CROs industrywide

- ▶ Number of virtual companies

- ▶ Focus on specialty drugs

- ▶ Funding for translational research

- ▶ Reorganizations of pharma into smaller business units

- ▶ Use of incubators and shared facilities for start-ups

---

entrepreneurialism to flourish. To use an earlier analogy, in this model the orchards of biotech are tended for their harvest rather than clear-cut. To many in the industry, this model makes a lot of sense, but whether its widespread adoption is imminent or not only time will tell.

One nagging question occurs to one perusing Dr Pisano's analysis: If drug discovery IP really resides in the minds of researchers and biotech's problem has to do with not being able to get enough of these minds together, why is big pharma, which employs them by the thousands, still having productivity problems? His answer: "Most major pharmaceutical companies have created their own islands of expertise inside their own corporate boundaries, a deeply problematic practice that probably explains their poor R&D productivity."[42] Again, organizational changes may well prove key to solving this problem. This and some other trends in the industry are summarized in Box 2.1.

## 2.4   New Technologies

### 2.4.1   S-Curves and Expectations

Most new technologies, like most new drug candidates, don't pan out, and even the ones that do will not necessarily earn back the money it cost to develop them. But fortunately this gloomy reality doesn't stop people from trying anyway, spurred on by factors including the large potential reward that a successful new drug or new technology can bring. We've already seen the tremendous impact recombinant DNA technology, for example, had on drug discovery, giving birth to a whole new industry and entirely novel types of drugs. Others like structure-based drug design (SBDD, discussed in Chapter 7) have proven to be valuable tools that thousands of medicinal chemists depend on each day. Many more, for example, surface plasmon resonance technology, have also found real, if more limited, applications.

But as with drugs, for each of these success stories there are probably hundreds more that just never worked out. As Chief Technology Officer at Vertex Pharmaceuticals, Dr Mark Murcko has been asked to evaluate quite a few new technology proposals. "What typically happens," he says, "is that if it works great on the first test case, some people will naively assume it's going to work just as well in every other situation. But in reality, when the technology is given to five other researchers and they try it on their problems, it doesn't work. That's because it isn't robust. Most technologies aren't battle-tested and they fail in real-world situations." He suggests that what's needed is "a 'toolbox' mentality and a skeptical approach."

Taken in perspective, the high attrition rate for new technologies shouldn't really be surprising. The legendary drug hunter and Nobel Laureate, Sir James Black pointed out that "in all the excitement of the scientific breakthroughs, we can easily forget *the iron law of development of new technologies, that the complexities of real problems can never be entirely predicted.* [Italics added] Time is needed to meet and solve unexpected problems. The great enemy is impatience."[74]

Time is definitely a critical factor in the development of new technologies, and the same temporal pattern emerges again and again with them as Dr David Brown recently pointed out.[75] He refers to the *S-curve theory* of new technologies and their business applications as discussed in a book by Richard Foster,[76] and explains how the benefits of a given successful new technology will typically develop over time, as illustrated in Figure 2.9.

"There is an induction phase for each new technology with no payback," he points out, "then a payback phase, then an obsolescence phase. The critical point is the length of the induction phase, i.e. the time from first development of a technology through to when it actually starts to earn money in the marketplace, the payback phase."[75] A company or an industry investing in such a technology too early will be doing so at a loss as money and manpower is consumed and no real benefit is obtained yet. On the other hand, getting in too late isn't a very good option either: Competitors have probably enjoyed most of the steep payback phase and there's little left to gain. So the key question is, "When will the payback phase begin?" What's the $EC_{50}$ going to be for this particular dose–response curve?

Most scientists whose careers haven't yet spanned such a complete cycle tend to think it's on the order of a few years. Among those with a longer perspective, estimates are larger but still diverge a bit. According to Dr Mark Murcko, "it varies of course, but I think it's often as much as a decade before the pharmaceutical industry can gain widespread value from new

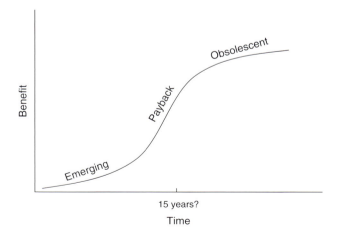

**Figure 2.9** ▶  Benefits versus time for a given new technology.

technologies. Look at combichem. Combichem was around in the early-to-mid 1990s, right? But many people in the field will tell you that it wasn't until the late 1990s, or even early into this century that they were starting to get real value out of it. Genomics? It's been around since the late 1990s, but it's only now that people seem to be getting much mileage out of such information. So it seems like it's always close to a decade before the widespread use of the technology can be robustly expected to add value. Of course there are always going to be isolated examples where on a particular problem set the technology may bear fruit more quickly."

Others point to examples of still longer timeframes. "The pattern of mAbs (monoclonal antibody therapeutics) suggests that it can be nearly 25 years before a key scientific innovation becomes effectively translated into new therapies, again suggesting that benefits are hard won."[77] Dr Brown agrees with these sorts of long estimates. "There is ample evidence," he says, "that new technologies take 15–25 years to mature and have an impact and pay you back. So if you want to ask what's going to impact in the next 5–10 years, well, okay, ask what technologies started 10–15 years ago." Table 2.1, which is illustrative rather than comprehensive, lists examples of a few new technologies that might help the reader get some perspective on this. It's often very difficult to put an exact date on when a given technology was born, so as a surrogate for this the table lists the earliest year in which this author could find a paper in PubMed bearing the technology's name in its title line.[78]

Of course, just as the $IC_{50}$ for a dose–response curve doesn't tell you what the maximal response will be, predicting the time-to-payback for a given new technology doesn't tell you the amount of payback to expect. "Whether it's a worthwhile impact, whether it's just another tool or whether it's going to be a major breakthrough is a different question to ask," says Dr Brown. But the timeframe required to get there is definitely long. "Look at HTS as an example. It was 1987 when Pfizer led the industry in developing HTS for the first time. We now see a few compounds that had their origin in HTS coming to the market—but only a few. It's taken a generation, 15–25 years."[79] For a promising new technology like *systems biology*,[80] "my view is that it's certainly going to take 20–25 years to pay back. I don't think

**TABLE 2.1 ▶** **Some Newer Technologies Potentially Impacting in Drug Discovery**

| Technology | Date[a] | Impact[b] |
|---|---|---|
| Antisense oligonucleotides | 1985 | TV, therapeutics |
| Combinatorial chemistry | 1991 | LID, LO |
| High throughput screening (HTS) | 1991 | LID |
| Proteomics | 1997 | TV, preclinical |
| Virtual screening | 1997 | LID |
| RNAi | 1998 | TID, TV, therapeutics |
| Systems biology | 1999 | TID, LO |
| Fragment screening | 2003 | LID |

[a] Earliest occurrence of the technology's name in the title of a paper listed in PubMed.
[b] LID is lead identification, LO lead optimization, TID target identification, TV target validation.

you'd expect a return before that. You know, that is the trendy thing in 2005/2006 with lots of academic papers. But it's going to be 2025 or 2030 before there's a real impact in the form of useful medicines saving lives in patients."

All of which presents a big problem. Unless a new technology like systems biology gets adequately financed for the next decade or two, it's all but guaranteed that no one will see any benefit from it. But consider the dilemma this poses for innovators and corporate executives. It's hard to imagine a pharma CEO rationalizing tens of millions of dollars spent on some new, unproven technology to investors by promising financial rewards that, with luck, might just be available in their *lifetimes*. And for smaller companies, using the carrot of profits 20 years away isn't likely to sway VCs who can see no further than three. Raising capital to fund the development of new technologies can only be done if the promised paybacks happen a lot quicker than that. So the most optimistic possible predictions are made and taken as reference points by others also seeking to fund the technology who believe they can develop it even faster. This seems the only possible response, given "the need for innovators and their sponsors to create high expectations to get access to the very considerable resources (money, people, and intellectual property) required to develop new medical technologies."[77]

Add to this the societal factor that Americans seem to favor new, high-tech things, and you've got a formula for unrealistic expectations or even, to use Alan Greenspan's famous phrase, "irrational exuberance." This can be charted on a new curve that's perhaps every bit as important as the benefit-versus-time S-curve, one of expectations-versus-time. This is shown in Figure 2.10. Dr Murcko explains: "In the beginning there's a lot of hoopla, a lot

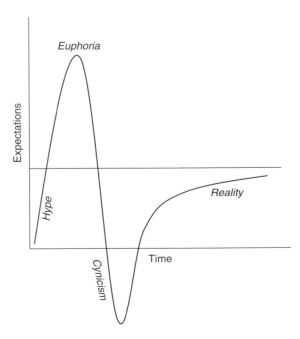

**Figure 2.10 ▶** Plot of how expectations for successful new technologies often change with time. (See Bezdek, J. *IEEE Fuzzy Systems* **1993**, *1*, 1–5.)

of hype, and expectations are very, very high. But of course the new technologies never work out as well as people had thought. So then people get disenfranchised—they lose all interest in the technology—and so expectations crash. And that's where the story for most technologies ends, because most of them really don't work. But for some technologies, there really is some value, and so asymptotically the world approaches a realistic understanding of what the technologies are good for."

Almost inevitably, early expectations for a given new technology are unrealistic in anticipating not only how soon they'll come to fruition but also how broadly applicable they'll be. The ones that prove successful by and large are tools that contribute to some part of the drug discovery process, but don't result in a major paradigm shift for the whole process. Again, biotechnology is the glaring exception. Tools are worth having and normally justify the expense of their purchase. But having, say, a nice new electric screwdriver, one shouldn't work only on projects involving a lot of screws, or worse, try and use it to pound nails in. It's an important addition to the toolkit, but not the whole thing.

After years of struggle, uncertainty, and bipolar opinions, after overcoming problems no one could have foreseen, a few new technologies will prove their merit. But the need for more new drugs and technologies will remain. What then? One interesting idea is that real progress depends on the summation of a number of new technology S-curves, constantly jumping from one that's been through its payback phase to the next one just entering it.[75] Figure 2.11 illustrates this.

Timing the necessary leaps is tricky, though. One needs to be sure that the next curve is close enough, goes higher, and won't give way before jumping. Dr Brown argues that one factor in today's suboptimal industry productivity might be just such an ill-timed leap that drug discovery took some years ago. In the earlier paradigm, projects were built around lead compounds that were active in physiological systems, even if the mechanism of action wasn't

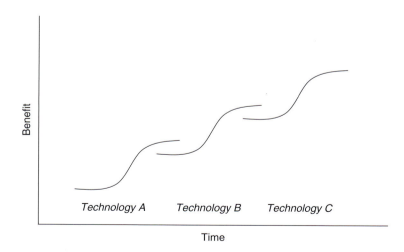

**Figure 2.11** ▶ The benefits of additive S-curves. (See Brown, D. Target selection and pharma industry productivity: What can we learn from technology S-curve theory. *Curr. Opin. Drug. Discov. Devel.* **2006**, *9*, 414–418.)

known. Compounds were tested and optimized on cells, tissues, and whole animals. The modern researcher would think of this as *phenotype screening* (as discussed in Chapter 5) or the *observation-led approach*. These days the word "blind" might come to mind. This approach had its problems, but drugs were obtained. Then advances in molecular biology and genomics led to the new S-curve one might call the *hypothesis-led* or *target-driven approach*. Although more intellectually satisfying than the earlier approach, this added in a new dependence on things like cell permeability, metabolic stability, and an understanding of the target's role in vivo—factors already taken into account by the observation-led approach—which could make the most potent, picomolar lead become a total dud in animals or man.

Should pharma have taken a better look before it leapt to that particular next S-curve? Probably. Dr Brown's editorial concludes that "it appears that there are unsolved problems with both the 'observation led' and the 'hypothesis led' approaches and these are holding back the discovery of new small molecule medicines . . . Resolution of these problems could lead to a more satisfactory 'payback' phase of the new S-curve. And we must hope that this new S-curve, which appears to have started at a performance level below that of the old one, does indeed have the potential to overtake its predecessor."[75]

Many in the industry have decried the over-reliance on new technologies in general, pointing out that "from combinatorial chemistry to genomics, new concepts or technologies that claim to help accelerate drug development have arguably been too rapidly embraced without true validation," and feel that pinning one's hopes on technologies detracts from what the real focus in drug discovery should be. "Innovation . . . springs less from technologies and more from placing scientists in a stimulating environment, and providing them with the platform to explore all hypotheses and results."[81] All of this should be seen as a warning not to put the cart before the horse, not a suggestion that none is needed.

### 2.4.2 Genomics Redux

As we've seen in the last chapter and now understand better in light of the expectations phenomenon highlighted by Figure 2.10, the heady days of the race to sequence the human genome gave way to an enormous letdown when the expected masses of new drugs and targets proved not to be forthcoming. But in keeping with the S-curve timetable of a decade or two before payback, it seems that the cynicism observed in its wake is finally giving rise to more realistic appreciation of its usefulness. Genomic technology has provided a tool for understanding inherited contributions to disease, classifying them by their genomic signatures, identifying and targeting infectious organisms,[82] and many more things besides.[83]

Being able to sequence all or part of the genome for a number of individuals allows their sequences to be lined up for comparison so that sites where nucleotides vary called *SNPs* (*single nucleotide polymorphism*) can be found. SNPs are referred to as "common" if a particular variant is found in more than 1% of the population. Such SNPs represent most of the 0.5% of the genetic differences that occur between different people.[84] They aren't the only way that genomes vary. Non-SNP mutations include insertions and deletions of nucleotides (collectively known as *indel*s) which can have profound effects if they occur within coding sequences. This is especially true when they don't occur in multiples of three base

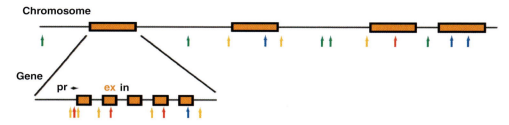

**Figure 2.12 ▶**   Arrangement of SNPs along a chromosome. Green arrows indicate rSNPs (random) which are distant from genes, yellow arrows denote gSNPs (gene-associated), situated alongside genes or in introns, blue arrows denote cSNPs (coding) situated in exons, and red arrows indicate pSNPs (phenotype-relevant), which are gSNPs or cSNPs that affect phenotype. (Reproduced from the Roche Science Publication *Genes and Health*, third edition, available at www.roche.com/pages/facets/22/gene2_e.pdf, copyright 2007, F. Hoffmann-La Roche Ltd.)

pairs, thus making for *frameshift mutations* that garble and/or prematurely terminate the messages of subsequent *codons*, the DNA triplets that specify the mRNA and ultimately the amino acid to be placed in a given position. But let's start by considering SNPs.

SNPs, of which there are millions, are found all along the genome in introns, exons, promoter regions, and between genes, as Figure 2.12 illustrates. SNPs that code for a different amino acid being built into a protein (a *missense mutation*) at an important spot, such as in the active site of an enzyme, can obviously make a difference in the protein that might result in a phenotype change, but lots of other effects are possible too.[85] SNPs can have absolutely no effect or be responsible for changes in protein expression, *loss of function* (*LOF*), or *gain of function* (*GOF*). SNPs that result in a codon for an amino acid being mutated into a stop codon (a *nonsense mutation*) can result in a truncated form of the protein being produced. Even the seemingly innocuous and heretofore largely ignored *synonymous SNPs* or *silent mutations*, meaning different nucleotides that end up coding for the same amino acid through the redundancy built into the genetic code, might surprisingly result in changes in protein expression,[86] truncation, or conformation and thereby have an effect on function.[87] So understandably a lot of work has focused on differences that key SNPs might make in diseases as well as in response to drugs.

A frequently cited example of the contribution of SNPs to disease is in sickle cell anemia, which occurs in about 1 out of 500 African Americans. Here a simple change in one nucleotide in the hemoglobin beta gene (HbB) as shown in Figure 2.13 eventually leads to the morphological abnormalities in red blood cells that characterize the disease. This being an autosomal recessive disease, both alleles need to have the mutation in order for someone to have it, but this particular mutation didn't evolve to reduce the life span of those who inherit it. Instead, it has a positive side: having just one mutant allele confers significant resistance to malaria. Other *monogenic diseases*, those traceable to problems with a single gene (although not necessarily with a single SNP), seem to have been selected to confer protective advantages as well: those individuals heterozygous for one cystic fibrosis allele have a survival advantage against cholera, and people with one copy of the Tay-Sachs allele are unusually resistant to tuberculosis. It seems that Mother Nature's way of dealing with diseases comes with its own set of serious adverse effects.

**Normal hemoglobin sequence**

| DNA | | CTG | ACT | CCT | G**A**G | GAG | AAG | TCT |
|---|---|---|---|---|---|---|---|---|
| Amino acid | | Leu | Thr | Pro | *Glu* | Glu | Lys | Ser |
| Position | | 3 | | | 6 | | | 9 |

**Mutant hemoglobin sequence**

| DNA | | CTG | ACT | CCT | G**T**G | GAG | AAG | TCT |
|---|---|---|---|---|---|---|---|---|
| Amino acid | | Leu | Thr | Pro | *Val* | Glu | Lys | Ser |
| Position | | 3 | | | 6 | | | 9 |

**Figure 2.13** ▸   A single cSNP makes for a deadly disease. Inheriting both alleles of this A→T mutation results in sickle cell anemia.

The hunt for SNPs associated with diseases is made easier by the well-known fact of genetics that pieces of DNA located near each other on a chromosome are much less likely to be separated by recombination during crossover than pieces farther apart, which is called *linkage disequilibrium* (*LD*). The net effect is that nearby SNPs tend to be inherited together in groups, and the resulting gene variations in a particular allele are called *haplotypes*. So a single *tag SNP* could identify the whole set. This is important because it greatly reduces the number of computations required to do *whole genome association studies*, which scan the genomes of individuals to look for variations that correlate with particular diseases. Thanks to the work done by the International *HapMap* consortium[88] it may be possible to see such associations using only about 300,000 tag SNPs rather than the full set of 10 million SNPs. In many cases this should make it unnecessary to narrow the search down to a biased list of likely suspects, or *candidate genes*.

But by and large neither one SNP nor one haplotype can explain everything about a disease. Diseases attributable to a single SNP or some other defect in a single gene tend to be rare Mendelian disorders, not the kind of thing most doctors see on a daily basis or most patients will ever experience. Most genetic variants that can be associated with more common diseases aren't causal the way the sickle cell SNP is, but only increase susceptibility to the disease instead. Even "smoking gun" mutations which do seem to be causal, such as the deletion of a critical Phe 508 in the cystic fibrosis transmembrane receptor that's present in 70% of Caucasian CF patients, don't necessarily allow one to predict when the disease will appear or how severe it will be.[89] This is partly explained by the fact that other genes, such as TGFβ1 in this case, can modify physiological response to the changes induced.[90] In fact most diseases are *polygenic*, not monogenic, and some may involve dozens of different genes, which makes a genome-based understanding of their etiologies hard to come by. The search for the multiple sets of genes that are expected to be associated with diseases like schizophrenia, asthma, and type II diabetes has been underway for years and will likely continue for many more.

To make matters worse, a persistent problem in gene association studies is the lack of reproducibility of findings.[91] "It is disturbing," says one editorial review, "that only 16–20% of initially reported significant associations have been consistently replicated without any evidence of between-study heterogeneity or bias."[92] Much of the problem can be minimized by increasing the size of the studies, but as the same review points out, and in a rare example of a potential benefit of continued over-population, in some cases that might require data to be collected from more patients than currently live on Planet Earth!

To keep things in perspective, consider the fact that identical twins share an identical genome but don't always get the same diseases. This gives a strong hint—as if any were needed—that environment, itself divisible into many complex components, may be equally important if not more so. And the biggest real-world problem of all is that even having a good genetic understanding of a given disease in a specific patient doesn't necessarily mean that one can *do* anything about it once it's there. Until the technology exists to go in and correct the SNPs, indels, etc., via a more sophisticated form of *gene therapy* than we now know about or some other method that doesn't exist today, "fixes" will be indirect and difficult. But understanding the genetic factors involved is at least an important first step.

For the present, though, the benefits of genomics have arguably accrued more from its use in predicting individual responses to drugs and aiding in the selection of treatment regimens for patients. In this way, it's beginning to prove its utility in the field of *personalized medicine*, discussed below.

### 2.4.3  Personalized Medicine

The idea behind personalized medicine is to get the right drug at the right dose to the right patient at the right time. There's nothing particularly recent about this idea: it's always been a primary aim of medicine. More than a century ago Sir William Osler noted that "If it were not for the great variability among individuals, medicine might be a science, not an art."[93] The problem up until now has been that this variability means that prescription drugs are just not safe and effective for every patient. Since part of the variability is genetic, genomic technology holds the prospect of better matching patient to drug. The holy grail of this field would be to be able to predict which of three groups a patient belongs to—the one in which a given drug is likely to be efficacious, one that it won't work for, or a third in which it's likely to be toxic.

Measuring gene expression levels via messenger RNA, a field known as *transcriptomics*, has shown some real-world applications to personalized medicine already. A new assay called Oncotype DX® uses expression profiling of 21 gene products to quantify the likelihood of recurrence and probable magnitude of chemotherapy benefits for many patients with early-stage breast cancer.[94] Before this was available, they faced the difficult choice of whether or not to have chemotherapy to prevent later recurrence that would happen in only a few percent of patients who skipped it.[95] A similar approach to predicting the risk of recurrence ($\sim$25% overall) for patients with stage 1A non-small-cell lung cancer (NSCLC) called the lung metagene predictor has demonstrated positive results in the clinic so far.[96] And the study of "genomic signatures" holds the future promise of selecting which chemotherapeutics to use for a given patient as well.[97] So the contributions of genomics, transcriptomics, and other "omics" shown in Table 2.2 to personalized medicine are real and growing. But the biggest factor so far has been the application of genomics to the study of individual variations in drug response, the field that will be referred to as *pharmacogenomics*.

**TABLE 2.2 ▶   Some Popular "Omics" and What they Involve**

| Omic | Related terms | Study of | Analytical methods |
|------|---------------|----------|--------------------|
| *Genomics* | | DNA | DNA sequencers |
| *Transcriptomics* | | mRNA | Microarrays |
| *Proteomics* | | Proteins | 2D Gels, Protein microarrays |
| *Pharmacogenomics* | Pharmacogenetics | Genotype effects on drug response | Various |
| *Metabonomics* | Metabolomics | Endogenous metabolites | NMR, MS |

## 2.4.4   Pharmacogenomics

Definitions and usage of many of the "omics" terms are amorphous and user-dependent. Strictly speaking, *pharmacogenetics* is to pharmacogenomics what genetics is to genomics. In each set the latter is much broader (and of more recent vintage) than the former. But in an example of polymorphisms in what might be called vocabularomics, definitions of the "P-words" seem to vary even more than individual drug responses do. Often the terms are used interchangeably. Sometimes the non-committal abbreviation *PGx* is used as well.[98] In this book, the distinction between the two will be ignored to avoid confusion, and the term "pharmacogenomics" will be used to when discussing genetic variations in both broad and narrow contexts. Box 2.2 outlines some of the things that pharmacogenomics hopes to accomplish.

Table 2.3 shows some broad estimates of the failure to achieve efficacy in patient populations for some common classes of drugs. In light of these kinds of numbers, the best that can be done, for example, to get a particular patient's hypertension under control involves a lot of trial and error. "Upon initial treatment, only approximately 40 to 50% of patients respond sufficiently to a randomized choice of a single therapeutic drug. However, this can be increased by the use of sequential monotherapy, with the subsequent addition of each major drug class or rotation through the various alternatives until the most effective single therapeutic or combination is identified."[99]

Worse still, drugs can cause toxic effects, called *Adverse Drug Reactions* (*ADR*s) or *Adverse Drug Events* (*ADE*s). By one estimate (which is occasionally disputed) more than

---

**Box 2.2   Some Goals of Pharmacogenomics**

▶ Identify the right patients and dosing to maximize efficacy of a drug

▶ Minimize ADRs for new and existing drugs by identifying those at risk

▶ Guide patient selection for smaller, cheaper, more successful clinical trials

▶ Rescue failing clinical drugs by identifying responder subgroups

**TABLE 2.3 ►     Some Classes of Drugs Along with Estimates of the Percentages of Patients who Experience Poor Efficacy when Using Them**

| Drug class | Poor efficacy rate (%) |
|---|---|
| ACE inhibitors | 10–30 |
| Angiotensin II antagonists | 10–25 |
| Beta 2 antagonists | 40–70 |
| Beta blockers | 15–25 |
| SSRIs | 10–25 |
| Statins | 30–70 |
| Tricyclic antidepressants | 20–50 |

Source: Roche Science Publication *Genes and Health*, available at www.roche.com/pages/facets/22/gene2_e.pdf.

2 million hospitalized patients in the United States experienced serious ADRs in 1994 and 106,000 died as a result, making ADRs about the fifth leading cause of death that year.[100] A more recent study found that adverse drug events lead to about 700,000 emergency room visits per year in the United States.[101] Both failure to respond to a drug and adverse response to it have serious personal and economic consequences. Being able to predict, without trial-and-error, which drug will work or at least which won't cause toxicity in a given patient would therefore be a boon.

There are three main ways through which genetic factors can alter an individual's response to a drug, thereby making it either inefficacious or toxic. They can affect the **metabolism** of the drug, its **transport**, or the **drug target** itself. Of these, the effects of genetic polymorphisms on their metabolism by *drug metabolizing enzymes* (*DMEs*) have been the most widely studied. They're the most frequently encountered and most easily examined pharmacogenomic effects. That's because many genetic variations in drug metabolism can be traced to corresponding variations in only a handful of enzymes, the *Cytochrome P450* family known as *CYPs*, and a few others.[102] And conveniently, in many cases the only thing needed to study drug/DME interactions once the alleles have been genotyped for a given patient is the plasma concentration of the drug at various timepoints, the bread-and-butter of the science of *pharmacokinetics* (*PK*).

CYPs, which are discussed in more detail in later chapters, are major players in the body's mechanism for eliminating hydrophobic substances, including most drugs. These heme-containing proteins use oxidative mechanisms to make their substrates more hydrophilic and often provide a handle like a hydroxyl group for further transformations by other enzymes. As they're mostly found in the liver, genomics provides a convenient way to assess CYP activity without having to do a liver biopsy or administer a *probe drug*. This is a substance thought to be exclusively metabolized by the same enzyme as the drug of interest so that measuring its plasma concentration can be used to determine that enzyme's activity in a particular patient.

The activity of a given CYP (or any other protein, for that matter) can be affected by SNPs, indels, *gene deletion*, and *gene duplication* by which individuals can have anywhere from zero to more than a dozen copies of the gene coding for one of these enzymes.

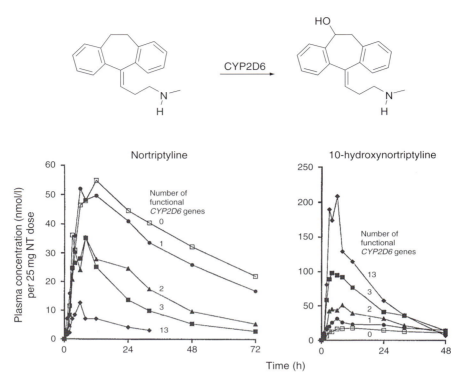

**Figure 2.14 ▶** Mean plasma concentrations of nortriptyline (left) and 10-hydroxy-nortryptaline (right) in different CYP2D6 genotype groups after a single oral dose of nortryptyline (Reprinted by permission from Dalen, P., et al. 10-Hydroxylation of nortryptyline in white persons with 0, 1, 2, 3, and 13 functional CYP2D6 genes. *Clin. Pharmacol. Ther.* **1998**, *63*, 444–452, copyright 1998, Macmillan Publishers Ltd.)

Figure 2.14 shows just how much of a difference this can make in plasma levels of a drug, in this case nortriptyline, one of the many CNS drugs that CYP2D6 acts upon. The net effect of inherited variations in DMEs will depend, among other things, on whether a given one results in loss or gain of activity as well as whether the particular metabolite it produces is more active or less active than the drug. Most metabolites are less, or sometimes equally active, but some drugs, called *prodrugs*, are only precursors, requiring some in vivo chemical reaction(s) to convert them to the active component. If a DME happens to be involved in prodrug activation,[103] polymorphisms that inhibited its function would produce not higher, but *lower* levels of the active species.

So by making drug concentrations too high or too low, the Scylla and Charybdis of dosing, polymorphisms in DMEs can profoundly influence observed drug response. And things can get complex for other reasons as well. A genetic impairment of one metabolic path might force a given drug to be processed by other enzymes that don't normally contribute much to its metabolism. This is called *metabolic switching*, and it can give rise to odd metabolites with different toxicological properties than the usual breakdown products.

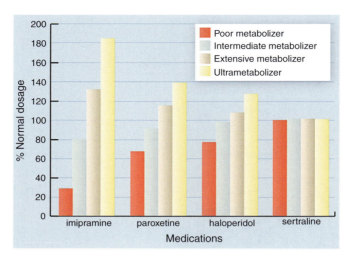

**Figure 2.15** ▶   Tolerance of patients with different CYP2D6 status to normal dosage levels of various antidepressants. (Reprinted with permission from Service, R.F. Going from genome to pill. *Science* **2005**, *308*, 1858–1860, AAAS.)

Inherited differences in CYP2D6 are responsible for patients running the gamut from *poor metabolizers* (*PM*) through *intermediate metabolizers* (*IM*) and *extensive metabolizers* (*EM*) to *ultrametabolizers* (*UM*). How much this will affect plasma levels of a drug they take will depend on the extent of its metabolism by the enzyme. Figure 2.15 shows how the tolerated doses for four different antidepressants can vary depending on the patient's CYP2D6 activity.[104] This enzyme plays a major role in the metabolism of Tofranil (impramine, Figure 2.16),[105] while it's only a minor player in Zoloft (sertraline, Figure 2.16) metabolism,[106] so variations in its activity make little difference to plasma levels of the latter.

The economic advantages of testing certain patients for CYP2D6 and CYP2C19 metabolism have been clear enough for one company, Roche, to develop an FDA-approved microarray chip, called AmpliChip, which tests for 27 alleles of the former and 3 of the latter. This enables most patients to be characterized into PM, IM, EM, and UM phenotypes.[107] And these days for those willing to pay a few thousand dollars for non-FDA

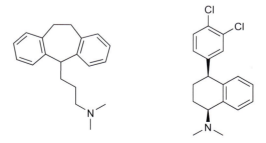

**Figure 2.16** ▶   Left: Tofranil (imipramine). Right: Zoloft (sertraline).

**Figure 2.17 ▶** The immunosuppressant Imuran is converted by glutathione (GSH) into 6MP, itself a prodrug for the active immunosuppressive, antineoplastic 6-thioguanosine nucleotides. DME thiopurine methyltransferase (TPMT) S-methylates 6MP to an inactive metabolite.

approved testing, more extensive genotyping of one's DMEs can even be arranged through companies that advertise on the Internet, which isn't surprising considering that DNA testing kits have now become available at your local chain store.

CYPs aren't the only DMEs currently worth genotyping for certain therapies. Groundbreaking work by George Hitchings and Gertrude Elion at Wellcome Research Labs in the 1950s resulted in the discovery of two important antineoplastic, immunosuppressive compounds, Imuran (azathioprine, Figure 2.17) and 6-mercaptopurine (6MP, Figure 2.17).[108] Imuran acts as a prodrug for 6MP, which itself requires multiple in vivo reactions to convert it to the active components, 6-thioguanosine nucleotides. Imuran and 6MP are still used today to treat inflammatory bowel disease (IBD) and acute lymphoblastic leukemia (ALL), the most commonly occurring childhood cancer. A cytosolic enzyme called thiopurine methyltransferase (TPMT) S-methylates 6MP, and is one major route for its clearance. At least 20 variant alleles that reduce or eliminate the activity of this enzyme, allowing higher than normal levels of 6MP to build up, have been found.[109] The most common one in Caucasians, TPMT*3A, has two non-synonymous SNPs that produce an enzyme that's less effective because it's degraded too rapidly, making its concentration unusually low.[110] About 89% of Caucasians have both copies of the more active gene, around 11% carry one high- and one low-activity allele, while 0.3% are homozygous for the low-activity variants and in these patients standard doses of 6MP can be fatal.[111]

As a result, testing for TPMT activity before administering 6MP has been standard practice for some time. If poor TPMT activity is found, only 5–10% of the normal dose allows those patients to reach standard levels of the active components, whereas patients homozygous for the more active enzyme get up to four times the normal dose.[112] Either an enzyme activity assay or a genotyping assay is run on a blood sample prior to dosing, assuming there's been no recent transfusion to muddle the results. Although the odds of a particular patient being severely deficient in TPMT activity are only about 1 in 300, considering the severity of the ADR and the ease of testing to predict it, this particular example of the use of personalized medicine to prevent adverse effects has been widely adopted.

The second way in which genetics can influence drug response is via transporters. To reach the right target, drugs not interacting with extracellular or cell surface receptors need to get inside cells. Hence the modern medchemist's preoccupation with cell permeability. As Chapter 9 explains, in most cases passive diffusion, which shouldn't depend on any variant alleles, is the main mechanism by which small molecules enter cells, but even drugs reaching the cytosol passively can be quickly shot out of it by active transporters such as *P-glycoprotein* (*P-gp*), an important efflux pump that's encoded by the *MDR1* (also called *ABCB1*) multidrug-resistance gene. Furthermore, P-gp in the small intestine can be involved in limiting an oral drug's absorption so that it never reaches the right cells in the first place. P-gp is notorious for its overexpression in many cancers which helps make them refractory to a broad variety of chemotherapeutics, but it also plays a crucial role in keeping foreign substances (*xenobiotics*) out of various tissues, including those in the brain.[113] So it seems almost a given that allelic variants in MDR1 or other transporter genes could lead to higher- or lower-than-normal drug concentrations in cells and/or plasma thus having an effect on drug efficacy and safety. To date, however, less work has been done on genetic polymorphisms affecting these transporters and their resulting effects on drug concentrations than on CYP effects, and the results are a bit less clear.[114–115]

A study in 2000, reported an SNP (C3435T) in MDR1 that correlated with alterations in P-gp expression and function. For a given dose of the cardiotonic agent digoxin (digitalis glycoside, Figure 2.18), plasma levels in patients having two T alleles were significantly higher than in other patients.[116] A number of clinical studies done afterward have either reported an effect of this mutation on patient response to *HAART* (*Highly Active Anti-Retroviral Therapy*) for HIV[117] or found no association at all between the two.[118] Other SNPs in MDR1[119] and SNPs in MRP2, another drug efflux transporter,[120] have been reported. Further work establishing the possible roles of these and other transporters in individual variations to drugs, especially drugs having low therapeutic ratios, could conceivably result in genomic testing to avoid toxicity *àla* TPMT.

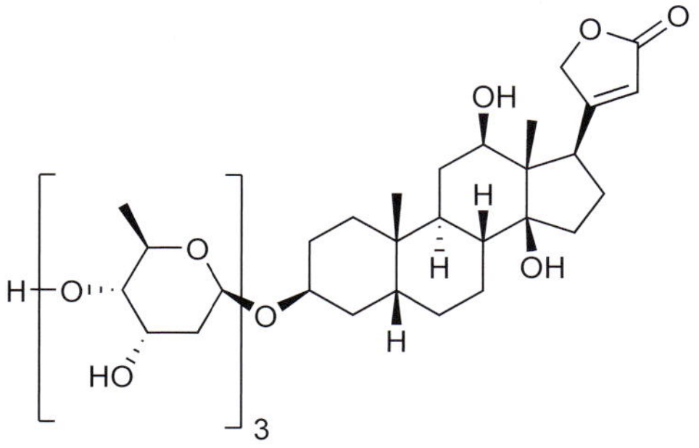

**Figure 2.18 ▶**   Digoxin.

**Figure 2.19 ▶**   Left: R = Me, epinephrine (adrenaline); R = H, norepinephrine. Right: bucindolol.

More work has been done in the area of polymorphisms affecting drug–target interactions. One example involves $\beta_1$-adrenergic receptors ($\beta_1$AR), 7-transmembrane-spanning GPCRs found on cardiomyocytes that act to increase heart rate and cardiac contractility upon binding to their native agonists, epinephrine (adrenaline, Figure 2.19) and norepinephrine (Figure 2.19). A G→C SNP means that amino acid 389 of this protein, which is in the cytosolic loop close to where the G-protein binds, can be either Gly or Arg.[121] About half the population is homozygous for the Arg alleles, 20% are homozygous for the Gly alleles, and the rest carry one of each.

In cellular and tissue experiments, the Arg-$\beta_1$AR seemed to be more responsive to the effects of bucindolol (Figure 2.19), a $\beta$-blocker, than the Gly-$\beta_1$AR. This drug had earlier failed to demonstrate any survival benefit in a clinical trial for heart failure (HF).[122] No genotyping was done at the time but samples had been saved and were available for analysis. When genotyping was done on them, it was found that treatment of the Arg-389 homozygous patients with bucindolol had resulted in a statistically significant 38% reduction in mortality whereas those homozygous for Gly-389 showed no significant clinical response.[123] Had the trial been run exclusively on patients homozygous for Arg-$\beta_1$AR, the results presumably would have looked fantastic and the drug might have become the standard treatment for such patients. Instead, bucindolol was never approved. Perhaps not surprisingly, one of the original researchers co-founded a new company called Arca Discovery that's aiming for bucindolol approval, along with a companion genetic test, which would make it the first genetically targeted cardiovascular drug.[124]

This sort of combination of a drug targeting a defined patient-subgroup with a diagnostic assay to identify the right patients to use it on is an example of the growing trend toward *theranostics*, which is sometimes given the more arreligious-sounding name *theragnostics*. Genentech's Herceptin (trastuzumab) is a powerful example of what theranostics at their best can do to personalize medicine and also an example of the success of a drug that's only relevant to a subset of patients. Somewhere between 15 and 30% of patients with breast cancer have tumors with an overabundance of the *human epidermal growth factor receptor 2* (*HER2*), a member of the EGFR-TK family that's involved in regulating cell proliferation. These tumors are generally aggressive, expressing few hormone receptors that would make them susceptible to drugs like tamoxifen (Figure 2.20). They tend to respond poorly to standard chemotherapy.[125] Consequently the prognosis for HER2-positive breast cancer has not been good. Herceptin, a humanized monoclonal antibody that binds to the extracellular domain of the HER2 receptor, has literally proven to be a lifesaver for many of these patients, providing a response rate of 35% when used as a first-line, single-agent therapeutic

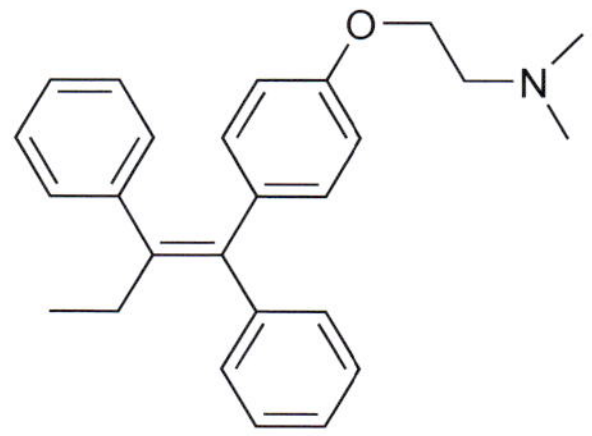

**Figure 2.20** ▶   Tamoxifen.

for metastatic, HER2-positive breast cancer.[126] Results of a series of recent trials testing its use as an adjuvant after surgery "are simply stunning," demonstrating "highly significant reductions in the risk of recurrence, of a magnitude seldom observed in oncology trials," according to an editorial in the *New England Journal of Medicine*, not normally given to hyperbole.[127]

The availability of a number of different FDA-approved tests for HER2 overexpression, using either *immunohistochemistry* (*IHC*) or *fluorescent in situ hybridization* (*FISH*) makes the selection of the patients who can benefit from this therapy possible. And big pharma may well take note of Genentech's sales of $1.2 billion worth of Herceptin in 2006 despite the *market segmentation* involved in leaving the "one size fits all" drug development paradigm. "The ultimate market segmentation," says Dr Samuel Broder, Executive Vice President of Celera Genomics and former Director of NCI, "is a failure in the clinic, which gives you a market segment of zero."[128] Limiting a drug's market size can be a lot smarter than losing it altogether, and in an increasingly accepted approach, companies like Perlegen of Mountain View, California now frequently work with pharma to identify genetic markers involved in drug response during clinical trials. The recent publication of FDA guidance on pharmaco-genomic data submission will likely accelerate the trend toward collection of this information, as submissions which have so far been voluntary become more greatly encouraged or even, someday, required.[129]

The greatest success story to date for a small molecule in personalized medicine—which sounds a bit like an Academy Award nomination—belongs to Novartis's Gleevec (imatinib mesylate, Figure 1.17). Almost all patients with chronic myeloid leukemia (CML) have a characteristic abnormality called the *Philadelphia chromosome* caused by a chromosomal translocation. This results in two proteins being fused together to produce the constitutively activated enzyme BCR-ABL tyrosine kinase, which plays a key role in too many white blood cells being made.[130] In early clinical trials Gleevec, a molecule specifically designed to inhibit this unusual enzyme, not only proved remarkably non-toxic for a cancer drug but at doses of more than 300 mg resulted in a complete hematologic response in 98% of the patients taking it.[131] Later, larger trials showed that Gleevec therapy resulted in improvements in survival with many complete remissions.

Figure 2.21, a gel electrophoresis immunoblot that can be read much like a TLC plate, shows its dose-dependent effect on preventing phosphorylation of one of BCR-ABL's sub-strates, the protein CRKL, in data from the earlier trial. This is an example of using a protein *biomarker* to assess what a drug does in the body, part of the science of *pharmacodynamics* (*PD*), which is a key feature of an increasing number of clinical trials.[132] A biomarker can be

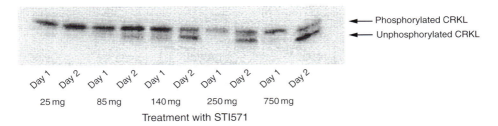

**Figure 2.21** ▶  Immunoblot assay demonstrating the degree of phosphorylation of the BCR-ABL substrate CRKL in individual patients. (Reproduced from Druker, B.J., et al. Efficacy and safety of a specific inhibitor of the BCR-ABL tyrosine kinase in a chronic myeloid leukemia. *N. Engl. J. Med.* **2001**, *344*, 1031–1037, copyright 2001, Massachusetts Medical Society. All rights reserved.)

loosely defined as a quantifiable characteristic (anything from blood pressure to levels of a specific molecule) that can serve as an indicator of disease progression or response to therapy. The kind of application of proteomics shown in Figure 2.21 is one many researchers will become accustomed to, as such biomarkers are even more frequently obtained in animal models of experimental drug efficacy these days, providing a real readout of whether or not the compound is hitting its target in vivo. After all, the usual targets of drugs are proteins, not genes, so measuring levels of the proteins affected is a good way of finding out what's going on.

Gleevec is interesting in another way too. Although designed to selectively inhibit one kinase, with about 500 other kinases existing in the human genome—this being a big family— it, not surprisingly, has activity against several of the others. But rather than being a drawback this turned out to be a bonus! The kinase c-KIT, which Gleevec inhibits about as well as it inhibits BCR-ABL in cells, is believed to be responsible for the fact that Gleevec works in another kind of cancer too: gastrointestinal stromal tumors (GIST). The drug is not without its problems: in CML, resistance can develop through a variety of mutations in BCR-ABL kinase, which newer inhibitors like BMS's Sprycel (dasatinib, Figure 2.22) and Genzyme's diagnostic kit for the mutations hope to address.[133] But with cancer patients living longer, higher-quality lives and Novartis bringing in about $2.2 billion in Gleevec sales in 2005 even with the relatively limited market size, it's hard to imagine that the future for personalized medicine doesn't look rosy.

But as the poster child for personalized medicine, Gleevec isn't exactly typical. Few diseases result from any single genetic defect the way CML does, and even in the ones that do, it's uncommon for the exactly same variant to be present in >95% of those affected.

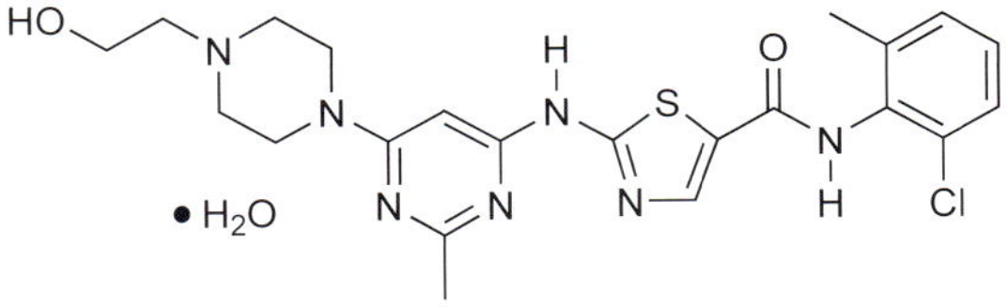

**Figure 2.22** ▶  Sprycel (dasatinib).

Recall that 30% of Caucasian CF patients *don't* have the most common mutation, but instead have a wide variety of others that affect the same protein, perhaps in different ways.

Gleevec has other advantages too. Most cancers involve solid tumors. Getting a drug to penetrate these and measuring its effectiveness can be a lot more difficult than hitting a target in the plasma accessible to drugs and then just counting cells in a blood sample. And, in general, mutations are more likely to result in loss-of-function (LOF) rather than in gain-of-function (GOF) that's seen with BCR-ABL kinase. Medicinal chemists know it's generally easier to inhibit or antagonize something, making up for GOF, than it is to increase function or agonize it to deal with LOF. There may be more bark than bite to this objection, though. Drugs for type II diabetes, for example, include glitazones that are PPARγ *agonists* and a newer class, DPPIV inhibitors that increase concentrations of an endogenous beneficial protein, GLP1, by inhibiting the protease that destroys it, a sort of double negative approach. So increasing function is not impossible.

Considering all the favorable circumstances involved, the development of drugs as successful as Gleevec for personalized medicine is far more likely to be the exception than the rule. To balance out our view of what role pharmacogenomics is likely to play in future years, here's a devil's advocate summary of many of the principal scientific limitations on it:

▶ Most diseases are polygenic, perhaps requiring more than one target to be addressed.[134]

▶ Most mutations cause LOF, not the easier-to-handle GOF.

▶ There are many "common SNPs" (frequency >1%) and many more "rare SNPs" (frequency <1%) and these vary by geography and ethnicity: those common in one population may not be in another. Besides raising concerns about the wisdom of sending clinical trials overseas, this means that diagnostics might miss the less common ones. Also, medicines aren't likely to be personalized for those in the developing world.

▶ Demonstrating efficacy in targeted trials might be done with a smaller number of patients, but, especially for non-life-threatening diseases, large numbers may still be required to demonstrate sufficient *safety*.[134]

▶ Response to drugs depends on other factors besides genetics. These include but—for readers who may be attorneys—are not limited to: gender, age, weight, smoking, disease state (e.g. renal or hepatic impairment), diet (e.g. grapefruit juice, which inhibits CYP3A4), alcohol consumption (ethanol being a CYP2E1 inducer and liver destroyer), other drugs (e.g. ciprofloxacin, a CYP1A2 inhibitor), herbal remedies (e.g. St. John's Wort, a CYP3A4 inducer), and—an especially big factor in the real world—*patient compliance* (taking the medicine on schedule).[135]

### 2.4.5  Other "Omics"

Since it doesn't take into account environmental effects, which vary between individuals, no one expects pharmacogenomics to be the whole story behind personalized medicine. Some say that another new "omics" technology that combines classical biochemistry with modern analytical techniques and sophisticated pattern recognition algorithms has the potential to overcome this problem and provide a more real-world readout: *metabonomics*. The simplest definition of this is "measurement of the complete metabolic response of an organism to an environmental stimulus or genetic modification."[136] As usual, a competing term, *metabolomics*, also exists which is either a subset of the former term or includes it, depending on whom you believe.[137] For our purposes "metabonomics", which seems to be more popular, will be used.

The theory behind metabonomics is that drug administration will cause changes in levels of endogenous metabolites as part of their physiological effects. These are called "fluxes" which sounds a bit Aristotelian. So looking at levels of endogenous metabolites—and not *drug* metabolites, which would be a different sort of study—as found in a substance like urine, which "carries information on almost everything because [the kidney is] your ultimate excretory organ, where homeostasis is maintained," can get you "a tremendous amount of information . . . if you can analyze all the thousands of metabolites that are in there,' according to Dr Jeremy Nicholson of Imperial College.[136]

This is often accomplished using proton NMR spectroscopy. But the resulting NMRs are understandably complex and difficult to interpret, not to mention time-dependent like the underlying processes involved. A given endogenous metabolite can be involved in dozens of in vivo reactions,[138] and there are lots of them. So it's not surprising that any interpretation of metabonomic data will seem "too true to be good." Computer pattern recognition software is essential. A method called *principal component analysis* (*PCA*) that attempts to find patterns among spectra based on two or three principal components is often used. Each spectrum is reduced to a datapoint on a 2D or 3D graph. Those clustered together display similar patterns, those farther apart do not. Further analysis can provide information on specific endogenous metabolites (probably a combination of several, rather than a single one) that might serve as biomarkers for animal studies and might even be of interest for clinical drug studies.

To date, metabonomics has most often been used for preclinical toxicology, as with the recently completed COMET study, where it was used to generate a database and a predictive expert system for target organ toxicity for consortium members, which included five pharma companies.[139] But there are those who foresee its wider use in the clinic in future years. Like any other technology, the field is not without problems. Metabolites like citrate, hippuric acid, and taurine are so frequently noted to change concentrations in so many studies that a list of such "usual suspects" has been published. And as with any new data-rich technology, "our ability to generate masses of data far outstrips our capacity to understand it."[137] In drug discovery, finding patterns doesn't mean much unless you can rationalize them and then use them in a predictive sense.

But that's exactly what an extension of metabonomic technology dubbed *pharmacometabonomics* (with or without a hyphen) was able to do in at least one notable case recently. The definition of this new "omics" is "the prediction of the outcome (for example, efficacy or toxicity) of a drug or xenobiotics intervention in an individual based on a mathematical model of pre-intervention metabolite signatures" according to Dr Nicholson, whose group has focused on metabonomic research.[140]

In a recent study, a single large does of paracetamol (acetaminophen, the active ingredient in many analgesics) was given to rats. This compound has an unusually low therapeutic ratio for such a widely used OTC drug and acetaminophen overdoses account for many cases of liver failure each year. Dr Nicholson's group found that both the extent of liver damage in

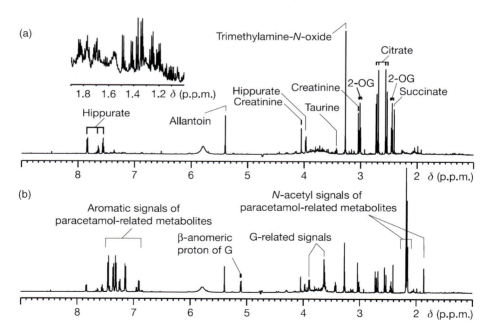

**Figure 2.23 ▶**    Paracetamol (acetaminophen) and its glucuronidated metabolite.

the rats as determined by histology and the extent of *glucuronidation* (an important drug-clearing pathway, see Figure 2.23) of the drug could be predicted by the pre-dose urinary metabolite profiles for the individual rats.[141]

The most important factor in predicting glucuronidation was the pre-dose integral around δ5.1 in the proton NMR spectrum (Figure 2.24), which probably represents the anomeric protons of other glucuronides produced in vivo. Individual rats with bigger (but still hard to see!) integrals here would seem to be more likely to be good glucuronidators. The extent of liver damage correlated inversely with pre-dose concentrations of the metabolite taurine and directly with pre-dose levels of trimethylamine-N-oxide and betaine, for which some theories were offered.[141] Although rats overdosing on acetaminophen wouldn't seem to be much of a concern to pharma, being able to predict patient response in clinical trials by simply analyzing a urine sample in advance would be much more intriguing. Perhaps a new S-curve is underway here.

**Figure 2.24 ▶**    Representative pre-dose (a) and post-dose (b) proton NMR spectra of urine samples from a rat dosed with paracetamol (600 mpk). Inset in (a) is an expansion. 2-OG, 2-oxoglutarate; G, paracetamol glucuronide. (Reprinted by permission from Clayton, T.A., et al. Pharmaco-metabonomic phenotyping and personalized drug treatment. *Nature* **2006**, *440*, 1073–1077, copyright 2006, Macmillan Publishers Ltd.)

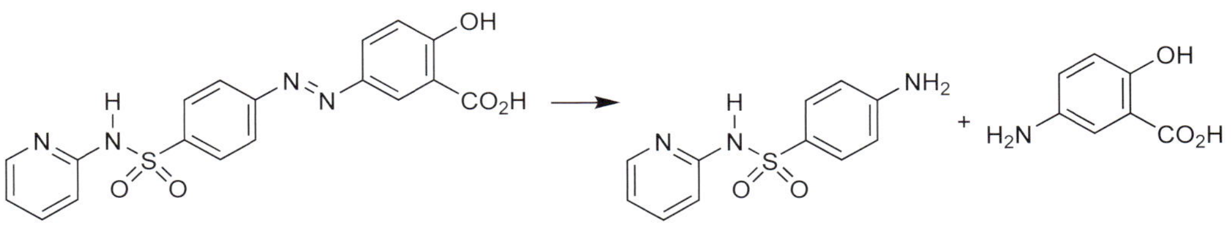

**Figure 2.25 ▶**  Intestinal bacteria break down the prodrug sulfasalazine into sulfapyridine and the active anti-inflammatory component 5-aminosalicylic acid (5ASA).

The same paper also mentions one variable that's usually been given short shrift when it comes to individual response to a drug. As we'll see in later chapters, drug metabolism can occur in many parts of the body. It's most often studied in liver (the major metabolism hotspot), but metabolism in other organs is often considered as well. Still, all of these reflect a kind of speciocentricism. Human cells aren't the only ones involved in metabolism. Like it or not, literally trillions of microorganisms live in our intestines and outnumber our human cells 10 to 1: the *microbiome* (or *microbiolome*), of course.[142] In fact, "humans are superorganisms whose metabolism represents an amalgamation of microbial and human attributes."[143] These bacteria are beneficial, helping to produce vitamins and amino acids as well as allowing their hosts (us) to absorb nutrients not otherwise available. Rats raised without them, in a germfree environment, are even known to exhibit morphological abnormalities.[144]

Some papers dealing with the role of intestinal microbes on drug metabolism can be found, many dating back to the 1960s or 1970s.[145] Intestinal bacteria have been reported to be important for the reduction of azo bonds, for example in the metabolism of sulfasalazine, a prodrug used in the treatment of ulcerative colitis and rheumatoid arthritis, into its active component, 5-aminosalicylic acid (5ASA; Figure 2.25).[146] For the most part, though, gut flora have been "a forgotten organ."[147]

But evidence that individual gut microbiomes metabolize differently would mean that this is a factor that can't be ignored in medicine. That kind of effect has already been seen in rats. Metabonomic profiling of Sprague–Dawley rats identified a new subset of animals that differed from all others of the same strain from the same supplier in how they metabolized chlorogenic acid, a dietary component. The difference was traced to a single room at the animal supplier from where the unusual metabolizers came. When housed with the normal metabolizers, they reverted to that phenotype. The investigators concluded that "it is likely that the two phenotypes are related to distinct populations of gut flora that particularly impact the metabolism of aromatic molecules."[148] Not only might this mean that another variable exists to confound the interpretation of animal experiments, but if it happens in humans it might represent yet another variable in clinical drug response.

And human gut microbial populations are known to vary. Recent work at Washington University in St. Louis has shown that the relative proportion of the two main groups of bacteria in the gut, *Bacteroidetes* and *Firmicutes*, correlates with obesity in man and metabolic potential in mouse.[149] More work can be expected along these lines, and if major correlations with drug metabolism are proven, the day may come when clinical medicine gets still more personal.

### 2.4.6  The Adoption of Personalized Medicine

A number of factors that aren't particularly scientific are also important to the progress of personalized medicine and these need to be briefly considered. One of these—big pharma's fear of market segmentation and the possible loss of the blockbuster model—has already been mentioned. Another one has to do with patient apprehensions. This is the fear that, once obtained, a person's genetic data could be used for *genetic discrimination*, meaning that she might be denied health (and other) insurance because a strong genetic predisposition to a disease like breast cancer or Alzheimer's disease might be considered a "preexisting condition" by insurers. Employers, afraid of the negative effects on their own costs, might also refuse to hire her. Cases involving allegations of genetic discrimination have already been reported.[150] The fear of genetic discrimination has probably kept many people away from genetic-based tests or from giving consent for their samples to be used for genetic analysis in clinical trials. It's even been reported that there are patients paying for genetic tests out of their own pockets so that insurance companies won't know the results, in some cases even using false names.[151]

In the words of Dr Francis S. Collins, "Without legislative protections against genetic discrimination in health insurance and the workplace, many people will be reluctant to undergo potentially life-saving genetic tests or to participate in the clinical trials needed to develop genetically targeted therapies. We need this legislation."[152] A seemingly simple fix for this problem would be to pass a law prohibiting the use of genetic data for insurance or employment eligibility. In fact, most states now have these, but they vary from state to state. A uniform federal law would do much to alleviate patient concerns and benefit the field of personalized medicine. In 2000, President Clinton issued an executive order banning the use of genetic data in federal hiring. Both houses of Congress and two Presidents have spoken in favor of a broad anti-genetic discrimination law ever since. Such bills have repeatedly been introduced, always with broad, popular support, and then have mysteriously evaporated—until now. In May 2008, after months of delays, Congress approved the *Genetic Information Nondiscrimination Act* (*GINA*), which applies to health insurance companies and employers.[153] President Bush is expected to sign the bill, thus removing one obstacle to the collection and beneficial use of genetic information in medicine.

Laws and lawyers may have another role to play in the adoption of personalized medicine as well, this one a curious benefit of living in a litigious society. "Legal liability has the potential to be a powerful driver pushing implementation of personalized medicine," concludes one review. "Individuals injured by adverse drug effects are increasingly likely to bring lawsuits alleging that they have a polymorphism or biomarker conferring susceptibility to the drug that should have been identified and used to alter their drug treatment."[154] In fact, one analysis attributes the widespread adoption of TPMT testing mentioned earlier to fears about lawsuits. "Use of the TPMT test has been implicitly driven by the threat of medical liability," says that report. "Oncologists do not want to risk prescribing Purinethol [6MP] to a child with leukemia without ensuring that the child is not at risk of bone marrow toxicity from a TPMT gene defect."[155]

Economic factors are likely to be key to the adoption of personalized medicine as well. In light of its continuing high-profile coverage and occasional success stories, the general public is likely to demand access to it. But the potentially high price tags involved aren't likely to convince payers (mostly health insurance companies and government agencies) to welcome this development and issue some sort of blanket coverage. Instead, it's likely to go on a case-by-case basis depending on the cost-effectiveness of the drug and/or diagnostic involved. No one could be more enthusiastic about new, cost-effective treatments than payers: if personalized medicine should prove to be cost-effective overall they would not only roll out the welcome wagon but carry it along on their own backs. And personalized medicine does have some potential economic advantages. "Payers would prefer to support highly effective medicines for a premium in an identifiable proportion of the affected population in preference to paying for a medicine for 100% of patients that would be expected to work in only 10%."[156]

But is personalized medicine really cost-effective? A report on it by the Royal Society concluded that "Although the introduction of pharmacogenetic testing has the potential to reduce costs through improved interventions, greater efficacy, less inappropriate prescribing and fewer ADRs, it is not clear whether or not the tests will increase or decrease overall health costs."[135] As the report goes on to note, expenses like medical personnel training and additional clinical time need to be factored in. And not unexpectedly, the economics will vary with each case. Even crunching the numbers isn't easy. One review illustrates "the complexity of cost-effectiveness analyses of PGx tests" which necessitates "estimates of the prevalence of genetic variation among the relevant populations; the impact of testing on non-response as well as adverse drug reactions; the availability of alternative diagnostic and therapeutic approaches; the availability of effective interventions that can be implemented on the basis of genetic information" and enough other data to induce vertigo in anyone who isn't a card-carrying pharmacoeconomist.[157]

The relatively few studies done to date that examined particular PMs for cost-effectiveness are generally encouraging but a mixed bag. Two studies which looked at the cost-effectiveness of CYP genotyping, one for CYP2D6[157] and the other for CYP2C19,[158] which happen to be the two DMEs that Roche's Amplichip analyzes, concluded that each was cost-effective, at least within certain patient groups. For example the latter pointed out, "Our calculations suggest that genotyping for CYP2C19 could save approximately $5000 for every 100 Asians tested, but none for Caucasian patients."[158] A review that evaluated the cost-effectiveness of 11 different pharmacogenomic interventions found that 7 were cost-effective, 2 weren't cost-effective, and 2 were equivocal.[159]

So payers would like to "cherry pick" which PGx tests to pay for in which patients. They might even go so far as to sponsor the development of ones that could save them money. In an example of this, "United Healthcare is conducting a study with Interleukin Genetics to develop a PGx test that could exclude patients who may be non-responsive to certain high-cost rheumatoid arthritis drugs."[155] The use of personalized medicine to tell someone what she *can't* have is likely to be alarming both to the patient and to pharma, but the cloud has a silver lining. Properly implemented, the chance for ADRs would be reduced, greater efficacy in the smaller patient-base can be expected, and payers won't have any excuse not to reimburse those who do fit the right profile. For drug companies to effectively deal with this kind of "formulary control," concludes one report, they need to "fight fire with fire: only by incorporating genetic tests during the drug development process can manufacturers prevent managed care from defining their markets for them."[160]

Right now, reimbursement for personalized medicine is spotty. Patients find that some things are reimbursed by some insurers and others aren't. Managed care organizations seem to be waiting to see what the biggest payer of them all, the federal government's *Centers for Medicare and Medicaid Services* (*CMS*) will do. And for pharma, "there's no point in investing in developing personalized drug therapies if payors won't cover them" so that many in the industry "are understandably wary of committing to drug development plans that assume prices an order of magnitude higher than what is typical today—at least until the government says that higher prices can be justified."[161]

Uncertainties about the effects of market segmentation and allowable pricing have definitely slowed pharma's adoption of personalized medicine. So, what's the prognosis for the next 10 or 20 years? The report by the Royal Society, hardly a forum for Pollyanna predictions, notes that pharmacogenomic sampling in clinical trials "is being conducted in 20–30% of current early-stage programmes of drug development, with a higher proportion in oncology" and anticipates that "a proportion of these...will progress to Phase III clinical trials, where the therapy is tested with an associated diagnostic." "Some companies," it notes, "have estimated that within five years [by 2010] this will be the case for up to 10% of their late-stage portfolio, perhaps rising to 20% in the next 10 years. Again, the proportion will tend to be higher for companies with a stronger portfolio in oncology."[135]

"Certainly I'm sure we're going to have personalized medicine," says Dr David Brown. "It will cost, and it will tend to be used for wealthy people in wealthy countries. But it's going to happen, so the industry has to face up to it."

## 2.5  Summary

The world of industrial drug discovery today is a dynamic one, constantly adapting to a wide variety of changing business, legal, and scientific factors. Big pharma has retained its position as the major source of marketed prescription drugs, but the in-licensing of many drug candidates and outsourcing of numerous sections of the R&D process represent breaks from its vertically integrated model of a few decades ago,[162] allowing input from biotech, universities, and CROs to become increasingly important and the number of drug discovery scientists in those organizations to grow. Organizational size is now seen as an impediment to innovation at many big pharma companies and a number are experimenting with reorganization into smaller business units. Another trend that will likely continue is the increasing emphasis on narrowly targeted specialty drugs, especially for cancer and inflammatory diseases, versus more broadly prescribed medications. Pressures on drug pricing due to generic competition, the proliferation of tiered formularies and mandated discounts, and the possibility of US government price control have made cost-effectiveness an important new addition to the properties of efficacy and safety required for any new drug.

Academia has seen a major shift toward the production of clinical candidates via translational research aided by universities themselves, government and foundation funding, and, in some cases, even outsourcing-for-equity use of CROs. The biotech industry now supplies an increasing number of clinical candidates to pharma via collaboration or acquisition, the value of such deals having gone up tremendously in recent years as pharma looks farther abroad for potential new blockbusters. Genentech, the "California model" of biotech success, has now become a different sort of model: the financial stability afforded by big pharma (Roche) majority ownership combined with its relative autonomy has resulted in innovative new therapies as well as profits that would not have been attainable by either party alone.

At the other end of the size spectrum, start-up biotechs, for some years severely pinched by financial pressures and short timeframes, have been working to reduce infrastructure costs via outsourcing, sometimes to the point of becoming virtual companies, and the use of incubators or shared facilities to do early studies. All indications are that the greatest challenge to innovative drug discovery lies neither in universities, in biotech, in CROs, nor yet in big pharma, but instead in how well the contributions of each can be brought together in coming years.

Promising new technologies for drug discovery abound, but how many of them will live up to the lofty summits of their expectation curves, or at least provide good, usable tools, remains to be seen, perhaps 15 or so years after their introduction. After its failure to meet particularly stratospheric expectations circa 2000, genomics has begun to prove its mettle. Much of its impact on future drug discovery depends on its utility for predicting drug response via pharmacogenomics, an important component of the inevitable advance toward personalized medicine, the matching of medicine to patient. Other "omics" are expected to contribute as well, many of which (e.g. cytomics, immunomics, phenomics, biomics, fluxomics) this author will prudently make no attempt to describe or define. Obstacles to the widespread implementation of personalized medicine that go beyond science and into the realm of law (genetic discrimination) and economics (insurance reimbursement) are currently being grappled with. Of course, new technologies impacting drug discovery aren't all "omics" and don't apply only to personalized medicine; some of them, like systems biology and *diversity oriented synthesis* (*DOS*), will be visited later.

So, having gone from coal tar to Gleevec by way of Hatch–Waxman, virtual companies, and the ineluctable gut microbiome, all in a mere two chapters and all by way of background, it's now high time to turn our attention to what the real-world drug researcher will find in industry today.

# Notes

1. Robbins-Roth, C. Broken? readers comment on biotechnology hurdles. *BioWorld Today*, May 31, 2005. Available at www.bioventureconsultants.com/5-31-05_Article.html.

2. Clinton, P. A year of threats. *Phar. Exec.*, December 1, 2005. Available at http://www.pharmexec.com/pharmexec/article/articleDetail.jsp?id=256553&pageID=2.

3. Drews, J. *In Quest of Tomorrow's Medicines* (New York, Springer-Verlag, 1999).

4. Wiederrecht, G.J., et al. Parnership between small biotech and big pharma. *IDrugs* **2006**, *9*, 560–564.

5. Jarvis, L. P&G to drop pharma R&D. *Chem. Eng. News*, February 27, 2006, p. 6.

6. Huston, L., Sakkab, N. Connect and develop: Inside Procter & Gamble's new model for innovation. *Harv. Bus. Rev.* **2006**, *84*, 1–8.

7. Forest Labs 2006 Annual Report. Available at http://ir.frx.com/phoenix.zhtml?c=83198&p=irol-reportsannual.

8. See their website: http://www.ranbaxy.com/rnd_activities.htm.

9. Although it's not clear whether their definition of "R&D" is the same as big pharma's, for reference, PhRMA member companies channel almost 16% of sales revenues into their R&D efforts.

10. Dr Reddy's Labs 2005–2006 Annual Report, available at http://www.drreddys.com/investors/pdf/annualreport2006.pdf.

11. Drews, J. *In Quest of Tomorrow's Medicines* (New York, Springer-Verlag, 1999), Reference 3, p. 165 and 208.

12. Gilbert, J., et al. Rebuilding big pharma's business model. *In Vivo* **2003**, *21*, 73–82.

13. GlaxoSmithKline 2002 Annual Report, pp. 19–20.

14. "Pfizer Announces Priorities to Drive Improved Performance, Position Company for Future Success and Enhance Total Shareholder Return", Pfizer press release January 22, 2007. Available at http://mediaroom. pfizer.com/index.php?s=press_releases&item=142.

15. Frantz, S. Pipeline problems are increasing the urge to merge. *Nat. Rev. Drug Discov.* **2006**, 5, 977–979.

16. Fontanez, C., et al. *Specialty Pharmacy Management Guide & Trend Report* (St. Louis, MO, CuraScript Pharmacy, June 2005).

17. "IMS Health Forecasts 5 to 6 Percent Growth for Global Pharmaceutical Market in 2007", press release October 24, 2006. Available at www.imshealth.com/ims/portal/front/articleC/ 0,2777,6599_3665_79210022,00.html.

18. Gottlieb, S. The Great Shift to Specialty Drugs. Available at www.aei.org/publications/pubID.22748/ pub_detail.asp.

19. Kaiser Family Foundation, Prescription Drug Trends, June 2006. Available at www.kff.org/rxdrugs/upload/ 3057-05.pdf.

20. In addition, it isn't clear whether the growing trend among managed care organizations to limit prescription quantities—say, requiring twelve 30-day prescriptions for Lipitor per year instead of four 90-day prescriptions, a net gain of eight prescriptions—is reflected in these numbers.

21. Kaiser Family Foundation/Harvard School of Public Health, The Public's Health Care Agenda for the New Congress and Presidential Campaign, December 2006. Available at www.kff.org/kaiserpolls/upload/7597.pdf.

22. ABC News/Kaiser Family Foundation/USA Today, Health Care in America 2006 Survey. Available at www.kff.org/kaiserpolls/upload/7572.pdf.

23. Kaiser Family Foundation, Health Poll Report, January/February 2005. Available at www.kff.org/kaiserpolls/ upload/January-February-2005-Health-Poll-Report-Survey-Toplines.pdf.

24. Lichtenberg, F.R. Do (more and better) drugs keep people out of hospitals? *Am. Econ. Rev.* **1996**, *86*, 384–388.

25. Faulding, S. The National Drugs Bill—Impact of New Price Controls. Available at www.ic.nhs.uk/psu/pdf/ pharmacy_management.pdf.

26. See Grimley, J. Pharma challenged. *Chem. Eng. News*, December 4, 2006, pp. 17–28.

27. Herper, M. Top of the cancer market? *Forbes*, October 12, 2006. Available at www.forbes.com/business/2006/ 10/12/cancer-drug-pricing-biz-cz_mh_1012drugs.html.

28. See Hileman, B. Regulatory trends. *Chem. Eng. News*, June 19, 2006, pp. 80–99.

29. Pipes, S.C. Price controls dry up drug discovery. *San Francisco Chronicle*, August 24, 2006, p. B7. See also its companion article in favor of this plan, Clayton, E., Alonzo-Diaz, L. "Make affordable medicine the law", ibid.

30. Farkas, C., Henske, P. Reference pricing for drugs. *Forbes*, April 13, 2006. Available at www.forbes.com/ columnists/2006/04/13/pharma-reference-pricing-cx_cf_0414pharma.html.

31. Taurel, S. "The Campaign Against Innovation", Chapter 18 in *Ethics and the Pharmaceutical Industry*, Santoro, M.A., Gorrie, T.M., Eds. (New York, Cambridge University Press, 2005).

32. Ellison, S.F., Mullin, W.P. Gradual incorporation of information: Pharmaceutical stocks and the evolution of President Clinton's Health Care Reform. *J. Law Econ.* **2001**, *XLIV*, 89–129.

33. Anon. "BIO Year by Year" 1993. Available at www.bio.org/speeches/pubs/milestone03/yearbyyear.asp.

34. Ellison, S.F., Wolfram, C. Pharmaceutical Prices and Political Activity (unpublished manuscript, Massachusetts Inst. Tech., April 2000).

35. Syntex (now Roche Palo Alto) for example, once sold dental chairs, x-ray equipment, and hair care products. Many pharma companies were involved in cosmetics, agricultural chemicals, and veterinary products before deciding to focus once more on their "core competency".

36. Quoted in Anon. An Audience with...Joseph DiMasi. *Nat. Rev. Drug Discov.* **2007**, *6*, 512.

37. This is not meant to imply that economic considerations weren't a concern in drug development until now, but is simply meant to emphasize just how critical they've recently become.

38. Loffler, A., Stern, S. Kellogg School of Management report, "The Future of the Biomedical Industry in an Era of Globalization", www.baybio.org/pdf/future_biomedical_industry.pdf.

39. Anyone interested in finding out more about pharmacoeconomics and outcome analysis should see Bootman, J.L., Townsend, R.J., McGhan, W.F. *Principles of Pharmacoeconomics* (Cincinnati, OH, Harvey Whitney Books Co., 1991).

40. See Building biotechnology but not via biotech companies, Cynthia Robbins-Roth, *BioWorld Today*, August 2, 2005. Available at www.bioventureconsultants.com/8-2-05_Article.html.

41. Quoted in Bailey, R. The new global villains: Drug companies and "Obscene Profits" *Policy* **2001**, *17*, 6–12.

42. Pisano, G.P. Can science be a business? *Harv. Bus. Rev.* **2006**, *84*, 114–124.

43. Robbins-Roth, C. The list: Innovative models on which to build biotech. *BioWorld Today*, August 3, 2005. Available at www.bioventureconsultants.com/8-3-05_Article.html.

44. See their website at http://nihroadmap.nih.gov/.

45. Birmingham, K. What is translational research? *Nat. Med.* **2002**, *8*, 647.

46. "HHMI Awards $10 Million to Graduate Programs That Combine Science and Medicine", available at www.hhmi.org/news/20060215.html.

47. See www.leukemia-lymphoma.org/all_page?item_id=11620.

48. Arnaud, C.H. Scripps takes a new approach *Chem. Eng. News*, December 4, 2006, pp. 64–67.

49. See their website at www.broad.mit.edu.

50. Alper, J. Biotech thinking comes to academic medical centers. *Science* **2003**, *299*, 1303–1304.

51. Kozikowski, A.P., et al. "Why academic drug discovery makes sense" Letter to *Science* **2006**, *313*, 1235–1236.

52. For a real-world example of some of the difficulties involved in translational research, see Couzin, J. Magnificent obsession. *Science* **2005**, *307*, 1712–1715.

53. For a glimpse into how Genzyme did this, see Chapter 4 of Cynthia Robbins-Roth's *From Alchemy to IPO* (Cambridge, MA, Basic Books, 2000).

54. Milken Institute, Financial Innovations for Accelerating Medical Solutions, October 2006. Available at www.milkeninstitute.org/pdf/fin_innov_vol2.pdf.

55. See Nagle, M. Synosis therapeutics revives roche CNS drugs. *DrugResearcher.com*, October 1, 2007. Available at www.drugresearcher.com/news/ng.asp?n=73233.

56. Anon. Creating a company. *Nat. Rev. Drug Discov.* **2007**, *6*, 93.

57. Quoted in Finance: Good alternatives. *Pharm. Exec.*, September 1, 2006. Available at www.pharmexec.com/pharmexec/article/articleDetail.jsp?id=369266.

58. Davies, K. Biogen-Idec's innovation incubator. *Bio-IT World*, January 18, 2007. Available at www.bio-itworld.com/newsitems/2007/01-18-07-biogen-idec?

59. Levine, D.S. Biotech babies born in "Garage". *San Francisco Bus Times*, September 15, 2006. Available at http://sanfrancisco.bizjournals.com/sanfrancisco/stories/2006/09/18/story3.html.

60. Baker, M. A bench to call your own. *Nature* **2006**, *444*, 120.

61. See their website at www.mmrx.org.

62. See Robbins-Roth, C.A kinder, gentler place for biotech hallmark: Start-ups. *BioWorld Today*, March 13, 2006. Available at www.bioventureconsultants.com/3-13-06_Article.html.

63. Robbins-Roth, C. When Frank talks, everyone listens but will they Act? *BioWorld Today*, July 5, 2006. Available at www.bioventureconsultants.com/7-5-06_Article.html.

64. See www.ott.nih.gov/pdfs/VelcadeCS.pdf.

65. Isis press release, Isis Pharmaceuticals and Symphony Genisis Enter into $75 Million Product Development Corporation, April 7, 2006.

66. Ransom, J. A reprise for collaborative financing in biotech. *Nat. Biotechnol.* **2006**, *24*, 873–874.

67. See the press release, Isis Acquires Symphony GenIsis. Available at http://ir.isispharm.com/releasedetail.cfm?releaseid=266423.

68. The term actually refers to the corporate philosophy of swamping prescribing physicians with multiple sales representatives, an approach which may now be on the wane, and was used by industry analyst Les Funtleyder in Alex Berenson's article New chief at Pfizer will reduce sales force. *New York Times*, November 29, 2006.

69. For insights into just how many services some of the larger CROs provides, check out Ricerca's website at www.ricerca.com or Covance's website at www.covance.com.

70. Bock-Marquette, I., et al. Thymosin β4 activates integrin-linked kinase and promotes cardiac cell migration and cardiac repair. *Science* **2004**, *432*, 466–472. But see also more recent encouraging results: Smart, N., et al. Thymosin β4 induces adult epicardial progenitor mobilization and neovascularization. *Nature* **2007**, *445*, 177–182.

71. Quoted in Company Interview, J.J. Finkelstein, RegeneRx Biopharmaceuticals, Inc. *The Wall Street Transcript*, February 27, 2006. Excerpts available at www.regenerx.com/img/JJFTWSTInterview.pdf.

72. Quoted in Thoughtleader: Small Cost, Big Win. *Pharm. Exec.*, September 1, 2006. Available at www.pharmexec.com/pharmexec/article/articleDetail.jsp?id=369265.

73. Pisano, G.P. *Science Business: The Promise, the Reality, and the Future of Biotech* (Boston, MA, Harvard Business School Press, 2006).

74. Black, J. Future perspectives in pharmaceutical research. *Pharm. Policy Law* **1999**, *1*, 85–92.

75. Brown, D. Target selection and pharma industry productivity: What can we learn from technology S-curve theory. *Curr. Opin. Drug. Discov. Devel.* **2006**, *9*, 414–418.

76. Foster, R. *Innovation: The Attacker's Advantage* (New York, Summit Books, 1986).

77. Nightingale, P., Martin, P. The myth of the biotech revolution. *Trends Biotech.* **2004**, *22*, 564–569.

78. For a more detailed discussion of the time course of some new drug discovery technologies, see Gershell, L.J., Atkins, J.H. A brief history of novel discovery technologies. *Nat. Rev. Drug Discov.* **2003**, *2*, 321–327.

79. Another book Dr Brown cites agrees with this estimate for new technologies in other industries. See Christensen, C.M. *The Innovator's Dilemma: When New Technologies Cause Great Firms to Fail* (New York, Harvard Business School Press, 1997).

80. For a review, see Rajasethupathy, P., et al. Systems modeling: A pathway to drug discovery. *Curr. Opin. Chem. Biol.* **2005**, *9*, 400–406.

81. Anon. Five years on . . . and four challenges for the pharmaceutical industry. *Nat. Rev. Drug Discov.* **2007**, *6*, 3.

82. One example of this, the genotyping of HIV strains to predict drug resistance and thereby enable better chemotherapy, is reviewed in Blum, R.A., et al. HIV resistance testing in the USA—a model for the application of pharmacogenomics in the clinical setting. *Pharmacogenomics* **2005**, *6*, 169–179.

83. For a great introduction to the medical applications of genomics, see Guttmacher, A.E., Collins, F.S. Genomic medicine—a primer. *N. Engl. J. Med.* **2002**, *347*, 1512–1520.

84. Previously, genetic variations between humans was estimated to be 0.1%, but this ws revised by the recent sequencing of J. Craig Venter's diploid genome (both alleles). See Levy, S., et al. The diploid genome sequence of an individual human. *PLoS Biol.* **2007**, *5*, 2113–2144.

85. For another well written, short introduction to the study of SNPs and a rare example of a patent actually teaching rather than obscuring something, see the "Background of the Invention" section of Venter, et al. Polymorphisms in known genes associated with human disease, methods of detection and uses thereof, US 6812339 (November 2, 2004). This and other patents are readily available via a website search of www.uspto.gov or Google Patents.

86. Nackley, A.G., et al. Human catechol-O-methyltransferase haplotypes modulate protein expression by altering mRNA secondary structure. *Science* **2006**, *314*, 1930–1933.

87. Kimchi-Sarfaty, C., et al. A "Silent" polymorphism in the MDR1 gene changes substrate specificity. *Science* **2007**, *315*, 525–528.

88. See www.hapmap.org.

89. Burke, W. Genomics as a probe for disease biology. *N. Engl. J. Med.* **2003**, *349*, 969–974.

90. Drumm, M.L., et al. Genetic modifiers of lung disease in cystic fibrosis. *N. Engl. J. Med.* **2005**, *353*, 1443–1453.

91. Ioannidis, J.P.A., et al. Replication validity of genetic association studies. *Nat. Genet.* **2001**, *29*, 306–309.

92. Page, G.P., et al. "Are we there yet?": Deciding when one has demonstrated specific genetic causation in complex diseases and quantitative traits. *Am. J. Hum. Genet.* **2003**, *73*, 711–719. References to the source of those percentages are given therein.

93. Osler, W. *The Principles and Practice of Medicine* (New York, D. Appleton, 1892).

94. Paik, S., et al. Gene expression and benefit of chemotherapy in women with node-negative, estrogen receptor-positive breast cancer. *J. Clin. Oncol.* **2006**, *24*, 3726–3734.

95. See Neergaard, L. Gene Test May Show Who Can Skip Chemo. *Associated Press*, June 12, 2006.

96. Potti, A., et al. A genomic strategy to refine prognosis in early-stage non-small-cell lung cancer. *N. Engl. J. Med.* **2006**, *355*, 570–580.

97. Potti, A., et al. Genomic signatures to guide the use of chemotherapeutics. *Nat. Med.* **2006**, *12*, 1294–1300.

98. Hopkins, M.M., et al. Putting pharmacogenetics into practice. *Nat. Biotechnol.* **2006**, *24*, 403–410.

99. Cowsert, L.M. Genomics, proteomics and pharmacogenetics—experimental approaches. *Curr. Opin. Mol. Ther.* **2005**, *7*, 200–201.

100. Lazarou, J., et al. Incidence of adverse drug reactions in hospitalized patients. *J. Am. Med. Assoc.* **1998**, *279*, 1200–1205.

101. Budnitz, D.S., et al. National surveillance of emergency department visits for outpatient adverse drug events. *J. Am. Med. Assoc.* **2006**, *296*, 1858–1866.

102. For a review of CYP metabolism and drug response, see Wilkinson, G.R. Drug metabolism and variability among patients in drug response. *N. Engl. J. Med.* **2005**, *352*, 2211–2221.

103. For example, CYP3As are involved in unmasking Metabasis Therapeutics' HepDirect prodrugs. See Erion, M.E., et al. Liver targeted drug delivery using HepDirect prodrugs. *J. Pharmacol. Exp. Ther.* **2005**, *312*, 554–560.

104. Service, R.F. Going from genome to pill. *Science* **2005**, *308*, 1858–1860.

105. Niwa, T., et al. Contribution of human hepatic cytochrome P450 isoforms to the metabolism of psychotropic drugs. *Biol. Pharm. Bull.* **2005**, *28*, 1711–1716.

106. Obach, S., et al. Sertraline is metabolized by multiple P450 enzymes, monoamine oxidases, and glucuronyl transferases in human: An in vitro study. *Drug Metab. Dispos.* **2005**, *33*, 262–270.

107. de Leon, J., et al. The AmpliChip CYP450 genotyping test: Integrating a new clinical tool. *Mol. Diagn. Ther.* **2006**, *10*, 135–151.

108. See Elion, G.B. The purine path to chemotherapy. *Science* **1989**, *244*, 41–47.

109. Stanulla, M., et al. Thiopurine methyltransferase (TPMT) genotype and early treatment response to mercaptopurine in childhood acute lymphoblastic leukemia. *J. Am. Med. Assoc.* **2005**, *293*, 1485–1489.

110. Tai, J.-L., et al. Enhanced proteolyisis of thiopurine S-methyltransferase (TPMT) encoded by mutant alleles in humans (TPMT*3A, TPMT*2): Mechanisms for the genetic polymorphism of TPMT activity. *Proc. Natl. Acad. Sci. U.S.A.* **1997**, *94*, 6444–6449.

111. Collie-Duguid, E.S., et al. The frequency and distribution of thiopurine methyltransferase alleles in Caucasian and Asian populations. *Pharmacogenetics* **1999**, *9*, 37–42.

112. Nebert, D.W., et al. Pharmacogenomics and "Individualized Drug Therapy": High expectations and disappointing achievements. *Am. J. Pharmacogenomics* **2003**, *3*, 361–370.

113. Fromm, M.F. Importance of P-glycoprotein at blood—tissue barriers. *Trends Pharmacol. Sci.* **2004**, *25*, 423–429.

114. For a general review see Kerb, R. Implications of genetic polymorphisms in drug transporters for pharmacotherapy. *Cancer Lett.* **2006**, *234*, 4–33.

115. For a review of P-gp polymorphisms and their effects, see Dey, S. Single nucleotide polymorphisms in human P-glycoprotein: Its impact on drug delivery and disposition. *Expert Opin. Drug Deliv.* **2006**, *3*, 23–35.

116. Hoffmeyer, S. Functional polymorphisms of the human multidrug-resistance gene: Multiple sequence variations and correlation of one allele with P-glycoprotein expression and activity *in vivo. Proc. Natl. Acad. Sci. U.S.A.* **2000**, *97*, 3473–3478.

117. Fellay, J., et al. Response to antiretroviral treatment in HIV-1-infected individuals with allelic variants of the multidrug resistance transporter 1: A pharmacogenetics study. *Lancet* **2002**, *359*, 30–36.

118. Verstuyft, C., et al. Absence of association between MDR1 genetic polymorphisms, indinavir pharmacokinetics and response to highly active antiretroviral therapy. *AIDS* **2005**, *19*, 2127–2131.

119. Woodahl, E.L., et al. MDR1 G1199A polymorphism alters permeability of HIV protease inhibitors across P-glycoprotein-expressing epithelial cells. *AIDS* **2005**, *19*, 1617–1625.

120. Hulot, J.-S. A mutation in the drug transporter gene ABCC2 associated with impaired methotrexate elimination. *Pharmacogenet. Genomics* **2005**, *15*, 277–285.

121. Mason, D.A., et al. A gain-of-function polymorphism in a G-protein coupling domain of the human $\beta_1$-adrenergic receptor. *J. Biol. Chem.* **1999**, *274*, 12670–12674.

122. Eichhorn, E.J., et al. A Trial of the beta-blocker bucindolol in patients with advanced chronic heart failure. *N. Engl. J. Med.* **2001**, *344*, 1659–1667.

123. Liggett, S.B., et al. A polymorphism within a conserved $\beta_1$-adrenergic receptor motif alters cardiac function and $\beta$-blocker response in human heart failure. *Proc. Natl. Acad. Sci. U.S.A.* **2006**, *103*, 11288–11293.

124. See Arca's website, www.arcadiscovery.com.

125. Burstein, H.J. The distinctive nature of HER2-positive breast cancers. *N. Engl. J. Med.* **2005**, *353*, 1652–1654.

126. Vogel, C.L., et al. Efficacy and safety of trastuzumab as a single agent in first-line treatment of HER2-overexpressing metastatic breast cancer. *J. Clin. Oncol.* **2002**, *20*, 719–726.

127. Hortobagyi, M.D. Trastuzumab in the treatment of breast cancer.*N. Engl. J. Med.* **2005**, *353*, 1734–1736.

128. Quoted in Tolchin, E. Targeted medicine fraught with challenges. *Genomics Proteomics* November 15, 2004.

129. "Guidance for Industry: Pharmacogenomic Data Submissions". Available at www.fda.gov/cber/gdlns/pharmdtasub.pdf.

130. Capdeville, R., et al. Glivec (STI571, imatinib) a rationally developed, targeted anticancer drug. *Nat. Rev. Drug Discov.* **2002**, *1*, 493–502.

131. Druker, B.J., et al. Efficacy and safety of a specific inhibitor of the BCR-ABL tyrosine kinase in a chronic myeloid leukemia. *N. Engl. J. Med.* **2001**, *344*, 1031–1037.

132. One estimate is that over 80% of all clinical trials will make use of biomarkers by 2011. See Jain, K.K. *Biomarkers: Technologies, Markets, and Companies* (Basel, Switzerland, Jain PharmaBiotech Publications, 2006).

133. See "Dasatinib (BMS-354825): Oncology Drug Advisory Committee (ODAC) Briefing Document", NDA 21-986, June 2, 2006. Available at www.fda.gov/ohrms/dockets/AC/06/briefing/2006-4220-B1-01Bristol-MyersSquibb-Background.pdf.

134. Horrobin, D.F. Realism in drug discovery—could Cassandra be right? *Nat. Biotechnol.* **2001**, *19*, 1099–1100.

135. The Royal Society Report, Personalised medicines: Hopes and realities, September 21, 2005. Available at www.royalsoc.ac.uk/displaypagedoc.asp?id=23244.

136. Henry, C.M. New "Ome" in town *Chem. Eng. News*, December 2, 2002, pp. 66–70.

137. Robertson, D.G. Metabonomics in toxicology: A review. *Toxicol. Sci.* **2005**, *85*, 809–822.

138. Nielsen, J., Oliver, S. The next wave in metabolome analysis. *Trends Biotechnol.* **2005**, *23*, 544–546.

139. Lindon, J.C., et al. The consortium for metabonomic toxicology (COMET): Aims, activities, and achievements. *Pharmacogenomics* **2005**, *6*, 691–699.

140. Quoted in Owens, J. Predicting predisposition. *Nat. Rev. Drug Discov.* **2006**, *5*, 455.

141. Clayton, T.A., et al. Pharmaco-metabonomic phenotyping and personalized drug treatment. *Nature* **2006**, *440*, 1073–1077.

142. Bäckhed, F., et al. Host-bacterial mutualism in the human intestine. *Science* **2005**, *307*, 1915–1920.

143. Gill, S.R., et al. Metagenomic analysis of the human distal gut microbiome.*Science* **2006**, *312*, 1355–1359.

144. Gordon, H.A. Characteristics of the germfree rat. *Acta Anat.* **1966**, *64*, 367–389.

145. See, for example, Scheline, R.R. Metabolism of foreign compounds by gastrointestinal microorganisms. *Pharmacol. Rev.* **1973**, *25*, 451–523.

146. Goldman, P., et al. Metabolism of drugs by microorganisms in the intestine. *Am. J. Clin. Nutr.* **1974**, *27*, 1348–1355.

147. O'Hara, A.M., Shanahan, F. The gut flora as a forgotten organ. *EMBO Rep.* **2006**, *7*, 688–693.

148. Robosky, L.C., et al. Metabonomic identification of two distinct phenotypes in Sprague-Dawley (Crl;CD(SD)) rats. *Toxicol. Sci.* **2005**, *87*, 277–284.

149. See Ley, R.E., et al. Microbial ecology: Human gut microbes associated with obesity. *Nature* **2006**, *444*, 1022–1023, and Turnbaugh, P.J., et al. An obesity-associated gut microbiome with increased capacity for energy harvest. *Nature* **2006**, *444*, 1027–1031.

150. See Lewin, T. Commission sues railroad to end genetic testing in work injury cases. *New York Times*, February 10, 2001.

151. Glabman, M. Genetic Testing: Major Opportunity, Major Problems. *Managed Care*, November 2006. Available at www.managedcaremag.com/archives/0611/0611.genetest.html.

152. Collins, F.S. Personalized medicine: A new approach to staying well. *Boston Globe*, July 17, 2005.

153. Harmon, A. Congress Passes Bill to Bar Bias Based on Genes. *New York Times*, May 2, 2008.

154. Marchant, G.E. Legal pressures and incentives for personalized medicine. *Future Med.* **2006**, *3*, 391–397.

155. Bernard, S. The 5 Myths of Pharmacogenomics. *Pharm. Exec.*, October 1, 2003. Available at www.pharmexec.com/pharmexec/content/printContentPopup.jsp?id=72796.

156. Roses, A.D. Genome-based pharmacogenetics and the pharmaceutical industry. *Nat. Rev. Drug Discov.* **2002**, *1*, 541–549.

157. Phillips, K.A., Van Bebber, S.L. Measuring the value of pharmacogenomics. *Nat. Rev. Drug Discov.* **2005**, *4*, 500–509.

158. Desta, Z., et al. Clinical significance of the cytochrome P450 2C19 genetic polymorphism. *Clin. Pharmacokinet.* **2002**, *41*, 913–958.

159. Phillips, K.A., Van Bebber, S.L. A systematic review of cost-effectiveness analyses of pharmacogenomic interventions. *Pharmacogenomics* **2004**, *5*, 1139–1149.

160. Rawson, K. The next coverage hurdle: Pharmacogenomics as formulary control tool. *RPM Report* **2006**, *1*, No. 11. Available at www.windhover.com/ezine/pdf/Payors0611.pdf.

161. Rawson, K. Reimbursing designer drugs. *RPM Report* **2006**, *1*, No. 10.

162. For example, in a move that would have been inconceivable 20 years ago, Eli Lilly recently announced its planned transformation from a FIPCO to a " 'network structure' of outside contractors, service providers, and others." See Russell, J. Lilly shifting work to outsiders. *Indianapolis Star*, December 7, 2007.

**Chapter 3**

# Industrial Considerations

## 3.1 Intellectual Property

*Disclaimer: Nothing in this chapter or this book can substitute for the advice or guidance that a corporate attorney or other qualified legal professional can provide, and no attempt should be made to use it in lieu thereof. Issues can be complex and case-specific, laws can change or vary, and strategies can diverge. Always consult with the appropriate authority when advice is needed or concerns arise regarding intellectual property or other corporate legal issues. In perhaps, no other area of drug discovery is specific professional expertise so important.*

### 3.1.1 The Value of New Ideas

New ideas are the lifeblood of drug discovery. Without a way of profiting from them, the tremendous cost of new drug development couldn't be borne, investor capital couldn't be raised, and industrial drug discovery as we know it wouldn't be viable. So a way of defining, documenting, and protecting such *intellectual property* (*IP*) is crucial to drug discovery at any company. New ideas aren't confined to industry, of course. They're every bit as important to academic research groups, but there the emphasis has traditionally been on the rapid publication of the results of cutting-edge research. This made for a very different atmosphere in academia than in industry. In the former, the "publish or perish" model held sway. But in industry, the phrase "publish *and* perish" almost seems to apply, research results being doled out almost grudgingly, often through a cryptic medium seemingly more the domain of lawyers than scientists, the patent.

The reason for industry's reluctance to shoot out fresh new discoveries to *Science* or *The Journal of Medicinal Chemistry* is a compelling one: as we'll see later, publicly revealing these ideas before filing a patent application can make them difficult or impossible to patent, in effect throwing away valuable IP rights. With the passage of the Bayh–Dole Act (discussed in Chapter 2) and many demonstrations of just how marketable and profitable academic inventions can be, university research groups are now more careful about early disclosure of potentially patentable ideas. Still, the graduate or postdoc going into industry won't fail to notice the emphasis placed on the proper care of IP and new researchers will need to be especially careful about following company policies when it comes to IP.

The particulars will vary, and if in doubt the reader, again, needs to check with the appropriate authority such as a supervisor or member of the legal department. But some things are common to almost every company. Every drug discovery scientist in industry will be expected to know the correct procedures for dealing with at least three critical aspects of corporate IP protection:

▶ Documenting a new idea by filing an *invention disclosure* (sometimes also known by other names, like *record of conception of invention*),

▶ Properly recording data in a legally acceptable form in a lab notebook, and,

▶ Protecting corporate IP by guarding against inappropriate public *disclosure*.

All of these things are important for obtaining a patent, the major mechanism by which IP is defined and protected in drug discovery.

### 3.1.1.1 Invention Disclosures

New ideas need to be documented in writing. Not only might they form the basis for a patent application later, but also in addition they might be used to establish *priority*, a way a company can prove that its employees thought of the idea first, should that ever become an issue. Employers will specify how invention disclosures should be written up. This usually includes at least a description of the idea (such as the structures of novel compounds and how they might be made), why the idea might be useful, the name(s) of the inventor(s), and the date conceived. It should also be read, signed, and dated by someone else, preferably a knowledgeable scientist from within the company who isn't involved in the project. Most companies have either paper or web-based forms to be filled out. Others might ask the inventor to write it down in her lab book—either her regular lab book or a separate one kept just for this purpose—using a particular format. These disclosures will be submitted to a supervisor and/or the legal department for evaluation and recordkeeping purposes. Ultimately, an invention disclosure might become the basis for a valuable patent and establish priority of inventorship, which explains the concern about them being written up properly and witnessed.

So who would own the resulting potential invention? If you said "the employer," you're correct. If a patent based on your new idea ends up being issued you'll be listed on the first page as "inventor" as discussed later, but the "assignee" will be your company. In industrial research, people in research are "employed to invent" and generally sign an employment contract stating that all rights to their inventions will be assigned to the employer. As property of the company, these ideas aren't something one can disclose at one's own discretion or take with him when he leaves unless they've already been published, and even then only what's revealed in the publication—often much less than everything that's been learned—can be discussed. And, unfortunately, extra compensation doesn't necessarily accompany a profitable invention, at least in the US (5% of profits would be nice, thank you very much!). In some countries, however, notably Germany, Japan, and the UK, the inventor may legally be entitled to some.[1] Sometimes, a US company will later reward particularly valuable inventions with a bonus, but it doesn't have to. Most of the time rewards will be indirect, like another patent to list on one's resume, which *is* a sort of currency among drug discovery researchers, especially those looking for jobs.

In writing up an invention disclosure to document an idea, how can a scientist, especially one new to a project, know that the idea is "new?" A place to start is to ask your supervisor or someone who's been working on the project for a long time. Searching an in-house database can establish whether the proposed compounds were made and registered there before, but of course can't tell you if (a) someone has tried, but failed, to make them in-house before, (b) someone else is currently working on their synthesis, or (c) someone at another company or university has made such compounds. Talking to in-house project scientists can shed some light on all of these, and a quick search using CrossFire® or SciFinder® can turn up much of what's known about them outside the company, if they are known compounds. The researcher shouldn't get discouraged, if it turns out that the compounds he proposes to make are described in the literature unless they were previously made for exactly the same target, so that the whole idea has already been tried and published. Even then, although an invention disclosure might not be in order, the compounds might still be worth making as standards, if they're active.

For simple invention disclosures, elaborate searching isn't required and when in doubt just write it down. It's better to repeat a known idea than to fail to document an original one. Most invention disclosures, alas, never end up resulting in patent applications, but if it turns out that the compounds are fantastic and your idea may be worth applying for a patent on, the first thing the legal department will probably do is ask you, the inventor, what you know about the *prior art* (previous papers, posters, patents, etc.) and then at some point do a thorough search of databases like CAS and Derwent, an important one for patents.

Corporate legal departments like it best if invention disclosures are always written up before work on the idea starts, but many times chemists don't do it this way, preferring to do the experiment first and write up an invention disclosure only later if good compounds are obtained and someone twists their arm to do so. This is a bad way of doing things, but it stems partly from the fact that most of us prefer science to law and experiment to theory, and partly from the fact that many of us don't really appreciate how important it is to document the early stages of an invention. Always write down your ideas, even if you can't do so in legal-sounding prose. Once in a while a problem arises when two different scientists at the same company each claim to have come up with an idea first, especially if it's a successful one. Having a written invention disclosure can help prove that you came up with it first.

Wayne Montgomery is a patent attorney and Vice President of Patents at Exelixis with firsthand industrial research experience stemming from his years as a pharmacologist before going into law. "It's really crucial," he points out, "to get that entry in quickly and to have it witnessed. If you don't have something entered in the notebook and someone else says that he or she came up with the idea and has a piece of paper or an e-mail message to back it up, if there's no record that you thought of it first, you're going to get scooped." Writing up invention disclosures is really a way of protecting your *own* IP.

It's also a sign of creativity and commitment that bodes well for future evaluations and career advancement. Like anything else, though, it can be overdone or misused. Trying to maximize your output of invention disclosures by writing obvious or incremental ones (e.g. methyl esters as prodrugs of carboxylic acids followed by ethyl esters as prodrugs . . .) or "scoop" your co-workers by concentrating on ideas for their parts of the project instead of your own is a losing strategy. Research projects are exercises in teamwork, not one-upsmanship.

### 3.1.1.2 Notebooks and Data Recording

As with invention disclosures, most companies will have specific requirements about keeping records of experimental data and these need to be followed. Some common rules usually apply here too. Writing things down on paper towels or loose sheets of paper, however convenient, is taboo as these are easy to lose or discard, and they establish no chronological trail of experiments. Bound lab notebooks with numbered pages or sometimes, these days, electronic notebooks are used instead. With normal lab notebooks, experiments are written up in ink and if mistakes are made these are indicated by crossing them out with a single line and then initialing. Any blank spaces on a page are typically crossed out with a diagonal line. Data for compounds are usually printed out in shrunken form, then taped into the notebook with the researcher's signature across the data, tape and regular notebook page to allay potential suspicions that things were tampered with. Alternately (again, check on how it's done at your company), it may be acceptable to store the *nuclear magnetic resonance* (*NMR*) spectra, *liquid chromatography–mass spectrometry* (*LCMS*) spectra, etc., in a binder separately and refer to their location in the lab notebook. One commonsense note: labeling an NMR spectrum or file "1036-120A" if it's for the first product mentioned on page 120 of notebook 1036 turns out to be a lot more practical than calling it "P3 cyclohexyl ethyl ester" or "High $R_f$ spot" when trying to find the right data weeks or months later.

Like invention disclosures, notebook entries are sometimes used to establish priority or to document *diligence*, the fact that an idea wasn't just abandoned but was actively pursued instead, so having dates on experiments is important. All entries need to be dated and when the experiment is done it's signed and dated again by the researcher. Being a legal record, lab books also need to be witnessed, and "in a timely manner," which is often taken to mean within a week or two of completion. This means that another researcher needs to look it over, then sign and date at the bottom of the page where it says "Read and understood" or "Witnessed by." As with invention disclosures, someone who understands the technical aspects (i.e. another medicinal chemist), preferably one not involved in the project, is required. In practice, this usually occurs through a kind of "lab book buddy system:" you witness mine and I'll witness yours. New employees should find out who's available to do this. Only later will they find out who actually follows up and does it quickly. These kinds of buddies are always in demand. Having a notebook witnessed slowly can be excruciating: while your "buddy" has the book you can't enter any new experiments in it.

Dating, signing, and witnessing experiments shouldn't be hard to do, but if you multiply the number of scientists doing research by the number of experiments they do every day you'll see that thousands of pages might need to be dated, signed, and witnessed at a research organization every month. Evaluations rarely hinge on how promptly lab experiments were recorded and witnessed, while many depend on ideas and compounds. Trying to get compounds made quickly, chemists can fall into the trap of poor record keeping and incredibly infrequent witnessing. Shameful notebooks filled only with bare sketches of reactions and numbers of milligrams of reagents are known to exist. Don't let it happen to you. And keep in mind that lab book witnessing is a nice Friday afternoon activity.

Every so often lab notebooks-in-progress will be microfilmed or scanned in, so that a copy can be kept offsite in case the original should be lost, destroyed in a flood or fire, or perhaps eaten through by $POCl_3$ (which happened to the author once). Completed notebooks are routinely collected for microfilming and are sometimes stored thereafter at another location, with subsequent access only to digital images of them, not the original notebooks, being given to the scientist.

### 3.1.1.3  Avoiding Inappropriate Disclosure

"With very few exceptions, the work product of a drug discovery organization is not research. The work product is a patentable chemical compound that becomes a drug."[2] Anything that interferes with a company's ability to patent its compounds so that it can turn them into profitable drugs detracts from its value and needs to be avoided at all costs. Patents, as we'll see, can only be issued for new inventions, not to ones that are already known to the public, so revealing your great new idea through a journal article, a talk at an *American Chemical Society* (*ACS*) meeting, or an impromptu exposition to anyone willing to listen at the neighborhood bar before a patent application has been filed can destroy your company's ability to get a patent on it. As eager as the scientist might be to let the world know his results, he needs to "keep a cork on it" until corporate attorneys say they're OK to divulge.

The specific effects of public disclosure on patentability vary by country. The *US Patent and Trademark Office* (*PTO*) allows for a 12-month *grace period* after public disclosure, use, sale, or offer for sale of an invention during which time a patent application can be filed. This means that, if on January 1, 2007 a journal article revealing the invention is published either on the web or in print, as long as a US patent application is filed before January 1, 2008, it won't have an adverse effect on its patentability. Outside the US, some countries likewise have grace periods (Canada also has one of 12 months), but most don't, or may have grace periods tied to conditions that probably won't apply.[3] Importantly, the *European Patent Office* (*EPO*) does not have a grace period except for very specific conditions, such as disclosure at an officially recognized international exhibition, of which there aren't many. The bottom line, though, is this: if the invention is publicly disclosed, publicly used, sold, or offered for sale before a patent application is filed for it, in many countries it won't be patentable.

Note that "public" disclosure isn't the same as talking about it to your co-workers or consultants who will have signed a *confidentiality agreement* (*CDA*). Companies that have ventured into the realm of IP paranoia sufficient to avoid revealing new ideas to their own consultants are fortunately few in number.[4] Note, however that, even in cases where public disclosure doesn't affect patentability, as in the US 12-month grace period, it still tips one's hand that much earlier to potential competitors.

Corporate legal departments serve as the guardians of IP, and if any issues relating to disclosure come up they need to be consulted and their advice followed. Mechanisms exist within each company to regulate what information can go out in posters, talks, and publications. Often an online or paper permission request needs to be filled out and approved by managers and corporate attorneys before presentations can be given or manuscripts sent out to journals, and this can take some time, so potential speakers and authors need to learn the system and apply early.

It's axiomatic that the less useful or important the corporate research is, the more likely it is to be published early. A company will just be less zealous in guarding ideas it's not interested in patenting. In fact, it might not mind spoiling the water for those who *are* interested: early publication can make ideas and compounds unpatentable for them too, while at the same time making in-house scientists happy and adding to the publication record of the company. But at the other extreme, the structures and syntheses of clinical candidates constitute incredibly valuable IP, with drug exclusivity potentially being worth millions of dollars *per day*. Understandably, to avoid aiding the competition, clinical candidate

---

**Box 3.1    Key Points About IP Protection**

▶ The researcher needs to properly document new ideas as invention disclosures

▶ Employees can be inventors, but companies are assigned the rights to their inventions

▶ Notebooks are important legal records and need to be signed and witnessed frequently

▶ Publication, public use, sale, or offer for sale of an invention prior to patent application can result in the loss of valuable IP rights

▶ Legal approval will be required before information can go out via journal articles, posters, or meeting talks

---

structures will be kept secret for as long as possible. Even though they may be thoroughly patented by the time of late-stage trials, companies will play it close to the vest. Chemists know just how much information can be conveyed and how many ideas can be engendered upon seeing a single structure in a journal.

Concern for IP protection involves other things as well. Under no circumstances, should a scientist take a lab book with him at the end of the day in order to work on it at home. Although the intention is good, it can be lost or—employing just a *little* paranoia here—seen by others. Access to corporate databases is highly regulated too. Employers don't like the idea that proprietary information can be readily removed on a CD or thumb drive and may have rules against it. Many companies allow employees to have remote access to their e-mails, but few allow easy offsite access to sensitive databases. Sometimes, it's very convenient to be able to work on a paper or presentation at home, and this is often allowed, but one needs to check and abide by the rules.

To the author's knowledge, not a single biotech or pharma company tries to restrict researchers from coming up with new ideas at home, as long as they're kept confidential. Many fine ideas have come from notepads filled with structures and synthetic routes drawn in fits of postprandial or insomniac inspiration that result in invention disclosures the next day. It's also a sign that you enjoy your work and are probably in the right business. Box 3.1 summarizes some points the researcher should keep in mind when it comes to intellectual property.

## 3.1.2  Patents

*"Obtaining a patent and market exclusivity is not rocket science. If it were rocket science, the process would be much simpler."*[5]

Judge David M. Gersten, Third District Court of Appeal

### 3.1.2.1  Introduction and Definition

Researchers coming out of academia these days will probably already be familiar with good notebook practices and might have dealt with a few patent procedures or maybe even invention disclosures before. But working in industrial drug discovery will give them something of an education in utility patents. At times these can seem to be as inscrutable as if written in Sanskrit.

They're complex enough that even after many years of experience with them, scientists probably won't be capable of properly drafting their own patent applications. But, of course, no one expects them to do that! Companies employ *patent agents* (individuals who have passed the patent bar exam but aren't attorneys) and *patent attorneys*, probably at considerably higher expense, to do this instead. Smaller companies may not keep such professionals on their payroll, but instead might opt to work with IP law firms as needed.

Large companies, sometimes, employ not only patent agents and attorneys, but also another type of specialist, liaisons that interface between science and law. In attorney Wayne Montgomery's experience this is "a real luxury that it's nice to have. They do things like audit notebooks, help draft examples, and go back and forth between the chemists and the attorneys and facilitate that interaction."

So, the research scientist won't be expected to master patent law, but should be capable of gleaning necessary information from patents and patent applications and efficiently providing information to the specialists who draw them up. There are really five practical reasons why industrial drug discovery researchers need to be concerned with patents:

- ▶ To see whether a compound or series of compounds have already been described and/or patented

- ▶ To locate compound data and synthetic procedures of interest

- ▶ To keep abreast of the latest developments in the field

- ▶ To become familiar enough with some basic IP issues to be able to work well with patent attorneys or agents

- ▶ Occasionally, as an introduction to an alternative career for scientists

At its simplest, a patent represents an agreement between a governmental authority (in the US, the PTO) and an inventor or assignee. Each party gives up something and, in return, gets something. The inventor gives up her secret by disclosing the invention sufficiently well that "a person of ordinary skill in the art" would be able to "practice the invention." In addition, in the US (but not in other countries) the patent must describe the *best mode* for practicing the invention. In return for all of this, the inventor or assignee is granted the legal right, usually for 20 years from the date of filing,[6] to exclude others from practicing the invention she's just spelled out to them. This can create a long, lucrative monopoly if the invention has commercial value. It becomes actual property and if another party uses the invention without licensing it from the owner first, it can be taken to court for *patent infringement*. Hence the oft-repeated, unofficial definition of a patent is "*a license to sue*." From the government's point of view, granting such a monopoly is justified by the useful public knowledge made available upon patent expiration and by the advances (new medicines, cheaper goods, etc.) resulting from the new ideas that the process encourages.

Note the important fact that a patent *does not* confer on the inventor the right to practice his invention, only the right to keep others from doing so. Many scientists are surprised to find that a given compound can be covered by two or more valid patents, often belonging to different companies. Suppose, a broad array of *N*-substituted tetrahydroisoquinoline-8-carboxylic acids is covered by patent *A*. Other researchers might later discover that a specific subset of these compounds having a very narrowly defined side chain at the 3-position has unexpectedly good activity or selectivity not found for other compounds described by patent *A*. These might still be patentable as a so-called *selection invention*, a kind of novel improvement on an existing invention. If you're granted such a patent and want to bring a drug of this type to market, but another company owns patent *A*, you won't be able do so without a licensing agreement with them. Your patent is said to be *dominated* by *A*. At the same time, whoever owns patent *A* can't market a drug described by your patent *B* without such an agreement either.

Coverage by multiple patents can apply to other situations like polymorphs and enantiomers as well. So, although the real estate analogy is often used in describing patents (marking off claims, right to defend against trespass, etc.), sometimes a particular patch of dirt can actually belong to more than one party with neither one being able to develop it without working out an agreement with the other owner.

"Intellectual property" covers more than just patents. Things like copyrights and trademarks are ways of protecting IP too, but don't usually concern the drug discovery researcher very much. As shown in Figure 3.1, types of patents include plant, design, and utility patents, the last of which includes *composition of matter* patents, the most frequently utilized way of protecting new drug discoveries. These patents may claim a compound (*new chemical entity*, *NCE*), a polymorphic form, or some unique formulation. *Process patents* describing an industrially feasible way of making a drug can be valuable too, but if that's the only type of patent one has, resourceful chemists might just come up with another, non-infringing way of making it. Sometimes *method of use* patents, like the use of inhibitors of enzyme *X* to treat patients with disease *Y*, can be lucrative too, but these may well involve a drug *Z* that another company already holds patents on. And as we've seen in the section on repurposing, in recent years composition of matter patents tend to mention many different diseases in the context of potential utility, largely to prevent just this sort of patenting by another party. So out of all these types, composition of matter patents are the most robust, and when promising new compounds are discovered the push will be onto file an application for this type of patent first.

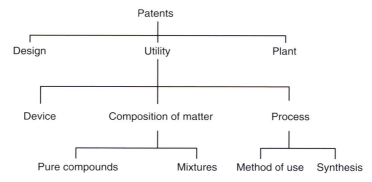

**Figure 3.1** ▶ Partial breakdown of the types of patents available, showing those of greatest interest for drug discovery researcher at the bottom.

## 3.1.2.2 Patent Requirements

There are some things that are inherently unpatentable, like mathematical equations, the human body, perpetual motion machines, and ways of making, say, a better letter bomb. Here, society has decided that some things just shouldn't be patented. For everything else, there are three fundamental requirements that must be met in order for an invention to be patentable:

▶ Novelty

▶ Unobviousness

▶ Utility

Describing the invention well enough for "one skilled in the art" to practice it, as previously stated, is also important. And plenty of other things (proper attribution of inventorship, fee payment, etc.) are needed to get a patent as well, but patentability is founded on these three key concepts.

### 3.1.2.2.1 Novelty, Priority, and Prior Art

If old ideas could be patented, someone would have made a killing "inventing" the wheel in recent times.[7] Patents are granted in return for the disclosure of *new* inventions and aren't intended to hand out monopolies on known ones, thus forcing the public to pay for something to which it long had free access.[8] As a result, in applying for a patent it's important to establish that the invention isn't already known, e.g. it hasn't already been disclosed in a paper, at a non-confidential talk (like a chemistry conference), or in the existing patent literature, all of which can be considered *prior art*. This is the driving force behind the need to avoid inappropriate disclosure mentioned earlier. To recap, although many researchers might enjoy publishing new research results as quickly as possible, doing so can result in the loss of valuable patent rights. In the US, a 12-month grace period is allowed between such disclosure and patent filing, but most of the rest of the world makes no such provision: there public disclosure prior to patent application makes inventions non-novel and therefore unpatentable.

Another major difference, between the US patent system and others that exist as of the date of this writing (November, 2007) has to do with who should be granted a patent, if more than one applicant should claim the same invention. Should *priority* be given to the party that files the application first (*first to file*) or to the party that actually came up with the idea and reduced it to practice first (*first to invent*)? The "first to invent" system currently applies in the US and the Philippines, while "first to file" reigns everywhere else.

Many American scientists will instinctively feel that potential rewards should be given on the basis of invention date, not paperwork-filing date. A kind of David and Goliath scenario is imagined, where an independent inventor comes up with a valuable invention but, being better at science than at patent law, is slow in filing a patent application and as a result is beaten to filing by a large multinational corporation with its army of Armani-clad patent attorneys. It seems only fair that, he should at least have a *chance* of being awarded the patent by proving in an *interference proceeding* that he invented it first. Under the current US system he is, in fact, given that opportunity, but the odds are against him.

In an interference proceeding, the party with the earlier filing date is called the *senior* party and the *junior* party has the later filing date. The junior party then needs to prove that it actually invented first. Most interference proceedings end up being decided in favor of the senior party, giving them a kind of home-court advantage, but, more importantly, the median cost of such legal procedures is now somewhere around $600,000 and they can take up to 10 years. Thus, less-than-independently-wealthy inventors are at a tremendous disadvantage when facing off against well-financed organizations regardless of whether they're first or second to file. In reality, most junior parties that do win are corporations anyway, not individual inventors. Considering all of this as well as the fact that somewhere around 99.9% of US patents don't involve any interference, much less ones concerning small businesses or independent inventors, many have concluded that first to invent in practice confers no real advantage to the "little guy" and that much trouble and expense could be saved by switching the US to first to file.[9] This idea has been around, since at least the 1960s, but it now looks increasingly likely to be adopted within the next few years. The Patent Reform Act of 2007 (S. 1145, H.R. 1908) would, under certain conditions, switch the US to a first to file system, among other things.

Whether or not this adversely affects small companies and independent inventors, it would bring about one change that many drug discovery researchers would welcome. Currently, as mentioned earlier, lab notebooks and invention disclosures need to be scrupulously signed, dated, and witnessed, largely because they might need to be used as evidence in the interference proceedings that can arise under the US's first to invent system. This is the case even for researchers outside the US because their work might still result in a US patent. Although it wouldn't eliminate the need to keep good records, if the US changed to a first to file system, the consequences of failing to sign or witness experiments in a timely way would be less dire and one might at least expect less of an emphasis on it. As a caveat, though, remember that this may not happen and that relevant company rules will need to be followed either way.

Since an invention is considered novel if it doesn't form part of the prior art, this is a good time to consider some differences between patents and journal articles, which make up much of the prior art. Both scientific papers and patents refer to work that's been done before, but to a different extent and for different purposes. Journal articles tend to use a lot of references to build up a foundation that supports the newly reported results, which are then viewed as a logical extension of established science. The emphasis is on consistency and continuity with an existing body of knowledge, a sort of new twig on the acknowledged tree of knowledge. Sometimes the new research calls into question accepted ideas and hypotheses, but even then citations will be made to earlier references that support the new theory if such exist.

In contrast, patents need to meet the novelty requirement, so emphasizing an invention's continuity with a prior body of knowledge would be counterproductive, to say the least. All material prior art known to the applicant (and those associated with it filing a patent application) *do* need to be disclosed, and failure to do so can result in a patent being ruled invalid or unenforceable in subsequent litigation. An honest, but minimalist approach is therefore generally taken. Compared to journal articles, citations tend to be few. This is one reason why patents and patent applications have a completely different "feel" to them than scientific journal articles. Dr Aaron Filler, a scientist experienced with both, notes that "in a patent, the argument is 'There was a need, but no one knew where to go and nothing was done that was very relevant. And out of the blue, by surprise'—the key thing is surprise—'exploring...we stumbled on this completely surprising thing...and...created something of industrial use out of it.'"[10]

### 3.1.2.2.2  Unobviousness

Surprise is a wonderful thing in the context of patentable drug discovery. That's because something surprising isn't obvious and things that are obvious can't be patented. The problem, though, lies in defining what exactly the word "obvious" means. Defining it, for example, as something that would be apparent to a theoretical, rather unimaginative "person of ordinary skill in the art" simply shifts the difficulty over to defining what "apparent" means, while raising the additional point that what's apparent to one person may not be to the next, even at the same skill level. Like many other concepts, "obviousness" seems easier to recognize than to define, and even there people will disagree about it. It's no wonder then that so much effort always goes into trying to determine and define the meaning of that word in the context of patents.[11] A recent Supreme Court case (*KSR* v. *Teleflex*),[12] for example, addressed the standards used to determine obviousness, with the Court's ruling in effect making it easier for patents to be rejected as obvious, at least in some cases.[13]

One thing is for sure, though: the word "obvious" (which should be avoided in lab notebooks) tends to be used quite differently by scientists than it does by patent attorneys. Scientists assume for it a very high standard by which most inventions (especially those of others!) might well be unpatentable. This is partly due to the healthy skepticism that seems to be required of a good scientist. But, we also tend to forget that, as the magician unwise enough to reveal the secret of his fantastic trick learns, once disclosed, *everything* becomes "obvious" in retrospect. "It is often very easy to reconstruct an invention with the benefit of hindsight, as a series of logical steps from the prior art, but it does not necessarily follow that the invention was obvious, especially if there is evidence that the invention was commercially successful, or supplied a need. The question 'if the invention was obvious, why did no one do it before?' is usually a relevant one to ask."[8]

The success of an approach can sometimes be used to argue against its obviousness. "If you take a single point change," says patent attorney Wayne Montgomery, "like changing an ethyl group to a methyl group or changing one atom for a similar type of atom, that single point change is obvious. But then the applicant has the opportunity to say, 'Well, alright, maybe it is *per se* obvious, but it wasn't obvious that if you do that you'll come up with a compound that's one hundred times more potent than anything that's described in the art!' "

Corporate IP people would rather not shoot down valuable potential patents by adopting a much stricter standard of obviousness than a patent examiner will. Because of this, in conversations between a patent attorney and a scientist, when the word comes up they'll likely be "talking apples and oranges." It's unfortunate that there aren't completely different words in the English language to describe *scientifically obvious* and *legally obvious*, which might save a lot of confusion. Suffice it to say that when it comes down to it, it's the patent attorney, not the scientist, whose evaluation of obviousness will count in determining whether a patent should be applied for.

### 3.1.2.2.3  Utility

Of the three major requirements for patents, utility or usefulness is generally the easiest one to satisfy in drug discovery research. In some fields, inventions can be easier to come by than

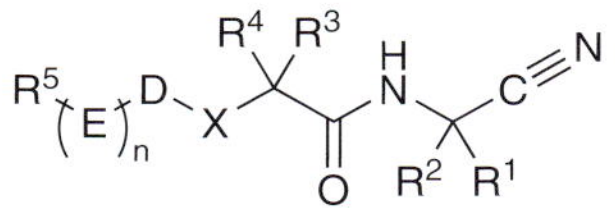

wherein $R^1$ is hydrogen, $C_{1-6}$ alkyl or $C_{1-6}$ alkenyl wherein
said alkyl and alkenyl groups are optionally substituted
with halo;

$R^2$ is hydrogen, $C_{1-6}$ alkyl or $C_{1-6}$ alkenyl wherein said alkyl
and alkenyl groups are optionally substituted with halo;

or $R^1$ and $R^2$ can be taken together with the carbon atom to
which they are attached to form a $C_{3-8}$ cycloalkyl ring
wherein said ring  system is optionally  substituted with
$C_{1-6}$ alkyl, hydroxyalkyl or halo;

$R^3$ is hydrogen, $C_{1-6}$ alkyl or $C_{1-6}$ alkenyl wherein said alkyl
and alkenyl groups are optionally substituted with $C_{3-6}$
cycloalkyl or halo;

$R^4$ is hydrogen, $C_{1-6}$ alkyl or $C_{1-6}$ alkenyl wherein said alkyl
and alkenyl groups are optionally substituted with $C_{3-6}$
cycloalkyl or halo;

or $R^3$ and $R^4$ can be taken together with the carbon atom to
which they are attached to form a $C_{3-8}$ cycloallcyl ring,

**Figure 3.2** ▶  Example of a Markush structure and part of the description thereof. (From Prasit, et al., US 7,012,075 (to Merck & Co. and Axys Pharmaceuticals).)

and require varying amounts of expertise to use properly, although most chemists will probably get plenty of hands-on experience with the first two these days. Searching composition of matter patents will introduce the researcher to *Markush structures*, sort of generic structures with multiple allowable substituents like the one shown in Figure 3.2. Descriptions of all the allowable substituents in a Markush structure can run for many pages and encompass billions of compounds. Although general structures are used in journal articles, Markushes tend to have far more substituents, each with more possible structures than one would ever see in a journal. The idea, in fact, is for them to be as broadly inclusive as possible, thereby pushing out the bounds of that valuable IP real estate and keeping would-be neighbors at bay.

### 3.1.2.3.2  Patent Anatomy

Published patents and patent applications start off with a *front page*, these days bearing a bar code at the upper right. Here, basic information like the title, inventors, assignee, priority date, and foreign filings is spelled out. A listing of references, both patents and journal articles, will be given as will a brief abstract, sometimes containing a Markush structure. Not much is required of the title and abstract, and they sometimes seem to be written to be as uninformative as possible, again a major deviation from what one finds in journal articles.

In US patents and application, the front page is sometimes followed by *drawings* (figures, graphs, etc.), which instead are found later on in PCT applications. After that, a short *field of the invention* section which may not be terribly different from the abstract can sometimes be found. Both US PTO and PCT publications will include a *background of the invention* section.

This presents a need that exists or a problem to be solved for which the invention to be described affords a novel solution. For example, for compounds to be used to prevent cancer metastasis the background of the invention section would probably highlight the facts that cancer is a major source of mortality, metastasis is major cause of cancer mortality, and that few effective antimetastatic drugs exist, all of the current ones having their drawbacks. In contrast to journal articles, patents generally give scant coverage to mechanisms and theory, but this section will often cite a target's role in the disease process in order to explain why the proposed compounds should make for effective drugs. Before drafting this section, the patent attorney involved will normally talk to project scientists familiar with existing treatments (the competition) and their shortcomings, so that the advantages of the compounds presented can be properly highlighted.

The *summary of the invention* section, which is next, should "set out the exact nature, operation, and purpose of the invention" by describing it in "one or more clear, concise sentences or paragraphs."[14] A *detailed description of the invention* will likely follow. There's some variation in which of these two will show the Markush describing compounds of the invention and the laundry list defining substituents thereof, but one or the other will always have this important information. In composition of matter patents, the summary may also describe the use of such compounds in pharmaceutical compositions and the use of such compositions to treat one or more diseases.

The detailed description section defines the terms used (e.g. "heteroaryl" or "aralkyl"), so there's no possible misunderstanding, although, again, they're likely to encompass a broad array of molecules. This section, should also provide an explanation of how to make each aspect of the claimed invention. *Examples* of specific compounds and their syntheses will be given to satisfy the requirement for disclosure, making this largely equivalent to the experimental section of a journal article. One difference between the two is that, like activity data, less compound data tend to be reported in patents than in journal articles. Final compounds are sometimes characterized solely by a parent ion peak in the mass spec, which is hardly up to *Journal of Chemistry* (*JOC*) standards. Intermediates might be characterized by melting points or not at all. In many cases, more data than this is given, and proton NMR spectra for important compounds can often be found, but in general the requirement for "hard data" is fairly modest, and to patent attorneys it makes little sense to include more than is necessary. Chemists putting together procedures and data for attorneys to use in application filings will find this a relief, but when searching for data in someone else's patent, minimalism won't seem such a virtue. Biological assays used in determining activity will be described in the detailed description section as well, while biological data for the specified compounds are generally only given to the minimum extent needed to support patentability.

If you really want to know what a patent is about, skip to the end. The heart of a patent or application lies in the *claims*. This last section normally begins with a phrase like "What is claimed is" or "The subject matter claimed is" and proceeds to a numbered list spelling out the actual compounds the invention encompasses. In the end, exclusivity is granted only for claims that appear in the issued patent. These can differ markedly from the universe of compounds that fit the Markush structure in the earlier sections, and may not even include many of the example compounds. For instance, US 7,071,184 includes in the detailed description section a Markush structure with more than five pages worth of definitions of substituent groups and lists 64 specific compounds as examples, but only one single compound is allowed in the claims.[15]

Claims listed usually follow a specific pattern. Where multiple compounds are covered, the first claim usually has a Markush structure with the broadest description of substituents. Claim #2 will be slightly narrower, perhaps, "A compound of claim 1 wherein $R^1$ is fluoro." Claim #3 might be a different subset of claim #1 or it might be an even narrower subset of claim #2, like "A compound of claim 2 wherein $R^2$ is a morpholino group attached by nitrogen." Claims are thus nested from broad to narrow as fallbacks, so that if earlier claims are disallowed (often for being too broad) the broadest *allowable* range of compounds will still be covered. Not surprisingly, *the compounds of highest interest (e.g. clinical candidates) in an issued patent tend to be the most narrowly defined, deeply nested ones, which will usually be specifically named or drawn in the claims.*

Some claims use the format "A compound of claim 1 selected from the following group:" followed by a list of chemical names, requiring the chemist to reach for a pencil and paper (or ChemDraw®). Composition of matter patents and applications usually also contain claims to pharmaceutical compositions containing the compounds and their use to treat different diseases toward the end of the list. Sometimes a process for preparing the compounds is claimed as well.

Claims that are carved out of other claims, like #2 and #3 above, are called *dependent claims*, while in this example #1 is an *independent claim*. An unlimited number of either used to be allowed in US patent applications, but very recently a limit of no more than five independent claims or 25 total claims per application (unless an "Examination Support Document" is filed)[16] was supposed to be put into effect, but a lawsuit by GSK has currently put this on hold.[17]

### 3.1.2.3.3  Locating Information in Patents

Patents and patent applications are a frequently neglected source of useful information. The patent literature sometimes contains the best available procedures for a given synthetic transformation or information on compounds that can't be found anywhere else. Finding this might require a little more detective work than locating it in a journal article, but it's well worth the effort. Being able to determine, if a particular compound is "covered" or a particular synthesis is described in a patent or patent application is often part of the industrial drug researcher's job.

Synthetic procedures, again, can be found in the detailed description of the invention section. Sometimes, the schemes are written up out of sequence so that the final step is described first, followed by the first, and then the intermediate steps. The researcher needs to know, when perusing a patent or patent application that sometimes *a listed synthetic procedure might not have actually been carried out and the product may never have been made!* Inclusion of such procedures in a journal article would be considered outright fraud, and many academics consider it a kind of industrial science fiction and fantasy, but when done properly it's an acceptable practice in filing a US patent application. The acceptance stems from the problems involved in applying for a patent before all of the relevant SAR can be fully explored. Sometimes scientists will suspect that particular compounds, although not yet made, can be synthesized and will be active. They may include such *paper examples* (also called *prophetic examples*) in the patent application as long as they don't use the past tense in describing those experiments, as described below. Examples that represent work that's really been carried out instead are called *working examples*. Paper examples aren't (or shouldn't be) written with any intent to deceive. They represent viable routes to compounds reasonably anticipated to be good.

Although not specifically labeled as such (which would be much better from a researcher's point of view), paper examples should be distinguishable by the verbs used to describe them—tenses as in "the reagent *is* added . . ." versus "the reagent *was* added . . ." or "the compound *can be* made" as opposed to "the compound *was* made." According to a recent appellate court ruling, using the past tense in prophetic examples, which makes them sound as if they had been carried out, may constitute *inequitable conduct*, which can render patents unenforceable.[18] Some older patents and applications, however, have been known to use the past tense for paper examples, so one needs to be careful. Another characteristic of paper examples worth noting is that, since the compounds haven't yet been made no data should be listed for them.

Actual *reduction to practice* is not required for filing a patent application. They only need to teach one skilled in the art how to "make and use" the invention. The bottom line is that "so long as the paper examples are scientifically reasonable (that is, they do not offend the chemist's chemical sense), the PTO will consider them to have been performed."[19] One caveat is that should an interference proceeding arise, it might be necessary to show that paper example compounds can really be produced as described, but again, the odds of an interference occurring are very low and good product yields wouldn't be required even then.

A chemist will often want to know whether a compound he's interested in is "patented."[20] Searching through just the claims of an issued patent (*not* patent application) for specific structures, named compounds, and Markush structures which would include the compound in question can determine whether or not it's the one for which rights have been granted in that particular patent. This can involve a bit of work, especially when Markush substituents can themselves contain other substituents chosen from a long list. Everything can hinge on the fine points of how a word like "heteroaryl" is defined, and even sometimes the exact placement of a, comma so if it's important to the project one might want to bring a resident IP professional (e.g. patent agent or attorney) into such a search. Keep in mind that even if a claim does include the compound of interest, patent rights may not apply in a different country or if the patent has expired or been abandoned, which sometimes happens. For US patents, checking the "status" link at the website shown in Box 3.2 can help determine this.

But even without current local patent rights, a compound that "falls within the scope" of a patent or a patent application can be considered part of the prior art, and this too has implications. To see if this is the case, the entire document, not just the claims, needs to be searched, again looking for specific drawings, named compounds, and inspecting Markush structures to see if the compound in question would be described by them.

As mentioned earlier, activity data are rather sparsely reported in patents, which often leave one scratching one's head and wondering which of the millions of compounds a patent might encompass are the best ones. Searching for a *preferred embodiment* or just the word "preferred" in the detailed description of the invention (easy to do by searching the pdf file), might help, but be warned that so many things might be "preferred" that it's hard to see what *isn't* preferred. Terms like "more preferred," "still more preferred," and "yet more preferred" will often be found, but here it's apparent which of these should be best.

Searching for the "best" compound isn't always easy. Says attorney Wayne Mongtomery, "if it's a patent, you're more than likely going to have the compound specifically claimed.

You want to have at least one claim that addresses just your compound. If it's a patent *application*, people tend to try to hide it, because those applications publish. In time, you can amend your claims for the particular compound you're most interested in. Some companies don't like to do that, but they'll still try to hide it in a claim, say, covering a dozen or so compounds. In fact a lot of companies will try to camouflage what the real compound is and they'll have several claims to compounds that they're not that interested in." So, searching for the innermost nested claim or the claim specifying the fewest compounds is more likely to work for a patent than a patent application, but it's always worth a try. For practice and amusement, you might try going through patents online and picking out the best compounds, which will at least give you a feeling for what real-world patents are like.

After the patent application is published, of course, data might become available in a journal article, but that sometimes doesn't happen for months, years, or perhaps ever. Still, once a series of interests have been identified, setting up a regular, automated search by compound type, target, or inventor using a service like NERAC or SciFinder® in order to get newly released information ASAP makes a lot of sense.

### 3.1.2.4   Inventorship

The difference between a patent inventor (say, employee), who comes up with an invention but assigns her rights to a patent assignee (say, employer), who therefore owns it, has already been noted. Although remuneration, if any, overwhelmingly goes to the assignee, which probably financed the research and sent the inventor regular paychecks even during the many times that she wasn't inventing, being an inventor on a patent brings with it a bit of prestige. Patents and published applications are nice to add to one's resume, especially if the patent eventually results in a marketed drug (most don't). Needless to say, it provides a sense of accomplishment as well, so researchers are keen to get their names on patents and can become particularly embittered if they feel they've been unjustly denied a place among the inventors.

Inventorship itself has been described as "one of the muddiest concepts in the muddy metaphysics of the patent law."[21] Determining who the inventors are on a given patent application, which frequently involves having to tell people who genuinely believe they're inventors that in fact they're not, is the unenviable job of the patent attorney. "It's one of the hardest parts of the job," admits Wayne Montgomery. For the researcher, reviewing a few basic points about patent inventorship can clear up some misconceptions and hopefully help avoid or minimize future disappointment, bewilderment, and bad feelings.

Inventorship rightly belongs to someone who made an inventive contribution to at least one of the claims. As some of the claims are often removed before a patent is issued, if an inventor was responsible only for those claims and not any of the surviving ones, her name would need to be removed from the list of inventors too. There can be, and in fact usually are more than one inventor per patent, about three–five co-inventors being typical. A much larger number is hard to justify, but courts have proven even more critical of patents that omit an inventor. All of this is particularly important, because under US patent law, incorrect attribution of inventorship can invalidate a patent. As long as no intent to deceive was involved, it's possible to correct inventorship during the examination process or even after a patent is issued, but it's far better to just get it right in the first place.

A remarkably widespread and persistent but erroneous belief among scientists is that synthesizing a compound gives one a certain amount of inventorship rights. This isn't necessarily so. Patent law is replete with statements like "insofar as defining an inventor is concerned, reduction to practice, *per se*, is irrelevant,"[22] "one must contribute to the conception to be an inventor,"[23] and "invention requires conception."[24] Perhaps, the best way to put is to say that "to be a co-inventor one must contribute to the idea, rather than to the work."[25]

If someone else comes up with the idea for some new compounds, but you're the one who actually synthesize them, he's the inventor and you're not. It is possible that, if the synthesis proved unexpectedly difficult and those compounds could only be made by a very different route or with a critical reaction modification that *you* came up with, you may well have made an inventive contribution that qualifies you to be listed as a co-inventor, but that's *if and only if* the process you developed ends up being claimed. Although you can try presenting your case, once again it's the patent attorney who determines inventorship.

In doing this, attorneys will look at the compounds that might be included in a patent application and relevant invention disclosures, then talk to the researchers involved. At this point, the wisdom of having a written invention disclosure to point to in establishing inventorship becomes apparent. Trying to remember who suggested what first and when is a surefire route to frustration. The patent attorney needs to be able to refer to written evidence.

Mr Montgomery suggests one additional consideration for those to whom inventorship is important: "The best advice I can give to anyone who wants to be an inventor is to get in on the ground floor early. Write up your inventions and when they generate ideas about scaffolds that they're going to develop and the kinds of modifications that could be made, those people who come up with those ideas early on are the ones who are most likely going to be inventors. The people who come into the game later will find it harder and harder to be inventors, because as the invention gets better defined it's going to be difficult for them to get outside of the box. They're going to keep making things that are in the box. You have to go beyond just making foreseeable or easily recognizable variations of what's already been made."

Inventorship decisions can be particularly tricky in collaborations, where not only are multiple scientists at the same company involved but multiple companies are as well. Here, contentiousness about inventorship can lead to suspicion, loss of cooperation, and worse. Another delicate situation for a patent attorney involves facing down, as would-be inventors, the occasional high-level manager who feels that his position gives him the right to have a thumb in every pie. These are the kinds of situations that make a researcher's job seem easy indeed.

Another thing that might not be apparent to scientists is the fact that filing a patent application is only the beginning of the process for the patent attorney. Prosecuting the patent can be a long and complicated undertaking.[26] "There's a lot more involved in the actual prosecution of the case," according to Mr Montgomery, "because then we have to follow up with the patent office, present our arguments, and fight for the broadest invention that we could possibly get." And of course even that isn't the end of a corporate IP professional's job responsibilities. "There are other issues involved," he says. "With collaborative partners, are we segregating the intellectual property properly? Is there freedom to

---

**Box 3.3   Some Key Points About Patents and Patent Applications**

▶ Patents provide government-granted exclusivity rights

▶ Sometimes, though, more than one patent can cover the same compound

▶ Compared to journal articles, patents contain less theory, fewer citations, less extensive data, and may include "prophetic" procedures

▶ U.S. and PCT patent applications usually are published 18 months after filing

▶ Most patents and applications can be readily viewed online

▶ Composition of matter patents are the most desirable type in drug discovery

▶ To get a patent, inventions must be novel, legally unobvious, and useful

▶ In most countries the first party to file a patent will get it if more than one apply

▶ Currently in the U.S., the first to invent, as determined in an interference proceeding, will get it instead

▶ Only the claims of an issued patent have exclusivity rights

▶ The best compounds in a patent will likely be the ones in the most highly nested claim or contained in the claim which names the fewest compounds

▶ Inventorship depends on contributing ideas—just making a compound doesn't count

▶ Patent attorneys decide on who to list as inventors before an application is filed

---

operate? Do we have issues with publications that may overlap with what we're trying to patent? So there's more under the surface than just preparing, filing, and prosecuting a patent."

While some of us might shy away from such things, other scientists see it as a challenge and as an alternate way to put their scientific knowledge to work. Patent law is probably the most popular career for scientists in pharma and biotech to switch into. The job requires a detailed understanding of the technical aspects, and in fact technical competency needs to be demonstrated by a science degree or sufficient course credits before an applicant can take the patent bar exam. All the same, previous *industrial* research experience is especially useful. Those considering such a switch can find helpful information on the *American Intellectual Property Law Association* (*AIPLA*) website, www.aipla.org. A summary of key patent issues that impact industrial drug discovery is shown in Box 3.3.

## 3.2  Outside Resources

### 3.2.1  Consultants

Every new drug discovery project will run into the occasional roadblock. Unexpected problems in chemical synthesis crop up that can't be overcome. Rather than following an expected trend, cellular or in vivo results turn out to be erratic or irreproducible. Specialized expertise beyond the experience of anyone at the company can become crucial to project decision-making.

The correct first response to these irritating but unavoidable situations, of course, is to make sure that all available in-house expertises have been properly called up. In a very small company that may not even be a consideration: you might *be* the entire company's in-house expertise. At the other end of the spectrum, in a large organization communication may require not only talking to scientists in the next lab or on a different project, but also going to a different department or different company site. It's human nature not to relish presenting one's own project's "dirty laundry" to others. After all, the person presenting the problem is going to look … well, like he has *problems*, while the casual observer sitting back in the audience stands to lose little by making suggestions off the top of his head that won't require *him* to do one iota of work to test out. Possibly adding to the reluctance to air the issue may be the fact that in graduate school, students and postdocs are usually encouraged to present progress, not problems at seminars.

Still, in corporate research, allegiance to the project and ultimately to patients with a given disease requires that problems be faced head on and presented to the people who may be able to help solve them. Don't be afraid to go to your in-house experts. Companies have established ways of doing this, and you may be pleasantly surprised at what you find.

When the combined efforts of the best minds in the company can't resolve a particularly thorny problem, it's time to speak to a consultant. Consultants specializing in every aspect of the drug discovery and development process are frequently employed by companies large and small. Their backgrounds vary. For example, organic chemistry consultants are often Professors at large universities who are well known in the literature and bring with them a particularly impressive string of credentials. Consultants in less thoroughly academic fields such as formulations, radiation safety, or toxicology are more likely to have had previous full-time employment in industry. Consultants have proliferated to the extent that there are probably now metaconsultants, consultants who give advice on which consultants to hire!

Although it's possible for a new problem to arise that calls for the hiring of a new consultant, a process which might take weeks or months to complete, it's more likely that a given problem can be presented to a consultant already available to the company. A word about the mechanics of consulting contracts: all corporate consultants will have signed a *CDA*, a legal contract requiring the consultant to maintain the secrecy of any information revealed to them in the course of their consultation. They'll also have a consulting agreement that generally guaranteed a certain minimum number of hours per year at a given rate of pay. Of course, the payment varies greatly and can be quite high for the most prestigious and desirable consultants. Such agreements can be exclusive, meaning that the consultant isn't allowed to consult for any other company, but usually they're non-exclusive. Most of the better known ones are under contract to multiple companies and may be members of *Science Advisory Boards* (*SAB*s) for others still. Maintaining confidentiality for all these companies and not applying proprietary information they've learned from one of them to another must be particularly challenging, but it goes with the job.

Before presenting anything to any outside party, just make sure that he or she *is* a consultant with a CDA and not just, say, a seminar speaker specializing in an area of interest to the company. As we've seen, presenting your results non-confidentially can cause a loss of

IP rights, besides which seminar speakers may well resent perceived attempts to use them as unpaid consultants.

You might initially think that consultants should be "on call," that is, available to contact immediately, whenever a problem arises. In fact, this is about as likely as your physician is to drop everything and make a house call whenever you get sick. In rare circumstances (possibly with consultants who are less busy than most) it might almost work that way, with perhaps phone consultations available, but most of the time your chance to present your problem to the consultant will occur during a regularly scheduled consulting visit. These vary in frequency, from about monthly to just a few times per year. During these visits, everyone who has anything worth asking about (and often many people who don't) will be encouraged to avail themselves of the expert advice available. And make no mistake about it, these are opportunities for valuable input and good suggestions by people who know what they're talking about. Don't expect all your problems to be magically solved, but if they're not you'll at least be assured that you haven't overlooked something that's apparent to the best minds in the field.

In most cases, your consultant will not review your problems before you present them at the consulting visit. The disadvantage here for the consultant is that he or she comes in "cold," has only a few minutes to analyze a new situation, and is then expected to provide insight and suggestions that put the Oracle at Delphi to shame. In the meantime you, as the scientist personally involved with the problem, have been able to spend hours thinking about it, have gotten other people's suggestions (see above), and have gone to the literature to thoroughly familiarize yourself with everything known that might bear upon it. It's a tribute to the knowledge and foresight of consultants that, despite coming in at such a disadvantage, they still frequently come up with good ideas.

Some consultants do solicit problems in advance of their visit. The only possible disadvantage to you in this case is that you need to make up and send over slides early and things may have changed by the time the consultant shows up at the company's door. Still, no one's averse to hearing the latest news, if new findings have come to light.

The actual format for presenting your problems to the consultant will vary from informal one-on-one sessions to hour-long sessions in a small group setting to somewhat intimidating, long formal sessions which include the consultant, almost every member of the department, and other observers who just want to see the show. It depends entirely on the company's own customs. Although one-on-ones seem easiest and most direct, some companies feel that their own in-house expertise is best brought in during a big group session. If that's the case, you may find that there's an intimidation factor involved in presenting before a lecture hall filled with experienced PhDs as well as a visiting Nobel laureate or someone who authored your college text. In truth, even in these situations the intimidation level lessens with experience.

You may at times find yourself in the position of presenting work on your ongoing project to consultants even when there aren't any real problems for them to address, or where the consultant doesn't feel obliged to actually make suggestions. This happens for at least two reasons. The first is that, as stated before, the consultant's contract includes the number of hours of consulting time and is laid out in advance. What's the point in paying for something and then not using it? At times when there are few departmental difficulties for a given consultant to address (a good situation!), management may consider that using his time to inform him of project progress, look over the programs, and possibly make suggestions is better than letting those hours evaporate. Which leads into the second reason, a critical fact that's usually not apparent to the novice: *in many cases supplying expert advice to project scientists is only part of a consultant's job.*

To understand the other part, you need to imagine yourself as the VP or the Head of the Research. Like most pharma and biotech executives, you have quite a bit of scientific knowledge and experience, and you've done some laboratory research yourself a long time ago, but nowadays your real expertise and the issues you deal with extend into the realms of business and management. Between planning strategies, attending endless meetings, hiring in and evaluating managers, etc. you simply don't have time to read scientific journals and familiarize yourself with the current state-of-the-art in each and every subdiscipline anymore. As a result, can you really be sure they're right, when your scientists tell you they can't initiate *high-throughput screening* (*HTS*) for a given new target because it's somehow not amenable to that format? Are your chemists really doing all they can to synthesize that hard-to-make compound that's expected to provide the company with a new and patentable compound for oncology? And is it really wise to shift resources away from that ongoing fast follower project to pursue a new target you've never heard of before, and that no other company seems to be interested in? You might ask these questions to the scientific managers who report to you, but what you really want is an unbiased outside opinion, particularly that of someone well-known in the field and with broad experience. So it's, "Hey, isn't that consultant of ours going to be here next week?"

The other, less visible part of many consultants' jobs, then, consists of providing upper management with an independent, expert opinion on the company's ongoing efforts in its programs, how they might be improved, and perhaps how they stack up against others. The more well-known and experienced a consultant is, the more likely this is to be a significant part of the job. At a large company, the CSO may not be interested in hearing the details of that particular asymmetric reduction that gave you a poor *enantiomeric excess* (*ee*) in the eighth step of your synthesis, but he may well be interested in hearing the medicinal chemistry consultant's opinion on the level of technical expertise to be found in your department or project and maybe, if he happens to mention it, whether more chemists are needed to work on that series.

This doesn't mean that one will be singled out and lambasted by the upper management for bringing up problems instead of brilliant accomplishments at consultant's meetings, thereby defeating the main purpose of having them in the first place. It really doesn't work that way. General project and departmental observations, the need, or lack of need for more resources in a given place, and a perspective on the research portfolio are the kinds of things that consultants might provide to management, when they do. Implicit reassurance that nothing has been missed and no wrong turns have been made often goes with the report. Finally, a summary of this and other consultant-related issues is shown in Box 3.4.

---

**Box 3.4   Key Points About Consultants**

▶ Consultants can be useful when in-house expertise can't solve a problem

▶ They sometimes also provide management with feedback on projects, capabilities, and other big picture issues

▶ Researchers normally present problems and progress at regularly scheduled consulting meetings

## 3.2.2  Academic or Government Research Agreements

As we've seen, there are now many different types of organizations involved in the various aspects of drug discovery research including CROs (both foreign and domestic), big pharma, biotech companies ranging in size from just a few people to that of Amgen, university groups sometimes with ties to translational research initiatives, non-profit research institutes, and government labs. The permutations on how two or more such organizations might collaborate on a project and the different models for doing so are too numerous and complex to try and list here. Instead, just a few aspects of the types of agreements an industrial researcher might commonly see will be briefly discussed.

It's a popular perception that academic ideas end up being developed into drugs by industry, the flow being completely unidirectional. A few years of industrial experience are likely to change this view. Besides frequently coming up with its own ideas, industry often provides the tool compounds to academia that enable their theories to evolve and be tested. According to an estimate by Dr Chris Lipinski, in 2005 alone Pfizer provided about 2300 tool compounds to outside (mostly academic) investigators.[2] Industry supplies these compounds, unavailable elsewhere, to interested academic researchers for free via legal contracts called *material transfer agreements* (*MTAs*).

Such agreements are necessary because of a company's fiduciary responsibility not to squander valuable IP by giving it away to another party, but instead to ensure that the company can share in the profits in the unlikely case that any should eventually result. This gives rise to two of the major "strings attached" to MTAs. The first is a delay, often of 90 days, in submission of publications resulting from the research, so that the disclosure can be reviewed by the company and patent applications filed, if necessary. Sometimes, no time limit is given and corporate approval is simply required before any resulting manuscript can go out. The second is the company's right to license inventions resulting from the research. Other conditions will probably apply too. Often, only experiments listed in the agreement can be carried out, and in fact protocols detailing what's to be done, especially for in vivo experiments, may be required to ensure that experiments meet all legal and ethical standards before the compounds can be sent out. It makes little sense to send them out if no results ever come back to the company (a surprisingly frequent outcome), so MTAs will often specify that a report listing data obtained for the compounds needs to be sent within, say, a year.

The clause specifying that the company has the right to license inventions resulting from the experiments is frequently a source of disagreement between university and corporate legal departments, but a modification in the details and wording that's acceptable to both parties can often be worked out. The publication delay can be irritating to academic investigators too, but isn't terribly different from an issue involved in publishing research results in the wake of Bayh–Dole that they now face anyway. "With or without an alliance to industry, the academic scientist now has a responsibility to address disclosure issues prior to *any* public dissemination of information."[27] These kinds of contractual obligations sometimes surprise new faculty members working on their first MTAs, but those with more experience know that all of these things are standard.

Intellectual property concerns can make publishing difficult in another way too. If a company should decide to send out proprietary compounds, for which structures haven't previously been disclosed in papers or patents, they probably won't reveal their structures to the academic investigator, but only use a code number instead. The agreement will probably also specify that the investigator must not attempt to determine the structure in this case. This makes it difficult or impossible to publish the results obtained in many journals. All the investigator might be told is that XYZ-2236 is a selective, reversible 2 nM inhibitor of enzyme *X* with a molecular weight of 431.3. However exciting the results may be, *Nature Medicine* is unlikely to publish them without a structure for the wonder compound, and only when the company is ready to reveal it can that happen.

Academic investigators usually request tool compounds from companies after seeing them in publications, or they might know that a given company is working in their area of interest and make contact to see if they have anything useful available for them to experiment with. Trips out to present results and give seminars, perhaps even financial research support or consultancy, can sometimes result. For companies with back-burner projects, which they lack the will and/or the resources to pursue, often at an early *target validation* (*TV*) stage, providing compounds to an academic group can be quite advantageous. In general, enabling an academic group to investigate biology that could result in new uses for their compounds or resurrect a program that "has reached a motivational dead end within a company,"[28] has little downside other than the potential for IP loss that the MTA is intended to minimize. "Rather than feeling challenged, pharmaceutical and biotechnology companies say they welcome anything academia can do to validate drug targets and identify promising drug leads early on, thereby reducing corporate risks."[29] Likely barriers to more widespread use of industry compounds in this way include insufficient networking to match academic investigators with the corporate compounds available (there being no matchmaking service with "compounds wanted" and "compounds available" ads) and—a very real-world consideration—the need for someone within the company with enough time and interest to push the paperwork through.

The multiple myeloma drug, bortezomib, provides an example of just how useful this kind of strategy can be when properly employed. Originally developed by a company called Myogenics that was later renamed ProScript, this drug was apparently viewed as a long-shot and at least 50 companies sought out as collaborative partners turned it down.[30] The fact that proteasome inhibition was a novel (and therefore unproven, as we'll see in Chapter 5) mechanism of action and that the molecule contained a boronic acid, a functional group rarely seen in the pharmacopoeia, certainly didn't help. This being the case, "the company distributed bortezomib from early on . . . because the company did not have the resources to carry out all the studies, and high-profile academic groups gave the drug more creditability."[31] Academic findings in this case contributed to the success of the compound, which ended up being the Millennium Pharmaceuticals' drug Velcade that made it to market in near-record time.

There's obviously a difference between simply sending out a compound via MTA and a true collaboration between a company and an academic group involving financial sponsorship, frequent communication of data and ideas, and the divvying up of work. These kinds of partnerships won't be discussed here, but can sometimes begin with an MTA.

Not only academic labs, but also government ones can be involved. Sometimes, researchers at the *National Institutes of Health* (*NIH*) can be interested in the experimental use of tool compounds that a biotech or pharma company may have. Rather than an MTA, the contract covering such a transfer will probably be what's called a *materials cooperative research and development agreement*, or *materials CRADA*.[32] As in the case of academia, more of a true

---

**Box 3.5   Key Points About University and Government Research Agreements**

▶ Contracts called MTAs and materials CRADAs allow thousands of corporate tool compounds to go out to academic and government labs, respectively, each year

▶ The results obtained can help validate a target without using company resources as well as build momentum for a project

▶ More extensive collaborations with academia and government labs are also possible

---

collaboration is also possible, in this case through a *standard CRADA*. The development of bortezomib mentioned above, for example, made good use of such governmental agreements in addition to academic ones. One should realize too that the roles of the NIH in industrial drug research go far beyond CRADAs. For example, it's also responsible for drug discovery-directed *Small Business Innovation Research* (*SBIR*) grants, an important source of funding for early stage research.[33] Box 3.5 offers a summary of things the researcher should keep in mind when it comes to external research agreements.

### 3.2.3  Big Company–Small Company Collaborations

That collaborations between big pharma (or sometimes, these days, big biotech) and smaller biotech or pharma companies, which are becoming increasingly popular, was remarked on in Chapter 2. These can take different forms and occur at different stages of the drug discovery and development process. They might involve a licensing agreement for a clinical candidate or a co-marketing agreement for a drug, neither of which will necessarily involve any discovery research. Often, however, particularly when the demand for biotech clinical candidates outstrips the available supply, collaborations involving discovery stage or preclinical projects proliferate, and these generally involve joint research efforts, many of them being directed at small molecules.

Why should big pharma, with thousands of its own amply qualified researchers, be interested in such partnerships? Back in Chapter 2, Dr David Brown alluded to the fact that "you've only got a very small amount of global research going on at any one company" so that access to external innovation is important. Another analysis by Merck agrees: "With approximately 10,000 employees in its R&D division and its US $4 billion annual research budget, Merck is estimated to perform only 1 to 2% of the global biomedical research that is relevant to its selected therapeutic targets. Given that approximately 100-fold more medical research is performed outside the company, Merck scientists are encouraged to expend as much effort monitoring the activities of, or working with, outside research groups as they do on their own internal research."[34]

From big pharma's point of view, accessing innovation through discovery-stage alliances can "offer an alternative to building expertise from scratch, and present a degree of flexibility, cost advantage, and risk-sharing not afforded by in-house programmes."[35] From the biotech point of view, such collaborations provide needed cash and resources, while also furnishing a sort of unspoken validation for their research as well as a more straightforward path to market.

In such collaborations, each company's responsibilities will be clearly laid out in an agreement negotiated by both parties that specifies who will do what. Synthesizing project compounds and running them through primary screening, for example, might be done at the smaller company, while the big pharma partner carries out secondary assays and further development work on compounds of interest. In practice, compound synthesis might be done at either company or at both, say, if each is pursuing a different lead series. Normally, each partner will designate a project leader or program director to lead the collaboration onsite and coordinate the activities between them.

Once the agreement is in place and the collaboration has begun, good communication becomes incredibly important, both within each company and between partners. Although it's not common, in a few cases employees of one of the companies might even temporarily work onsite at the other.[36] E-mails, phone calls, faxes, teleconferences, and videoconferences will link the two companies, and joint, face-to-face meetings will be held on a regular basis, often quarterly. Not everyone involved in the project will attend every meeting, especially the ones that require travel, but good collaborations ensure that results are shared and ideas are solicited from all parties. "Get everybody involved" is probably the best advice.[37] In a very real sense, the likely success of a given collaboration can be gauged by how open and inclusive it feels to the scientists working on it.

How will working on this kind of collaborative project differ from working on an in-house project, from the point of view of a scientist at the smaller company? The sudden wealth of new resources, excitement, and expectations involved can combine to make it seem like a normal, in-house project "on steroids!" A big pharma or big biotech collaborative partner brings with it an incredible range of expertise, abilities, assays, and considerations you may not have previously known even existed. Particularly at the point where optimized leads are evaluated for potential suitability in the clinic, pharma-sponsored projects typically entail a wider, more comprehensive range of studies than most small biotechs would have carried out on their own, largely because they already have the mechanisms, resources, and experience in place to do so. Even in early discovery research, routine use of specialized ex vivo assays or studies on things like metabolite identification, which provide information not readily available without big company resources, can accelerate compound optimization. Bottlenecks can disappear. Expensive, difficult assays which before would have been contracted out and run one-at-a-time on a few selected compounds might now be run in parallel on every compound made. Experienced industrial scientists who know what it takes to make a drug and can steer a project clear of potential obstacles will come onboard.

None of which is meant to detract from the importance of the ideas, abilities, and unbridled entrepreneurship that the smaller company brings to the table, or to say that such a company couldn't develop a drug without them: they can and do. But the popularity of such collaborations owes much to the complementarity of what each side makes available, and in a good collaboration the whole can even be greater than the sum of its parts. Working on such projects at either company can be an exciting and rewarding experience. With the right partners and collaborative agreement it can really represent a win–win situation.

There's a potential downside too, of course. When downstream project bottlenecks do disappear, one needs to make sure that he himself hasn't become the new one: expectations for finding good compounds and subsequent pressure to produce them quickly, thus enabling important milestones to be met, can be very high. So, rather than things "settling down" after the rush to make good compounds that might attract a corporate partner, the pressure might be cranked up a notch instead after one is found. This is ultimately driven by the nature of the collaborative agreement. Often, the bigger company can decide to terminate it at any time if it's unhappy with the progress or for any other reason just by giving a couple of months' notice. And collaborations might need to be renewed—or not—on an annual basis. Much of this depends on the specific agreement involved.

Similarly, changes in the management, research portfolios, or mergers and acquisitions at the bigger company can result in sudden death of a partnership unless some sort of protection is provided by the original agreement. In many cases, collaborations will compete for resources with in-house projects at a big pharma company. The company might be pursuing one or more other programs for the same indication, and the *NIH syndrome* (*Not Invented Here*) can be a factor, if the respective projects need to be prioritized. Sometimes, the biotech company might perceive that for all of its promises and resources the bigger company is moving with great deliberation and might even want its compounds back, feeling it could progress them to the clinic faster without its partner's help. Not all collaborations are "made in heaven," and when corporate cultures collide or, to use a cliché, "the chemistry isn't right" they might seem to come from a very different place.

In the real world, most collaborations fall somewhere between these two extremes and with their growing popularity, it's likely that most drug discovery scientists at companies large and small will end up contributing to a number of them in the course of a career. Regardless of their "feel" to the scientist and even their ultimate success, they can play a vital role in introducing the researcher at a small company to many aspects and details of what's required to get a drug into the clinic. Researchers at the larger company may benefit principally from access to the innovative and exciting projects their smaller partners will bring, and perhaps have their horizons broadened by exposure to a different way of doing things. Both stand to benefit (Box 3.6).

One final note: IP protection is particularly critical in collaborations. Before releasing any information through journals, meeting talks, etc., the appropriate release form will probably need to be signed off by both parties, and this may be slower and more difficult than

---

**Box 3.6   Key Points About Big Company–Small Company Collaborations**

▶ Collaborations can match creative ideas and projects from small companies with big pharma resources and expertise

▶ With increasing resources comes increasing expectations and the need to meet project milestones quickly

▶ Good collaborations rely on well-defined agreements, inclusiveness, open communications, and ultimately the people involved

▶ Some special IP considerations may apply in partnered projects

for in-house projects. Another tricky situation can arise too. In many collaborations, discoveries—not only just discoveries ultimately resulting in patents but also synthetic tricks, interesting pharmacophores, useful bioisosteres, and the like—are considered proprietary to the collaboration and can't legally be used for other projects. If a collaboration between your biotech company and big pharma turns up a new kind of, say, kinase inhibitor or a novel way of making some known ones, you may not be free to use that knowledge in other projects, either in-house or those involving a third party. This can make for a difficult situation. Trying to "forget" something like this is like trying to *not* think about blue elephants, but it's just part of the job. If in doubt, play it safe and check with those responsible for carrying out the collaborative agreement.

## 3.3  The New Drug R&D Process

Turning concepts into new drugs requires an incredible number of activities involving many different disciplines. The whole process tends to follow an overall pattern, which can be divided into a number of different stages. Nomenclature and definitions, however, aren't completely standardized. Sometimes, slightly different names will be used for the same stage, and stages may be broken down slightly differently. A general scheme that agrees with most is shown in Figures 3.3 and 3.4. In keeping with the previous section, note that IP issues need to be considered at each of these stages.

### 3.3.1  Target Identification

These days most drug discovery projects are directed at specific molecular targets. Some targets (e.g. β-adrenergic receptors, HMG-CoA reductase, etc.) have long been known to be good ones to aim at. For such "established targets," there's little doubt that a compound interacting with them appropriately would be efficacious in man. For such projects neither the earliest stage of drug discovery, called *target identification* (*TID*, alternately known as

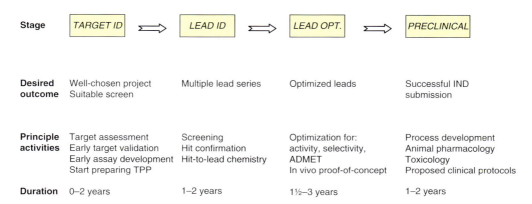

**Figure 3.3 ▶** Early stages of the new drug process. Target identification, LID, and LO are considered "research" or "discovery stage" activities, while preclinical development marks the beginning of the "D" part of "R&D."

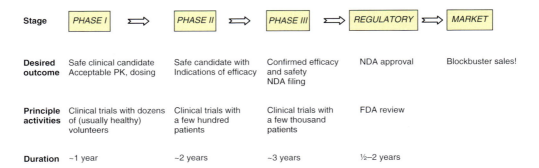

**Figure 3.4 ▶**   Stages in clinical development and beyond.

*target assessment*) nor the process called *target validation* (*TV*) would be necessary. Only a business decision as to whether the company wants to compete with known drugs for a particular market and some input from scientific departments on the feasibility of assays and models, suitability of available compound libraries, etc. would be required. Hopefully, even bureaucratic companies could do this in a matter of months.

For novel targets, which lack the ultimate validation that known efficacious drugs provide, there's always the question of whether the company wants to spend hundreds of millions of dollars to develop a drug that turns out not to work because the target proves irrelevant to the disease state. This is a very reasonable concern and explains the failure of many projects and drug candidates. So here, TV becomes crucial to the project and one of the first things that needs to be addressed during TID. As we'll see in Chapter 5, *TV is an ongoing process, not a one-time experiment*. It's perhaps the most difficult and uncertain part of the drug discovery process and is best thought of as having different successive levels, including cellular, whole animal, and clinical validation. Although sometimes TV is listed as a discrete stage of drug discovery in a flowchart, doing a single successful TV experiment early on and considering the target "validated" is somewhat like taking a bath once and considering oneself "clean" forever after. But the bottom line is that for novel targets, the TID stage must involve early TV experiments, and only proceed further as long as the experiments (perhaps, cellular experiments using RNAi reagents or results obtained from knockout mice) prove positive.

If the project is to proceed, assays capable of determining activities at a sufficient throughput need to be available, and so assay development work typically begins in TID. And for a project to succeed, regardless of whether the target is established or novel, the ultimate goal of the project needs to be defined, and the earlier the better. It isn't enough to say that you want to have a new drug for, say, lowering blood sugar in Type II diabetics by interacting with a certain target. You need to define whether the drug needs to be an oral one or if i.v. dosing is acceptable, how often (once a day? twice a day?) it needs to be administered, how safe it needs to be in terms of interacting with other drugs that might commonly be used along with it, etc. It's especially important to define what your drug should be able do that others can't, that is, what advantage it will have over existing therapeutic agents, as drugs with no advantage will be hard to sell. All of this information and more make up the *target product profile* (*TPP*), which will be used to guide the discovery and development process. The more thought and detail that goes into a TPP, the better off a project will be. The sorts of issues it might address can be seen by perusing the listings for known drugs in the *Physicians Desk Reference* (published annually by Thomson/PDR) or online drug label information from *Food and Drug Administration* (*FDA*).[38] Looking through either of these is a good way of getting familiar with real-world drugs too.

### 3.3.2  Lead Identification

Having an incipient project that fits with a company's business plan, including its patent strategy, has at least an initial amount of validation and a working screen allows for entry into the next stage of drug discovery called *Lead identification* (*LID*). Although, the earlier TID stage often involves a limited chemistry effort—chemical tools often being necessary even to do assays or early validation experiments—it's really at LID that chemistry starts becoming a driver for the project. Screens, often these days high-throughput ones run on thousands of compounds, will be carried out and, as described in Chapter 6, "actives" will be found. These require some follow-up experiments to rule out artifacts like compound fluorescence and aggregation, and this might take longer than it did to run the initial screen. In the end, though, a list of confirmed hits that represents a starting point for medicinal chemistry efforts will be obtained. Other, non-HTS-based projects will arrive at a starting point by screening fewer compounds but with more information about the target and its ligands.

The process by which medicinal chemists convert confirmed hits, which generally have little to recommend them other than some defined level of activity against the target, into leads, which have better activity and additional properties like good selectivity and perhaps cell permeability, is known as *hit-to-lead* (*HTL*). The criteria for what constitutes a lead will depend on the company and the project. Ideally, multiple confirmed hits are turned into multiple lead series. In reality, hits may be few to begin with, and many of them can peter out for various reasons, like flat, un-optimizable *SAR*, lack of patentability, insolubility, etc. Ending up with multiple, diverse lead series, though, is a tremendous advantage and something that projects strive for. Good lead compounds, may well be in demand for ongoing TV experiments too.

### 3.3.3  Lead Optimization

Medicinal chemistry efforts really peak in the *lead optimization* (*LO*) stage. Although leads entering this phase of drug discovery may have good in vitro activity and selectivity and might even work in cells (especially, if they came out of *phenotype screening*, discussed in Chapter 5), this turns out to be a far cry from all that's required for a compound to enter clinical trials. Lead optimization tends to focus on the properties necessary to achieve in vivo results, such as bioavailability, metabolic stability, and the avoidance of toxicity due to known liabilities like hERG binding. In addition, further optimization of properties like potency—of which one can rarely get enough—and selectivity will probably still be needed. Success is usually determined through animal testing, which provides readouts of *pharmacokinetics* (*PK*, "what the body does to the drug"), *pharmacodynamics* (*PD*, "what the drug does to the body") and, ultimately, efficacy. Good activity in a relevant animal model of the disease of interest will be required.

Lead optimization might require hundreds or thousands of compounds, many of them not being amenable to high-throughput synthesis, and this takes manpower. Often 10 or more chemists will work full-time on a given project in LO, far more than one will find involved in HTL chemistry at the previous stage. Even so, this is often the most time-consuming part of the new drug process prior to lengthy clinical trials. Typically, LO might take several years, making it, at times, a favorite target for high-level executives wishing to speed things up. Unfortunately, attempts to treat it as little more than a factory exercise haven't worked well and too much pressure can result in lowering the bar for "optimized leads" that are eventually produced, thus contributing to failure in the next phase.

### 3.3.4   Preclinical

Whether or not an optimized lead ever becomes a candidate going into the clinic depends on what happens in the next stage, called *preclinical* or maybe *clinical development candidate selection*. Although, the term "preclinical" can be taken to broadly mean any new drug activity done prior to clinical trials (TID, LID, etc.), it normally refers specifically to this stage, which marks the transition to the "D" in "R&D." As such, it's the point where many things change.

During HTL and LO, medicinal chemistry concentrates on making tens or hundreds of milligrams of many analogs. Purity requirements are modest (often 90–95%) and any available reagents or processes can be used. Good yields, although desirable, aren't mandatory. In contrast, preclinical chemistry demands the synthesis of large amounts (100 g to multikilogram quantities) of a single, very pure potential clinical candidate using cheap, safe, and relatively non-toxic reagents. Yields are much more important and things like lengthy routes, chromatography-requiring steps, and even rotary evaporation need to be avoided whenever possible. Many chemists are surprised to learn that some of the most elegant synthetic routes to be found in industry stem from work done by process chemistry groups, and it isn't unusual for the final manufacturing route to a drug to be completely different from, and far superior to, the original route by which it was synthesized. In fact, for the chemist interested primarily in synthetic organic chemistry who enjoys optimizing routes and scaling up syntheses, process chemistry can make a rewarding career, as many have found.

But, it is different from research stage chemistry. Even apart from differences in scale, HTL or LO chemists are unlikely to, for instance, write down the lot number of solvents used in each experiment or take the trouble to identify and quantitate minor impurities, activities that will probably seem like a waste of time to them. But these and many other factors (like reproducibly getting the most desirable polymorphic form) which don't seem that important in earlier stages become so as compounds approach clinical status, where their syntheses require that GMP procedures be followed. The development of compounds at this stage can lead to valuable new IP such as process improvements or unique polymorphic forms.

If the "flavor" of chemistry changes in the preclinical stage, so does that of biology. By the time they become optimized leads, compounds will already have gone through at least preliminary screening for factors that might contribute to toxicity like the binding to hERG receptors or CYP inhibition, as described in later chapters. But during the preclinical stage, their safety will be much more thoroughly investigated and toxicology testing in multiple animal species will be carried out. Some long-term toxicology experiments begun here will even extend well into the clinical development stage. Needless to say, extensive experiments which address efficacy will be involved as well, with many PK studies being done in multiple species with the aim of predicting the compound's PK properties in man by a process called *allometric interspecies scaling*.

All clinical activities, of course, will depend on a suitable optimized lead being chosen for development. In the lucky event that a number of optimized leads are available, only one of them will be probably be developed at a time. Each clinical candidate costs many millions of dollars to develop, so a selection needs to be made if more than one of them make the cut. Having others available as backups, however, is very desirable. If only, a single optimized lead compound enters preclinical development, however, only a GO/NO GO decision is required. Either way, further testing as described above is in order. The actual criteria for taking a compound on will depend on the project (for instance, how likely it is to fit the desired TPP) and the company. Many detailed and strict requirements will need to be met before any compound can be considered for clinical testing. When a compound is found that meets all of them, a new title and acronym will be bestowed on it. These vary a bit, but something like *clinical development candidate* (*CDC*) might be typical. But keep in mind that the ultimate goal of the preclinical stage isn't just nominating such a compound, but actually getting the approval required to take it into the clinic.

In order to do this, in America an *Investigational New Drug* (*IND*) application will need to be filed with FDA.[39] This, of course, includes plenty of safety information from toxicology studies as well as massive amounts of in vitro and in vivo data and proposed clinical protocols that have been developed by the clinical investigators now busily preparing for the project. *Institutional Review Boards* (*IRBs*), committees of physicians, scientists, statisticians, and others who oversee clinical studies, begin to get involved as well. After submission, an acknowledgement of the filing along with an IND number is sent back to the company and if no hold is placed by FDA within 30 days of its receipt of the completed application, the study can begin and a major milestone has been reached: a true clinical candidate or IND is now in hand.

You can probably see already that drug development is every bit as complex and specialized as drug research. Two implications of this will be discussed here. First, under the system pharmaceutical companies used a few decades ago, "R" didn't really talk to "D" very much. The phrase often used was that compounds were "thrown over the wall" into development, sometimes with concerns about any property other than potency being just an afterthought on the part of research. Some of these afterthoughts, though, like bioavailability and formulatability, could go onto kill the compound either at the preclinical or clinical stage. These days, needless to say, that doesn't happen. Parallel screening for many properties beside potency is the rule at every stage of research, and communications between research chemists and their development counterparts are extensive, particularly when it comes to transferring synthetic routes and knowledge. There's definitely more foresight and more of a team approach involved.

The second implication of the complexity of development is that it's a big subject, with too much additional material for this book to attempt to cover, so *Real World Drug Discovery* will remain true to its title and concentrate on only discovery instead. But, as the ultimate goal of the combined process is a new drug, a very short review of the rest of that long path follows, if only to assuage the reader's curiosity and provide a little perspective. Readers interested in a more complete explanation of the clinical development and drug approval process are referred to several outstanding references.[40]

### 3.3.5  Stages in Clinical Development

The progression from IND to approved drug is a long and expensive one. Much of research is focused on trying to ensure the highest possible success rate for such compounds in clinical development, the costliest part of coming up with a new drug. Although no one wants failure at any stage, failure in the research stages is better than failure in development for this reason.

The path from clinical candidate to marketed drug is outlined in Figure 3.4. Note that, this is the "classical" clinical path. In practice, these days two new phases of clinical trials exist. After the preclinical stage but before *Phase I* studies, clinical candidates might enter first-in-man studies via *Phase 0*, where subtherapeutic *microdoses* of the compound are administered to a small number of healthy volunteers to give a preliminary indication of whether human PK is as predicted from animal studies. Valuable information about PD and biomarkers that might be useful for subsequent clinical trials might also be obtained. The other stage not shown above is called *Phase IV*. These studies take place after drug approval and are generally either required by a regulatory agency like FDA as part of drug safety surveillance (*pharmacovigilance*) or might be done voluntarily by the company to expand the market for the drug by adding a new indication.

In the standard scheme of things, though, INDs first enter the clinic in *Phase I* studies. Here, a relatively small number of healthy volunteers (perhaps 20–100) are given the investigational drug. For some very serious diseases like cancer, actual patients may be enrolled instead. The focus will be on establishing safe dosing levels and understanding the compound's PK properties. Other information on PD, biomarkers, the type of toxicity that might be associated with the compound, and maybe even an early indication of efficacy (if patients and not healthy volunteers are given the drug) can also be obtained. If the company is encouraged by the results of a Phase I trial and FDA hasn't imposed a clinical hold, which it can do at any time in clinical development if it has concerns about safety or other factors, it's onto the next stage.

In *Phase II* studies, a few hundred patients (not healthy volunteers) will be treated with the investigational drug, usually in *double-blind randomized controlled trials*. "Double-blind" implies that neither patient nor clinical investigator knows which treatment group the patient is in, "randomized" means that patients are randomly assigned to treatment groups, while "controlled" means that the investigational drug is given to some patients while others get either placebo (in most trials), or, sometimes, a known drug instead to allow for comparison. Industry critics frequently call for the use of comparator drugs in every trial where they exist, on the theory that new drugs should have to be better than old ones, but this approach isn't without its own problems.

The first true, if preliminary, indication that an investigational drug is efficacious usually comes at Phase II. Safety/toxicity data are also a focus. Sometimes, these trials are now broken down into *Phase IIa*, done on a smaller number of patients (perhaps fewer than 100) and *Phase IIb*, later done on a slightly larger group. Phase combinations like "Phase I/II" sometimes occur to make things more complicated still.[41] Whether thus subdivided or homologated or not, statistically significant efficacy in Phase II trials isn't conclusive. If it was, a lot of money and trouble could be saved, but unfortunately sometimes after a Phase II trial that successfully met its endpoint in a statistically significant way another Phase II or a Phase III trial using the same investigational drug in the same indication won't reproduce this efficacy, leaving everyone scratching their heads and wondering what went wrong.

Real proof of efficacy and safety comes from large, expensive *Phase III* trials. These usually involve up to several thousand patients at multiple sites around the world. These too are likely to be double-blind randomized controlled studies. By the time they're over, all the types of information—dosage, indication, efficacy, PK (including *food effects*, which can affect bioavailability), metabolism, side effects, etc.—required for a package insert or product label (and hopefully one that fits with the original TPP) will be available from the combined trials.

If the sponsoring company feels that the data warrant it, all of this and more will be put together and a *New Drug Application* (*NDA*) will be filed with FDA's *Center for Drug Evaluation and Research* (*CDER*). These can run well over 100,000 pages, and in years gone by photographs of rooms crammed with boxes full of papers for an NDA submission were commonly used at pharma companies to illustrate just how much was involved. These days, though, filing is electronic. If the NDA submission is deemed incomplete or unacceptable by FDA, they can indicate that they won't waste their time reviewing it by issuing a *refusal to file* (*RTF*) letter, which throws a monkey wrench in the works.[42] Otherwise, the NDA is successfully *filed* and the regulatory stage of drug development (Figure 3.4) has begun.

FDA can respond to an NDA in three ways. It can send a *non-approvable letter*, which is pretty self-descriptive, an *approval letter*, the much longed-for green light to market and sell, or an *approvable letter*, which is somewhere in between. An approvable letter means there are issues to be resolved. These might not be too hard to work out, like labeling changes, or they might involve running lengthy new clinical trials.

Just how long it takes to get a response to an NDA can vary quite a bit. Some get priority review, if they're felt to represent "breakthrough" drugs. Gleevec (imatinib) was approved in only 2½ months and after only Phase II trials (three of these), with the understanding that further clinical studies would continue. More commonly, it can take over a year to get a decision, which is actually quicker than used to be the case. Review times (and also the length of the clinical trial stages) can vary by therapeutic area, with HIV drugs and antiinfectives, for example, being reviewed more quickly on average than, say, CNS drugs.[43] Recall, though that US patents for new drugs can be extended to make up for the time it took to get regulatory approval.

## 3.3.6   What are the Odds?

Approval to market and sell a new drug represents the consummation of everything that went before. The creativity, hard work, and resolve of hundreds of individuals in many different disciplines, an investment of hundreds of millions of dollars, and a commitment probably lasting more than a decade finally come to fruition with an NCE that gives patients a medicine that didn't exist before. But, how likely is this to happen? And if a project doesn't work out, where is it most likely to fail? What's the weakest link in the chain?

| | Research | | | | Development | | | | |
|---|---|---|---|---|---|---|---|---|---|
| | Target assess | Lead ID | Lead opt | Pre-clinical | Phase I | Phase II | Phase III | Reg | On market |
| Rate (%, median) | 63 | 60 | 63 | 57 | 58 | 45 | 58 | 85 | |
| Cumulative success rate (%) | 1.7 | 2.7 | 4.6 | 7.3 | 13 | 22 | 49 | 85 | |
| projects (for one approval) | 57 | 36 | 22 | 14 | 8 | 4.5 | 2 | 1.2 | 1 |

*Drug Discovery Today*

**Figure 3.5** ▶ Success rates at various stages of new drug R&D process as assembled from the experience of several leading pharma companies. The top row shows median success rate at each stage, the middle row reflects cumulative success rates for projects by stage, and the bottom row indicates the number of projects required to get a single drug to the market at each stage. (Reprinted with permission from Brown, D., Superti-Furga, G. Rediscovering the sweet spot in drug discovery. *Drug Discov. Today*, 8, 1067–1077, copyright 2003, Elsevier.).

To answer these important questions statistics have been gathered on the attrition rates for projects at each stage. One such collection of data is shown in Figure 3.5. The top row indicates the success rate for each stage, which ranges from 45% in Phase II to 85% in the regulatory stage. The weakest link, then, turns out to be Phase II clinical trials, where the majority of compounds going in never make it out to Phase III, a bitter outcome so late in the game. You can also see that fewer than one-quarter (24%) of all newly minted research projects are able to successfully deliver optimized leads into the preclinical stage, and that the cumulative attrition rate in development alone exceeds 90%, meaning that projects are even more likely to fail in development ($$$$) than in research ($$). On average, a staggering 57 projects will be needed to get one new drug, and some of us might even consider this an underestimate.

All of which goes to show that industrial drug R&D is not for the faint of heart. Success may be elusive rather than guaranteed, but if scientists allowed themselves to get too discouraged by the odds, it would only ensure that no new drugs would be found. Some new drug projects can and do succeed, and it's now time to take a nuts-and-bolts look at how they operate in real-world drug discovery.

# Notes

1. For more about how this works in Germany, see Harhoff, D., Hoisl, K. Institutionalized incentives for ingenuity—patent value and the German Employee's Invention Act. *Res. Policy* **2007**, *36*, 1143–1162.

2. Lipinski, C.A. The anti-intellectual effects of intellectual property. *Curr. Opin. Chem. Biol.* **2006**, *10*, 380–383.

3. Soames, C. Grace periods and patentability. *Nat. Rev. Drug Discov.* **2006**, *5*, 275.

4. Here the concern isn't that telling the consultant bound by a CDA could be construed as public disclosure (it can't), but that he might not follow the terms of the CDA and instead might leak information.

5. Gersten, D.M. The quest for market exclusivity in biotechnology: Navigating the patent minefield. *NeuroRx®: J. Am. Soc. Exp. NeuroTher.* **2005**, *2*, 572–578. Available at www.pubmedcentral.nih.gov/picrender.fcgi?artid=1201316&blobtype=pdf.

6. As discussed in Chapter 1, though, patent extensions are sometimes granted for drugs. In addition, before 1995 U.S. patents were in effect for 17 years from the date of approval.

7. Although a few years back some Australian patent protection was given to a "circular transportation facilitation device" (aka wheel) by a patent attorney wishing to point out the then-current "rubber stamp" policy of IP Australia. See Knight, W. Wheel patented in Australia. *New Scientist*, July 3, 2001. Available at www.newscientist.com/article/dn965.html.

8. For more about novelty and many other aspects of biotech and pharmaceutical patents, see Chapter 4, "What Can Be Patented", in *Patents for Chemicals, Pharmaceuticals, and Biotechnology* by Philip W. Grubb (Oxford, Oxford University Press, Fourth Edition, 2004). This book represents a particularly up-to-date and well-written resource for those seeking further information.

9. For a good discussion of the merits and demerits of first to file versus first to invent, see "A Patent System for the 21st Century", Merrill, S.A., Levin, R.C., Myers, M.B., Eds. (Washington, DC, National Academies Press, 2004). Available at www.nap.edu/html/patentsystem/0309089107.pdf.

10. Filler, A.G. quoted in Myers, G. From discovery to invention: The writing and rewriting of two patents. *Soc. Stud. Sci.* **1995**, *25*, 57–105. The reader interested in differences between science journal articles and patents is encouraged to consult this fine case study.

11. One particularly blunt attempt at a definition of "obvious" was "so easy that any fool could do it". *Edison Bell* v. *Smith* (1894) 11 RPC 457 at 497 (CA). But note that this definition conflicts with the observation attributed to Edsel Murphy (of Murphy's Law fame) that "it is impossible to make anything foolproof because fools are so ingenious."

12. See www.supremecourtus.gov/opinions/06pdf/04-1350.pdf.

13. See Wingfield, B. Getting a patent just got harder. *Forbes*, April 30, 2007. Available at www.forbes.com/businessinthebeltway/2007/04/30/scotus-patent-microsoft-biz-wash-cx_bw_0430scotus.html.

14. 37 CFR 1.73. See www.uspto.gov/web/offices/pac/mpep/documents/0600_608_01_d.htm.

15. Cummings, M.D., et al. US 7,071,184, "Protease Inhibitors" (to SmithKline Beecham).

16. See www.uspto.gov/web/offices/pac/dapp/opla/presentation/esdguidelines090607.pdf.

17. The proposed new regulations affect other things as well, including the number of continuations and divisional patent applications (not discussed herein) that are allowed. See "Court Temporarily Halts New Patent Rules", Associated Press, October 31, 2007. Available at www.forbes.com/feeds/ap/2007/10/31/ap4285576.html.

18. This was *Hoffman LaRoche, Inc.* v. *Promega*, 323 F .3d 1354 (C.A.F.C., 2003). See www.dwt.com/practc/life_sciences/bulletins/07-03_patentapps.htm.

19. Maynard, J.T., Peters, H.M. *Understanding Chemical Patents: A Guide for the Inventor* (Washington, DC, American Chemical Society, 1991), 2nd edition, p. 52. This is a particularly useful book for chemists interested in patents, but unfortunately it's long out of print and no new, updated edition has come out.

20. Another, slightly more detailed set of instructions on searching for compounds in patents can be found in Friary, R. *Job$ in the Drug Industry* (San Diego, California, Academic Press, 2000), p. 131.

21. *Mueller Brass Co.* v. *Reading Industries, Inc., 352* F. Suppl. 1357, 1372 (ED. Pa. 1972), actually describes the concept of joint inventorship, but the phrase is just too good not to use!

22. *Fiers* v. *Revel*, 984 F .2d 1164, 1168, 25 USPQ2d 1601.

23. *In re Hardee*, 223 USPQ 1122, 1123.

24. *Ex parte Smernoff*, 215 USPQ 545, 547 (Bd. App. 1982).

25. Grubb, Reference 8, p. 383.

26. If you don't believe that, check the "USPTO status" link shown in Box 3.2, type in the number of a recent U.S. patent, then click on the "Transaction History" and "Image File Wrapper" tabs to see a bit of what's involved post filing.

27. Chin-Dusting, J., et al. Finding improved medicines: The role of academic-industrial collaborations. *Nat. Rev. Drug Discov.* **2005**, *4*, 891–897.

28. Gray, N. Drug discovery through industry-academic partnerships. *Nat. Chem. Biol.* **2006**, *2*, 649–653.

29. Lengauer, C., et al. Cancer drug discovery through collaboration. *Nat. Rev. Drug Discov.* **2005**, *4*, 375–380.

30. Sánchez-Serrano, I. Success in translational research: Lessons from the development of bortezomib. *Nat. Rev. Drug Discov.* **2006**, *5*, 107. This article provides a fascinating introduction to the unexpected paths from concept to market that drugs can take in the real world.

31. Sánchez-Serrano, ibid. Attributed therein to a personal communication from P. Elliott.

32. An example of what a materials CRADA form looks like can be found online at www.cit.nih.gov/NR/ rdonlyres/8E8EA8B7-0938-4518-A208-CC0D3AC3F371/0/MatCRADA.pdf.

33. For more about the roles of NIH in commercial drug discovery research, see Ben-Menachem, G., et al. Doing business with the NIH. *Nat. Biotechnol.* **2006**, *24*, 17–20.

34. Wiederrecht, G.J., et al. Partnership between small biotech and big pharma. *IDrugs* **2006**, *9*, 560–564.

35. Jones, A., Clifford, L. Drug discovery alliances. *Nat. Rev. Drug Discov.* **2005**, *4*, 807–808.

36. Bayer, for example, had some of its employees working onsite at Millennium Pharmaceuticals to facilitate communications during their collaboration. See Ziegelbauer, K., Farquhar, R. Strategic alliance management: Lessons learned from the Bayer-Millennium collaboration. *Drug Discov. Today* **2004**, *9*, 864–868.

37. See Ziegelbauer & Farquhar, ibid.

38. Currently viewable at www.accessdata.fda.gov/scripts/cder/drugsatfda/index.cfm.

39. To see exactly what's required, check out www.fda.gov/opacom/morechoices/fdaforms/FDA-1571.doc.

40. Schacter, B. *The New Medicine$* (Westport, CT, Praeger Publishers, 2006) gives a very good description of the clinical development process using examples like that of celecoxib. More succinct outlines of what goes on in the clinic and after can be found in Campbell, J.J. *Understanding Pharma* (Raleigh, NC, Pharmaceutical Institute, 2005) and Eckstein, J. *ISOA/ARF Drug Development Tutorial*, available from the Alzheimer Research Forum at www.alzforum.org/drg/tut/ISOATutorial.pdf.

41. See Piercey, L. The two-headed beast. *Signals Magazine*, January 20, 2004. Available at www.signalsmag.com/ signalsmag.nsf/657b06742b5748e888256570005cba01/ 8b347a8cdbc59c2088256de5005db0dc?OpenDocument&Highlight=0,piercey.

42. This happened, for example, with the biologic cetuximab (Erbitux) when its BLA was submitted to CBER, the biologics equivalent of an NDA submission to CDER.

43. For NCEs approved by FDA between 1999 and 2001, the mean approval times were 0.4 years for HIV drugs, 1.0 year for antiinfectives, and 1.8 years for neuropharmacologics. See Kaitin, K.I., Cairns, C. New Drug Approvals of 1999, 2000, and 2001: Drug Development Trends a Decade After Passage of PDUFA, in *Parexel's Bio/Pharmaceutical R&D Statistical Sourcebook 2006/2007*, Mathieu, M.P. Ed. (Boston, Parexel International, 2006).

# Chapter 4

# How Things Get Done: The Project Team

## 4.1 Introduction

Scientist just entering the field of industrial drug discovery research probably won't be familiar with project teams. Understanding what they are, how they function, and the roles that scientists play on them turns out to be as important to the biotech or pharma researcher as understanding the structure of the atom is to the chemist. A simple explanation can save a lot of head-scratching and allow for a quicker, more seamless integration into the flow of the project, which is something both employer and employee will appreciate. That's because projects and project teams lie at the heart of drug discovery research. Projects are conceived, proposed, staffed, carried out, reported on, assessed, scrutinized, and ultimately either come to fruition in a new drug or die. Either way, the scientists move onto new projects and begin the life cycle all over again. So a general understanding of project teams and how they work is probably the single most immediately useful piece of knowledge a new researcher can learn.

A major difference between academic and industrial research lies in the organizational units used to get things done. In university research, this unit tends to be the research group, headed by the professor whose name it bears, and consisting of the postdocs, graduate students, and possibly other people reporting to him or her. Organization along these lines allows for deep research in a narrowly defined field, say, new synthetic techniques for macrolide synthesis in a chemistry group or caspase activation pathways in cellular apoptosis (programmed cell death) in a biochemistry group. Sometimes these two groups may decide to collaborate to see if macrolides produced in the first group can influence apoptosis in the second. For the biochemistry group this means they have compounds to work with and publish on, while the chemistry group might benefit from biochemical data to add to presentations and papers that, if interesting activity is found, could make it easier to get grant funding. This kind of collaboration usually just involves a transfer of compounds from the synthetic group to the biochemistry group for the latter's research. This is normally done by the two principals without extensive input from the rest of the groups and without some sort of overall management by university administrators, who wisely leave research matters to the research faculty members.

And, importantly, this kind of collaboration only involves a couple of groups, not all the conceivable specialties that could be involved in a task as complex as bringing a new drug to market. Right now even these limited collaborations are the exception: most of a professor's research will stand as independent of other groups. This is perfectly understandable as academic work has always tended to be . . . well, "academic"—that is, deep and narrow, the ultimate aim being an expansion of knowledge, not product line.

Although this model has been around for centuries, change is in the wind, as we've seen in Chapter 2. Increasingly, universities have begun to emphasize seemingly more lucrative *applied* research that is more likely to require either internal or external interdisciplinary collaborations. So the situation has begun to change, but at present a lot of academic research still goes on as it always had, focused on one particular discipline and carried out within the confines of a given research group. In the words of Dr Frank L. Douglas, former Executive Vice President of Aventis and currently Executive Director of the MIT Center for Biomedical Innovation, "In academia each researcher tends to see the problem through his or her own paradigm. The need for addressing real-world projects and having project milestones is much more honed in industry than it is in academia."[1] Let's now look at how these project issues are addressed.

## 4.2   The Project Team

In industry, the ability to bring together different groups from various disciplines is an absolute prerequisite for success. Here, there's no less need for deep knowledge, but such knowledge is required in many fields simultaneously. No one person, no matter how brilliant, can have it all. A thorough understanding of just a part of what's required is difficult enough, which means that multiple individuals in many specialties will need to work together. People being more complex than any scientific model, getting this to happen efficiently requires a certain kind of genius as well.

Because of this absolute requirement for interdisciplinary research, most real-world drug discovery is carried out by the *project team*. This is a group of individuals brought together to function as a unit and representing various disciplines, scientific and otherwise (business, law, etc.), required to accomplish a defined *project goal* in a given period of time. Just how interdisciplinary a project team is by nature is revealed by its composition. Often it includes synthetic chemists and lawyers, molecular biologists, and clinicians. The composition of a project team will vary with the type and stage of a given project, with, for example, more clinical representation as it gets closer to having a clinical candidate. But the project team usually includes at least one representative from each of many different disciplines.

Project teams aren't unique to drug discovery but are a key part of any industry that depends on new products for revenue, be it software, airplanes, or TV movies. It is hard to believe, though, that the sheer diversity of specialties—not to mention the time and expense of bringing a drug "product" to market and the resulting requirement for a concerted interdisciplinary effort—really has a parallel in many other fields. The project team system is used in drug discovery R&D for one very simple reason: years of pharmaceutical experience have shown it to be an efficient way of combining these many skills and getting a new drug to market.

### 4.2.1   The Project Goal

In a sense, the project goal defines the project team. With just a few words it states what needs to be accomplished, defines where and when the project begins and ends, supplies a starting point for deciding on resources, and provides the unifying force for everyone on the team. It's the context within which the project team exists. If you ask any project team member what the project goal is and he or she can't tell you, there's a big problem. Understanding and working toward the project goal is a prerequisite for the professional industrial researcher. Upon entering into a new project, determining what the project goal *is* has to be the first order of business.

In drug discovery, research project teams are typically organized around particular biological targets, although sometimes this can vary, as with screening for a new drug that gives a particular phenotype effect, as we'll see in the next chapter. The ultimate goal in either case is to bring a new drug to market. Because of the complexity of that task, however, goals for new projects just coming online need to be more limited, covering just a few early steps in the process. If those steps are successfully completed the project team will be redefined to bring in whatever new resources (chemical development, toxicology, clinical, etc.) may be required in the next stage, and the next few steps will be taken, beginning with the writing of the new project goal.

So, although the ultimate goal is going to be to bring a new drug to market, as a project goal that's far too long term and complex, too big to swallow whole. Project goals will constitute small, distinct sections of the overall new drug goal. They'll tend to span a timescale of perhaps six months to a year, while the time required to go from concept to approved drug, as we've seen, typically exceeds 10 years.

In light of the probable failure rate for new projects, which is over 90%, project goals also need to be realistic, meaning that if what the project team seeks is for some reason unachievable, they can then report that fact rather than bury their heads in the sand and waste resources on what clearly cannot be done. Determining that the project has reached a dead end would be considered a victory nonetheless. According to Dr Cynthia Robbins-Roth, an industry consultant and keen observer, "The need to kill projects in a timely fashion is as crucial as recognizing success."[2]

Think of it as a kind of project Darwinism, an evolution toward the survivable projects that can uncover successful new drugs. Most projects may not make the cut and the project team members should be rewarded, not punished, for finding an answer—even if that answer was "No"—that couldn't have been predicted in advance to a question that senior corporate management agreed was well worth asking. Project killing is never fun, and a certain amount of patience and forbearance need to go into making the decision. Managers should know that all the necessary resources have been properly brought to bear on the issue, and project team members should know that they aren't being punished. But when a project *is* killed, the decision should be made on the basis of predefined criteria. This again leads us back to the project goal.

### 4.2.1.1  Compound Validation Goals

Most projects in industry focus on targets already known to have in vivo effects in man, or at least in a number of animal models of the disease of interest. Projects having this kind of validation are less risky than those aimed at previously unexplored targets, as we'll see in the next chapter. A typical early stage goal for such a project might be the following:

> To discover a novel, reversible selective inhibitor of BCR-ABL kinase that demonstrates target inhibition and an antiproliferative effect in both wild type and imatinib-resistant Ba/F3 cells with an $IC_{50} < 20$ nM by the third quarter of 2008.

Although what's meant by "selective" needs to be further defined, here the project team needs to find a "novel" (read *patentable*) new drug for imatinib-resistant CML that at least works in cells. A target date has been set. As you can see, although medicinal chemistry will be central to this compound validation goal, other departments such as legal, cell biology, and molecular biology will also need to be involved. So the goal begins to define the project team. Note that outside events can impact such a project goal. Newly found mutations in Gleevec-resistant CML may need to be added to the assays, and newly approved drugs for the same indication, like Sprycel (dasatinib, Figure 2.22), that come to market years ahead of any drug that might come out of this new project might call for the project to be re-evaluated. These kinds of things can't be predicted in advance, though: the goal can only be written on the basis of what's known at the time.

Successfully meeting this project goal would lead to new project goals, rewritten to reflect additional hurdles to be overcome such as PK properties (oral bioavailability, compound half-life, etc.) and efficacy in animal models. Failure to meet the project goal, on the other hand, although not an outcome people will want to see, can happen even with an outstanding project team. It could just be that the bar was initially set a little too high so that goal couldn't be reached with the given resources in the timeframe allotted. Sometimes a lead series just couldn't be sufficiently improved or make it outside the scope of another company's patent, and so further efforts would have to go back to square one. As is the norm in scientific research, a lot of things can go wrong. If it really is just a matter of a few more FTEs or another few months, management will often be willing to give a little on these. On the other hand, if it looks like the goal is just not achievable, then the project is likely to be terminated, freeing up team members for other projects. In this tough business it happens all the time. As long as the project team made every effort to meet the project goal, no blame should befall anyone. Unfortunately, though, the warm, fuzzy glow of success will be lacking too.

Regardless of the ultimate outcome, the writing of the project goal has laid the groundwork for this project team and the corporate management has bought into it, in effect gambling a considerable chunk of the company's money on it. It's not surprising, then, that they'll keep a sharp eye on the team's progress.

### 4.2.1.2   Target Validation Goals

A slightly different type of project goal is usually called for when it comes to working on a novel target that isn't already known to have an effect in the clinic or at least the vivarium, as we'll see in the next chapter. Here, it's not certain that discovering an ideal inhibitor/agonist/antagonist with perfect drug-like properties is likely to result in either animal or human efficacy. There's an excellent chance that in the end hitting the target will not give the desired effect in vivo. So the goal for this type of project is not yet to find a lead that can be developed into a new drug but to simply assess whether or not the target is a relevant one in cells and animals. An example would be the following:

> To determine whether or not inhibition of cathepsin H causes antiproliferative effects in non-small cell lung cancer cell lines using RNAi reagents and sub-100 nM, cell permeable inhibitors by the third quarter of 2008.

Notice how loose the criteria are here compared with those required in the previous goal. The bar for potency is lower, good selectivity and reversible may not even be required, and compounds will not necessarily have to be novel. Medchem may or may not be required yet. Commercially available or existing in-house compounds may be good enough to lay the groundwork at this early point. It could be that the company has an optimized compound or series in mind from another project and wants to investigate a potential new indication for it with just a few experiments in cell biology. Perhaps a paper recently appeared implicating a target the company already works on in a new and unexpected indication.

If appropriate compounds are not available, it is unusual for a company to begin screening for hits or begin compound optimization chemistry for this kind of speculative project absent a particularly strong biological rationale, although it can happen at times. Instead, some other validation methodology like RNAi or knockout mouse (see Chapter 5) is likely to be used, with small molecules to be made later if the results look promising. Whatever method may be used, compounds and targets either progress or do not, together.

Whatever the outcome, you can see that for goals directed only at *target* validation, if it turns out that hitting the target has no effect in cells or animals, the project team, with or without medchemists, is not going to suffer. Instead they'll be congratulated for demonstrating that fact quickly and efficiently. Again, pharma wisdom says that killing projects that would drain resources away from more viable ones is the next best thing to getting a drug.

## 4.2.2 Project Team Organization

### 4.2.2.1 The Matrix Management System

As the project team functions as a largely autonomous unit focused on a particular goal, it acts as a kind of company-within-a-company. So its hierarchy must by definition be different than the normal ("functional" or "line") reporting relationships within the company, a fact that adds some complexity to everyone's job situation. First of all, this means that the person you report to and who gives you your job evaluations, aka "the boss," is probably not the person in charge of the project team, which is headed by the *project team leader* or *co-leaders*. The resulting system is known as *matrix management* and Figure 4.1 illustrates this principle.

In this fictitious scenario, seven individuals at the scientist job level, namely Jill, Morgan, Susan, Kristof, Harold, Andrea, and Lisa report to Bob, their supervisor. In this management role, Bob is the one responsible for drawing up their goals, checking on their progress, helping them when they run into chemistry problems, ensuring their effective management of people who may be reporting to them, discussing, writing, and eventually presenting their annual assessments and working with them as a mentor for their career development, among other things. At Bob's group meetings, all of them may be asked to report on their progress and "brainstorm" when they run into problems. You can see that Bob has a lot to do and that good supervision on his part is important for things to work smoothly.

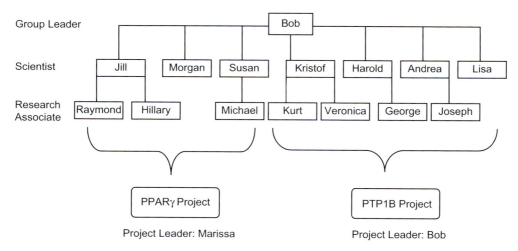

**Figure 4.1** ▶   Matrix management.

You might notice that some of the scientists in Figure 4.1 have research associates reporting to them and some do not. There's no universal rule on this. It varies with corporate policy, the manpower situation, and even the individuals involved. Research organizations have a diverse collection of reporting structures, and smaller companies tend to have smaller, flatter, less hierarchical ones so that, for example, the research associates might report to Bob directly. But the situation shown in Figure 4.1 is a typical one for a medium-to-large size biotech company these days.

For the research associates, who are really the backbone of any given project and a fertile source of routes and ideas themselves, their functional, "line managers" bear the same responsibilities to them that *their* supervisor, Bob, in turn has toward them. These, of course, take the form of dealing with day-to-day problems, checking progress, doing annual evaluations, etc. In all of these situations, keep in mind that neither degree levels nor project knowledge necessarily correlates with supervisory ability, an important factor in reporting relationships, and the last time this author checked no management or "people skills" courses were required for a chemistry degree at any level. Talent in dealing with people and experience count for a lot here and the best companies take this into account in deciding on reporting relationships.

Everything discussed about this fictitious group so far deals with the *reporting* or *functional* or *line management* relationships involved, which are reflected in the job titles "group leader," "scientist," and "research associate." Now the other dimension of the matrix needs to be considered: the *project* relationships.

In this case, as you can see from Figure 4.1 that people in Bob's research group are split between two projects, with six of them working on a PPARγ agonist project and the remaining nine, including Bob himself, working on a PTP1B project. There are a couple of implications to this to consider here. First, almost half of Bob's group works on a project he has no direct involvement in, meaning that a fair and accurate assessment of their contributions to the project will require the input of the PPARγ project team leader. And although he isn't on the PPARγ project team himself, three of the people he supervises and three of the people they supervise are. Clearly, he'll need a working knowledge of the project plan in order to help his reports along, this kind of multitasking being one hallmark of group leadership.

The splitting of a given group, particularly between vastly different projects (receptor binders versus inhibitors or between two different therapeutic areas) involves a lot of mental gymnastics and shifting of gears. It's less difficult and more common for a given group, if split, to work on similar targets (say, different kinases) or at least the same therapeutic area (say, diabetes as in this example). Group splitting is best avoided if possible, but sometimes it is just not possible. It happens when groups are big so that all group members can't fit in a single project or in cases where there are too many projects and too few scientists.

It's almost always the case with non-medchem groups like ADME, X-ray crystallography/SBDD, or other resources where the entire department or group can't work exclusively on just one project. For people in this position, matching their available time and facilities with probably much greater demands from many projects can become a problem unless things are strictly regulated by scheduling and priorities that everyone can see. It isn't easy to break the news to a project leader that an important PK study of their lead compound can't be done for a few more weeks because the company has given top priority to another project. But, if that's the case at least having a transparent prioritization process is better than either giving no good explanation or, worse, promising something that can't be delivered. The bottom line is that good "project splitting" in resource groups demands good organization.

A second implication of the arrangement shown in Figure 4.1—and this is where the matrix comes in—is that even though Jill, Raymond, Hillary, Morgan, Susan, and Michael are part of Bob's group, when it comes to their roles in the PPARγ project they're accountable to the project chemistry leader, Marissa. This is what matrix management is all about. The project leader can be from a totally different group and is almost as likely to be a pharmacologist or cell biologist as a chemist. Also, as we'll see, the project leader doesn't necessarily have to be at a higher job level than everyone else on the project, nor does she necessarily bear the job title "project leader." But the key to understanding matrix management is to grasp the fact that line supervisors and project leaders have different, discreet responsibilities and to have a clear understanding of what these differences are. Some typical guidelines for this are summarized in Table 4.1.

### 4.2.2.1.1 Day-to-Day Supervision

No project team leader in his right mind would attempt direct supervision of project personnel who don't report to him. That's why there are line supervisors. Any attempt to do so would give rise to what's been called the "two boss" syndrome: "He told me to do this but you're telling me to do that. Do I have *two* bosses now?" This classic, dreaded line introduces a demoralizing potential pitfall of matrix management but one that's easily avoided given reasonable people and a clear demarcation of responsibilities. It can happen at times as a result of short project timelines and pressure from the top, which might make a project leader not used to operating in this environment tend to demand frequent status reports direct from the source. This can end up looking like dual supervision. This kind of

**TABLE 4.1 ►   Some Issues and Responsibilities in the Matrix Management System**

| Issue | Responsible party |
|---|---|
| Routine supervision, day-to-day problems | Line supervisor |
| Target compounds | Sources vary (see text), but project team leader has final approval |
| Written and verbal progress reports | Typically technical reports (e.g. synthetic chemistry) due to line supervisor and group leader, project progress reports (e.g. SAR) to project team/leader, but varies |
| Performance evaluation | Line supervisor with input from project team leader |

over-the-line behavior calls for a little understanding on the part of project team members and line supervisors combined with a short, friendly talk between all parties involved. Making sure that everyone understands how the system is supposed to work is the best way to nip this problem in the bud. And fortunately, it's a surprisingly infrequent side effect of matrix management, especially when experienced supervisors and project leaders are involved.

One thing worth noting is that the two boss problem obviously doesn't apply when the project leader and the line supervisor are both the same person, as in the case for Bob's employees on the PTP1B program. Here, the two-dimensional matrix has been reduced to one dimension for the people involved and there's no question of who's responsible for any kind of supervisory activity: it's Bob.

#### 4.2.2.1.2   Target Compounds

The second topic in Table 4.1, that of who designs the target compounds, is an interesting one. In our example group, neither Marissa nor Bob will dictate which target compounds team members make. Target compounds could have originated with them or they may have originated with the individual researchers themselves, with someone in senior management who just wanted particular compounds made (not an unusual occurrence), or with anyone at the company who participated in a "brainstorming session" and had a good idea. But regardless of their sources, they will have been agreed upon and prioritized by the PPAR$\gamma$ project team and particularly its project leader, Marissa, who's ultimately responsible for the direction of that project's chemistry. If she's not a chemist herself and there's no chemistry co-leader, she'll be acting on the advice of others who are. As project leader, she in turn is accountable to senior management, so they've probably had to buy into the target compounds as well. In general, though, how much oversight people this high up in the corporate hierarchy have at this level of detail is a function of the company and the people involved. As they say, "Your results may vary."

Of course, another factor worth considering is having the buy-in of the person making the compound at the bench. It doesn't require an MBA degree to realize that when a chemist is making a compound that originated in his own head, or at least when he's making a compound he believes will be good, he'll work harder and will be more likely to find his way around synthetic obstacles than if the idea originated elsewhere. In the real world, of course, it isn't always possible to have chemists working on what they like versus what the project needs at a given point. For example, it's frequently necessary for a number of chemists to participate in the scale-up of a synthetic intermediate that few of them will actually use, which a lot of non-process chemists regard as necessary but not a particularly rewarding task. So it's all the more important to let researchers work on their own ideas when it's possible to do so.

Partly because working on one's own ideas is such a strong motivating factor, some companies will even allow the researchers to dedicate 10–20% of their time to working on interesting medchem ideas they may have that aren't necessarily even part of any existing corporate project. It's even been suggested that this can function as a reward for reaching a project goal, as a kind of sabbatical.[3] The companies that currently allow this kind of "free research" typically require chemists to inform them of what they're trying to make and why, and then get corporate approval (or at least avoid corporate condemnation) so that they know what to expect. Come to think of it, asking whether or not a company has such a policy when interviewing is probably worthwhile, but be prepared to answer the question of what it is that you're so eager to make! A candidate who has the motivation and the ability to do this is someone a lot of employers would favor, and not someone they run across every day.

### 4.2.2.1.3   Progress Reporting

The subject of whom a researcher reports what to when is, as this very twisted grammar would indicate, a tricky one. Organizations tend to have different reporting mechanisms (group meetings, monthly reports, project reports, etc.) at different intervals, so there isn't much one can say about it all that's universally true, and a lot of variation will occur. Since the researcher's daily supervision comes from a line supervisor, while project matters fall under the purview of the project leader, splitting up the reporting this way is a good place to begin.

Using a little common sense we can see that for a medicinal chemist, reaction schemes, yields, problem steps, and as such belong to the supervisor and will probably be reported at group meetings, while giving the same sort of presentation at project team meetings would be far too detailed for the non-chemistry crowd. There the interest will be limited to the effect of structural variations on activity, cell permeability, etc. So the situation might seem clear-cut, but it really isn't. The people at the chemistry meetings are going to want to see SAR as well: after all that's what they do. And getting stuck on a key synthetic step for a couple of weeks might leave the researcher with no SAR to talk about at the project meeting. So the reality is that some duplication in reporting is going to be inevitable. At best it can be kept to a minimum. But as an employee and as a project team member, make sure you understand the system and the rules that apply to your particular situation. When in doubt, ask.

### 4.2.2.1.4   Performance Evaluations

It isn't the intention here to go over the mechanics or details of performance evaluations—how often they're done, what kinds of systems are used, etc. Much of this comes up in Chapter 11. Here, we'll concentrate only on how it fits in with the matrix management system.

As alluded to above, the person responsible for doing performance evaluations is really "the boss," the line supervisor, who knows you and your work most closely from interactions in the lab, probably on a daily basis. There's no one more appropriate. But even he can't know everything about your performance. For one thing, your line supervisor may be involved in a different project (as in the case of Jill and Bob in Figure 4.1). He may not attend all of the project team meetings to see what ideas and suggestions you've been contributing there or to sense how much you are contributing to the overall effort. If he's not there, he won't know how your presentations went and how well you answered questions that involved knowledge outside your area of core expertise, a surrogate marker for career development in this business. And even if he did attend all the meetings his observations might well be different from those of the person assigned to oversee the project, the project team leader.

One could argue that, with the exception of spacewalking astronauts, people don't work in a vacuum and it's going to be difficult for a particularly stellar or really abysmal performance by an employee in a given project not to be noticed by his or her line supervisor, who is aware of the chemistry going on each day. But it would be foolish to rely on this kind of information osmosis in rating how well a person has done on the job. The input of the person who eats, drinks, and sleeps the project, the project team leader, is generally an important component of an evaluation. This input could be formal (like a written form to be filled out) or informal (like a short phone conversation) and its place in the evaluations could be similarly formalized (weighted along with the line supervisor's input) or simply "taken into consideration" at the discretion of the line manager. It's a necessary part of an objective evaluation system.

There's another reason for this as well. Putting the line supervisor entirely in charge of evaluations and the subsequent raises, bonuses, and possible promotions would mean locking the project leader out of them. And if that's the case, especially if she lacks high rank within the functional hierarchy, then why should any team member do what she asks for? In a perfect world, of course, everything is done "for the common good" and this kind of financial motivation wouldn't be required. But just take a glance at the title of this book again! Appeals to altruism and inspirational speeches can only go so far. In the real world, monetary motivation is required and that's what the project leader's input into employee evaluations helps provide. So as a member of a project team keep in mind that although your evaluation is probably given to you and signed by your line supervisor, the project team leader's observations on your performance as a team member will likely be an important part of what you end up with.

### 4.2.2.2   Project Team Roles

#### 4.2.2.2.1   Project Team Leader

A closer look at the different roles that exist on a project team is outlined in Table 4.2. The team will be led by one or two *project leaders* (sometimes confusingly called *project managers* instead, although the latter refer to a different more administrative role in this book). Because, even early stage drug research involves such a broad spectrum of issues and so many types of arcane knowledge, expecting one single scientist, regardless of discipline, to have an understanding of all these sufficient to run the whole project smoothly is asking a lot. For this reason in many cases two project co-leaders, one from the chemistry side, and one from the biology side will be chosen. They then need to work together to present a unified plan for reaching the project goal.

The position of project leader is always an important one within the team, but within the company the actual amount of power it confers can range from a little to a lot. First of all, the title of "project leader" doesn't usually imply an actual job level in the matrix management system. It's often not a job title the way "research associate" or "group leader" is. So the

**TABLE 4.2 ▶ Titles and roles in the project team**

| Title | Synonym | Responsibilities |
|---|---|---|
| Team leader | Team co-leader | Ensuring the team meets project goals and milestones, coordinating team activities and planning, reporting progress to senior management, ensuring effective communications throughout the team |
| Team member | | Carrying out assigned activities, presenting data, contributing ideas, effectively interacting with other team members |
| Team representative | Council member | In large teams, represents all project members from a given department. Facilitates transfer of information back and forth between them and the the rest of the project team |
| Project manager | Team administrator | Optional administration delegate to project teams, usually sent to ensure adherence to timelines and critical path activities in larger companies |

project leader doesn't necessarily get a higher salary or have more reports than any other project team member the way that, say, a group leader normally would. There may or may not be some kind of extra compensation or bonus scheme associated with it, and at some, generally large companies there may really be a discrete job position titled "project leader" that's relatively high up in the hierarchy. But most of the smaller companies don't do it that way, so the bottom line is that extra compensation and perks don't necessarily accompany the role.

Of course, even if it doesn't involve extra cash it's nice to have the words "project leader" in your personal portfolio as a sign of ability and experience, and for the upwardly mobile researcher the position comes with lots of exposure to higher management. The fact that, in our example scheme, Marissa is a project leader means that management is confident that she has the knowledge, experience, and drive it takes to get the job done. She might have been the *project champion* who took the initiative to propose the project in the first place and has been pushing for it ever since. But it doesn't mean that she has a higher salary or a lot of people reporting to her. In theory, at least, as the two dimensions of matrix management are totally independent of each other, she could even have the lowest job level of anyone on the project.

Just as compensation and authority given to the project leader vary with corporate *mores*, who exactly the project team leader is accountable to varies as well. It can range from a project manager or VP all the way to the corporate board of directors in a distant state or country. At regular intervals the project team leader may need to provide updates on the progress of the team to senior management to ensure that the team is accomplishing its *critical path* activities, the necessary sequence of activities key to reaching the project goal on time. And you can be sure that a few weeks before each of those meetings there will be a push to get lots of results, and good ones, to populate the PowerPoint® presentation! A statistical analysis would probably show that at least half of all project results are obtained in a two- to three-week window before these meetings. Said meetings will typically tie in with milestones that senior management will use to evaluate the team's progress, and if this is the case a good project leader will ensure that the milestones and dates are more easily recalled by team members than their own birthdays.

Project review meetings with senior management will probably not be complete unless the project headcount (full time employees, or FTEs), both current and projected, is discussed and approved, although project managers, not leaders, generally deal with this when they're used. It turns out that there are two kinds of FTEs, the projected ones relating to how many people are *supposed to be* working on a project and the real ones based upon timesheets everyone turns in specifying how much time they've actually worked on a given project in the last week or month. The latter frequently exceeds the former if people aren't scrupulously accurate about keeping track of hours or if things like increased project demand on resource groups like ADME and SBDD is not taken into account. Keep in mind too that in determining planned headcount the project leader or manager needs to work with the heads of each of the departments involved.

A word about budgeting: the assignment of a budget, which dictates available resources in terms of new equipment, contract work, and the number of FTEs, can either be done by project, where the team leader or manager works with a budgeting authority, or it can be done by department. Either way the project leadership will be part of the process but might not ultimately have a lot of control over it. Of course, as the project's situation changes and new problems or new opportunities loom on the horizon, resource needs will change too. Regardless of how and when the budgeting process is done, it's their responsibility to make the changing needs known to management and push for additional resources when necessary. Communicating the need for more resources to meet changing situations is part of the job.

### 4.2.2.2.2   Project Team Member

Most people working on a given project will be classified as *project members*. As stated before, this will include scientists and a few non-scientists with expertise in all of the fields required to make the project go. Each one of these brings unique talents to the table that the team needs in order to meet its goal. Only through their joint, interconnected efforts can this happen. Acting separately and in isolation from one another it won't.

The task of seeing to it that everyone's efforts are interconnected and coordinated belongs to—you guessed it!—the project leader. But as every project leader will tell you, how easy it is to get people to work together depends entirely on the individuals involved. Scientists, who represent the majority of project team members, tend to be independent, critical thinkers. They probably came out of a university system that emphasized individual efforts, not collaborative work. So at one extreme, attempting to get researchers to act as a team has been likened to "herding cats" or "pushing limp spaghetti."[4] Although there's something to be said for the independence of cats, which can't be herded over a cliff like lemmings[5] and limp spaghetti that never breaks, the essence of a project team is cooperation. The more of this there is, the more likely it is that the team and all its members will succeed.

---

**Box 4.1   Some Desirable Qualities for Project Team Members**

▶ Motivation

    Gets the job done

    Volunteers for necessary but mundane duties

    Suggests ideas

▶ Cooperation

    Actively participates at meetings

    Gives clear, concise presentations when asked to

    Available for discussions outside of meetings

▶ Openness

    Admits it when problems crop up or help is needed

    Shares credit and does not assign blame

    Stays objective

---

We've already gone over many aspects of what a project member does and the divvying up of responsibilities between project and line reporting. For the reader who may want to know what's important to a project member's performance, some suggestions follow in Box 4.1.

### 4.2.2.2.3   Project Team Representative

Depending on the size of a project team, a third classification may exist: that of *project team representative*. As discussed further below, when project teams get big enough, for various reasons it's just difficult to have everyone participate in all aspects of project team meetings. For instance, if there are 15 medicinal chemists on a project, which sometimes happens, and only one or two of them can present at each meeting it might be a matter of months before they all get through the rotation. Even then, listening to one or two individual presentations at any given meeting wouldn't give team members a very good picture of global progress on the project. To circumvent this and other problems, the role of *project team representative* was created.

The project team representative serves as the liaison between the team members in his department and the overall project team. Instead of 15 chemists slowly presenting their individual progress, one of them, the team representative, will report their overall progress at each meeting. When this kind of system is used it's generally used for all departments involved even if they don't have large contingents. So there will be one representative from chemistry, one from cell biology, one from enzymology, etc. The job will be the same for all of them: to communicate departmental results to the project team and to ensure effective communication of what's going on in the project team to all departmental team members.

The assignment of team representative for a given department is usually made by the department managers with the input and approval of the project team leader. It isn't an easy job and we'll take a closer look at what it involves in the team meeting section below.

#### 4.2.2.2.4  Project Team Manager

Finally, some companies (generally big pharma) provide high-level administrative help to project teams. This takes the form of a *project administrator* or *project manager* (not to be confused with the project team leader), who attends project meetings to make sure things stay on track business-wise. While team leaders tend to be scientists, project team managers often come from business and administrative backgrounds. Although having someone in direct contact with administration present, who will be keen to know about the status of various things, may not sound like much of a favor from management ("the suits"), these people may actually be quite valuable when it comes to planning and coordinating studies, justifying additional resources the project may need, organizing team meetings, etc. So keep an open mind. Again, you can refer to Table 4.2 that summarizes the various project titles, synonyms, and what the responsibilities are at each level.

### 4.2.3  Project Team Meetings

Regular project team meetings are the heart and soul of a project. There information is exchanged, decisions are reached, plans are made, and duties are assigned. In order for all of this to happen, and happen smoothly, the cooperation of every project team member and a lot of effort on the part of the project leader are required. And, as usual, the devil is in the details—logistics!

There's a world of difference between a relatively small, informal project team meeting at a start-up company and a large, highly structured one on a late stage project at big pharma. As with the projects themselves, there's no such thing as "one size fits all." A lot depends on the particular people and the organization involved. But the biggest difference in what meetings will be like breaks down to the number of people involved in the project. Big project teams have to deal with some problems that smaller project teams don't.

To understand this, we need to look more closely at what happens when project teams are particularly large. All-inclusive team meetings in such projects might mean having more people there who really don't want to be there, especially if they aren't presenting anything. These scientists may be entirely dedicated to doing their part to reach the project goal but see that part as working in the lab, not pouring over someone else's histology slides or staring at an endless succession of Gantt charts. They may see their role more as that of a specialist rather than as a generalist and might even resent being forced to go to too many meetings and participate in discussions and decision-making outside of the area they feel most comfortable in. These "hardcore" scientists have made the decision to "go narrow" rather than "go broad" and that decision needs to be respected. A small but significant fraction of all chemists probably fall into this category.

A second factor that looms large in project team meetings, or for that matter, in any meeting with more than 20 or 30 people, is that a curious thing happens. Not only the logistics but also the dynamics of regular meetings become problematic. Think of it this way: if you have one hour to spend on a meeting for a team with 30 members—and believe me, just trying to get 30 people to meet in the same room at the same time is no small feat in itself— there just won't be much time for individual interaction, "thinking out loud," etc. Quite apart from this time constraint, meetings in large groups have well-known tendency to get side-tracked on less essential subjects. Many people—often the ones who need to be heard from— become reluctant to speak up in large groups, while in others the audience size inspires new and time-consuming oratorical heights. Differences of opinion seem to become more pro-nounced and consensus becomes much harder to reach.

The dynamics of a large project meeting are just very different from those of a smaller group, and under some circumstances divisiveness, indifference, or worse can take over. One theory is that there's some sort of natural tendency for people to associate in units of a few families or "tribes" of no more than a couple of dozen people at work here, and that in larger groups an inexorable kind of social mitosis takes over to split off a second tribe. This kind of division, however, would spell ruination for the project team, where unity is strength.

In reality, there are at least two possible solutions to this "critical mass" problem that don't involve increasing the project team hierarchy, but both depend on exceptional situa-tions. With an extraordinarily broad-minded, open, and cooperative project team large yet functional meetings are sometimes possible, but it's not something one can always count on. Alternately, an autocratic project team leader can run effective large meetings, but if he's that much of an autocrat, why have a project "team" at all?

In order to avoid these and other problems, many companies prefer "closed," "core team," or "project council" meetings to limit the number of individuals involved in the decision-making part of project team meetings. Under this system, only the project leader(s) and one project representative, sometimes alternatively called a "council member," from each department will attend regular decision-making meetings. The representative, as we've seen, is charged with keeping the other, non-attending project members of her department current about what's going on, as well as representing the research results of that group to the core team. The representative will generally organize regular meetings of all the project team members in her department where, for example, medicinal chemistry progress can be reviewed in detail. In addition, special, less frequent, more informational meetings meant to fill all the non-council members in on what's going on in the project more broadly will probably be scheduled as well.

Dr Mark Murcko of Vertex Pharmaceuticals explains how the system works there. "A project team may be 25 people or more. To function properly, different kinds of meetings will be required. Different subsets of the project team will meet as needed—for example, a weekly chemistry meeting where everybody on the chemistry team and maybe a few other folks (maybe the modelers, crystallographers, a cell biologist) will meet in a very nuts-and-bolts, detail-oriented meeting. So we'll have a meeting like that, and the biologists may have a similar sort of meeting. For broader discussions and decision-making, representatives of the different functional groups form the project council. We try to keep this core team to eight or 10 people. If you bring 25 people together all at once and you're hoping to make a decision that's probably not a good format.

"But, the various sub-team meetings like the chemistry and biology meetings I just mentioned are also important, so that if there were important topics that were discussed at the project council, those messages are relayed efficiently to all team members. If something happens at the core team meeting that's relevant, the minutes from that meeting go out to everybody on the team. Then whoever was in the room knows it's their responsibility to make sure that everyone in their neck of the woods is informed and can ask questions. So there's a lot of attention given to how information gets communicated throughout an entire project team."

In summary, the role of project representative has evolved to allow for more efficient meetings in large projects and to enable decisions to be made in the more manageable confines of a smaller group. The project representative is in a sense the middleman between the team members in his department and the decision-making body and needs to do a good job of always keeping the lines of communication open.

There are some disadvantages that come with this system. Unless the project representative is particularly enthusiastic and committed to crediting the individual project members in his or her area, these may have considerably less visibility to project leadership. Also, yet another title and position have been added to the matrix management system while project members become one step removed from real-time discussions as well as decision-making. Although not ideal, such is the price that's paid at many companies to avoid rambling, unproductive, and indecisive large project team meetings.

As the size of the project team is often directly proportional to the size of the company, researchers who prefer direct involvement and smaller teams may want to consider working at a smaller company. At the very least, a candidate interviewing in industry should ask about how the project team system works there so that he or she knows what to expect. This is a question not often asked at interviews, especially by those just starting in industry, and the fact that a candidate would even be aware of the project team concept will probably impress the interviewer.

### 4.2.3.1   Meeting Scheduling

Regular project team meetings will typically be held at the same time and place every two to four weeks. The project leader, co-leader, or manager often does the scheduling, sometimes with the help of a departmental administrative assistant. Although scheduling a meeting doesn't sound like too demanding of a task, with the busy schedules of most scientists these days, finding an acceptable time and place is often a challenge. Many a project leader has sent out a proposed meeting date and time only to receive equal sets of responses saying it's fine, it needs to be moved back an hour or two, or it needs to be moved forward by an hour or two. Software packages like Microsoft Outlook Calendar® do facilitate finding a time when most people can make it, assuming everyone involved actually uses it to schedule their appointments, vacations, etc.

At larger companies with more than one site involved in a project, getting together by web link or teleconference requires a consideration of the time difference and the additional encumbrance of getting the teleconferencing or video equipment to work at both ends as well, which can be a hassle. But, all in all, with a little diligence and by scheduling well in advance of the actual date a good, if not always perfect, time will be found. Making it a regular, recurring event on everyone's calendar will get team members into the flow of things.

### 4.2.3.2 The Meeting Agenda

As mentioned previously, corporate customs will dictate whether all project team members or just a subset of them called project team representatives will attend these meetings. As a general rule, before anyone shows up at the meeting, they should have a reasonable idea of what will be presented, discussed, and decided there. Hence the meeting agenda, which will normally be prepared and sent out by the project leader a day or two in advance of the meeting. A typical one might read as follows:

Feedback from management presentation (Ernie – 15 min)
Medchem progress (Susan – 10 min)
Analysis of new lead series patentability (Vincent – 10 min)
Enzymology – new antitarget assays (Joe – 10 min)
PK results for XYZ-1421 (Jocelyn – 10 min)
Discussion – revised criteria for POC compound (all)
Action item update (all)

In order to prepare the above agenda the project leader would have polled the various members or representatives to see who needed to present what and for how long. In the not-too-distant past overhead transparencies were preferred for presentations, but these days the ubiquitous PowerPoint® slide is the rule. The advantage over the older method is that everything can be put on a common server, allowing all project team members to have instant access to a wealth of information whenever they might need it. The disadvantage is that while bulb burnouts, the only thing that could really go wrong with overhead projectors, were infrequent, server and network difficulties are not as rare.

In any case, the responsibility for getting presenters to post their slides to the server and possibly ensuring that the audiovisual setup works lies once again with the project leader, who at this point has spent time trying to schedule the meeting, asked around to find out who's prepared to present, reminded members to post their slides, sent out an agenda, reserved the room, checked the projector and with the looming possibility of having to take meeting notes and keep track of action items himself is beginning to feel more like an overworked administrative assistant than some kind of junior executive. Some of these responsibilities actually will fall on the shoulders of a project manager at larger companies but if such a resource is not available they're the responsibility of the industrious project leader.

### 4.2.3.3 Meeting Notes

Meeting notes are a touchy subject. The best person to take meeting notes is actually someone dedicated to doing specifically this, which does happen sometimes but not always. In the absence of such a valuable person much experimentation with different systems can go on. Making it the sole domain of the project leader is unfair, considering all the other responsibilities that the person already has, and in a practical sense it's just difficult to do while you're trying to run a meeting. Rotating the responsibility alphabetically or otherwise to every person at the meeting is a great way to bring out the many and varied reasons why a particular person can't do this on any given day. It also highlights the difficulties inherent in a multidisciplinary team: a synthetic organic chemist, especially one new to the project team, may not understand, say, the concepts of experimental pharmacology involved well enough to be able to take good notes on it, and the same in reverse may apply to the pharmacologist taking chemistry notes.

A better system is to have each of the presenters provide his or her own written synopsis of less than 200 words or a single page of structures for chemical presenters, which can then be gathered up and pasted together with a few other lines from the project leader to make meeting notes. These can then be posted on a server accessible to all team members. Getting these synopses requires, once again, gentle reminders, persuasion, and occasional arm-twisting by the project leader. But in the absence of an attendee dedicated to taking notes, which, you can see by this point is a very valuable person to have, it probably is the best way to disseminate important information for the benefit of those who couldn't attend. It's also a good reference even for those who were there.

### 4.2.3.4   Action Items

Looking back to the first page of this chapter, you'll notice that it's entitled "How Things Get Done: The Project Team." Strangely enough, the primary, day-to-day driver for getting things done is actually not an aggressive project leader or tough-minded project review committee or even the very necessary self-motivated scientists on the team. Instead, it's an inconspicuous listing of often mundane tasks and necessary completion dates that's usually updated at or shortly after every project team meeting. This would be the action item list.

Just as project goals break down the drug development paradigm into smaller, more easily manageable pieces, in turn each of those pieces can be broken down into a distinct series of actions required of each of the departments involved. In our relatively simple example BCR-ABL goal above you can see that a number of departments including medicinal chemistry (which makes inhibitors), enzymology (which screens for in vitro kinase inhibition), protein expression (which produces the kinases for them to screen), molecular biology (involved in producing the engineered cells), cell biology (which probably runs the cellular assays), and the legal department (which decides how patentable the inhibitors might be) will need to be involved. Each will have a series of defined tasks that may be represented as action items. For medicinal chemistry, at least a rough quantitation of the number and types of compounds to be made as well as their projected completion dates would be nice. Of course, everyone knows that in synthetic organic chemistry, as in research in general, it's often impossible to predict in advance whether routes which look perfectly reasonable on paper will actually go to deliver target compounds, much less by when. The best we can do is to make an educated guess, but at least that's a place to start.

A typical medchem action item might be "Prepare several analogs of XYZ-13459 substituted with ester bioisosteres at the 5-position for improved plasma stability—RR." A cell biology action item might be "Assay 10 selected project compounds for antiproliferative effects in T47D and one other breast cancer cell line—KP." In all these cases, action items should be specific without being overly detailed and one needs the capacity to make reasonable guesses about quantities, times, and what can and can't be done.

Action items are normally reviewed and updated at project team meetings, often as the last meeting item. The usual suspects (project team leader or manager, designated team member, or departmental admin) will probably be entrusted with recording and distributing these, and the project team leader with overseeing them. A good way to keep track of the current status of all action items is by using a server-accessible spreadsheet with separate tabs for completed and ongoing action items. Other formats such as distributed paper copies can be used as well. Experience has shown that a widely distributed list showing just how long ago RR was supposed to have had those bioisosteres made is a powerful incentive for getting that action item moved over to the "completed" tab, a basic tenet for a successful career being "don't be the bottleneck." And besides such motivational factors, both the meeting notes and the action item lists should be available to everyone on the project as these are the keys to direction and performance for the whole team.

### 4.2.3.5 Project Planning Tools

The timing of tasks and activities needs to be coordinated across all the different departments involved in the project. Rushing to make compounds for which no assay has yet been brought up or, conversely, rushing to get an assay in place months before compounds are available for testing is, to say the least, inefficient. This sort of lack of coordination wastes both money and goodwill. Animal testing needs to be particularly well coordinated so that the necessary animals, personnel, written protocols, vivarium space, reagents, etc., are all available at the right time. Beware the wrath of the pharmacologist who's ordered special animals for a study and has then been forced to watch them grow old because the compound needed for testing isn't ready. To avoid these kinds of problems as much as possible projects use a planning tool known as the *Gantt chart* (Figure 4.2).

Henry Gantt (1861–1919) was an engineer who collaborated with legendary efficiency expert Frederick W. Taylor. He also was a proponent of using positive reinforcement to motivate workers, probably an unpopular idea at the time. Back in 1917, Gantt came up with the idea of breaking down any project into a series of separate tasks, and then representing each of those tasks as a horizontal bar on the same calendar. This kind of visual representation allows one to see what's supposed to be going on at any given date, when each particular task is supposed to start and end, and also makes plain the "dependency" that a given task may have on a previous one so that the cart can't be put before the horse.

So from Figure 4.2, you can see among other things that on 28 April the efficacy model for XYZ-3032 will be running while a critical PK study of a second compound will be going on, the PK department will need to fit the studies in late April and early May into their schedule, the pharmacology department will need to find the resources to have two efficacy studies

**Figure 4.2** ▶ A Gantt chart.

running in tandem during early May, and the chemists scaling up the second compound will need to make sure that it's done by 26 April for everything to go smoothly. In Gantt charts, a thin line or arrow connecting the horizontal bars indicates dependency of the later task on the earlier one. The diamonds represent "milestones," significant events such as project review meetings or a declaration that proof of concept has been obtained, which would of course be celebrated with a new project goal and new Gantt charts.

Although for many scientists preparing, trying to stick to, and eventually "nudging" the horizontal bars of Gantt charts to the right are about as enjoyable as working on their tax returns, it's really hard to imagine how all of the necessary interdepartmental planning and coordination could be done without them. Dozens of frantic phone calls, e-mails, and scribbled-up scraps of paper would not be enough to replace a single Gantt chart. At the risk of sounding like an advertisement for a certain Seattle-based company, Gantt charts are usually prepared using Microsoft Project® although other programs, including some open source ones, can be used. The world of Gantt charts in project management is vast and subtle indeed, and additional abilities and nuances exist to which only MBAs are privy and which will not be covered here. One thing you might see occasionally, in place of solid bars, are hollow rectangles filled in to represent percent of the task that's been completed, like halfway for 50%, etc. Also, the level of detail can vary from one chart to another, with a requirement for multipage Gantt chart when given tasks are broken down into particularly detailed components. But whatever kind of Gantt chart is used, once you've been through a few projects and seen several of them they become a familiar and easily understood project planning tool.

Another type of project planning graphic that all project team members will eventually encounter is the flow chart. Figure 4.3 illustrates one of these, a graphic (although G-rated) description of how an important project decision is to be made. In this case, the question is whether the experimental compound should be administered orally (p.o., which stands for per oral) or intraperitoneally (i.p., an injection into the peritoneal cavity) in an animal model. Obviously, the former would be the best way to dose *patients* with a drug, assuming they're not unconscious, uncooperative, vomiting, or unable to swallow.[6] In animal models, though, other routes are often acceptable, at least for proof of concept.

We won't go into the details here, but will only point out that the diamonds represent key experiments,[7] the outcomes of which will determine where to go from there. For example, an oral bioavailability exceeding 20% will lead to further evaluation as to whether or not to use an oral formulation in animal models, while less than that will mean that other experiments will need to be done to determine if i.p. dosing can be used instead. Although Figure 4.3 might seem a little complicated at first it's really not rocket science. Flowcharts similar to these are also often used to outline "funnels" for taking a large number of available compounds, running them through various primary assays (easy or cheap tests), and then deciding which are the best candidates to put into more resource intensive, expensive assays like efficacy tests, toxicity testing, or broad pharmacology antitarget screening. In such cases, the greater the number of decision points that are successfully passed, the closer the team is to having a drug, the more valuable the survivors become, and the greater the allowable expenses will be for their subsequent testing.

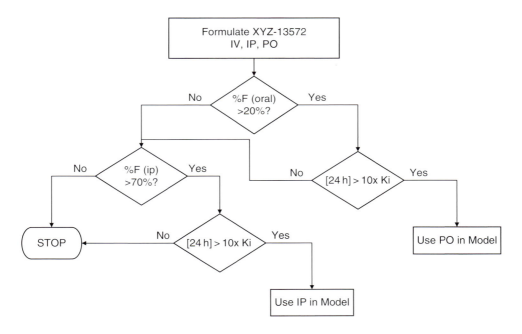

**Figure 4.3** ▶   A flow chart for in vivo drug administration.

## 4.3  Conclusions

### 4.3.1  Summing Up...

A drug discovery project team is a group of experts in many of the disciplines required to bring a new drug to market, brought together for the purpose of achieving a defined project goal, which represents a distinct segment of that lengthy process. They're usually organized according to the matrix management system, which imposes a second set of organizational relationships onto the normal reporting relationships found in the industry. Good communication and clear demarcation of duties and responsibilities are necessary to prevent the confusion and problems that might otherwise result. Project teams are headed by a project leader or, frequently, by two project co-leaders who report the team's progress on the road to milestones set out by executive management. They run regular team meetings, are ultimately responsible for project team decisions, oversee the direction of work done by project team members, and usually have input into their job evaluations. On most of the large project teams a project team representative from each department involved will be the intermediate between the department's team members and decision-making part of the team and will need to ensure the unobstructed flow of information between the two. At larger companies the position of project manager will probably exist as well.

At regular team meetings scheduled by the project leader and preceded by a meeting agenda, team members or team representatives will present progress, discuss options, and be assigned action items for the swift completion of critical path tasks. Planning aids such as

Gantt charts and flow chart will be used at that time. Information on meetings and project issues as well as a mechanism for all members to contribute their input and ideas to the project needs to be freely available.

## 4.3.2    Is it Really Best?

The system that's been described here represents the current industry paradigm for drug discovery that the industrial researcher is likely to interact with on a daily basis. But it also needs to be pointed out that it wasn't always so and that this model is not without its critics. No one doubts that bringing together all the various disciplines involved in an environment where communications are open and honest is a necessary part of drug discovery. But some see the modern management of such teams via such a structured, milestone-driven process replete with GO/NO GO decisions (key, pre-specified points at which a project can be discontinued if sufficient progress hasn't been made) as a product of business school thinking that ignores or at least minimizes some hard-earned lessons about what it takes to find a new drug.

Although it's now ubiquitous in industry, this model is of a relatively recent vintage. An earlier generation of researchers often felt that if there was a medical need sufficient to justify a project that project should be carried out until a drug was arrived at.[8] By today's industry standards this philosophy might seem positively medieval, but it may serve as a reminder that a sort of balance is needed. While no one wants to pour efforts and resources into a black hole from which nothing will emerge, the overall goal of research *isn't* to kill projects, but to select and pursue the most promising ones from a set. Every one of them will run into problems at times though a few will, at great effort and expense, defy the odds, overcome the obstacles, and lead to approved drugs.

Which brings us to a necessary ingredient for drug discovery research that hasn't changed over the years, the *project champion*. It's an admittedly unofficial position, but experience has shown that if it's vacant a project has little chance of succeeding. "Trying to invent new drugs is no picnic," according to Sir James Black, himself an outstanding example of a project champion.[9] Most likely not a single approved drug has ever resulted from a project that went as planned, met all deadlines, never had to overcome unexpected and damning problems, and never gave researchers reason for despair. Someone to lead the charge and not let these stand in the way seems to be a prerequisite for success. If this involves a bit of pigheadedness at times and opens one up to accusations of a lack of objectivity (something decried early on in this chapter), these seem to be necessary in small quantities, the spices that one needs to properly season the new drug recipe.

Traditionally a project champion would become involved early on, often being the one who proposed the project in the first place, the *drug hunter*, and would stick with it all the way to either approval or to the bitter end. Today, however, they may not end up as the project leaders and even if they do, their tenure may be short. For one thing, most employees no longer remain at a company for 10- or 15-year span that it takes for a new drug to be born. There's also a tendency to reorganize project teams around the preclinical development stage to reflect new needs, and as part of the "R" faction the project champion may be left out or given only a minor role. Reassignment to other projects happens frequently too. Whatever the cause, the lack of project champions or drug hunters—call them what you will, those committed to a project and dedicated to shepherding it on through to a drug—has been proposed as a possible factor behind the declining industrial productivity. "Rarely, if ever, does an individual with an original idea for a drug get to lead the project, or stay involved from the lab to the clinic. Indeed, targets and compounds are just as likely to be plucked from a database."[10]

Ultimately, drug discovery requires things that no database can supply: passion and commitment. Companies that bring this kind of project leadership to the fore are far more likely to succeed than those which view it as a calculated mechanical exercise.

### 4.3.3 The Benefits

What the project team experience will be like for the new industrial researcher depends a lot on the people involved. Being on a given project team can be quite an enjoyable experience or it can be a difficult one. And the difference between the two doesn't necessarily relate to how successful it's been in meeting the project goal. Instead, it's the people on the project team who make the difference. All other things being equal, the difference between working with an open, cooperative group of people and working with the "cooperatively challenged" can be profound. With a cooperative group and a good team leader, both project goals and project team members benefit. Besides enabling difficult, multidisciplinary goals to be achieved there are some added advantages for everyone involved.

Project teams, as semi-autonomous units, allow for a kind of flexibility and even egalitarianism which would be impossible within the much larger hierarchy of a big company. There may be a hundred or more medicinal chemists at your company, but on your project there are probably fewer than a dozen, making you one of "the few, the proud!" At a big company you may not normally have much involvement with departments like cell biology or pharmacology, but on the project team you'll likely get to know at least one or two people from those departments, and there you're equals working together for a common goal, interacting, and learning new science outside your core area of expertise that benefits both the company and yourself in the end.

Dr Matthew S. Bogyo of Stanford University, an expert in the field of proteomics, has had a firsthand view of the world of industrial research as a consultant and as a manager himself. He believes that this kind of interdisciplinary learning is one of the primary benefits of the project team, especially at big companies.

"In little biotechs you're probably exposed to biology as a chemist because the meetings are everywhere," he says. "So you probably sit down and listen to people presenting biological data in a small company. The problem is that you might not have that opportunity at a large company. This is where I think project teams are important in industry. I think it benefits you immensely to be able to link up with the biologists and understand where your compounds are going and what they're doing and what the assays are." His advice to project team chemists is, "You really have to open up your mind when being exposed to things that at first pass make no sense. Unless you do that you're not going to ever be able to appreciate the other side." And, increasingly, being able to appreciate the other side is the first step in becoming a more valuable contributor to the discovery of a new drug.

So if project teams, matrix management, and GO/NO GO decision points bring with them the potential for problems, compensation is provided by the opportunities for interdisciplinary learning that comes with them and by the unifying force of working toward common goal in a properly functioning project team. At the end of a project, even if a team member doesn't come away with a new drug, the ultimate goal, he can still take away something else: a kernel of knowledge, experience, and career growth that couldn't have been gotten any other way.

# Notes

1. Quoted in Anon, Frank L. Douglas *Nature Rev. Drug Discov.* **2006**, *5*, 368.

2. "Building Biotechnology But Not via Biotech Companies", Cynthia Robbins-Roth, *BioWorld Today* December 29, 2005.

3. Macdonald, S.; Smith, P.W. Lead optimization in 12 months? True confessions of a chemistry team. *Drug Discov. Today* **2001**, *6*, 947–953.

4. Robbins-Roth, Cynthia, ed. *Alternative Careers in Science.* Elsevier Academic Press, Amsterdam, 2006, p. 4.

5. Apparently, though, lemmings don't really follow each other over cliffs.

6. Other routes of administration, like i.v. injection, can, of course, sometimes be used in patients. Interestingly, a rare case where i.p. administration of drugs to human seems to be the best, if least convenient, route has recently been described for ovarian cancer. See Armstrong *et al.*, "Intraperitoneal Cisplatin and Paclitaxel in Ovarian Cancer". *N. Engl. J. Med.* 2005, *354*, 34–43.

7. In this fictional scenario, of course, both bioavailability and concentrations at 24 h could be determined in the same experiment.

8. See, for example Anon. A Call to Arms for Drug Discoverers. *Nature Rev. Drug Discov.* **2003**, *2*, 335.

9. Black, J. Future perspectives in pharmaceutical research. *Pharm. Policy Law* **1999**, *1*, 85–92.

10. Anon. Wanted: Drug Hunters. *In Vivo* **2003**, *17*, 45.

# Chapter 5

# Project Considerations

## 5.1 Introduction

There's really no such thing as a "typical" project. No two are alike. Each will have its own uncertainties, issues, and challenges that need to be addressed in a slightly different way. Taking a one-size-fits-all approach to projects would be an exercise in futility, while there aren't enough pages in this book to describe all the variations in project types that can and do occur. But projects do tend to have some commonalities that allow most of them to be divided up into several bins, if the bins are made large enough. Projects can be divvied up between these bins based on both the initial strategy (screening) and the kind of drug they're expected to result in (market strategy). Needless to say, these particular alphas and omegas will influence many of the letters in between.

As a scientist, project team member, or project proposer, a basic knowledge of these project types can help you fit the pieces of the puzzle together to answer questions like, "Why are we emphasizing these factors and not those?," "Why the short timelines?," "Why are we hung up at this stage?," "Why are we trying to make a drug when there are similar ones already on the market?," or even, "Why aren't we developing this perfectly good compound?" Knowledge is power, and being able to answer these questions is not only intellectually satisfying but also enables the researcher to do a better, more focused job as well.

Keeping this in mind we find that, much like Gaul, the landscape of drug discovery targets can be divided into three parts, as shown in Box 5.1. Each of these has its own strengths and weaknesses, advantages and disadvantages, and risks and benefits. Projects of these different types will have a different flavors and tend to run somewhat different courses. Figure 5.1 shows these schematically.

---

**Box 5.1   Categories of New Drug Discovery Projects**

▶ Established targets for which clinically effective drugs exist

▶ Novel targets which so far lack clinical validation

▶ Undetermined targets uncovered through phenotype screening

---

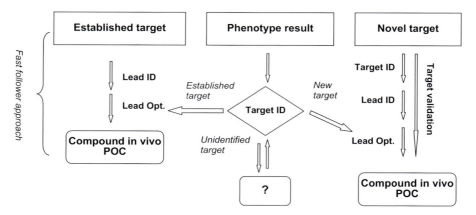

**Figure 5.1** ▶  Types of drug discovery projects and the major stages involved in their progression to POC in animals.

## 5.2  Established Targets

Doing drug discovery research on an established target implies a number of things of great importance to the resulting project, as described in Box 5.2. In this and the other boxes, what many would consider as the most important advantages and disadvantages have been italicized.

   *Target validation* (*TV*, discussed in more detail later), that is, systematically building up evidence that the interaction of a proposed target with an agent like a small molecule or a

---

**Box 5.2   Advantages and Disadvantages of Working on Established Targets**

**Advantages**

▶ Target ID and target (but not compound) validation can be skipped

▶ Known compounds and assays are available as tools

▶ *Higher than average probability of finding an efficacious agent*

**Disadvantages**

▶ Crowded market space

▶ Work needs to be fast: competition is far ahead

▶ Drug needs to display a clinical advantage to be viable

▶ Market may disappear due to newer targets or mechanistic toxicity

▶ Formularies or price regulation may limit profitability

▶ *Crowded IP space*

*monoclonal antibody* (*mAB*) can have a therapeutic effect in a human disease state, is often far from trivial as we'll see. It represents an ongoing, labor-intensive process that can be rate-limiting, but is necessary to avoid disastrous losses in time, money, and manpower that occur when a clinical drug fails to work because its target is not at all crucial to disease progression or symptoms. *Target identification* (*TID*), in this case the process whereby a new target is first proposed based on some minimal evidence, is less cumbersome but still requires devoted resources. The beauty of working on a project based upon an already established target is that both of these steps can be skipped.

Examples of established targets would be HMG-CoA reductase inhibitors for cholesterol lowering (e.g. atorvastatin, Figure 5.2), and PPARγ agonists for type II diabetes (e.g. rosiglitazone, Figure 5.3). A company going after either of these targets as of today would have the advantage of not having to do the expensive, messy work of first identifying and then validating it. This has already been done for them on somebody else's dime (or, more likely, hundreds of millions of dollars). Their new program could therefore begin with compound screening and as it progressed they wouldn't have to deal with the annoying question that people involved in financing such projects often ask, "How can we be sure this is a valid target?"

But the companies that originally worked out the preclinical and clinical TVs didn't do it out of humanitarian benevolence or to confer good assays and tools to future competitors. Instead, by definition they already have an approved drug or advanced clinical candidate well on its way to approval from which they hope to recoup the costs of this project as well as their less successful research and make a shareholder-pleasing profit to boot. And with the "blockbuster mentality" currently no less ensconced in big pharma management than it is in Hollywood, one can be sure that such a target is likely to draw whole flocks of companies to that particular watering hole. Trying to find enough room to squeeze in and enjoy may not be easy. After the first-in-class drug, a plethora of competing drugs will probably follow. And, as a recent study has shown, that profitable period of market exclusivity, when a single first-in-class approved drug exists for a given target, is becoming shorter and shorter, down to an average of 1.8 years in the late 1990s[1] and probably to only a matter of months today.

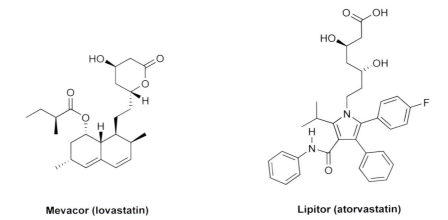

**Mevacor (lovastatin)**          **Lipitor (atorvastatin)**

**Figure 5.2 ▶**  Structures of two HMG-CoA reductase inhibitors. Lipitor is actually the calcium salt of the compound depicted.

That same study uses a statistical analysis of drug approvals by date to demonstrate the eye-opening fact that in recent years second- and third-in-class drugs were more likely to have been competitors vying for initial approval with the eventual first-in-class winner, than copycat compounds trying to shoehorn into its established market later. These competitors just didn't happen to win first place in the race, but they were very much in it, having started before the target was an "established" one. All of which begs the question of what exactly a "me-too" drug might be these days. Dr Albert Wertheimer of Temple University uses the following scenario: "Three different labs all start out within a month of each other and one is more successful with FDA or they had fewer questions or their CRO was more efficient or was luckier in getting more patients faster. So one of them got FDA approval first and we say, 'Ah, that's the originator!' But in France the one that came in second here could have been first there, and the American 'innovator' here is 'me-too' there. So what's a 'me-too'?"

Most likely, in the time it takes for a brand new drug project based on an established target to go from lead ID to approved drug, probably at least 8 years under the most favorable circumstances, any target for a major market which had Phase II clinical validation when it began will already be treated by several approved drugs. So starting that late raises a new annoying but incisive question, "Do you really want to bring the fourth- or fifth-in-class drug to market?"[2]

Now, most researchers whom I've ever met treasure innovation and are highly competitive. To them this prospect seems about as exciting as entering a race where the first three winners were announced before the gun was even fired. They won't be the only ones wondering how such a drug could get any significant market share when facing such competition. The research chemist, particularly if he ends up working with a close structural analog of the marketed drug, which is frequently the case in this kind of project, will inevitably think of it as a "*me-too*" project, while management will likely use a more elegant term like *innovative improvement*.[3] *Fast follower* is yet another tag sometimes used for this type of project, although, as mentioned above, "fast" here only applies on a geological time scale. There's a certain element of truth to all of these terms, but neither they nor any other two-word phrases can explain all that's really going on with such a project.

In pursuing a clinically established target, a company has made the corporate decision to minimize risk, at least the risk of target failure, which is responsible for a significant fraction of clinical drug failures. "Risk" here is relative. It's not quite the same as making the decision to buy a bank CD versus buying Internet stocks as an investment. In the drug discovery business there is no truly low risk strategy. Everyone involved is something of a gambler. But in this approach, everyone is confident that at least the target is a good one and it has been shown to lead to a salable therapeutic agent. Many useful, profitable drugs have come out of this approach. And this approach per se doesn't imply that chemists will be bored to tears generating compounds nearly identical to a known drug but falling into an IP loophole: *patent poaching* (another two-word epitaph). At least in theory, an established target can be approached with molecules resembling established drugs or it can be targeted with compounds looking nothing like any of them. The target doesn't *necessarily* define the chemistry, but, as we'll see, since using a known drug or pharmacophore as a starting point shortens the time to having an identifiable lead, it sometimes does for this type of project.

Having eliminated the TID and TV steps in a project based on an established target, keep in mind that *compound* validation is not dispensed with. The target may be a good one, but compounds can fail for plenty of other reasons (poor PK, toxicity, etc.). And although it might seem that, since the two stages of the drug discovery process have been amputated from the scheme, fewer resources should be required for this type of project, the need to work fast when starting from behind means that this isn't always so. All the *full time employees* (*FTE*s) and the budget saved in eliminating the first two steps and maybe even more could be required to get the resulting drug to market while there's still time. But, if the competition is so far ahead—assuming one isn't competing with oneself in a *second generation* project—that it won't even be remotely possible to beat them to the market, how can the company hope to sell the drug? Why should any physician prescribe it or any patient take it?

Because of some clinical advantage. It needs to have an advantage over existing drugs that will give people a reason to use it. This can take the form of greater efficacy (clinical studies prove that it just works better), better dosing regimen (say once daily as opposed to several times a day), or reduced toxicity (causing fewer or less severe side effects). What do these things mean to the medicinal chemist? They mean that while synthesizing and selecting the clinical candidate the emphasis will be on superiority to the existing drug in terms of some important and hopefully measurable property such as in vivo half-life in PK studies, where superiority could mean once (*q.d.*) versus thrice (*t.i.d.*) daily dosing, or selectivity against known antitargets implicated in observed toxicity for the existing drug. If no such advantage is found and the best that can be obtained is something equivalent to what's already out there or soon will be, the company may decide not to spend many millions of dollars required to try and bring it to the market. But the catch is that often advantages or disadvantages may not become evident until Phase II or Phase III clinical trials or even later, so who can really say short of that? A candidate with no obvious advantage might still be moved forward, but the project and the drug become a much harder sell, both internally and externally.

With a new drug that's only "just as good as" an old one, the best pharmaceutical detailers (sales reps) in the business will not be able to significantly "move the product" regardless of how many free samples they give away. But, if the existing drug causes some particular, significant toxicity in patients and yours doesn't, you stand a good chance of taking away a large amount of their market share. Or if the existing drug, like the oral anticoagulant warfarin, is difficult to dose correctly due to a narrow therapeutic window and lots of individual variability while your new vitamin K reductase inhibitor for some reason isn't, all other things being equal, your drug will likely take over that market.

A couple of myths that exist in the minds of many researchers need to be exploded here. The first is that "first-in-class" means "best-in-class" or at least "as-good-as-any-other-in-class." In some ways coming up with a new drug isn't terribly different from coming up with any other new product, say a new car. Does Henry Ford's Model T outperform the car in your driveway? Does Ford still sell a lot of them? The new drug business is an evolutionary process both in terms of targets for a given disease and drugs to treat any one of those targets, which means that "first-in-class" can eventually even become "worst-in-class." Prescription drugs, like anything else, are imperfect and there's plenty of room for improvement.

An example that can provide hope to those who labor to come up with a new drug for an established target is that of Lipitor (atorvastatin, Figure 5.2). Currently the #1 prescription drug sold anywhere, with over $13 billion in annual worldwide sales,[4] it was the *fifth* statin to be approved by the FDA, making it to market nine years after the first-in-class compound, Mevacor (lovastatin, Figure 5.2).[5] At least one major study has demonstrated improved effects on LDL-lowering for atorvastatin relative to other statins, and it has a low incidence of a nasty but fortunately rare side effect of compounds in this class, rhabdomyolysis. Other factors, including the marketing might of Pfizer are, no doubt, a big part of Lipitor's success story, but the point is that working on yet another drug for an established target is sometimes a legitimate formula for success, both from the point of view of the corporate executives and of the patients who take it.

It can even be exciting for the project's chemists. Consider the structures of lovastatin and atorvastatin, which are shown in Figure 5.2. Other than the lactone-containing side chain of the former, which upon hydrolysis generates the important pharmacophore contained in the side chain of the latter as well as all other drugs in this class (see Appendix, "The Periodic Table of Drugs"), there's really not much similarity. Lovastatin looks like, and is, a fungal metabolite, while atorvastatin doesn't look like a natural product because it isn't. Instead, it was the result of a determined synthetic effort by a group of chemists at Warner-Lambert who worked out an interesting, commercial scale enantioselective synthesis.[6] Here, indeed we have "innovative improvement" and one doubts that any of the chemists involved found this project trivial or boring.

Differences in the molecular structures of the drugs are key to the second myth that needs exploding. This is that finding a new drug for a given target that's structurally quite different from known drugs for this target provides some sort of advantage for the drug. As a synthetic chemist myself, I've never been fond of true "me-too" or "patent poaching" projects aimed at finding a "hole" in an existing patent via something like a one atom replacement. Only desperate corporate executives transfixed by Gantt charts could love such a project. Chemists would rather make novel—meaning highly dissimilar—structures. But dissimilar structures per se convey no advantage in the clinic. No physician is going to ask a patient, "Wouldn't you rather take an oxazole than a pyrimidine?" Once a drug has reached clinical testing, structure is largely irrelevant. Only two things really matter: safety and efficacy. That's where your compound needs to demonstrate its uniqueness. If the oxazole is less likely than the pyrimidine to be biotransformed into a toxic metabolite, allowing the question to become, "Wouldn't you rather take a safer compound than a more dangerous one?" then the clinical and marketing advantages are obvious. Structural dissimilarity probably will provide a basis for new chemistry, new papers, and patents, all of which really are benefits. But unless the resulting drug has greater efficacy and/or less toxicity than known drugs for the same target, it will fall flat.

Another potential disadvantage of using an established target strategy is that the market for drugs acting by a given target can suddenly disappear when another, better target comes along, as it did for older cholesterol lowering drugs with the introduction of the statins beginning with Mevacor. After all, how can one look at the market conditions as assessed a few years ago, and then project that the market for a given disease is going to be the same for another 10 or 15 years down the road? Few drug markets today are what they were in the early 1990s and crystal balls are hard to come by in this business. You might end up with a new and improved Edsel, a significant improvement that nobody will buy.

Dangerous, rare toxicity, the bane of the pharmaceutical industry these days, might also be encountered. Recall that we've mentioned several times that a real clinical advantage is required for this strategy to work. Just as a clinical advantage can't always be predicted in advance, neither can *idiosyncratic toxicity*. Should such toxicity be seen with a known drug

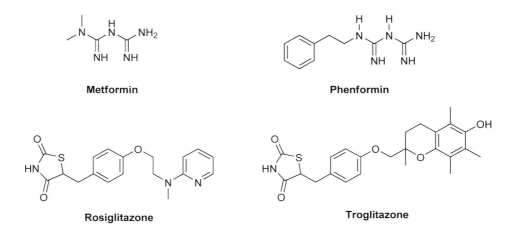

**Figure 5.3 ▶** Structures of four antidiabetic drugs. Glucophage (metformin) and Avandia (rosiglitazone) are currently marketed in the US, while phenformin and troglitazone are not due to safety issues.

acting on a given target, it isn't going to immediately be clear whether said toxicity was *compound toxicity* or *mechanistic toxicity*. The latter would doom all drugs acting on the target while the former might instead provide a better opportunity for competing drugs in the same class to grab market share.

An apparent example of compound-dependent toxicity can be seen in the metformin/phenformin story. Glucophage (metformin, Figure 5.3) is a biguanidine thought to act via multiple and not yet fully elucidated mechanisms to exert its blood glucose lowering effects. Another structurally similar drug, phenformin (Figure 5.3), is thought to utilize the same sort of mechanisms, but years of use by diabetic patients have shown an association with an unacceptably high-risk of life-threatening lactic acidosis, which fortunately metformin doesn't share. So in this case, although a theoretical basis for mechanistic toxicity might be postulated (buildup of lactic acid due to hepatic gluconeogenesis inhibition, one of the effects of both drugs), it appears that the toxicity was largely compound dependent. Phenformin was withdrawn from the US market while metformin still does brisk sales. Two other diabetes drugs, the PPARγ agonists Rezulin (troglitazone), which was withdrawn from the market due to liver toxicity, and Avandia (rosiglitazone) still an important drug today, although one recently highlighting another toxicity concern,[7] share a similar story with a different type of toxicity involved.

Because such situations can occur one might even wonder if fast follower or "me-too" approaches could count this type of possible scenario as a potential advantage. But that would be like setting up shop next door to your competitor in the fervent hope that lightning will strike and eliminate *his* business, only worse. It would be a ridiculous strategy.

To the risks that the market for a drug acting on an established target might disappear either because better targets have been found or unexpected mechanistic toxicity has been discovered, one can add the risk that in the future fewer drugs will compete in a given indication. This could be due to the increasingly restrictive formularies which are only looking to cover prescriptions for the cheapest drug in each class or it could be due to the increasing pressures on the US government to regulate drug prices. If that happens, and, as we've seen in Chapter 2, there are those who believe it will,[8] it would be very hard for pharma companies to justify high-initial pricing for a fourth-in-class or fifth-in-class drug regardless of its advantages. For these and other reasons many people believe that backing too many drugs based on established targets is just not a good idea and that the risk level of this approach isn't so low after all.

All of these objections are built on future scenarios that may or may not come to pass. But there's another difficulty with this approach that's seen every day in the new drug discovery business. It's just plain hard to find "IP space" for drugs acting on established targets. There are often already hundreds of existing composition-of-matter patents issued for compounds acting on such a target, each with a Markush structure implying more compounds than there are stars in the heavens. Gazing at page after page of descriptions of such structures brings the chemist closer to a true understanding of infinity, not to mention despair of ever finding something not covered. Add to that the fear that even if you do happen upon something not claimed in existing patents it doesn't mean that work isn't already underway somewhere to cover it in future ones. As we saw in Chapter 3, it takes 18 months from the date of filing for a patent application to be published. So whatever holes you find in their published ones may be in the process of being patched before you ever find them. Even leads which may have come out of screening and appeared to be proprietary may not be. All of this means that it might just be impossible to find a patentable compound of your own to develop when it comes to an established therapeutic target. And no patent generally means no project.

Studies of the industry consistently show that projects based on established targets are more likely to reach the clinical stage and are more likely to succeed in the clinic than projects based on non-established targets. In short, they're seen as being winners. So keep in mind that although disadvantages may outnumber advantages as shown in Box 5.2, the single advantage of "higher than average probability of finding an efficacious agent" is often thought to outweigh all the negatives. As a result, a large percentage of new drug projects will fall into this category and most research chemists can expect to get plenty of experience working on them.

## 5.3  Established "Tough Targets"

A second type of established target project exists where a number of these disadvantages are obviated. Dr Hans Maag, Vice President of Chemistry at Roche Palo Alto, refers to these as "tough targets where we don't know how to actually start chemistry, but where the biology is pretty clear cut."

"There are all these biologics—antibodies and so on—out there which are commercially successful. So if you want to make a small molecule, basically the target is already validated. There's no question. So it's not a matter of validation of the target. You just want to find some small molecule that interferes with that target either allosterically or otherwise," he explains.

An example of this kind of target is *tumor necrosis factor* (*TNFα*), an important cytokine shown to be a key player in many inflammatory disorders including rheumatoid arthritis, a painful, debilitating, and difficult-to-treat disease. The anti-TNFα antibodies Remicade (infliximab), Humira (adalimumab), and Enbrel (etanercept), a soluble form of the ligand binding region of the TNFα receptor, have proven beyond any doubt the validity of TNFα as a target. All of these, as biologics, require injection and carry the risk of immunogenicity, not to mention high-price tags.

A small molecule interfering with the binding of TNFα to its receptor wouldn't carry these liabilities and would likely become a billion dollar drug.[9] Although other small molecules are known to interact somewhere along the TNFα pathway, like Thalomid (thalidomide)[10] which somehow reduces TNFα levels, and thus have at least some sort of pathway validation, a molecule that could directly interfere with the binding of TNFα to its receptor, thereby mimicking the effects of Remicade and Humira, has proven hard to come by. Projects directed at this have encountered the usual difficulties of trying to mimic protein–protein interactions with small molecules, although recently a publication from Sunesis describes a small molecule inhibitor of this interaction which acts by essentially displacing one of the subunits of the normally trimeric ligand.[11,12]

For any such project directed at a small molecule interacting with a target that's already been validated in the marketplace by biologics, the drug's advantages are obvious, IP space is wide open (since probably no one else has such an agent either, although they may be looking for one), and the market for the resulting drug is almost guaranteed. The only stumbling block, which must not be underestimated, is finding such an agent—it's just not clear that it can be done. But the saving grace is that the failure to identify a good lead would at least happen early on in the project, once again as in a synthetic scheme where the trickiest step is one of the first ones and not the last step where you've invested much more time, energy, and hope. An oft-repeated aphorism in drug discovery is that if you're going to fail, fail early.

We've now looked at two "flavors" of projects based upon established targets. But what about the ones nearer and dearer to a chemist's heart and at the vanguard of biotech and pharmaceutical research, new targets?

## 5.4 Novel Targets

### 5.4.1 Identifying New Targets

Although the human genome contains slightly more than 20,000 genes which, due to post-translational modifications, alternative splicing, and protein complex association, might make for in excess of one million different proteins (not to mention other, non-protein targets) in the human "targetome,"[13] only a very small percentage of these are expected to provide good therapeutic targets. As we've seen in Chapter 2, genomics has till date proven to be far more capable of elucidating rare inherited disorders stemming from mutations in a single protein than in identifying new targets for common diseases, and the expected flood of new targets hasn't materialized. These days new TID has largely shifted to the science of proteomics.

The differential expression of a given protein in diseased versus normal cells or tissues is often the starting point for investigating whether or not it constitutes a good target for a drug. Potential targets can stem from other sources too, such as progress in understanding biochemical pathways and their rate-limiting steps, or new targets identified after phenotype screening, described later, but differential protein expression is probably the most common source of proposed new targets. Examples would include the overexpression of a given protein in a number of different tumor cell lines or its correlation with more aggressive or more highly metastatic tumors. Probably thousands of such proteins have been examined in industry and academia already, since such data raise the possibility that they might be crucial to tumorigenesis, tumor growth, or tumor spreading.

In such cases, a number of caveats need to be kept in mind. First of all, expression data obtained using microarrays[14] reflect only mRNA levels and don't necessarily correlate with protein concentrations which in turn don't necessarily correlate with protein activity levels. Concentrations of two proteins with the same relative mRNA levels might differ by more than 20-fold, showing that "transcript levels provide little predictive value with respect to the extent of protein expression."[15] And not all proteins are produced in active form. Some require further processing or association with other proteins or are modulated by, say, endogenous inhibitors whose levels might vary. So in interpreting such data, the scientist should keep in mind that "there's many a slip 'twixt the cup and the lip."

Second, the dynamic nature of protein expression means that it's always cell- and time-dependent. Expression levels in early stage tumor cells will probably vary considerably from those in later stage tumors, and both may have little in common with protein expression levels in, say, endothelial cells that line the tumor vasculature. A tumor cell from an aerobic environment (near blood vessels) will have a different proteomic profile from the one buried deep inside a tumor in an anaerobic environment. These kinds of issues can complicate the interpretation of protein expression in any type of tissue.

But most importantly of all, *correlation can't establish causation.* "Omics technologies are giving us correlative data while what we really need is causative data," says Dr David Szymkowski, Director of Biotherapeutics at Xencor of Monrovia, California.[16] He points out that the reasoning behind the assumption that a protein overexpressed in a given disease state is a valid therapeutic target is analogous to the following bit of (il)logic:

- ▶ *Firemen are found at burning houses*

- ▶ *Firemen are not found at normal houses*

- ▶ *Therefore, firemen cause house fire*

- ▶ *Therefore, eliminate firemen to prevent fires*[17]

Overexpression of protein in a disease state might actually be a defensive response and not something one wants to eliminate. A second possibility is especially worth considering:

- ▶ *More weddings occur in June than in any other month*

- ▶ *More suicides occur in June than in any other month*

- ▶ *Therefore, weddings cause suicides*

- ▶ *Therefore, eliminate weddings to prevent suicides*[17]

That is, the increase or decrease of protein expression in a given disease state can be entirely coincidental, the proteins in question being more like innocent bystanders than causative agents. The vast majority of the proteins differentially expressed in a given disease will fall into these two categories.[18]

Clearly, launching a new project based on such a target, however exciting, with the subsequent commitment of money and manpower, can only be justified by an equal commitment to find firmer evidence as soon as possible, and to drop the project quickly if no such evidence is forthcoming. Everything, then, depends on TV.

## 5.4.2 Target Validation

### 5.4.2.1 Levels of Validation

The term "target validation" can mean different things to different people. To a chemist proposing a new project it might refer to affirmative evidence from a mouse knockout or even just an *RNA interference* (*RNAi*) experiment in cells. To a clinician it probably means no less than proven statistically significant efficacy by a drug in large scale, controlled human trials. In the strictest sense, the clinician is right, but in the real world it's obviously impossible to put agents acting on every new proposed targets into the clinic to find out whether they're valid or not. Since, however, the need for new drugs acting on new targets is very real, a series of milestones which demonstrate a target's importance in cells and animals need to be met between the time that a target is identified and the time a drug acting on it can enter the clinic. The closer a compound gets to the clinic and the farther along a project advances, the higher the bar becomes. As mentioned earlier, this means that *TV doesn't consist of a single experiment but instead is an ongoing process*. It's best thought of as "target confidence building."[19] The closer a project gets to the clinic the more expensive it becomes and the more confidence that indeed the target is a valid one is required to satisfy scientists, managers, and investors.

A number of authorities have commented on the need for a staged TV and suggested some of the milestones that might be used to define each stage.[20] Table 5.1 shows one possible system that might be used, and is worth a good look.

"Level None" is included as a reminder that initially identified targets should in no way be assumed to be valid ones. Using our cancer analogy, Level I validation might consist of

**TABLE 5.1 ▶ Target Validation Levels – One Proposed System**

| Level | Milestone |
| --- | --- |
| **None** | Correlative evidence only |
| **I** | In vitro POC (e.g. works in a cellular disease model) |
| **II** | First in vivo POC (works in animal disease models) |
| **III** | Efficacy in one or more Phase II clinical studies |
| **IV** | Efficacy in Phase III trials sufficient for NDA approval |

evidence showing that blocking the function of Protein X induces apoptosis in various cancer cell lines but has less of an effect on normal cells. Tools for thus perturbing protein function (RNAi, knockout, antibodies, etc.) can vary, as discussed in the next section, and different results can sometimes be obtained with different tools. Early stage projects usually lack small molecules capable of selectively interfering with a single desired target, so at that point TV experiments often use other methods, but ultimately small molecules will provide the best validation when a small molecule drug is the goal.

Many proposed targets will drop out even at this lowest validation level. For those that don't, the push will be onto develop a tool (small molecule, antibody, etc.) capable of selectively perturbing the target in an animal model of disease. Achieving this first in vivo *proof-of-concept* (*POC*), thus reaching Level II validation in our scheme, is a major milestone and cause for celebration in any project, but it still doesn't prove that an efficacious drug is just around the corner. Level II validation in cancer might mean tumor growth inhibition, or, better still, regression in a *xenograft* (using human cancer cells in an immune compromised animal) or *syngeneic* (using a cancer native to the species) mouse model. For cancer especially, efficacy in a single model doesn't really predict human clinical success. But it's generally true that a compound that works in many different models has a better shot at clinical efficacy than the one that's only effective in a few.[21] The upshot is that for many indications Level II validation needs to be established in multiple models for it to mean much.

Once a program has produced a clinical candidate for a new target, the true test of its validation approaches. Failure to achieve a viable safety profile, pharmacokinetic properties, or withdrawal due to "business reasons" sometimes means that the new target ends up being neither validated nor invalidated. With proof of efficacy in a Phase II study, though, the first studies really designed to look for it, the first level of validation in man (Level III) has been achieved. Even here validation is not unequivocal. In some disappointing cases clinical efficacy can't be reproduced in a second Phase II study or a later, Phase III trial. Hence the highest level of target confidence, Level IV validation, is reserved for drugs and targets that have run the full gamut.

Successfully reaching each of these milestones to validate a new target is probably the biggest challenge to be met in all of drug discovery. Its difficulty is a major factor in the relative dearth of drugs aimed at new targets that get approved each year. So drug discovery projects directed at novel targets need all the "target confidence building" they can get along the way. Fortunately, a wide variety of validation tools are nowadays available, as discussed in the next section.

### 5.4.2.2   Target Validation Tools

#### 5.4.2.2.1   Knockouts and Knock-Ins

The ability to produce an animal lacking a particular gene product of interest allows researchers to look for disease-relevant phenotypes and has become a powerful tool for validating targets (Figure 5.4).[22] Because of their genetic similarities to humans, mice are generally used, although much information has also been gained from knockout experiments in lower organisms like *Caenorhabditis elegans* and drosophila. Three types of animals can be compared: wild type $(+/+)$, heterozygotes $(+/-)$, and homozygotes $(-/-)$ so that a crude sort of dose-dependence can sometimes be noted in phenotype.

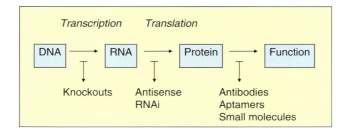

**Figure 5.4** ► Points of intervention utilized by different TV tools.

Knockout mice have proven to be useful in the validations of many drug targets, one good example of which is the cysteine protease Cathepsin K. This enzyme is mostly found in osteoclasts, a specialized type of cell involved in bone remodeling. Cathepsin K is a proteolytic enzyme whose purpose was thought to be the degradation of the collagen matrix that's left over after decalcification during the process of bone resorption. Osteoporosis, a disease characterized by low bone density and one that's particularly prevalent and problematic in aging populations, is believed to represent an imbalance between bone buildup and bone breakdown, so if Cathepsin K plays an important role in the latter, inhibiting its function might provide an effective way to treat the disease. When Cathepsin K −/− mice were generated, they were found to have abnormally dense bone, with severely impaired bone resorption.[23] This provided a measure of validation for this target in osteoporosis,[24] and further validations along with several clinical candidates aimed at it have since appeared.

A study by researchers at Lexicon Genomics (now Lexicon Pharmaceuticals) looked at the targets for the top 100 prescription drugs and found that for the targets where knockout models could be generated (most of them) in 85% of the cases the knockouts produced phenotypes that were "informative" for the therapeutic indication involved.[25] Not only Lexicon but also *National Institutes of Health* (*NIH*) has now embarked on projects aimed at knocking out most if not all of the genes found in mouse.[26]

Like any tool, though, knockouts aren't universally applicable and free of all problems. Although many knockout mice are now commercially available (but expensive), most require quite a bit of time and money to generate. In many cases (some say 20% of the time, others say more frequently) knockouts result in embryonic or neonatal lethality, the absent protein apparently being required for proper development whether or not they're needed later. When this happens, of course, information about the target protein's role in mature animals can't be obtained, although sometimes the lethality itself means something: finding embryonic lethality due to defective blood vessel formation in heterozygous *vascular endothelial growth factor* (*VEGF*) knockouts implied that nature didn't have any good ways of compensating for its loss during angiogenesis. This provided a key piece of information in developing anti-angiogenic anticancer compounds like Avastin, a VEGF antibody.[27]

In general, nature is adept at compensating for the loss of an incredible variety of proteins, and this presents a problem when phenotypes one expects to observe just aren't there because compensation by redundant mechanisms happened during the development. And, surprisingly, different phenotypes are sometimes observed for the same knockout when different strains of mice are used.[28] Furthermore, the phenotypes that are observed in knockout mice aren't always straightforward in predicting what will happen with small molecule inhibitors. For the important PPARγ receptor, heterozygous knockout mice demonstrated improved sensitivity to insulin when fed with a high-fat diet, which suggests the strategy of using PPARγ antagonists to treat diabetes.[29] However antidiabetic thiazolidinediones like rosiglitazone, which improve insulin sensitivity are PPARγ *agonists*. Needless to say, this took a bit of explaining.[30]

One big reason that knockout results may not correspond with results obtained using small molecules is that prospective target proteins may well be involved in more than the one particular activity that a small molecule might target. A small molecule might just interfere with its binding to a ligand at one site, but proteins are often multidomain and can form necessary complexes with other proteins via interactions not affected by the small molecule. This effect, illustrated graphically in Figure 5.5, means that small molecules are capable of tweaking proteins while knockouts, as the name implies, aren't exactly a light tap on the cheek.

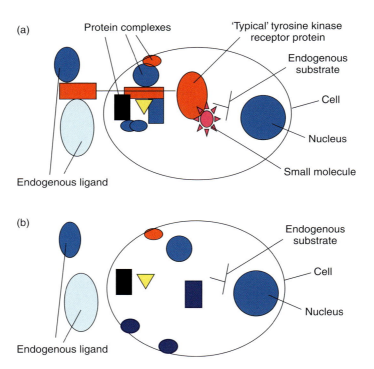

**Figure 5.5 ▶** A fundamental difference between pretranslational and posttranslational tools for TV, as illustrated for a receptor tyrosine kinase. (a) Posttranslational methods like inhibition by small molecules can affect target protein function without affecting its association with other proteins. (b) Pretranslational methods like knockout or RNAi knockdown result in the loss of target protein and all of its interactions with other proteins. (Reproduced with permission from Fitzgerald, K. RNAi versus small molecules: Different mechanisms and specificities can lead to different outcomes. *Curr. Opin. Drug Discov. Dev.* **2005**, 8, 557–566, copyright 2005, The Thomson Corporation.)

"Knockout of a gene results in complete loss of the target protein, which can disrupt protein complexes or impair functional domains that would be unaffected by a drug. These secondary effects can dominate the phenotype of a knockout, such that a knockout and a small molecule inhibitor targeting the same protein produce very different outcomes."[31] The authors of this quote recommend a less blunt-force transgenic method, replacing target proteins with mutants lacking catalytic activity (*knock-ins*) rather than deleting them altogether, as a more predictive alternative to knockouts. They provide an example (that of the kinase p110γ) where the two techniques produced opposite phenotypes. Along with other techniques like conditional knockouts, where the absence of a protein can be induced specifically in a desired tissue, such knock-ins provide additional useful tools for TV.

### 5.4.2.2.2  Antisense Oligonucleotides

In 1978, researchers at Harvard reported that a 13-base deoxynucleotide segment which was complementary to a section of Rous sarcoma virus mRNA could block virus production in cell culture.[32] Later work demonstrated that such *antisense oligonucleotides* (*AO*s), varying in length from 7-mers to about 30-mers, bind to their target mRNA sequences through Watson–Crick base pairing and either lead to cleavage of the mRNA by the enzyme RNAse H or prevent translation by other mechanisms.[33] The end result is the same, however: production of the target protein is shut down (see Figure 5.6).

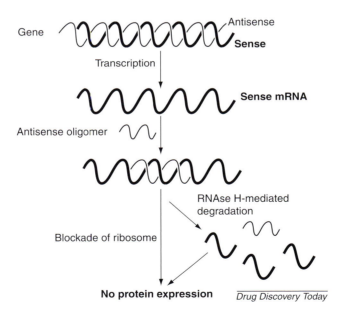

**Figure 5.6** ▶   How AOs block protein expression. An AO binds to the target sequence in the sense strand of the target mRNA. In some cases, RNAse H then cleaves the mRNA part of the resulting duplex, freeing up the AO for further cycles of binding and cleavage, while in other cases different mechanisms resulting in a blockade of the ribosome are induced. (Reprinted with permission from Taylor, M.F et al. Antisense oligonucleotides: A systematic high-throughput approach to target validation and gene function determination. *Drug Discov. Today* **1999**, *4*, 562–567, copyright 1999, Elsevier.)

"Naked" oligos like the 13-mer used for Rous sarcoma virus are rarely used nowadays. They normally aren't very stable because of the presence of nucleases, endogenous enzymes which can cleave them in minutes. This means that chemical modifications are normally required for them to stick around long enough to have a good effect. One of the earliest and the most consistently used modifications involves replacing a non-bridging oxygen atom in the phosphate linkage with a sulfur atom, resulting in phosphorothioate deoxyoligonucleotides. Others involve alkylation of the 2′ hydroxy groups of ribonucleotide oligomers with methyl or methoxyethyl groups (which also seems to increase their affinity for target mRNA), generation of methylphosphonates in place of phosphate linkages, and various modifications in the bases or in the ribose moiety, some even going so far as to replace the ribose rings with morpholines or polyamides. Sometimes chimeric oligos incorporating several of these modifications at different positions are used.[34]

Not every stretch of a given protein's mRNA sequence is a good site for AO binding, so some experimentation is required, but since only gene sequence information, not secondary or tertiary protein structural information is required to design such tool compounds, they can be generated much more quickly than small molecule inhibitors. Getting them into cells, however, isn't always as straightforward, and works better in some types of cells than in others. Methods for transfection include the use of polycationic reagents, liposomes, and special techniques like electroporation and microinjection. Antisense oligonucleotides have proven to be useful for TV both in cells and in animals, where the embryonic lethality often seen in knockouts is not an issue.[35]

Limitations, though, include possible non-specific effects due to interactions with various proteins and toxicity due to the transfection reagents that need to be used. As with the use of knockouts for TV, the absence (not just presence in inactive form) of target protein might invoke phenotypes that differ from those seen with small molecule inhibitors. Unlike knockout, however, the target protein will probably not be 100% absent when using antisense techniques, although >70% knockdown is often achieved. Importantly, since antisense reagents interact with cells or animals that already have the target proteins present and merely prevent future production of them, a lot depends on the half-life of the existing protein. If it's on the order of days (which can happen) it may take at least that long for antisense knockdown to have any effect on phenotype, yet another reason why AOs need to be very stable.

As is the case for every other TV tool mentioned in this section except for knockouts, AOs have human therapeutic potential as well. Vitravene (fomivirsen) is a phosphorothioate 21-mer from the pioneering antisense company Isis Pharmaceuticals that's been approved for use in CMV retinitis in patients with HIV. The drug is injected intravitreally (into the eye). Another Isis antisense drug, ISIS 30102, which targets apoB-100, has shown promise in Phase II clinical trials with a cholesterol lowering effect comparable to that obtained with atorvastatin.[36] This drug was given by *subcutaneous* (*SC*) injection. Other antisense drugs are in clinical trials as well.[37] Antisense oligonucleotides remain important for both validation and therapeutics, but in many ways they've recently been eclipsed in popularity by another technique directed at mRNA, called RNAi.

### 5.4.2.2.3 RNAi

Twenty years after the discovery that AOs could block the translation of mRNA into target proteins, another method of posttranscriptional target knockdown was found. In 1998, Dr Andrew Fire, then at the Carnegie Institute of Washington, Dr Craig Mello of the University of Massachusetts, and their co-workers published a paper demonstrating that injecting low levels of *double stranded RNA (dsRNA)* into *C. elegans* could inhibit the expression of the target protein in a sequence specific manner by a mechanism different from that observed for AOs.[38] This and subsequent work which uncovered the details of this new type of *RNA interference (RNAi)* eventually led to them being awarded the Nobel Prize for Medicine in 2006. Figure 5.7 outlines our current understanding of the processes involved.

Long dsRNAs like the one used in the above report are cleaved by the nuclease Dicer to produce double stranded 21–25mers with several unpaired nucleotides (overhangs) at each 3′-end. These also have 5′-phosphate groups and are called *small interfering RNAs (siRNAs)*. Unwinding of the strands and association with the *RNA-induced silencing complex (RISC)* allows the antisense strand to guide to the complex to the targeted mRNA by hybridization with the right mRNA sense sequence. Once bound, the mRNA is cleaved in the center of the cognate sequence by the

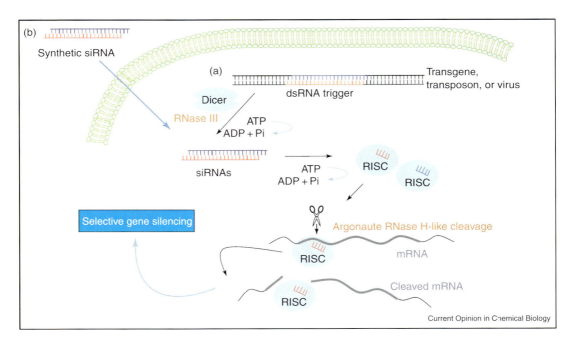

**Figure 5.7 ▶** The mechanism of RNAi. (a) Long dsRNA is cleaved by the ribonuclease Dicer to form double stranded 21–25 nucleotide fragments called siRNAs. (b) siRNAs can also be introduced directly into the cells. In either case, association of the antisense strand with the RISC guides the resulting complex to its cognate mRNA site, where the mRNA is cleaved, selectively blocking protein translation. (Reprinted with permission from Manoharan, M. RNA interference and chemically modified small interfering RNAs. *Curr. Opin. Chem. Biol.* **2004**, *8*, 570–579, copyright 2004, Elsevier.)

Argonaute component of RISC, thus selectively preventing protein translation, and the antisense strand of the siRNA is free to do it all again.

This catalytic mechanism is believed to have evolved as a defense against viruses in simple organisms lacking more complex immune systems, but it isn't used that way in mammals. Putting ds RNA segments longer than 30 nucleotides into mammalian cells is problematic: an interferon response is induced and the cells tend to die. Because of this RNAi is normally achieved using presynthesized 21–23 nucleotide siRNA reagents that are transfected into cells using methods (carriers, electroporation) similar to those used for AOs (Figure 5.7(b)), or by introducing a DNA vector for *short hairpin RNA* (*shRNA*), which is converted to the relevant siRNA reagent within the cells.[39]

"Naked" siRNAs, being double stranded, are more stable than "naked" AOs, and are, unlike the latter, normally fine for cellular experiments. But for in vivo use where they're added exogenously, their stabilities,[40] PK parameters, and perhaps affinities might need to be improved by the same sorts of chemical modification (phosphorothioates, etc.) used for antisense reagents.[41] Reagents for siRNA experiments in cells are probably the easiest of all TV tools to come by. They can be designed solely on the basis of gene sequence and ordered over the Internet. Vast siRNA libraries can now be purchased as well. As with AOs, getting the compound into the cells can be the trickiest part of the whole process, multiple controls containing some mismatches need to be used, and actual target protein levels need to be determined before any conclusions are arrived at. But compared to the efforts involved in finding a potent, selective small molecule inhibitor of the target, RNAi can constitute a much easier route to early TV, particularly for classes of targets where good small molecule tools are hard to come by. And most, but not all, studies comparing the use of siRNA with that of AOs have found the former to be effective at lower concentrations.[42]

These advantages go a long way toward explaining the increasing popularity of RNAi in TV. In a typical recent example, a commercially available siRNA targeting *matrix metalloprotease-11* (*MMP-11*) impeded cell proliferation, resulted in apoptosis in the mouse hepatocarcinoma cell line Hca-F, and attenuated the metastatic potential of these cells in a mouse model.[43] Researchers at Genentech took advantage of conditional RNAi, a newer technique[44] to show that knocking down *BRAF*, a serine–threonine kinase, slowed tumor growth in one xenograft mouse model of metastatic melanoma and caused regression in another, thus providing in vivo validation for *BRAF* as a cancer target.[45] Small molecule inhibitors of this novel target exist but are not selective.

Originally RNAi was thought to be exquisitely specific for the desired target, but recently this has been called into question, and off-target effects have been reported.[46,47] Some mismatches between the reagents and the target mRNA may thus be tolerated, and one possibility is that siRNAs might in some cases be acting like their cousins, *microRNAs*, important, newly discovered regulators of mammalian protein translation which seem to act non-specifically by recognizing as few as seven nucleotides.[48] Another possibility is that the sense strand of the siRNA duplex, normally assumed to be degraded, can still play a role. The sometimes non-specific effects of siRNA are an area of ongoing research, but things need to be kept in perspective: "While off-target activity of siRNA has received a large press coverage lately, the difficulty in obtaining small-molecule specificity has been an ongoing struggle within the pharmaceutical industry for many years."[49] This would make arguments against the use of siRNA reagents by chemists because of an alleged lack of specificity a little like the pot calling the kettle black. Specificity is an issue for *any* TV tool, or for that matter any therapeutic modality. Methods for demonstrating specificity, especially the use of *rescue experiments* (adding back the gene product via a transcript sequence that the siRNA doesn't target) and using multiple siRNA reagents targeting different parts of the transcript, have been suggested.[50]

Another assumption that dsRNA with less than 30 nucleotides (read, siRNAs) will never cause an interferon response has also been called into question.[51] These kinds of problems are to be expected for such a young technology, but in many ways the most remarkable thing is how far it's come in such a short time. RNAi therapeutics are already undergoing clinical trials and one of the companies most involved in the technology, Sirna Therapeutics, was recently acquired by Merck for more than $1 billion, which indicates just how promising the industry feels this technology ultimately is.

RNAi certainly deserves a seat at the TV table, its main advantage over small molecules being the ease of reagent design and construction.[52] No method is perfect, of course, and it's been suggested that much of the utility of RNAi or any other TV tool lies in combination screening with other tools. An example of this is the recent use of a genome-wide siRNA screen to identify genes that sensitize non-small cell lung cancer (NSCLC) cells to the effects of paclitaxel, an experiment that could help point the way to new therapeutic targets as well as more rational combination therapies for cancer.[53]

### 5.4.2.2.4 Antibodies

Antibodies represent an endogenous method for targeting extracellular proteins. Dozens of *monoclonal antibodies* (*mABs*), those derived from one clone and recognizing one antigen, have been approved for therapeutic use, mostly in cancer and inflammatory diseases, and literally hundreds more are in the clinic, which accounts for the massive amounts of research going on in this area. But they can also be used as exquisitely selective TV tools, *provided that the target is an extracellular one*. Figure 5.8(a) shows the structure of a full length antibody.

Their high molecular weights (~150 kDa) partly explain why they normally can't be used for intracellular targets: they're highly cell impermeable. And even introducing a full length

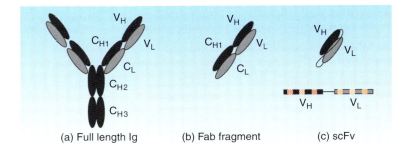

(a) Full length Ig    (b) Fab fragment    (c) scFv

**Figure 5.8** ▶ (a) A full length immunoglobulin, MW ~ 150 kDa. (b) Recombinant fab fragment, which lacks $C_{H2}$ and $C_{H3}$ domains, MW ~ 60 kDa. (c) Single chain Fv fragment containing the variable domain of heavy and light chains, with connection by a peptide linker, MW ~ 25 kDa. (Reproduced with kind permission from Mundt, K. Intrabodies—Valuable tools for target validation. *Eur. Pharm. Rev.* Winter 2002, 1–5, ESBATech AG (www.esbatech.com).)

mAB into cells through microinjection or having them expressed by the cells themselves is problematic. The reducing environment inside the cell doesn't allow for the disulfide bonds necessary for antibodies to maintain the right conformation and antibodies produced intra-cellularly don't fold properly.

For extracellular targets, although they aren't quite as simple to make as siRNAs, antibodies have proven to be useful in the early stage drug discovery via libraries generated through phage display[54] or as tools to study the physiological role of extracellular proteins like selectins.[55] One of the nice things about using them for TV is that if the target proves valid and useful, mABs used to establish this might themselves be developed into therapeutic agents. But generating mABs that bind selectively *and* have the desired effect on function (not a given) can be challenging.

Recent years have seen a growing interest in *intrabodies* (*intracellular antibodies*). As shown in Figure 5.8(c), these usually consist of the light chain ($V_L$) and the heavy chain ($V_H$) antibody domains containing binding recognition elements, held together by a peptide linker. These are referred to as *single chain Fv fragments* (*scFvs*). Genetic engineering is normally used to express them in cells, and their effectiveness sometimes depends on their ability to redirect target proteins to different locations within the cell via peptide sequences attached to them that function as "cellular Zip codes."[56] It's been observed that the biggest barriers to the effective-ness of scFvs are often stability and solubility, not binding affinity per se.[57] Despite these limitations, intrabodies have proven to be useful in validation experiments for the cysteine protease Cathepsin L in cancer,[58] and the hepatitis C serine protease NS3 in liver cancer,[59] among others.

Because of the size of these molecules, both intrabodies and antibodies have the exciting ability to affect protein–protein interactions, which small molecules are usually not good at, and because of their mode of action they're not limited by the time lag involved in slow protein turnover, but can take effect as soon as the reagent is there.

### 5.4.2.2.5 Aptamers

Besides their use as AOs and siRNA reagents, oligonucleotides find yet another use as *perturbogens* for TV studies as *aptamers*. The term refers to short (<100 nucleotide) stretches of DNA or RNA that fold in a way conferring shape complementarity with a target protein. *Peptide aptamers* exist as well, these having more in common with antibody fragments like those discussed above.[60] Oligonucleotide aptamers are selected from pools of up to $10^{15}$ candidates by an automated in vitro process called *systematic evolution of ligands by expo-nential enrichment* (*SELEX*)[61,62] that can now be completed in just a few days. Not only proteins but also small molecules and metal ions can be targeted: an RNA aptamer was found to bind to theophylline with 10,000-fold greater affinity than for its methyl analog, caffeine.[63] Aptamers don't seem to induce immunogenicity when given in vivo, which is often a major limitation for antibodies.

One recent example of the use of aptamers to study potential targets makes use of an RNA *Spiegelmer* ("spiegel" being German for mirror), a mirror image oligonucleotide made with L-ribose sugars in place of the normal D-ribose rings so that the resulting compound can resist degradation by nucleases for days.[64] This aptamer, called L-NOX-B11, binds to octanoyl-ghrelin with nanomolar potency. Ghrelin is a 28-amino acid peptide that's an endogenous ligand for the *growth hormone secretagogue receptor 1a* (*GHS-R1a*) and a potential obesity target that gets a lot of press coverage. Pegylated L-NOX-B11 was found to inhibit ghrelin's effects in *central nervous system* (*CNS*) in an in vivo experiment,[65] and a successor spiegelmer produced weight loss in a diet-induced mouse obesity model.[66]

Most aptamers find use against extracellular targets,[67] but, as with antibodies, an intracellular version of this technology also exists, in this case called *intramers*.[68] A good review of the use of aptamers for TV was recently published,[69] emphasizing the utility of aptamers in HTS assays as well. An aptamer drug, the VEGF antagonist Macugen was recently approved by the FDA for the treatment of *age-related macular degeneration* (*AMD*). Like the antiviral AO drug Vitravene, Macugen is injected into the eye. Several other aptamers are in the clinic as well.[70]

#### 5.4.2.2.6 Small Molecules

After reading the preceding sections, most readers probably won't need too much convincing that small molecules usually make the best tools to use in validating a target. Perhaps the best way to consider the advantages that small molecules bring is to look at the disadvantages they lack: embryonic lethality, compensation during development, structural effects due to the absence of normal protein–protein interactions, long delays between administration and effect, difficulties in determining dose–response, immunogenicity, interferon response, extreme challenges with intracellular delivery, and general unsuitability for oral administration.

They're not without problems of their own. Finding a selective small molecule inhibitor from scratch will probably be more difficult than finding a functional siRNA tool. A chemist can't look at the sequence of a gene, design a small molecule inhibitor, and order it online. Problems with selectivity, cell permeability, and oral bioavailability are no strangers to the medicinal chemist. But keep in mind that most drugs on the market are and will continue to be small molecules. Ultimately any project directed at such a compound will need to use them in validation experiments up to and including clinical studies. Although other validation tools can be helpful along the way, particularly when used in combination, for such projects only small molecules are "the real thing."

### 5.4.3 Working on Novel Target-Directed Projects

For purposes of this book, the line between an "established" target and a "novel" one has been drawn at clinical efficacy in Phase II, which represents Level III TV according to the scheme described earlier. This definition of target novelty therefore encompasses everything from targets completely lacking any validation through targets which have been proven to work in multiple animal models of a given disease. You can see that these represent widely varying degrees of "novelty," including targets only just now existing in the mind of a single scientist to targets for which positive animal model data have already been published in *Science* or *Nature*. How will such a project differ from those based on established targets?

Box 5.3 summarizes the advantages and disadvantages of projects directed at novel targets. First of all, when working on a project directed at a novel target, research scientists are generally going to be more excited and therefore more motivated, a factor that should not be underestimated. The prospect of designing and making with one's own hands a first-in-class

---

**Box 5.3   Advantages and Disadvantages of Working on Novel Targets**

**Advantages**

▶ More motivating to project scientists

▶ *Market space and potential market exclusivity*

▶ Potential for advantage over or synergy with other drugs

**Disadvantages**

▶ Requires validation

▶ May not have benchmark compounds or relevant assays

▶ *High attrition rate*

---

drug, especially a structurally novel one that fills an unmet need, will probably translate into greater efforts, more creativity, and more focus for even the most jaded veterans. Few are the medicinal chemists who haven't dreamt about coming up with the next Gleevec (imatinib), the breakthrough Bcr-Abl kinase inhibitor which has dramatically improved and extended the lives of patients with *chronic myeloid leukemia (CML)*.[71] Contributing to a first-in-class Hsp90 inhibitor for cancer or Cathepsin K inhibitor for osteoporosis is just more appealing to a scientist than coming up with a fourth-in-class drug. And as everyone knows, the difference between a motivated employee and a lackadaisical one is night-and-day.

One might think at first that new targets would bring with them wide open IP space. But novel targets do not, per se, imply novel compounds. A lot depends on just how different that target is from known ones. As we'll see later, leads for a new target that's a member of a known family, such as kinases or aspartyl proteases, can often be found in libraries of compounds active against other protein family members, as in the situation where screening an aminoquinazoline library turns up a hit for new kinase target. This class of compounds is extensively claimed as kinase inhibitors in the patent literature, so although the target may be new the inhibitor probably is old and there may not be much IP room. But if the novel target itself is significantly different from popular, well-explored targets, there's a better chance that a structurally novel lead that binds to it will be identified, if one can be found at all. So once again, although the target doesn't dictate the compound, it has a certain amount of "influence" over what it will be.

For novel targets, although patent *lebensraum* doesn't necessarily come with the territory, market space does. Even in a crowded therapeutic area where many effective drugs exist, a drug intervening at a different point is always welcome. The first-in-class renin inhibitor aliskiren (Figure 2.8), recently approved for the treatment of hypertension, is a good example of this. Renin, an aspartyl protease, is involved early on in the angiotensin pathway, which the well known *angiotensin converting enzyme (ACE)* is involved in later, and which largely regulates blood pressure. Renin inhibitors have been around for decades. But these inhibitors tended to be rather large, peptide-like compounds and problems with oral bioavailability kept them from becoming drugs. Before the appearance of aliskiren, renin inhibitor projects were most notable for having eventually provided leads for another aspartyl protease that came along much later, HIV protease. Not surprisingly, oral bioavailability always seemed to be the major obstacle to developing this class of drugs, but with great effort and after decades of work that obstacle has been overcome.

Three points are apparent from this story. First, targets in a given family do tend to prefer structurally related compounds. Second, even the most difficult problems can eventually be overcome given sufficient incentive. And third, our original point that market space does exist for drugs with novel *mechanisms of action* (*MOA*s). Now that aliskiren has made it to market, both single agent and combination use with other antihypertensives can be expected to provide a win–win situation for its maker, Novartis.

Why should there be a need for such a new drug, albeit one with a different target, when so many are available already? For the same two reasons we've encountered before: better efficacy and/or better safety, either with the new agent alone or in combination therapy with existing drugs. One could just as easily ask, "Why should an organic chemist need more than one way of putting on a silyl protecting group?" Because sometimes the "favorite" way either doesn't work or works too well and silylates other things. The answer breaks down to *reactivity* and *selectivity*, which are curiously similar to a drug's crucial attributes, *efficacy* and *safety*. As chemists we intuitively realize that transformations for which dozens of reagents and conditions exist are the most popular and/or the most difficult ones. An analogous situation applies to drugs.

The use of a drug acting against a new target in combination with another existing drug is a particularly attractive proposition, especially when it comes to difficult, life-threatening diseases, where therapeutics need all the help they can get, like *hepatitis C virus* (*HCV*), HIV, and cancer. While the hope remains that someday a single agent efficacious enough to at least keep such a disease in check or even cure it may be available, the reality is that no such drug exists today. Existing therapeutics can work but have limitations. So when different classes of them are available they're frequently used together in the kind of combination therapy as discussed in Chapter 1. Here, the beneficial effects can be additive or, better yet, synergistic, although studies will need to be carried out to make sure that their toxicities aren't synergistic too.

This approach is best exemplified by HIV drugs. The first target to be broached in the war on this fearsome virus was an enzyme called *reverse transcriptase* (*RT*), which the virus requires to convert its genomic RNA into DNA, and the first approved drug to target it was the nucleoside analog Retrovir (zidovudine, AZT, Figure 5.9), which had originally been looked at for cancer. Although this compound alone proved a lifesaver for many AIDS patients, efficacy was not ideal and toxicity in the form of neutropenia, severe anemia, and liver disorders was encountered in some patients. Furthermore, the fast and sloppy replication process the virus uses soon gave rise to mutations, some of which were resistant to the effects of the drug.

Other RT inhibitors soon reached the market including other nucleosides and later some *non-nucleosides* (*NNRTI*s), but all acting on the same target—incremental improvements but still far from ideal. A major advance in AIDS therapy came with the identification of another target, HIV protease where, as mentioned above, samples of renin inhibitors in the compound collections of pharma companies provided leads. Even before the first-in-class

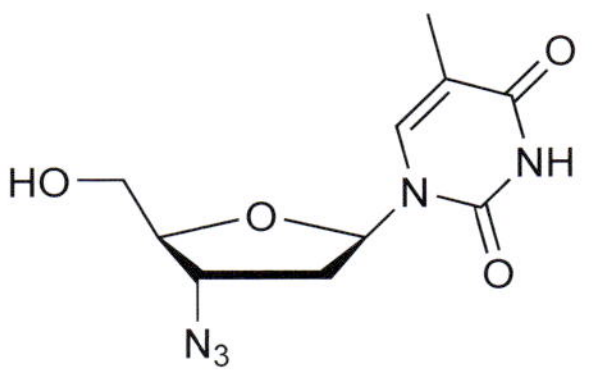

**Zidovudine** 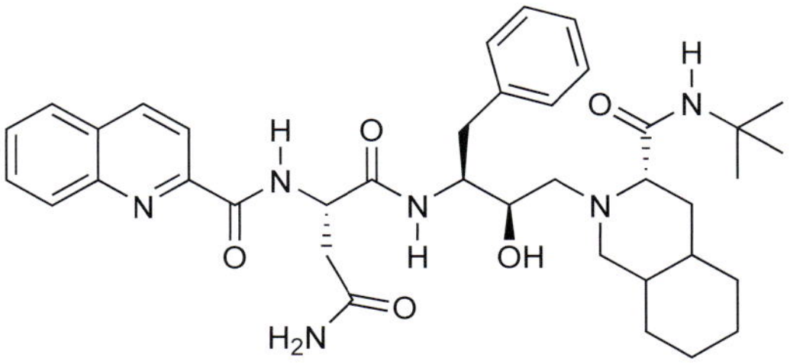                    **Saquinavir**

**Figure 5.9** ▶  The HIV RT inhibitor Retrovir (zidovudine, AZT) and the HIV protease inhibitor Invirase (saquinavir).

compound, Roche's Invirase (saquinavir), was approved for sale, it was general knowledge that if it worked, such a drug's real future would lie in combination with RT inhibitors, and so it has proven to be. Of course, additional targets for HIV (fusion inhibitors, integrase inhibitors, etc.) continue to be pursued,[72] with each new first-in-class drug expected to find a market for combination therapy of AIDS and each improving the lives of many patients.

If a drug acting on a new target can still find market space in a therapeutic area where other drugs already exist, imagine what it can do where there is no effective current therapy! This might include new diseases like *severe acute respiratory syndrome* (*SARS*) (assuming a company wanted to gamble on that eventually becoming a lucrative market), or old ones for which existing drugs only treat the symptoms. Finding something there would be a major accomplishment and would be valued as such.

Counterbalancing all of these advantages, of course, are the several major disadvantages. We've already seen that despite all the new, improved technologies that can be used, the process of preclinical TV is often slow, resource intensive, and, worse still, sometimes not predictive of clinical results. This last factor is the most discouraging of all, as late stage efficacy failure is the most expensive as well as discouraging sort of project failure.

A less severe, but still real limitation involved in pursuing novel targets is the fact that assays may need to be developed and reagents (things like the protein of interest or target substrates for proteolytic enzymes) may need to be made before validation can even begin. It might not be simply a matter of browsing through catalogs for reagents or kits or tracking down a procedure in *Methods in Enzymology*. Real work might need to go into it, which means FTEs will need to be approved by the management for a speculative project that hasn't even begun yet. Most likely, small molecules acting on the prospective target won't be known either, and as we've seen, small molecule inhibitors or agonists or antagonists are very fine tools for biologists to have at their disposal when doing TV. It could be that such a tool molecule wouldn't be available until the project is in full swing *lead optimization* (*LO*) mode, which a company is unlikely to enter until a target has been better validated, a sort of a Catch-22. Normally, this would mean that validation would need to progress without a small molecule tool. But there is one other way that's based on the preference of target families to bind to structurally similar compounds, as noted earlier. This is called *chemogenomics*.

Dr Mark Murcko of Vertex Pharmaceuticals provides an example of how this might work for a real-world project: "In the area of kinases or ion channels or GPCRs we've already worked on other members of the family," he notes. "We already have some know-how about how to set up an assay in the field. We may already have useful compounds—not drug candidates, perhaps—but pretty decent starting points. Let's say we read a paper in *Nature* that suggests that kinase X is a great target for cancer. But it's speculative.

"What are our choices? Our choices are, a) do nothing and wait a few years till there are more publications so there's more validation, b) put a lot of biologists on the project to try to make sense out of it, c) do a collaboration with somebody in academia who's an expert in that target and use them as our surrogate biologist—which is a sort of variation of b—or, d) we could say, 'Well, gee, in three months we can get an assay up and running, screen our kinase deck, see if we have a 100 nM inhibitor of kinase X, and then take that into primary cells for that cancer type and see whether there's anything interesting. If not, then we're done.' It's actually less work to do it that way than to do all the laborious biology upfront.

"And what if you're lucky and you find a molecule that's pretty good? Then you've given a tool to the biologists so that when they go do their basic biology studies of that target they now have a tool they wouldn't have had otherwise. If we pursue the target with no tool compounds, they can use RNAi or other biological tools, but it would be easier and faster for the biologists to do what they need to do if they had a tool compound. My view is that small molecules, if they're selective and cell permeable, are always better tools."

Add to this the fact that if successful your 100 nM tool compound might also be a 100 nM lead compound for which the chemistry is worked out, IP issues have been looked at, and some *structure–activity relationship* (*SAR*) may already exist if other analogs were also in the screen, and you've got quite a good strategy. But keep in mind that this relies on you having screening compounds directed at that family of targets already available. If you only have their second cousins (something that looks a little like something that might possibly be active) or would need to start making compounds from scratch it's a whole different story.

So if there's a great potential market for drugs acting at new targets, if delays due to the need for validation can be somewhat ameliorated by front loading, and if there's sometimes even a workaround for the problem of not having tool compounds, why doesn't everyone work on novel targets? If you remember the major advantage of working on established targets, the fact that resulting drugs have a higher-than-average probability of making it to market, you can probably predict what the major disadvantage of working on novel targets is.

By simply using readily available online resources like BioSpace (www.biospace.com) and DrugResearcher.com (www.drugresearcher.com), without delving any deeper it's likely that you'll be able to locate an average of at least one report of an exciting potential new target *every day*. Today, for example, my usual sources alerted me to *Heme oxygenase-1* (*HO-1*) and its potential role in Kaposi's sarcoma,[73] while yesterday brought reports on the role of *lysyl oxidase* (*LOX*) in cancer metastasis[74] and the promise of drugs acting on *metabotropic glutamate receptor* (*mGluR*) in Fragile X syndrome.[75] All three represent laudable accomplishments that hold promise for victims of cancer or the most common inherited form of mental impairment. But how many of these will eventually result in marketed drugs?

If we make the conservative estimate that such novel targets are reported at the rate of one per day and the guesstimate that for each of these another has been uncovered in industry but not talked about (secrecy being especially important in the early stages of a project) that gives us 730 new targets every year. But in any given year only a few drugs which act on novel targets are approved. Seven or eight such approvals would make for a banner year. Glossing over factors like the time lag between TV and drug approval, unpredictable changes in the numbers of approved drugs, etc., you can see that *at least 99% of all new targets do not lead to marketed drugs*. A recent estimate states that although 17% of projects based on established targets are likely to enter preclinical development (that is, result in optimized leads evaluated in the preclinical stage), only 3% of projects based on new targets will ever get that far.[76] All of this should in no way be seen as taking credit away from the dedicated and often brilliant researchers who did the hard work of uncovering them originally.

Instead, their "failure to launch" could be due to any number of things. As anyone who's ever made compounds described in the literature knows, experiments sometimes end up being difficult for other researchers to reproduce. Or no real follow-up might have been done because companies believed the markets to be too small or too difficult or it just didn't fit into their research portfolio or was lost in the shuffle of project competition and corporate mergers. Perhaps, they felt that the published work fell into the "NIH" (not invented here) category and wasn't up to their own, internal standards. Perhaps for a variety of reasons no suitable compounds could be found, as they couldn't for the then-new renin inhibitors for 30 years. For one reason or another, the proposed target just never was able to work its way up the rungs of the validation ladder. There might be many reasons for this, but the bottom line is that while novel targets are plentiful, novel approved drugs acting upon them are not. More's the pity.

## 5.5  Targets Arising from Phenotype or High-Content Screening

### 5.5.1  Phenotype Screening Versus Target Screening

Cellular assays play an important role in every project. For a project based upon a given target, initial in vitro assays are usually followed up by target assays which take place in cells. For example, a proteasome inhibitor project would probably begin by screening compounds for inhibition of isolated 20S proteasomes. This would be followed up by studies where the active compounds were incubated with cells and then the levels of some marker of cellular proteasome function like the protein p21, which tends to accumulate when proteasomes aren't actively degrading it, are measured. If compounds active in the first assay work in cells everyone will feel more confident that they're onto something. If not, the usual villains of cell impermeability or cellular efflux will have to be faced down if the project is to continue. Think of these kinds of simple cellular assays as a rite of passage for most target-driven projects.

But cellular assays can also be used in a totally different way. To understand this, we need to take a very brief look at genetics. A time-tested genetic technique for finding the roles of specific proteins in biochemical pathways is to use mutagenesis in some living organism to randomly interfere with gene products, to focus in on the mutations that cause observable effects (the phenotype), and then to go back and identify which genetic alterations and resulting perturbed proteins were responsible. This process goes by the name of *forward genetics*. Mutagenesis here is usually accomplished through the random mutations induced by a nasty chemical mutagen like ENU, ethylnitrosourea. If instead, we use as our

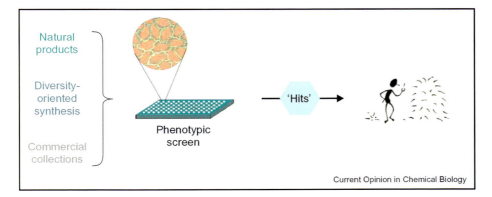

**Figure 5.10 ▶**    Forward chemical genetics. (Reprinted with permission from Lokey, RS. Forward chemical genetics: Progress and obstacles on the path to a new pharmacopoeia. *Curr. Opin. Chem. Biol.* **2003**, *7*, 91–96, copyright 2003, Elsevier.)

perturbogen[77] a library of small molecules which, rather than causing mutations, might interact with cellular proteins through inhibition, antagonism, or agonism, the process becomes what is known as *forward chemical genetics*.[78] Figure 5.10 is a simple schematic illustrating how this works.

Starting with a clean slate, that is, even before a target has been identified, cellular assays can be devised which measure observable phenotypes, like viability, morphological changes, or give some other readout like protein translocation which reflects a given disease state and not just the striking of one prespecified target. The most direct example of this would be for a type of pathogenic bacteria where compounds inducing death (bacteriocidal) or failure to proliferate (bacteriostatic) would be considered potentially useful, assuming they didn't do the same to human cells—a very big assumption. In this case it wouldn't matter, at least initially, what target was involved. Everything could be thrown into such an assay and the mechanism of action for interesting hits could be sorted out later. This general sort of strategy has been used for many decades in antibiotics, although not necessarily on a high-throughput basis, and is reminiscent in a way of those interesting Petri dishes of Sir Alexander Fleming which gave the world penicillin. Inhibition of proliferation in a cancer cell line, which can happen through any number of different mechanisms, is an example of a simple multitarget cellular assay such a strategy might use.

This approach stands the normal process completely on its head: instead of going from target to drug, it proceeds from "drug" (in the sense of hit or lead) to target. In a way, this represents a return to the older drug discovery methodology that prevailed in the days of broth extract screening and tissue or even whole animal primary assays for synthesized compounds, back before the new field of molecular biology shone the bright spotlight of scientific inquiry on specific molecular targets. Although nobody suggests that we turn that spotlight off, many people feel that illuminating the rest of the stage as well could only help. In that context, phenotype screening and other types of multitarget or *high-content screening* (*HCS*, which usually implies the use of subcellular fluorescence imaging) represent a valuable addition to the armamentarium of drug discovery.[79]

There's a lot of diversity in the kind of assays that can be carried out, some of them having only been enabled by recent technologies, such as the ability to detect cell morphology changes by microscopy in a high-throughput format. Cells (which of course don't constitute an in vivo system except in the case of single-celled organisms) can also be engineered to express things like known chemotherapy resistance factors and then compounds screened for their ability to restore the antiproliferative effects of a cancer drug like doxorubicin.[80] In some cases where traditional cellular models are hard to come by, such as HCV, which isn't amenable to simple infection and reproduction in cultured hepatocytes, cells can be engineered to express self-amplifying "minigenomes" called *replicons* containing the non-structural viral proteins, which represent most (but not all) conceivable antiviral targets.[81] A compound active in such an assay might be hitting any of them.[82]

True organisms, rather than cells, such as nematodes (*C. elegans*)[83] and zebrafish embryos (*Danio rerio*)[84] can be used in multitarget assays. Readouts can be as simple as fluorescence or as complex as a detailed analysis of modified protein pathways.[85] Even beginning to understand all the possible variations on this type of screening strategy is a daunting prospect, but however many there are and however complex they may be, they all have a couple of things in common. They allow one to screen for drugs in a complex living system, which is more "real world" than an isolated enzyme or receptor binding assay, and to "cast a much wider net than target-by-target approaches."[85]

An overall summary of the advantages and disadvantages of using this type of drug discovery methodology is shown in Box 5.4. The single biggest advantage of this method is that by definition hits give real effects in cells, rather than an effect on an isolated target that may or may not work in a cellular milieu when tested later. Note, however, that initial hits, which are defined by some arbitrary level of response in a high- or medium-throughput screen which has been validated by running a positive control, require a certain amount of follow-up before any conclusions can be drawn. It's not as simple as looking at the raw data for the screen. Artifacts must be ruled out, identity, purity, and good dose–response need to be established, all of which are further discussed in Chapter 6.

But if all these follow-up experiments have been done and hits are confirmed, especially if additional evidences like confirmation in a different cell line or apparent non-flat SAR for structurally related compounds in the screening set are present, then what's left at the end of the day (or month, more likely) is a hit that produces a biological effect in a disease-relevant cellular assay, and the value of that shouldn't be underestimated. Because the confirmatory experiments have ruled out insolubility (by visual examination, nephelometry, etc.), toxicity (by cell viability experiments), and other bad things, the genuine hits should provide unusually good starting points for compound optimization. Furthermore, if the target of the hits turns out to be intracellular, which many are, you can be sure that the compounds have a certain degree of cell permeability. All in all, what you end up with is a starting point that's far ahead of those you'd get in other types of projects.

Furthermore, with this approach it's possible to go in "blind," without prejudging which targets may be present and what type of compounds should hit them, which may be more likely to result in the discovery of new targets and novel compounds if it finds anything at all. In cases where such information isn't known this sort of phenomenological approach may be the only way to go. But if target information is available, bias can still be built in if desired,

---

**Box 5.4    Advantages and Disadvantages of Working on Cellular Phenotype or Multi-Target Projects**

**Advantages**

▶ *Compounds emerge with cellular validation*

▶ For all targets toxic, insoluble, and unstable compounds can be filtered out

▶ For intracellular targets, hits imply some cell permeability

▶ Leads can be uncovered in the absence of any target information

▶ Novel drugs acting on multiple targets can be identified

▶ Can identify unexpected prodrugs

**Disadvantages**

▶ Can involve expensive equipment, licenses, and lengthy assay development

▶ Not all diseases have relevant high throughput cellular models

▶ *Extensive TID work, probably involving chemistry, will be required*

▶ Target might be difficult to identify or undesirable

---

for example by selecting a library of fused nitrogen heterocycles which look like kinase inhibitors in a cancer cellular assay.

Another advantage of this approach that may not be obvious at first is that compounds whose effects depend on hitting not just one but several different targets, which may have been discarded in more "targocentric" approaches precisely because of non-selectivity or just because such combinations weren't considered, can still be identified. In fact, cellular screens may be the best way to study things like kinase inhibitors, where one recent study concludes that "cellular responses induced by multiplex protein kinase inhibitors may be an emergent property that cannot be understood fully by considering only the sum of individual inhibitor–kinase interactions."[86] And a number of such multitargeted compounds do populate the world of known drugs as we'll see later. A particularly appealing therapeutic area for such multitargeted compounds is oncology, where many different gene products are typically mutated and phenotype assays (antiproliferative, pro-apoptotic, etc.) are straightforward.

Yet another unique aspect of phenotype screening is that it might even detect prodrugs, where the active species is not the compound applied but some transformed version of it. Isoniazid (isonicotinyl hydrazide), a tuberculosis drug in widespread use for over half a century, requires reaction with a specific peroxidase followed by reaction with $NAD^+$ and $NADP^+$ to generate the nucleotide adducts that are the active species. A recent study using a chemical genetics approach identified one of the targets of a specific nucleotide adduct as bacterial dihydrofolate reductase.[87] Isoniazid originally came out of a simple bacteriocidal (that is, phenotype) assay, but had it been discovered today, elucidating the multiple MOAs of its several important metabolites would represent quite a challenge to the drug's development. More importantly, though, it would be virtually impossible to discover through a target-centered approach.

Using the phenotype or multitarget or high-content approach to drug discovery has its downside as well. First of all, the high-tech nature of many of the sophisticated assay techniques involved means that licenses need to be taken out and often expensive license fees paid before assays can even be done. Examples include the use of *green fluorescent protein* (*GFP*), a frequently employed reporter gene, and the previously mentioned HCV replicon assay. HCV is a particularly expensive area to do research in since fees will also need to be paid to Chiron (now Novartis) which, owing to the groundbreaking work done by its scientists, owns the viral genome patent. Other expenses associated with such a project will include specialized equipment as well as costs associated with the frequently lengthy assay development process itself. This can be trickier than the usual enzyme or receptor screens, cells being somewhat delicate and capricious.

The fact is that in some areas like CNS indications, good disease-relevant cellular assays may just not be available owing to the complex nature of the disorders, which may not be adequately modeled in a single, culturable cell type. Even in therapeutic areas where cellular models exist, like cancer, there's no guarantee that the transformed cell lines typically used (like the *NCI-60*, discussed later) will respond to a drug in the same way as an actual tumor, which is composed of many primary cell types living and dying in a variety of different aerobic and anaerobic environments. Even cancer cell lines themselves exhibit different behaviors in the presence of other cells, which can secrete factors to influence their proliferation or death.[88] So it should come as no surprise that a drug with Level I validation in such an assay might make it no further. It is, after all, just a model, not something you expect to get into and drive away.

But the biggest drawback in using this approach to drug discovery becomes obvious once hits have been identified and confirmed—what exactly are they hitting? There are widely used drugs (including, up until recent years, aspirin) which act upon unknown targets.[89] Drug candidates can still enter the clinic that way, a recent example being InterMune's Pirfenidone (Figure 5.11), an investigational new drug for *idiopathic pulmonary fibrosis* (*IPF*) for which a proposed MOA (p38$\gamma$ MAP kinase inhibition) was first suggested during Phase III testing. Such drugs might even make it to market without a known MOA. But the reality is that a drug going into the clinic without a known target is at a serious disadvantage, and many companies will be reluctant to pursue it for that reason. Everyone involved, including FDA, feels more comfortable if the MOA is known. Clinicians in particular need a feeling for how it would fit in with existing therapies based on its mechanism. Although drugs with unknown MOAs have been and continued to be approved, going in without one makes for a rougher road.

Knowing the MOA at an early stage would be far more preferable. That might allow the chemists to use *structure based drug design* (*SBDD*) for more rapid optimization. Knowing what the target is can also give scientists an idea of what the antitargets, related proteins possibly responsible for toxicity, might be. So in the real world identification of hits using phenotype screening will inevitably be followed by attempts to determine what target or targets are giving the effect. Target identification, in this case based on real cellular effects instead of correlative evidence and theory, has not been

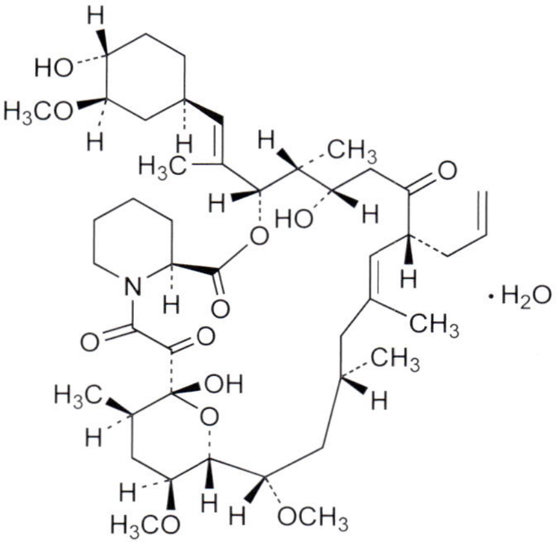

**Tacrolimus**

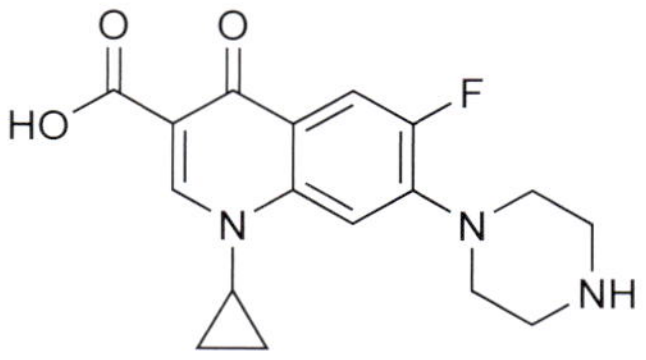

**Ciprofloxacin**

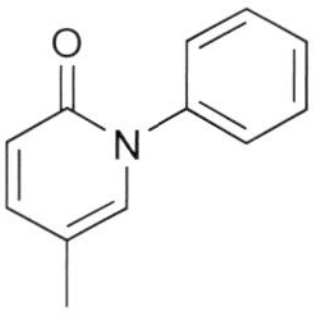

**Pirfenidone**

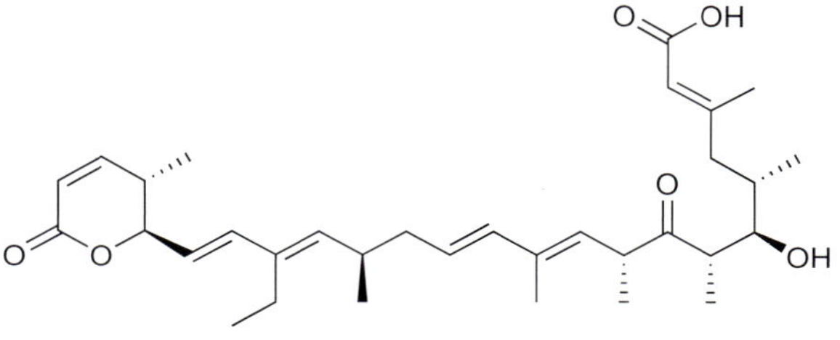

**Leptomycin B**

**Figure 5.11** ▸ Tacrolimus (FK506), Cipro (ciprofloxacin), Pirfenidone, and Leptomycin B.

avoided but only postponed. When it comes, it will be because an exciting new lead has been found from phenotype screening. Timelines will therefore be short and resources will need to be poured into it. It will be a far from trivial effort and may even call for more biologists than a small company has available. Additionally, chemists will probably be needed early on. This could be for several reasons. The initial hit can often be too weak (say 50 μM) to be useful in determining the target so that a certain amount of blind optimization needs to be done up front. Structural modifications involving linkers and/or labels may need to be carried out to fish out the target protein, as discussed below. So much excitement can be generated that chemists become eager to just start making analogs. All of these cases require the commitment of at least a few chemists to a still highly speculative preproject. These FTEs will need to be planned for in advance to avoid premature (and predictable) project termination for lack of resources later on.

When all is said and done, there are three possible outcomes for TID in this approach. If the target for a phenotype lead is discovered and found to fit into the established target bin, which is often the case, the company will need to decide whether or not it should be pursued. For example, uncovering an antibacterial agent that acted by inhibiting DNA gyrase, a target of fluoroquinolones like Cipro (ciprofloxacin) (Figure 5.11), would put the company in the position of having to go over the pros and cons mentioned in Section 5.2. A business decision would be called for, and it might or might not be a favorable one. If it was, though, they'd at least be starting out with a cell permeable and hopefully novel lead that could jumpstart the project in the LO stage.

The second possibility is that a novel target has been uncovered. Here all of the advantages of working on novel targets apply and one of the disadvantages, that of requiring validation, has at least been mitigated by starting out with a compound that works in cells. Although the higher risk of failure due to lack of further validation remains for such projects, they will at least excite everyone involved.

But there's a third possibility, too, as shown in Figure 5.1, that the timeline originally set for target determination runs out and the target has still not been identified. This is the most disturbing of the three outcomes, but is a real possibility owing to the often rate-limiting nature of the process. With all of the modern techniques (see below) and all the resources in terms of people, expertise, and funding that can be brought to bear it *should* be possible to identify the target or targets for any cellular leads. But that assumes unlimited resources including time, a situation that never exists in the real world. Determining the targets of compounds found through phenotype screening is never trivial and there's a real possibility that the budget and time requirements could end up being insupportable. The decision about whether to continue the investigation, increase the resources, and push out the timelines will need to be made. A second decision about whether or not to go into serious LO "blind," without that critical target information, will also need to follow. After all, it will be pointed out, drugs have made it to the clinic and even to approval this way!

### 5.5.2  Elucidation of Phenotype-Derived Targets

Whatever the outcome, we can see that TID plays a critical role in phenotype-based projects and it's been the subject of recent reviews.[90,91] It's now time to look at some of the more common techniques used in this important step, as shown in Box 5.5.

In many cases some idea of what the potential targets might be will exist before compounds have even been screened. There are always "the usual suspects" based on known pathways—tubulin binders in antiproliferative assays, MDR1 and MRP1 gene products in cancer drug resistance, etc. An educated guess and some follow-up can often prove successful in such cases. A more elegant example of this kind of knowledge-based approach was reported by Kau et al.[92] Library high-content screening for inhibitors of FOXO1a (a transcription factor) nuclear export turned up a number of small molecule hits that were Michael acceptors. A known compound, Leptomycin B (Figure 5.11), which is also a Michael acceptor, acts upon the nuclear export receptor CRM1, by forming a covalent adduct with the protein's Cys528 sulfhydryl group. The hypothesis that the new hits were also acting on this target via this mechanism was subsequently borne out by follow-up experiments.

A very direct method for identifying molecular targets of a phenotype screening hit involves *affinity purification*. A strong binding interaction between the target and the drug, at least in theory, means that if the drug is attached to a solid support through some part of the molecule unimportant to interaction with the target and an extract or lysate containing the target is passed over it, the target will bind sufficiently well that the many other cellular proteins in the "soup" can be washed away. Stronger conditions such as elution with buffers

---

**Box 5.5    Some Methods Used for Target Identification in Phenotype Screening**

▶ Hypotheses based on known pathways

▶ Affinity purifications

▶ Screening of expressed protein libraries

    ▶ Phage display

    ▶ 3-Hybrid systems

    ▶ Cellular microarrays

▶ Differential protein expression

▶ Transcriptional profiling

▶ Modulation by or comparison with genetic methods

---

of high-ionic strength or cleavage of the linker between the drug–protein complex and the solid support should allow for isolation of the drug–target complex which can then be characterized and analyzed, usually by LC/MS/MS, to reveal the identity of the target protein.

The poster child for this sort of approach was the groundbreaking work of Dr Stuart Schreiber and his group in determining the biological target of the important immune suppressant FK506 (Figure 5.11), the protein called FKBP.[93] Not only did this work elucidate the MOA for a drug (called Tacrolimus) which allows organ transplant patients to avoid transplant rejection and live for many years, but also, according to Dr Thomas Kodadek, Professor of Medicine at the University of Texas Southwest Medical Center, it "had a lot to do with convincing biologists that there really might be something to the idea of using chemistry as a key tool in molecular biology, while also suggesting to a recalcitrant synthetic chemistry community that it might be OK to work with real biological systems."

"The bad news," he continues, "is that it created the false impression among chemists without much classical biochemical experience that the target identification problem could be solved in most cases using this simple affinity chromatography approach. This view soon crashed upon the rocks of reality."[94] In the case of the FK506–FKBP complex the receptor was relatively abundant in cells and the ligand possessed high (single digit nanomolar) affinity, two things that are not true in many other cases.

Phenotype screening more commonly turns up leads in the double or perhaps single digit micromolar range, and the problem on using affinity purification then becomes one of signal-to-noise: washing away "sticky" non-target proteins without also washing away the relevant, low affinity target. Additionally, having to chemically modify the original hit so that it can be linked to a solid support isn't necessarily trivial, especially in the absence of knowledge as to what's important to target binding. A kind of preliminary SAR study may need to be done with linkers attached to various positions in the hit molecule in order to find an acceptable spot. One interesting approach is to build the linkers into the original screening library, as exemplified by recent work where the targets of a triazine hit in a zebrafish embryo morphology screen were identified as components of the 40S ribosome.[84] Another nice example of the use of affinity purification for phenotype TID has been reported by Tanaka et al., where a morphological phenotype screen revealed that a hydroxyl-containing analog of the kinase inhibitor PP gave rise to unique cellular effects, which affinity purification traced to a non-kinase target, CBR1.[95]

Thus affinity purification has proven to be successful from time to time but, like every other means of TID, it's by no means universally so. It's then not surprising that many different methodologies have been developed. Quite a few of these involve using the techniques of molecular biology to express large number of proteins, exposing these to tagged or tethered small molecule hits, detecting their interaction and determining the target involved. Among these are phage display,[96] and three hybrid systems. The latter, which is normally used in yeast has also been used in mammalian systems,[97] involves linking the hit to a known ligand for a protein receptor, then using this bifunctional molecule (the "bait") to bring together both domains required for the activation of a reporter gene but only in the presence of the hit's target.[98]

Another technique involves microarrays of cells engineered with different plasmids to overexpress various proteins.[99] Use of an appropriately labeled hit allows for TID, although so far literature examples involved only limited subsets of potential protein targets and not genome-wide arrays. The relatively more direct method, which makes use of arrays of proteins themselves bound to a glass slide, can also be used but likewise has its limitations.[100] As has been noted for years, immobilized arrays of proteins are more difficult to make and are less stable than DNA microarrays, but in this case one potential advantage they have over some of the other identification methods mentioned is that a detection technique called *surface plasmon resonance* (*SPR*), which obviates the need for hit modification via a label or linker, can be used.

In theory, one of the most direct ways of finding a target for a hit acting through an unknown mechanism is to simply look at the effects on cellular proteins by observing the differences in protein expression or composition in the presence and absence of the hit molecule using 2D gels. In practice, this is anything but simple owing to many factors such as the sheer number of proteins involved, normal differences in abundance (perhaps a million fold!) and activities, isolation difficulties, experimental variability, and massive data handling. So many proteins are likely to be differentially expressed that analyzing the data often results in a "haystack of needles."[101] The real target could be indistinguishable in the crowd of differentially expressed proteins, hiding in plain sight. That being said, this technique has had some successes. In one such instance, a comparison of 2D gels for an antiproliferative natural product, LAF389, showed a small shift in a protein which was traced to the incorporation of a single N-terminal methionine residue into the treated cells, suggesting that the compound was acting as an inhibitor of *methionine aminopeptidases* (*MetAP*s).[102]

A major simplification of this kind of proteomic technique involves limiting the number of proteins studied to a smaller subset of gene products such as cysteine proteases or metalloproteases where appropriate *activity based probes* (*ABP*s) exist. By virtue of their target binding mechanism, which depends on activity and not abundance, another obfuscating factor is eschewed as well.[103] The relevance of this kind of strategy is probably limited to cases where the gene family of interest has been targeted by a directed library in the first place and even then it couldn't rule out effects on the other, non-family targets per se. But within these tight constraints it's a powerful method for identifying targets.

A variation on the whole genome proteomics involves transcriptional profiling, a comparison of a hit's pattern of gene expression, or "signature," to that of other perturbogens such as known drugs, using DNA chips. Figure 5.12 illustrates this concept. A collection of known transcriptional profiles is required for comparison and this might be obtained from a public database or a library of several hundred to several thousand known drugs compiled and tested for this purpose.

An example involves the NCI-60, where researchers have analyzed the effects of more than 1400 compounds on protein expression in the 60 cell lines as determined by DNA microarrays for more than 8000 genes.[104] An analysis of the truly mind-boggling amount of data revealed clusters of drugs which acted by similar mechanisms (tubulin stabilizers, DHFR inhibitors, etc.). Profiling a hit from anticancer phenotype screening on such a gene chip and comparing it to the known patterns might place it in the appropriate MOA bin.

Another approach would be not to use transcriptomics or proteomics per se, but to do a direct comparison of growth inhibitory values in some or all of the NCI-60 cell lines with those of known drugs using the online program COMPARE. As its name implies, drugs acting through the same mechanism tend to display similar profiles here.[105] Figure 5.13

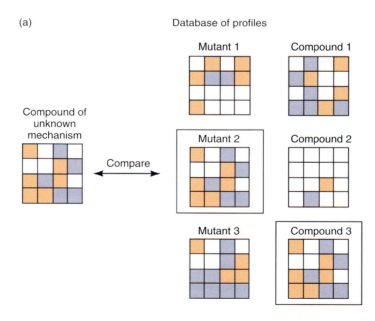

**Figure 5.12 ▶**  Identifying molecular targets using expression profiling. (Adapted with permission from Butcher, RA, Schreiber, SL. Using genome-wide transcriptional profiling to elucidate small-molecule mechanisms. *Curr. Opin. Chem. Biol.* **2005**, *9*, 25–30, copyright 2005, Elsevier.)

GI$_{50}$ mean graph for compound 67574
NCI cancer screen current data, June 2007
Average GI$_{50}$ over all cell lines is 4.46E-9

| Cell panel | Cell line | Log GI$_{50}$ | GI$_{50}$ |
|---|---|---|---|
| Leukemia | CCRF-CEM | −9.0 | |
| | HL-60(TB) | −9.0 | |
| | K-562 | −8.7 | |
| | MOLT-4 | −8.2 | |
| | RPMI-8226 | −9.0 | |
| | SR | −8.4 | |
| Non-small cell lung | A549/ATCC | −8.1 | |
| | EKVX | −8.3 | |
| | HOP-62 | −9.0 | |
| | HOP-92 | −8.1 | |
| | NCI-H226 | −8.2 | |
| | NCI-H23 | −9.0 | |
| | NCI-H322M | −7.3 | |
| | NCI-H460 | −9.0 | |
| | NCI-H522 | −8.2 | |
| Colon | COLO 205 | −8.8 | |
| | HCC-2998 | −9.0 | |
| | HCT-116 | −9.0 | |
| | HCT-15 | −7.4 | |
| | HT29 | −9.0 | |
| | KM12 | −8.7 | |
| | SW-620 | −9.0 | |
| Central nervous system | SF-268 | −9.0 | |
| | SF-295 | −8.6 | |
| | SF-539 | −9.0 | |
| | SNB-19 | −8.5 | |
| | SNB-75 | −9.0 | |
| | U251 | −9.0 | |
| Melanoma | LOX IMVI | −9.0 | |
| | MALME-3M | −8.5 | |
| | M14 | −9.0 | |
| | SK-MEL-2 | −7.4 | |
| | SK-MEL-28 | −8.3 | |
| | SK-MEL-5 | −9.0 | |
| | UACC-257 | −6.6 | |
| | UACC-62 | −9.0 | |
| Ovarian | IGROV1 | −8.2 | |
| | OVCAR-3 | −8.6 | |
| | OVCAR-4 | −7.5 | |
| | OVCAR-5 | −8.0 | |
| | OVCAR-8 | −8.3 | |
| | SK-OV-3 | −8.7 | |

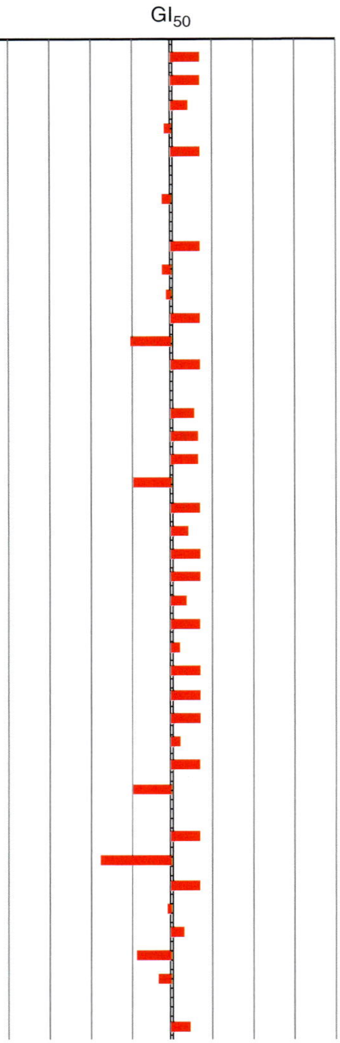

**Figure 5.13 ▶**    Partial NCI-60 profile for Vincristine.

shows part of a profile for one anticancer drug, Vincristine, with red bars to the left indicating less than average activity against that cell line, while those going to the right indicate greater than average potency in the corresponding cells. Regardless of the readout used, these signature-based approaches can provide valuable mechanistic insights, but have limited utility in identifying *novel* targets, which may have equally novel signatures. Still, a "unique" profile here is a data point well worth having.

Transcriptional profiling of an unknown against a custom-built signature database is illustrated by a study in which the profiles for a small set of antibacterial compounds were obtained and used to classify the relevant MOAs for others[106] and another in which a 2036-member library of small molecules was developed and used to elucidate MOAs for 85 antiproliferative agents.[107] Such comparison of databases are not only limited to small molecule perturbogens but can also make use of profiles generated through genetic methods such as deletion mutants in yeast, as an excellent recent review of the subject points out.[108]

In fact, in an ironic twist of fate, true genetic methods like gene-deletion mutants in yeast are often helpful in elucidating the targets of hits obtained through chemical genetics. Here the phenotype, not the expression profile as in the case above, is the observed variable, and collections where almost every gene product in yeast has been knocked out have been generated.[109] A mutation which mimics the phenotype induced by the hit might implicate its MOA. This is called *loss of function* (*LOF*) screening. Alternatively, a coded library of heterozygous yeast mutants can be treated with the hit to look for mutations causing sensitization to its effects—in effect a dose–response experiment where half the "dose" is induced by deletion of a known gene and the other half by the small molecule, a process called *haploinsufficiency profiling*.[110]

When this was done with Cincreasin (Figure 5.14), a simple small molecule identified through phenotype screening, one of the deletions conferring the greatest sensitivity was Mps1, the protein product of which has a spindle-checkpoint function. Follow-on experiments demonstrated that this was in fact the target of the hit.[111] The opposite kind of modification, making the cells less sensitive to drug which might indicate some compensatory or multidrug resistance mechanism, can also be useful. Yeast, of course, is not man, but it's a lot easier to manipulate, and the pathways involved can be very similar in both.[112]

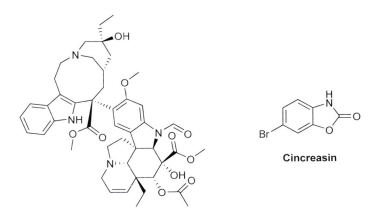

**Vincristine**

**Figure 5.14 ▶**   The cancer drug Vincristine and the Mps1 inhibitor Cincreasin.

## 5.6  In Conclusion

This chapter has presented a brief overview of how new drug discovery projects begin, some of the science and business decisions that go into them, what types of projects researchers can expect to see, and a little about how targets are identified and validated. Although this knowledge hasn't exactly been a state secret, trying to gather it up from several disciplines and a myriad of scattered information sources means that it's often difficult for those new to the industry to get the big picture. This overview obviously can't do justice to the depth and subtlety of knowledge involved in each of the fields mentioned. But the complexities of the drug discovery process are such that even with a profound, in depth understanding of all of these topics, which could only be the result of much effort over many years, the researcher would still understand only a part of the drug discovery process. For example, where exactly do screening compounds come from and what technologies are used to screen for hits?

## Notes

1. DiMasi, J.A., Paquette, C. The economics of follow-on drug research and development. Trends in entry rates and the timing of development. *Pharmacoeconomics* **2004**, 22, Suppl. 2, 1–14.

2. For more about the economics involved in drug entry order, see Cohen, F.J. Entry order as a consideration for innovation strategies. *Nat. Rev. Drug Discov.* **2006**, 5, 285–294.

3. Knowles, J, Gromo, G. Target selection in drug discovery. *Nat. Rev. Drug Discov.* **2003**, 2, 63–69.

4. Source: *MedAd News*. See Appendix: "The Periodic Table of Drugs".

5. Tobert, J.A. Lovastatin and beyond: The history of the HMG-CoA reductase inhibitors. *Nat. Rev. Drug Discov.* **2003**, 2, 517–526.

6. An interesting account of this is provided in Chapter 9, "Atorvastatin Calcium (Lipitor)" in *Contemporary Drug Synthesis*, Li, J.-J, Johnson, D.J., Sliskovic, D.R., Roth, B.D., Eds. (Wiley Interscience, Hoboken, NJ, 2004).

7. See Marris, E. Diabetes drugs under scrutiny in a post-Vioxx world. *Nat. Rev. Drug Discov.* **2007**, 6, 505–507.

8. Herrera, S. Price controls: preparing for the unthinkable. *Nat. Biotechnol.* **2006**, 24, 257–260. For an interesting discussion of price controls and other concerns impacting drug discovery research, see Santoro, M.A., Gorrie, T.M., Eds. *Ethics and the Pharmaceutical Industry* (Cambridge University Press, Cambridge, 2005).

9. For an excellent review, see Palladino, M.A. et al. Anti-TNF-α Therapies: The Next Generation. *Nat. Rev. Drug Discov.* **2003**, 2, 736–746.

10. For more on the thalidomide story, see Chapter 5 in Drews, J. *In Quest of Tomorrow's Medicines* (Springer-Verlag, New York), 1999.

11. He, M.M. et al. Small molecule inhibition of TNF-α. *Science* **2005**, 310, 1022–1025.

12. For more about small molecule TNF disruptors and small molecules interference with protein-protein interactions in general, see Wells, J.A., McClendon, C.L. Reaching for high-hanging fruit in drug discovery at protein-protein interfaces. *Nature* **2007**, 450, 1001–1009.

13. Kubinyi, H. Drug research: myth, hype, and reality. *Nat. Rev. Drug Discov.* **2003**, 2, 665–668.

14. Gerhold, D.L. et al. Better therapeutics through microarrays. *Nat. Genet. Suppl.* **2002**, 32, 547–552.

15. Gygi, S.P. et al. Correlation between protein and mRNA abundance in yeast. *Mol. Cell. Biol.* **1999**, 19, 1720–1730.

16. Quoted in Fitzgerald, K. Less correlation, more causation *Curr. Drug Discov.* January, 2002, 31–33.

17. Dr Szymkowski has presented these delightful analogies in slide presentations and in personal communications with this author.

18. See Szymkowski, D.E. Too many targets, not enough target validation. *Drug Discov. Today* **2001**, 6, 398–399.

19. Kopec, K.K. et al. Target identification and validation in drug discovery: the role of proteomics. *Biochem. Pharmacol.* **2005**, 69, 1133–1139.

20. See especially comments on target validation staging contained in Drews, J. Drug discovery: A historical perspective. *Science* **2000**, 287, 1960–1964, comments on the definition of target validation by Dr Christopher Miller in Szymkowski, DE. Target validation joins the pharma fold. *Drug Discov. Today* **2003**, 2, 8–9, and Winkler, H. Target validation requirements in the pharmaceutical industry *Drug Discov. Today* **2003**, 2, 69–71.

21. With the exception of targeted chemotherapeutic agents like imatinib, where efficacy in models not involving BCR-ABL kinase wouldn't be expected.

22. Zambrowicz, B.P. et al. Predicting drug efficacy: knockouts model pipeline drugs of the pharmaceutical industry. *Curr. Opin. Pharmacol.* **2003**, 3, 563–570.

23. Saftig, P. et al. Impaired osteoclastic bone resorption leads to osteopetrosis in cathepsin-K-deficient mice. *Proc. Natl. Acad. Sci. USA* **1998**, 95, 13453–13458.

24. Smith, W.W., Abdel-Meguid, SS. Cathepsin K as a target for the treatment of osteoporosis. *Expert Opin. Ther. Pat.* **1999**, 9, 683–694.

25. Zambrowicz, B.P., Sands, A.T. Knockouts model the 100 best-selling drugs—will they model the next 100? *Nature Rev. Drug Discov.* **2002**, 2, 38–51.

26. See www.lexpharma.com/discovery/genome500.htm for the Lexicon effort and www.nih.gov/science/models/mouse/knockout/ for the NIH project.

27. Ferrara, N. et al. Heterozygous embryonic lethality induced by targeted inactivation of the VEGF gene. *Nature* **1996**, 380, 439–442.

28. See Pearson, H. Surviving a knockout blow. *Nature* **2002**, 415, 8–9.

29. Kubota, N. et al. PPAR$\gamma$ mediates high-fat diet-induced adipocyte hypertrophy and insulin resistance. *Mol. Cell* **1999**, 4, 597–609.

30. Yamauchi, T. et al. The mechanism by which both heterozygous peroxisome proliferator-activated receptor $\gamma$ (PPAR$\gamma$) and PPAR$\gamma$ agonist improve insulin resistance. *J. Biol. Chem.* **2001**, 276, 41245–41254.

31. Knight, Z.A., Shokat KM. Chemical genetics: Where genetics and pharmacology meet. *Cell* **2007**, 128, 425–430.

32. Zamecnik, P.C., Stephenson, ML. Inhibition of Rous sarcoma virus replication and cell transformation by a specific oligodeoxynucleotide. *Proc. Natl. Acad. Sci. USA* **1978**, 75, 280–284.

33. Crooke, ST. Molecular mechanisms of action of antisense drugs. *Biochim. Biophys. Acta* **1999**, 1489, 31–44.

34. For a good review of AOs, see Dias, N., Stein, C.A. Antisense oligonucleotides: Basic concepts and mechanisms. *Mol. Cancer Ther.* **2002**, 1, 347–355.

35. Myers, K.J., Dean, N.M. Sensible use of antisense: how to use oligonucleotides as research tools. *Trends Pharmcol. Sci.* **2000**, 21, 19–23.

36. See www.isispharm.com/product_pipeline.html#301012.

37. See Rubinstein, M. et al. A review of antisense oligonucleotides in the treatment of human disease. *Drugs Future* **2004**, 29, 893–909, and Potera, C. Antisense—down, but not out *Nature Biotech.* **2007**, 25, 497–499.

38. Fire, A. et al. Potent and specific genetic interference by double-stranded RNA in *Caenorhabditis elegans*. *Nature* **1998**, 391, 806–811.

39. Arts, G.-J et al. Adenoviral vectors expressing siRNAs for discovery and validation of gene function. *Genome Res.* **2003**, 13, 2325–2332.

40. Paroo, Z., Corey, DR. Challenges for RNAi *in vivo*. *Trends Biotechol.* **2004**, 22, 390–394.

41. Chen, X. et al. Chemical modification of gene silencing oligonucleotides for drug discovery and development. *Drug Discov. Today* **2005**, *10*, 587-593.

42. For examples, see Grunweller, A. et al. Comparison of different antisense strategies in mammalian cells using locked nucleic acids, 2′-*O*-methyl RNA, phosphorothioates and small interfering RNA. *Nucleic Acid Res.* **2003**, 31, 3185–3193 and Vickers, T.A. Efficient reduction of target RNAs by small interfering RNA and RNAse H-dependent antisense agents. A comparative analysis. *J. Biol. Chem.* **2003**, 278, 7108–7118.

43. Jia, L. et al. siRNA targeted against matrix metalloproteinase 11 inhibits the metastatic capability of murine hepatocarcinoma cell Hca-F to lymph nodes. *Int. J. Biochem. Cell Biol*, 2007, 39, 2049–2062.

44. See Matsukura, S.et al. Establishment of conditional vectors for hairpin siRNA knockdowns *Nucleic Acid Res.* **2003**, 31, e71–e77.

45. Hoeflich, K.P. et al. Oncogenic BRAF is required for tumor growth and maintenance in melanoma models. *Cancer Res.* **2006**, 66, 999–1006.

46. Scacheri, P.C. et al. Short interfering RNAs can induce unexpected and divergent changes in the levels of untargeted proteins in mammalian cells. *Proc. Natl. Acad. Sci. USA* **2004**, 101, 1892–1897.

47. Jackson, A.L. et al. Expression profiling reveals off-target gene regulation by RNAi. *Nat. Biotechnol.* **2003**, 21, 635–637.

48. He, L., Hannon, G.J. MicroRNAs: small RNAs with a big role in gene regulation. *Nat. Rev. Genet.* **2004**, 5, 522–531.

49. Fitzgerald, K. RNAi versus small molecules: Different mechanisms and specificities can lead to different outcomes. *Curr. Opin. Drug Discov. Devl.* **2005**, 8, 557–566.

50. Cullen, B.R. Enhancing and confirming the specificity of RNAi experiments. *Nature Methods* **2006**, 3, 677–681.

51. Bridge, A.J. Induction of an interferon response by RNAi vectors in mammalian cells. *Nat. Genet* **2003**, 34, 263–264.

52. See Eggert, U.S. et al. Small molecules in an RNAi world. *Mol. BioSyst.* **2006**, 2, 93–96.

53. Whitehurst, A.W. et al. Synthetic lethal screen identification of chemosensitizer loci in cancer cells. *Nature* **2007**, 446, 815–819.

54. Rader, C. Antibody libraries in drug and target discovery. *Drug Discov. Today* **2001**, 6, 36–43.

55. Kneuer, C. et al. Selectins—potential pharmacological targets? *Drug Discov. Today* **2006**, 11, 1034–1040.

56. Biocca, S., Cattaneo, A. Intracellular immunization: antibody targeting to subcellular compartments. *Trends Cell Biol.* **1995**, 5, 248–252.

57. See Stocks, M.R. Intrabodies: production and promise. *Drug Discov. Today* **2004**, 9, 960–966, which is also an excellent review of this subject.

58. Rousselet, N. et al. Inhibition of tumorigenicity and metastasis of human melanoma cells by anti-cathepsin L single chain variable fragment. *Cancer Res.* **2004**, 64, 146–151.

59. Zemel, R. et al. Inhibition of hepatitis C virus NS3-mediated cell transformation by recombinant intracellular antibodies. *J. Hepatol.* **2004**, 40, 1000–1007.

60. See Baines, I.C., Colas, P. Peptide aptamers as guides for small-molecule drug discovery. *Drug Discov. Today* **2006**, 11, 334–341.

61. See Tuerk, C., Gold, L. Systematic evolution of ligands by exponential enrichment: RNA ligands to bacteriophage T4 DNA polymerase. *Science* **1990**, 249, 505–510 and Ellington, A.D., Szostak, J.W. In vitro selection of RNA molecules that bind specific ligands. *Nature* **1990**, 346, 818–822.

62. For a recent review, see Djordjevic, M. SELEX experiments: New prospects, applications and data analysis in inferring regulatory pathways. *Biomol. Eng.* **2007**, 24, 179–189.

63. Jenison, R.D. et al. High-resolution molecular discrimination by RNA. *Science* **1994**, 263, 1425–1429.

64. For a review of spiegelmers, see Eulberk, D., Klussmann, S. Spiegelmers: Biostable aptamers. *Chembiochem* **2003**, 4, 979–983.

65. Helmling, S. et al. Inhibition of ghrelin action in vitro and *in vivo* by an RNA-Spiegelmer. *Proc. Natl. Acad. Sci. USA* **2004**, 101, 13174–13179.

66. Shearman, L.P. et al. Ghrelin neutralization by a ribonucleic acid-SPM ameliorates obesity in diet-induced obese mice. *Endocrinology* **2006**, 147, 1517–1526.

67. Pestourie, C. et al. Aptamers against extracellular targets for in vivo applications. *Biochimie* **2005**, 87, 921–930.

68. Famulok, M. et al. Intramers as promising new tools in functional proteomics. *Chem. Biol.* **2001**, 8, 931–939.

69. Blank, M., Blind, M. Aptamers as tools for target validation. *Curr. Opin. Chem. Biol.* **2005**, 9, 336–342.

70. Lee, J.F. et al. Aptamer therapeutics advance. *Curr. Opin. Chem. Biol.* **2006**, 10, 282–289.

71. Capdeville, R et al. Glivec (STI571, Imatinib), a rationally developed, targeted anticancer drug. *Nature Rev. Drug Discov.* **2002**, 1, 493–502.

72. For a recent review of progress in HIV drugs, see DeClerq, E. New approaches toward anti-HIV chemotherapy. *J. Med. Chem.* **2005**, 48, 1297–1313.

73. Marinissen, M.J. et al. Inhibition of heme oxygenase-1 interferes with the transforming activity of kaposi sarcoma herpesvirus encoded G protein-coupled receptor. *J. Biol. Chem.* **2006**, 281, 11332–11346.

74. Erler, J.T. et al. Lysyl oxidase is essential for hypoxia-induced metastasis. *Nature* **2006**, 440, 1222–1226.

75. Miller, G. A fix for fragile X syndrome? *Science* **2006**, 312, 518–521.

76. Source: Accenture news release, January 19, 2005. Pharmaceutical Companies Must Re-think Innovation Strategies to Address Pipeline Challenges, Accenture and CMR Research Finds. http://accenture.com/xd/xd.asp?it=enweb&xd=_dyn\dynamicpressrelease_796.xml

77. Szymkowski, D. Chemical genomics versus orthodox drug development. *Drug Discov. Today* **2003**, 8, 157–159.

78. For timely and informative reviews of chemical genetics, see (a) Lokey, R.S. Forward chemical genetics: progress and obstacles on the path to a new pharmacopoeia. *Curr. Opin. Chem. Biol.* **2003**, 7, 91–96, and, (b) Smukste, I., Stockwell, B.R. Advances in Chemical Genetics. *Annu. Rev. Genomics Hum. Genet.* **2005**, 6, 261–286.

79. For more about HCS, see Kom, K., Krausz, E. Cell-based high-content screening of small-molecule libraries. *Curr. Opin. Chem. Biol.* **2007**, 11, 503–510.

80. Smukste, I. et al. Using small molecules to overcome drug resistance induced by a viral oncogene. *Cancer Cell* **2006**, 9, 133–146.

81. Bartenschlager, R., Lohmann, V. Novel cell culture systems for the hepatitis C virus. *Antiviral Res.* **2001**, 52, 1–17.

82. Bartenschlager, R. Hepatitis C virus replicons: potential role for drug development. *Nat. Rev. Drug Discov.* **2002**, 1, 911–916.

83. (a) Kaletta, T., Hengartner, M.O. Finding function in novel targets: *C. elegans* as a model organism. *Nature Rev. Drug Discov.* **2006**, 5, 387–399. (b) Kwok, TC. A small-molecule screen in *C. elegans* yields a new calcium channel antagonist. *Nature* **2006**, 441, 91–95.

84. Khersonsky, S.M. et al. Facilitated forward chemical genetics using a tagged library and zebrafish embryo screening. *J. Am. Chem. Soc.* **2003**, 125, 11804–11805.

85. Butcher, E.C. Can systems biology rescue drug discovery? *Nature Rev. Drug Discov.* **2005**, 4, 461–467.

86. Kung, C. et al. Chemical genomic profiling to identify intracellular targets of a multiplex kinase inhibitor. *Proc. Natl. Acad. Sci. USA* **2005**, 102, 3587–3592.

87. Argyrou, A. et al. *Mycobacterium tuberculosis* dihydrofolate reductase is a target for isoniazid. *Nat. Struct. Mol. Biol.* **2006**, 13, 408–413.

88. Khodarev, N.N. et al. Tumor-endothelium interactions in co-culture: coordinated changes of gene expression profiles and phenotypic properties of endothelial cells. *J. Cell Sci.* **2003**, 116, 1013–1022.

89. Drews, J., Ryser, S. Molecular drug targets and genomic sciences. *Nat. Biotechnol.* **1997**, 15, 1350–1359.

90. Hart, C.P. Finding the target after screening the phenotype. *Drug Discov. Today* **2005**, 7, 513–519.

91. Terstappen, G.C. et al. Target deconvolution strategies in drug discovery. *Nature Rev. Drug Discov.* **2007**, 6, 891–903.

92. Kau, T.R. et al. A chemical genetic screen identifies inhibitors of regulated nuclear export of a Forkhead transcription factor in PTEN-deficient tumor cells. *Cancer Cell* **2003**, 4, 463–476.

93. Harding, M.W. et al. A receptor for the immuno-suppressant FK506 is a *cis-trans* peptidyl-prolyl isomerase. *Nature* **1989**, 341, 758–760.

94. Burdine, L., Kodadek, T. Target identification in chemical genetics: The (often) missing link. *Chem. Biol.* **2004**, 11, 593–597.

95. Tanaka, M. et al. An unbiased cell morphology-based screen for new, biologically active small molecules. *PLoS Biol.* **2005**, 3, 764–776.

96. Smith, G.P., Petrenko, V.A. Phage display. *Chem. Rev.* **1997**, 97, 391–410.

97. Liberles, S.D. et al. Inducible gene expression and protein translocation using nontoxic ligands identified by a mammalian three-hybrid screen. *Proc. Natl. Acad. Sci. USA* **1997**, 94, 7825–2830.

98. (a) Licitra, E.J., Liu, J.O. A three-hybrid system for detecting small ligand-protein receptor interactions. *Proc. Natl. Acad. Sci. USA* **1996**, 93, 12817–12821. (b) Lefurgy, S., Cornish, V. Finding Cinderella after the ball: A three-hybrid approach to drug target identification. *Chem. Biol.* **2004**, 11, 151–153.

99. (a) Zlauddin, J., Sabatini, D.M. Microarrays of cells expressing defined cDNAs. *Nature* **2001**, 411, 107–110. (b) Bailey, S.N. et al. Applications of transfected cell microarrays in high-throughput drug discovery. *Drug Discov. Today* **2002**, 7(Suppl.), 1–6.

100. (a) MacBeath, G., Schreiber, S.L. Printing proteins as microarrays for high-throughput function determination. *Science* **2000**, 289, 1760–1763. (b) Mitchell, P. A perspective on protein microarrays. *Nat. Biotechnol.* **2002**, 20, 225–229.

101. Baak, J.P.A et al. Genomics and proteomics—the way forward. *Ann. Oncol.* **2005**, 16 (Suppl. 2), ii30–ii40.

102. Towbin, H. et al. Proteomics-based target identification: bengamides as a new class of methionine amino-peptidase inhibitors. *J. Biol. Chem.* **2003**, 278, 52964–52971.

103. (a) Phillips, C.I., Bogyo, M. Proteomics meets microbiology: technical advances in the global mapping of protein expression and function. *Cell. Microbiol.* **2005**, 7, 1061–1076. (b) Barglow, K.T., Cravatt, B.F. Discovering disease-associated enzymes by proteome reactivity profiling. *Chem. Biol.* **2004**, 11, 1523–1531.

104. Scherf, U. et al. A gene expression database for the molecular pharmacology of cancer. *Nat. Genet.* **2000**, 24, 236–244.

105. (a) Weinstein, J.N. et al. An information-intensive approach to the molecular pharmacology of cancer. *Science* **1997**, 275, 343–349. (b) Rabow, A.A. et al. Mining the national cancer institute's tumor-screening database: Identification of compounds with similar cellular activities. *J. Med. Chem.* **2002**, 45, 818–840.

106. Hutter, B. et al. Prediction of mechanisms of action of antibacterial compounds by gene expression profiling. *Antimicrob. Agents Chemother.* **2004**, 48, 2838–2844.

107. Root, D.E. et al. Biological mechanism profiling using an annotated compound library. *Chem. Biol.* **2003**, 10, 881–892.

108. Butcher, R.A., Schreiber, S.L. Using genome-wide transcriptional profiling to elucidate small-molecule mechanisms. *Curr. Opin. Chem. Biol.* **2005**, 9, 25–30.

109. Giaever, G. et al. Functional profiling of the *Saccharomyces cerevisiae* genome. *Nature* **2002**, 418, 387–391.

110. Lum, P.Y. et al. Discovering modes of action for therapeutic compounds using a genome-wide screen of yeast heterozygotes. *Cell* **2004**, 116, 121–137. For a brief, simplified explanation of the method in this paper, see Hughes, T. et al. *Cell* **2004**, 116, 5–7.

111. Dorer, R.K. et al. A small-molecule inhibitor of Mps1 blocks the spindle-checkpoint response to a lack of tension on mitotic chromosomes. *Curr. Biol.* **2005**, 15, 1070–1076.

112. For more on the utility of yeast in drug discovery, see Auerbach, D. et al. Drug discovery using yeast as a model system: A functional genomic and proteomic view. *Curr. Proteomics* **2005**, 2, 1–13.

# Hit Generation

## 6.1 Introduction

Connoisseurs of drug discovery often consider the pursuit of a good hit to be the most exciting and perhaps most challenging part of any new project. It's here that the drug hunter's determination, skill, and patience can pay off in that first fleeting glimpse of his elusive quarry. As a combination of hard work, scientific acumen, and random chance, there's just nothing else like it. Knowing which part of the woods to stake out and becoming familiar with the different techniques used can save many wasted hours and a tremendous amount of disappointment. Familiarizing the reader with these things is the purpose of this chapter.

## 6.2 Definitions

Most companies have slightly different definitions of what exactly constitutes a "hit" or a "lead," so it's best to ask and find out what the criteria are at your particular organization. Given below are some fairly minimalist but mainstream definitions as they'll be used in this text. They're used in the context of projects that begin with *high-throughput screening* (*HTS*), which is probably the most common, but not the only, approach used in drug discovery today, as we'll see.

To be called an *active* or *unconfirmed hit*, the compound present in a well need only demonstrate some predetermined level of activity at a given concentration. A *confirmed hit*, on the other hand, is an active that's gone through the *hit validation* (*HV*) process. This ensures that the apparent activity wasn't due to an artifact like the ones described later, that the compound present was what it was supposed to be, that it was pure enough, and that it could generate a good dose–response curve. The last of these implies an acceptable $IC_{50}$ or $K_i$ value, usually double-digit micromolar or below, and a curve that isn't too steep, meaning its *Hill slope* is around one. There are legitimate reasons why a desirable active compound could have a steeper slope,[1] but in the vast majority of cases a cliff-like dose–response curve should be treated like one of those flashing web ads congratulating you for having won a new laptop computer. It's almost always the mark of an artifact. Figure 6.1 compares what a desirable and an undesirable dose–response curve might look like. Also, aside from identity, purity, and dose-response, there may also be other criteria for a compound to be a confirmed hit, depending on corporate *mores*.

A *lead* is a compound or series believed to be a valid starting point for the development of a *clinical candidate*. Leads are generated from confirmed hits in the *hit-to-lead* (*HTL*) process,

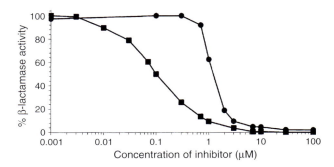

**Figure 6.1 ▶** Representative dose–response curves showing inhibition of the enzyme β-lactamase by a non-specific inhibitor, rottlerin (●) and by the specific and competitive inhibitor BZBTH2B (■). (Reprinted with permission from *J. Med. Chem.* **2003**, *46*, 1478–1483, copyright 2003, American Chemical Society).

described later, whereby fairly major changes to the scaffold and substituents can be made. Leads will need to meet certain criteria spelled out by the project team. These will include potency and selectivity requirements, ADME parameters like solubility and plasma stability, enough structural novelty to allow lead series members to be patented, and probably functional activity in a cellular screen.

Ultimately, the goal is to take leads through the *lead optimization* (*LO*) process, which typically involves the synthesis of very many analogs until tweaking of the structure results in an *optimized lead* suitable for more thorough investigation as a potential clinical candidate. Figure 6.2 shows the stages in this process, along with some typical requirements at each step.

Again, the above scheme applies to projects that begin with high-throughput screening. Other approaches, though, for example beginning with a small group of compounds known to interact with related targets (*chemogenomics*), might bypass the initial HTS stage and start out with a hit or maybe even a lead. But the same requirements will apply. Also note that not all the requirements for a lead will necessarily apply to *chemical tools*, compounds not intended to become drugs themselves but potentially useful in-house for studying targets and biological pathways. These, for example, may not need to be patentable.

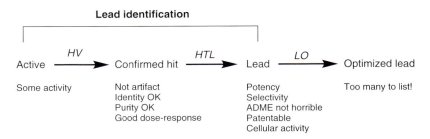

**Figure 6.2 ▶** Stages in the drug discovery process for projects beginning with HTS, and some typical requirements at each one. *HV*, hit validation, *HTL*, hit to lead, *LO* lead optimization.

# 6.3  Groups Involved in Hit-to-Lead

**Protein expression and purification** is a prerequisite for any target-directed project. Phenotype-screening projects will likely still require them farther down the road. Compound requirements can range from tens of milligrams to gram quantities. Not surprisingly, high-throughput screening of libraries will depend on the target protein being available. X-ray crystallography can also require substantial quantities of protein. Often, multiple anti-targets (e.g. other kinases), target proteins from other species (e.g. rat or mouse), protein tools, or mutagenized proteins will need to be made as well. Some proteins are covered by patents so that making them in a given country without a licensing agreement would constitute infringement. So the problems and requirements are many and the technology can be difficult: Expressing mammalian proteins in easier-to-work-with non-mammalian cells can be a challenge. Chemists tend to think of protein expression as a simple, well-precedented routine, exactly the same way that many biologists tend to view chemical syntheses. But a good protein expression group should be properly appreciated. Without them and the legacy of biotechnology they represent, most projects could never get off the ground.

**Assay development/HTS** plays a major role in most early-stage projects today. Converting a single-compound bioassay into a high-throughput format capable of detecting actives as quickly, accurately, and inexpensively as possible without being plagued by compounds which aren't truly active looking like they are (*false positives*) and those which are really active looking like they're not (*false negatives*) is quite a challenge. **Biochemistry/enzymology** will often be involved in pre- and post-HTS activities like target identification and secondary testing. At some companies assay development and enzymology will both be done by the same group.

**Medicinal chemistry** should really be involved from the project's conception, and for a very good reason. As pointed out by Dr Chris Lipinski, the term "good target" often means one thing to someone in biology and another thing to a medicinal chemist.[2] Understandably, biologists tend to judge targets by how likely it is that interfering with a pathway will have a therapeutically useful outcome, while chemists tend to scratch their heads and wonder how likely it is that a small molecule they can design will be able to carry out that interference. Targets vary in how difficult they are to address with small molecules. For example, many small molecules hit aminergic GPCRs, but few do well blocking protein–protein interactions. So both views of what constitutes a good target need to be considered carefully before a project can even begin. And once a project has the green light, medchem input should be an important part of deciding on what compounds to screen.

Later, after confirmed hits have been identified in HTS, medicinal chemists will get busy doing several things. One is deciding on whether or not a given hit is chemically tractable enough to be worth pursuing. They may decide, for example, that a complex natural product hit might be fun to synthesize, but would take more people and time than they have available. If the hit is deemed doable, the next step will probably be to make a small library of analogs to determine whether or not it should progress further. The object at that point isn't to turn a $10\,\mu M$ compound into a $10\,nM$ one (though that would be nice) but to see whether or not real SAR will be possible for it. This amounts to a preliminary assessment of optimizability. Many compounds with weak activity will only lead to a plethora of analogs that are similarly anemic, so it's best to rule out this kind of flat SAR as quickly as possible. If trends are apparent in the analogs, say, para substituents improve activity a bit while ortho substituents make for inactive compounds, it's a sign that it's a very reasonable hit to work with. One of the advantages of HTS screening of compound libraries is that this kind of SAR might even be observable from structural analogs run in the original screen, assuming the data is good enough for this, which isn't always the case.

These kinds of exercises, of course, are only the beginning of medchem's involvement. Further optimization of these and of many other properties will be needed as the project moves along from confirmed hits through hit-to-lead and finally through the long and arduous lead optimization stage. These will be discussed in the following chapters.

**Analytical chemistry** may be required for the identity- and purity-checks involved in hit confirmation. At smaller companies this might be part of the medicinal chemist's duties. **Structure-based drug design** groups can play an important role in hit optimization as we'll see later. They or a related computational group might also be involved in selecting (via pharmacophore searches or virtual screening) and maybe even obtaining the library compounds in the first place. **Cell biology** will be a very early player for cellular phenotype screening projects and will be required for just about every project during the HTL process. This will also bring the **ADME** group into the picture, as properties like solubility and cell permeability will need to be measured (not calculated!) to decide on whether a given compound should be considered a lead or not. **Legal** is likely to be involved in determining whether or not a proposed lead structure or series is novel enough to be patented, which is often not an easy thing to do.

As you can see, even for the first part of the drug discovery process, it really "takes a village."[3] Even early-stage projects will need good communications by all parties through an effective project team, and that need only increases as the project moves along.

## 6.4 High-Throughput Screening

### 6.4.1 History

Beginning sometime in the late 1980s a combination of technological improvements made it possible to screen compounds in a kind of automated, industrialized way. Rather than assaying just a few compounds at a time by hand, robotic systems became available that were designed to operate on 96-well plates, often measuring fluorescence as the readout. In this format, the first and last columns on the plate are usually reserved for controls, so that 80 new data points per plate can be obtained. Steady improvements in speed, reliability, and accuracy were followed by further miniaturization so that eventually 384- and even 1536-well plates could be used. Proponents and proud managers could, and did, toss around ever-increasing numbers of compounds that could be or that had been screened.

What this all meant in a practical sense was that for target-driven projects, screening capabilities were not going to be the bottleneck. On first seeing an HTS robotic system at work, the initial thought that came to many a medicinal chemist was, "Where are they going to get that many compounds?" Up until that time the entire compound collection at a mid-sized pharma company might only number in the tens of thousands, which seemed like a lot back then, representing thousands of chemist-years of work. A chemist could pick up a vial and admire the white crystals that represented the fruits of the creativity and perhaps several weeks' work by a respected scientist 30 years ago. But an HTS robot only saw a well that could be read in a few seconds. In practice, of course, HTS wasn't always simple or straightforward, and its results were and are inherently noisier than slower, more careful methods. Some tradeoff in quality can't be avoided in return for the gain in the quantity of data. But this new, high-tech method did run through compounds quickly.

So its development thus went hand-in-glove with the development of combinatorial chemistry, and it would be equally valid to say that the ability to produce large combi libraries drove the push to implement HTS. The theory behind their use was simply that screening more compounds faster should result in more drugs more quickly. The direction that high-throughput screening has taken and its current role in the real world of drug discovery are discussed in the section below. If presenting this in the format of a series of myths to be exploded seems a little "over the top;" the author can only argue that although no reasonable scientist is likely to have believed them all in the first place, many are common, unconscious assumptions made by those who haven't worked with HTS a lot.

## 6.4.2  Myths and Truths about HTS

*Myth #1: If it takes 10,000 compounds to arrive at one new drug, simply screening a library of 100,000 compounds will get you 10 of them.*
The statement that one out of five- or ten-thousand compounds made ends up being a drug has taken on a life of its own, a kind of *meme*. It usually comes up in the context of how difficult it is to find new drugs, often accompanying the argument that drugs are expensive because research is expensive.[4] It's probably true for analogs of leads, but certainly doesn't apply to the HTS step of drug discovery, where screening hundreds of thousands of compounds often produces few drug leads, much less drugs themselves.

Those wanting real-world insights into what to expect from HTS should see a recent article by Payne, et al. on some antibacterial screening campaigns run at GlaxoSmithKline (GSK).[5] Current antibacterials rely on only a few targets, and resistance is an increasing problem, so finding new targets is an attractive approach. A tremendous effort was put into finding new, genomics-derived targets for bacteria and this led to 67 HTS target-based campaigns being run on around 400,000 compounds over a 7-year period, each one of them costing roughly $1 million. Of the 67 target-based high-throughput screens, only 16 (24%) gave confirmed hits. All of those went into HTL, but after an average time of almost 1 year, only 5 of those hits (31%) had produced a lead. Two of those turned out not to have broad-spectrum properties, one was considered non-druglike, and another was too similar to a competitor's compounds, leaving a single candidate for lead optimization at the end of a long, expensive program.[5]

Keep in mind that the bar here was set higher than most. New targets are likely to be much harder to find hits and leads for than established ones, where knowledge about the kinds of structures that work is greater so that more directed libraries can be screened. And the requirement for broad-spectrum activity, difficult to meet because of the diversity of bacterial genomes, has no equivalent in most other therapeutic fields. Also, as the authors suggest, the diversity of the libraries screened may have been low. They suggest that campaigns in most other therapeutic areas succeeded in delivering a lead (not just a hit) about one-third of the time, versus the 7.5% rate observed here. But remember, a lead is only the starting point for all of those analogs required to get a drug.

*Truth #1: Screening massive libraries of compounds does not guarantee that a lead or drug will be forthcoming.*

*Myth #2: HTS should identify the same hits for a given target regardless of the type of assay used.*

To address this myth, we need to digress briefly to explain some of the more common types of assays used in HTS. A number of different types of readout can be used. Simple absorption measurements can be done, but given the short path lengths afforded by miniature wells and keeping Beer's Law in mind, it's obvious that it can often be a challenge to get a good signal-to-noise ratio here. All the same, this simplest of all techniques does find an occasional use.[6]

As mentioned above, assay readouts based on fluorescence are probably the most commonly used of all HTS detection techniques. A number of different "flavors" of fluorescence-based assays exist. The simplest one, called *fluorescence intensity* (*FLINT*) was used, for example, to assay for inhibitors of a proteolytic enzyme that releases a fluorescent component upon cleavage of a specially designed substrate, as shown in Figure 6.3.[7] In this case the enzyme of interest is Cathepsin K, a cysteine protease critical to bone resorption, inhibitors of which are of interest to fight osteoporosis.[8]

If other proteins don't interfere, the use of cell-permeable substrates allows these types of readouts to be obtained in whole cells as well, an example of this being a cell-permeable rhodamine-based substrate researchers at Maxim Pharmaceuticals used to find inducers of *apoptosis* (programmed cell death), which involves activation of another type of proteolytic enzyme, caspase-3.[9]

For another class of targets but along the same lines, cell-permeable fluorescent compounds can be used to detect intracellular increases in $Ca^{2+}$ levels caused by agonist effects on Gq-coupled GPCRs. Using the *Fluorometric Imaging Plate Reader* (*FLIPR®*)[10] and the calcium-sensitive dye Fluo-3, whose fluorescence increases about 100-fold upon chelation to calcium, all wells of a 96- or 384-well plate can be imaged simultaneously by a CCD camera at less than one-second intervals, allowing not only response, but response kinetics to be observed.[11] Release of intracellular calcium usually results in a 5- to 10-fold enhancement of signal relative to resting cells.[12] Screening of compounds in these cellular assays can be done directly to look for probable agonist response, or it can be done in competition with a

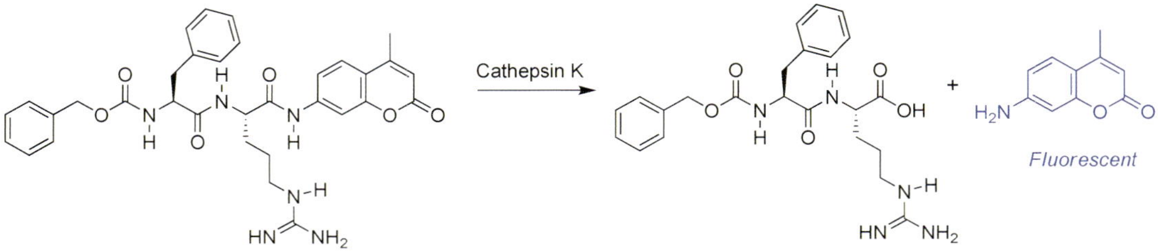

**Figure 6.3** ▶ Cathepsin K cleaves Z-Phe-Arg-AMC to release the fluorescent 7-amino-4-methylcoumarin (AMC). In the presence of inhibitors fluorescence is attenuated. (Palmer, J.T. et al. Design and synthesis of tri-ring P₃ benzamide-containing aminonitriles as potent, selective, orally effective inhibitors of cathepsin K. *J. Med. Chem.* **2005**, *48*, 7520–7534.)

known agonist to look for probable antagonism. Fluo-3 itself is introduced to the cells via incubation with its per-acyloxymethylated ester, which releases the free chelator (along with 5 equivalents of acetic acid and 5 equivalents of formaldehyde) intracellularly upon reaction with esterases, trapping the compound inside the cells, as shown in Figure 6.4. The FLIPR technique is useful for some ion channel screens as well.[13]

In a technique that differs slightly from FLINT, called *fluorescence polarization* (*FP*), polarized light is used to excite the fluorophore. If fluorophore isn't bound to receptor but is free in solution, it tends to be tumbling rapidly and fluorescence emission will therefore have low polarization. But binding to a macromolecular target will reduce the rotation rate and increase the measured polarization.[14] Fluorophores based on the fluoroscein or BODIPY (Figure 6.5), linked to ligands or substrates are frequently used. In this way, for example, a fluorescein-labelled DNA substrate was used to establish an HTS assay to identify small molecules interfering with its interaction with the potential anti-cancer target, replication protein A (RPA), which is involved in nucleotide excision repair.[15]

In *fluorescence resonance energy transfer*, also called *Förster resonance energy transfer* (*FRET*), an excited fluorophore in one part of a molecule (the donor) has its fluorescence

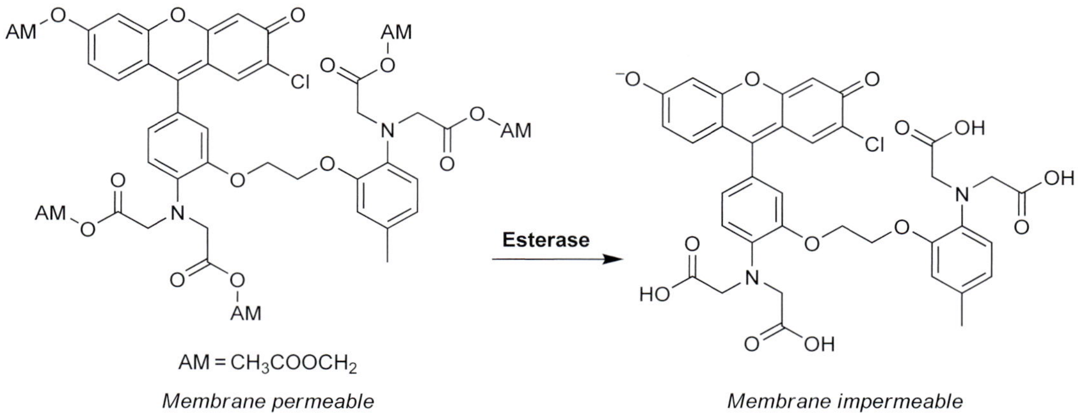

**Figure 6.4** ▶   Fluor-3 penta-acyloxymethyl ester passively diffuses into cells where it is cleaved by esterases and trapped as the parent compound, a calcium-sensitive fluorophore.

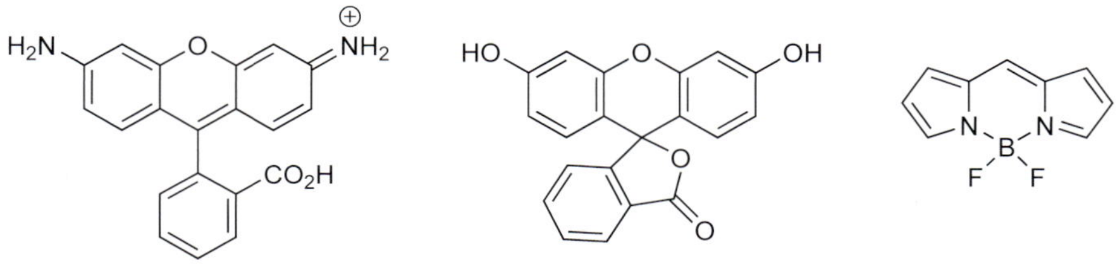

**Figure 6.5** ▶   Some common fluorophores, Rhodamine-110 (left), fluorescein (middle), and the BODIPY core scaffold (right).

internally transferred to a second label attached somewhere else on the molecule (the acceptor). The acceptor is often fluorescent itself, with its emission at a longer wavelength than that of the donor so that the two are easy to tell apart. As long as donor and acceptor are located within 10–75 Å of each other the transfer is efficient and relatively little donor fluorescence will be detectable. But if cleavage occurs anywhere in between, as in Figure 6.6, where the ring-opening action of the bacterial enzyme β-lactamase has separated the two, donor fluorescence will be observed. In this case, observing the change in fluorescence emission from 520 nM to 447 nM would indicate cellular expression of the enzyme, a reporter gene of interest for cellular high-throughput screens.[16]

Many other detection techniques rely on fluorescence readouts.[17] However, other types of measurements can be used too. Receptors have a long history of being amenable to

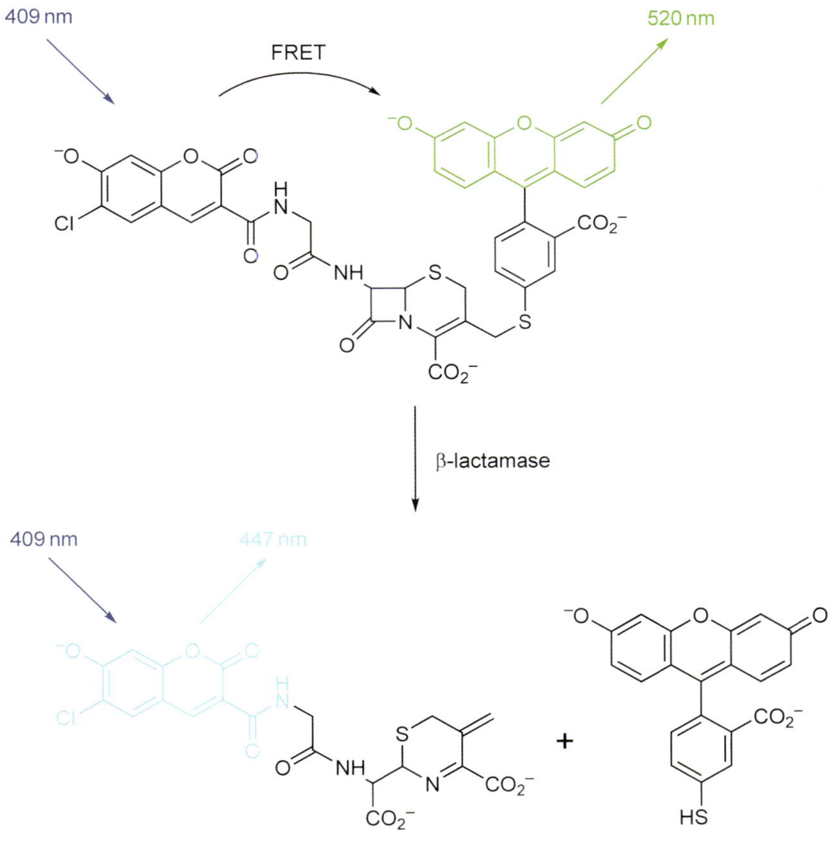

**Figure 6.6 ▶**  Compound CCF2 contains a fluorescent group which upon excitation at 409 nM transfers its energy to the acceptor group (green) that emits at 520 nM. Upon β-Lactamase-catalyzed hydrolysis of the lactam, spontaeneous decomposition of the intermediate results in loss of the acceptor group, and fluorescence at 447 nM instead of 520 nM is observed. (Reprinted with permission from Zlokarnik, G. et al. Quantitation of transcription and clonal selection of single living cells with β-lactamase as reporter. *Science*, **1998**, *279*, 84–88, AAAS.)

radioligand displacement assays. Typically the ligand, labeled with $^3$H or $^{125}$I, is incubated with a membrane-bound receptor preparation and its displacement by a screening compound is measured using liquid scintillation. Unbound ligand is separated from bound ligand by washing. These assays, once carried out by hand, have now been adapted for 96- and 384-well filter plates, but the washing steps complicate the screens and slow down the process. On the plus side, though, over the years this kind of assay has proven to be fairly rugged and reliable.

*Scintillation Proximity Assays* (*SPAs*) eliminate the need for filtering and washing to separate bound from unbound ligand. In this detection technique, *fluomicrospheres* (beads) which are coated with receptors or antibodies are able to emit light upon binding to a radioligand, while unbound ligand in solution has no effect, being too far away (see Figure 6.7).[18] In a recent example of its use, a team at Merck developed a high-throughput assay for inhibitors of a potential target for metabolic syndrome, 11β-hydroxysteroid dehydrogenase (11β-HSD1). This enzyme reduces cortisone to hydrocortisone (cortisol). Tritiated cortisone was therefore used in an SPA assay where the binding of [$^3$H]cortisol was detected via a monoclonal antibody attached to the beads that had 100-fold higher affinity for the alcohol than for the ketone.[19]

Radiometric techniques bear the distinct disadvantage of generating radioactive waste, which requires special handling and is increasingly expensive to dispose of. Another bead-based method for detecting molecular interactions that doesn't involve radioisotopes is called *AlphaScreen*™, which stands for *Amplified Luminescent Proximity Homogeneous Assay*. This uses two specialized types of bead, a donor and an acceptor. The donor bead, to which one molecule is conjugated, contains a photosensitizer (phthalocyanine) that produces large amounts of singlet oxygen upon laser excitation. Because $^1$O$_2$ has a very short lifetime, it can diffuse out only about 200 nM, so an acceptor bead, to which the second component is conjugated, will emit light if and only if it's been drawn close to the donor bead by binding interactions between the two components, as shown in Figure 6.8. This technique has found broad applicability to areas like second messengers, kinases, and protein–protein interactions.[20]

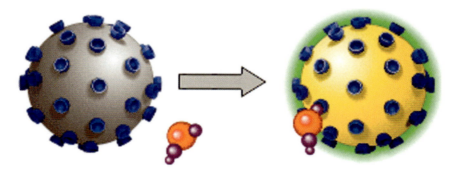

**Figure 6.7** ▶ Scintillation proximity assay concept diagram. Unbound radioligand (left) does not stimulate the bead, but binding brings it into close-enough proximity to cause the bead to emit light (right). (Courtesy of GE Healthcare Life Sciences. Copyright GE Healthcare.)

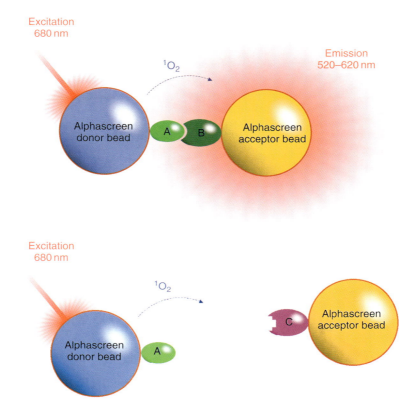

**Figure 6.8 ▶** AlphaScreen[TM] detection. Laser excitation of the donor bead causes it to convert ambient oxygen into singlet oxygen. If appended molecules A and B bind (top), the chemiluminescent acceptor bead is brought close enough to emit light. If a molecule C that does not bind A is attached to the acceptor bead (bottom) or if a small molecule interferes with the binding of A and B, then the donor and acceptor will be too far apart for light to be emitted. (Courtesy of PerkinElmer.)

With some detection technologies compounds can be screened in microarrays instead of plate wells. This requires that either the small molecule or the target be attached to a solid support, usually a glass slide, as shown in Figure 6.9.[21] The other potential binding partner, which can be labeled (say with a fluorescent dye if *fluorescence microscopy* is used) or unlabeled (if surface plasmon resonance detection, discussed later, is to be used), is then exposed to the slide and interactions are detected.

Box 6.1 lists the techniques described above, and the reader should keep in mind that many others exist. Another thing worth noting is that the versatility of many of them means that more than one method might be used to find hits for a given target. In the real world, only one method is normally used for a new HTS campaign, and its selection depends on factors like cost of reagents, assay development time, and previous experience in the department. In addition, whatever assay is used will be individually optimized to get the best Z-factor (a measure of goodness; 1.0 is ideal).[22] But, to finally return to our original myth, *will the same hits be found for a given target regardless of the method used?*

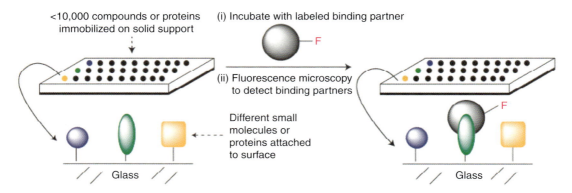

**Figure 6.9 ▶**  Small-molecule and protein microarrays: different small molecules or proteins are attached to a glass slide (usually $25 \times 75$ mm) and probed with a fluorescently labeled binding partner. Fluorescence microscopy is used to detect small-molecule–protein interactions. (Reprinted with permission from *ACS Chem. Biol.* **2007**, *2*, 24–30, copyright 2007, American Chemical Society.)

---

**Box 6.1    Some Types of Assays Suitable for High-Throughput Screening**

▶  Absorbance

▶  Fluorescence Intensity (FLINT)

▶  Fluorometric Imaging Plate Reader (FLIPR®)

▶  Fluorescence Polarization (FP)

▶  Fluorescence Resonance Energy Transfer (FRET)

▶  Radioligand displacement

▶  Scintillation Proximity Assay (SPA)

▶  Amplified Luminescent Proximity Assay (AlphaScreen™)

▶  Surface Plasmon Resonance (SPR)

---

A study at Novartis compared hits obtained by screening 30,000 compounds for tyrosine kinase activity in three different assays: a scintillation proximity assay (SPA), a homogeneous time-resolved fluorescence energy transfer assay (HT-FRET), and a fluorescence polarization (FP) assay.[23] Library compounds were screened as mixtures of five in a well, which meant that after active wells were identified, the five constituents would be *deconvoluted* by screening each compound individually or using other methods to establish which of the compounds was active.[24] When this was done SPA, HT-FRET, and FP turned up 30, 59, and 64 hits respectively. A low degree of overlap was observed between the sets. Only four or seven compounds, depending on where the activity bars were set, would have been identified as hits in all three assays, had they been run independently. The authors conclude that "the assumption that similar active compounds will always be identified in lead finding, regardless of the assay technology utilized, is not valid."[23] Indeed, another more recent paper from Boehringer Ingelheim comparing three other methods developed to screen for inhibitors of interleukin-2-inducible T cell kinase (ITK) reports that 57% of the hits overlapped between those assays, which is better, but still means that many would have been missed with any one of them.[25]

In another study, 42,000 compounds were screened in three different versions of a farnesoid X receptor (FXR) antagonism assay. AlphaScreen™, time resolved fluorescence (TRF), and time resolved FRET (TR-FRET) were used. These identified 104, 23, and 57 actives, respectively. Only 18 compounds were identified by all three methods. Further cellular testing showed that only about 30% of them were functionally active, with AlphaScreen identifying more of the functionally active compounds than the other two methods. This group also found that screening mixtures rather than discrete compounds was more likely to result in both false positives and false negatives.[26]

The medicinal chemist who keeps all of the above in mind is bound to have more respect for those involved in HTS, but will, paradoxically, probably be less willing to accept their initial results at face value!

*Truth #2: Actives and confirmed hits obtained for a given target do depend on the type of assay used.*

*Myth #3: Actives identified through a primary high-throughput screen are all bona fide hits.*

For project scientists, getting those first HTS campaign results back is exciting. Given half a chance, medicinal chemists will immediately pour through the results, judging structures, looking for trends, and dreaming up new compounds to make. Anyone who can look through this kind of data without at least some slight increase in heart rate is probably in the wrong line of work. The danger, though, is that too much of significance will be placed on initial, unconfirmed results, which will cause scientists with the best of intentions to launch off in new directions only to crash and burn later when they find that their new project is based on a false positive, a compound subsequently shown to be inactive. For this reason, primary screening data is sometimes withheld and information released only after the hits have gone through the validation process. Experience shows that most actives will, in fact, drop out. But why is that?

Many artifacts are specific to the detection method used. Assays using fluorescence readouts are most often fooled by test compounds that either have some fluorescence themselves (*autofluorescence*) or that quench fluorescence by absorbing at the observed emission wavelength. Depending on which of these effects a compound has and whether fluorescence indicates activity or lack of activity in a given assay, such compounds can account for many a false positive or false negative. "The situation is not improved by the fact that the fluorophores in many assays are used in the nanomolar range, whereas screened compounds are typically tested in the micromolar range," concludes one review of problems with fluorescence readouts in HTS.[27] To get around this problem, the actives from such a screen can be tested using a different assay not involving fluorescence in order to confirm or deny their activities. Sometimes a molecule that is fluorescent can also be a genuine hit, and for such compounds "only a secondary assay, using an alternative readout, can provide a definitive answer."[27] But if fluorescence or fluorescence-quenching caused a compound to be a false negative, that is, not picked up in the first place, just following up on the ones that *were* picked up by a fluorescence-based assay is like asking everyone who isn't present to raise their hands.

Aside from fluorescence artifacts, light scattering by compounds that aren't entirely soluble can also affect assays using an optical readout. Other assay-specific problems exist as well. AlphaScreen™, for example, is susceptible to interference from singlet oxygen quenchers. The FLIPR technique for monitoring intracellular release of calcium ion can't distinguish between $Ca^{2+}$ released from the ER (the desired effect), and $Ca^{2+}$ entering the cell either through a compound's effects on ion channels or through membrane permeabilization.

The last of these is the bane of many a cellular screen. In the massive GSK antibacterial screening work mentioned earlier, a cellular phenotype screen against *Staphylococcus aureus* turned up literally thousands of hits, the vast majority of which had detergent-like, non-specific effects on membranes. These were often lipophilic compounds with a positive charge and they could be picked out by secondary screening in a cell lysis assay. But the authors note that "Sorting through the high number of nuisance compounds to define whether the antibacterial activity was a result of either a nonspecific (and potentially cytotoxic) mechanism or a specific target was too burdensome."[5] And, of course, cytotoxicity induced by a compound through *any* mechanism can cause profound changes in biochemistry that might be mistaken for specific activity, so it's another thing than needs to be ruled out in cellular screens.

Compounds with redox activity can give false activity by affecting cysteine residues and the like. Reactive functional groups can often give hits that are rarely pursued owing to concerns about toxicity. Most chemists will instinctively realize that reactive compounds like alkylating and acylating agents are generally frowned upon as potential starting points for new drugs and as such are kept out of most screens. For example, Vertex Pharmaceuticals developed a program called *REOS*, which stand for *Rapid Elimination of Swill*, which can filter out reactive and other potentially problematic compounds so that screened libraries don't contain them in the first place.[28] Figure 6.10 shows some of these functional groups. Those familiar with such preselection know that caution needs to be exercised in using them, as genuine hits and leads can be filtered out along with many non-productive starting points. For example, Retrovir (AZT, zidovudine, Figure 5.9) contains one of the "bad" groups, an azide, and the anti-malarial Artemisinin is actually a peroxide.

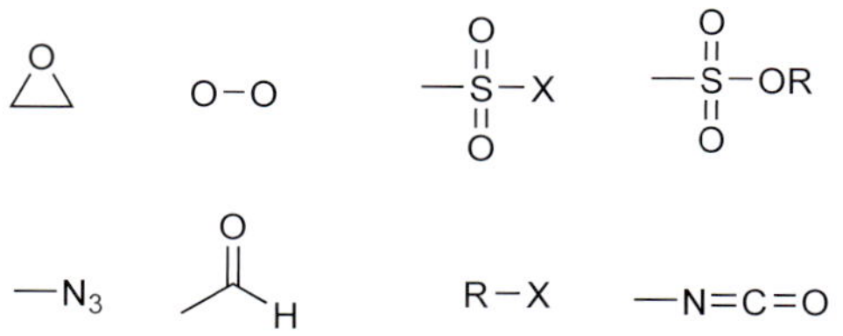

**Figure 6.10** ▶  Some functional groups flagged by REOS. (Walters, W.P.; Namchuk, M. Designing screens: How to make your hits a hit. *Nat. Rev. Drug Discov.* **2003**, *2*, 259–266.)[29]

Overall, keeping such reactive compounds out of screening, or at least flagging them as suspect, is probably a good idea. What's sometimes overlooked, though, is the fact that they can enter screens in a stealthy way as trace impurities, generally left over from a previous synthetic step (say a tosylation or reductive amination), and that certain screens can potentially be more sensitive to the presence of such impurities than any HPLC detector. Having clean compounds, as discussed below, can do wonders for cutting down on false positives and dead-end hits, and it's not without reason that some of the first things done after an active has been identified are checking its purity and often resynthesizing and retesting it.

Mechanical problems giving rise to false positives or false negatives can involve things as simple as pipette tips clogging up, solvent evaporating, or compound carryover from one well to the next. Partly because the first and last columns on a plate are normally reserved for controls, and compounds can easily be spilled or carried over, *edge effects*,[30] as shown in Figure 6.11, are often encountered.

All of these effects, fluorescence artifacts, edge effects, insolubility, etc., have been understood for some time. But even taking them all into account still wasn't enough to explain a mysterious problem that had probably been observed in many labs. Dr Brian Shoichet of UCSF elaborates: "We were looking for inhibitors of β-lactamase. We found all these hits and they were all noncompetitive, which was weird, and they were all time-dependent, which was weird. None of those things were impossible, but to have them all? We had twenty of these molecules and so we said, 'Well, let's just make sure they're not inhibiting chymotrypsin.' And they did. Then we said, 'Let's make sure they don't inhibit dihydrofolate reductase.' And they did. We said, 'Let's make sure they don't inhibit β-galactosidase.' And they did. We knew we were in trouble. Usually, when that happens in pharma, people drop the project."

Fortunately for screeners ever since, Dr Shoichet's group stuck with it. "I'd felt like I had seen these compounds before," he said. "You know, by then I had been in virtual screening

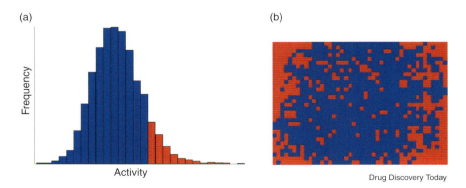

**Figure 6.11** ▶   A plate displaying edge effects. The highest measured activities are marked red in both (a) the histogram and (b) the plate map. It's clear that compounds with a high activity are clustered around the edge of the plate. (Reprinted with permission from Harper, G., Pickett, S.D. Methods for mining HTS data. *Drug Discov. Today* **2006**, *11*, 694–699, copyright 2006, Elsevier).

for 10 years and I'd been talking to people in high-throughput screening. Now, nobody would tell me this, but I knew there would be secret problems. People at one company told me that sometimes 90% of their hits had these funny properties. So that's what got us interested in it." What they found was that these and many other time-dependent, *promiscuous inhibitors* shared a common mode of action. They formed aggregates between 30 and 1000 nm in diameter that seemed to sequester the enzyme away from substrate, thereby reducing its apparent activity in the assay just as a genuine inhibitor would. Figure 6.12 illustrates what happens, by electron microscopy.

Compounds acting this way were found not only in combi library hits but also in some reagents sold as standards for kinase assays and even in 7 of 50 marketed drugs that were tested for this effect.[31] Aggregation seemed to go along with relatively high concentrations, and the papers point out that "compounds that form aggregates, and hence act promiscuously, at micromolar concentrations might well be specific at nanomolar concentrations, where aggregation is less likely."[32] One of the promiscuous inhibitors, quercetin (Figure 6.13), had even had its structure solved in the ATP site of two different kinases, so its activity against them must be real enough. Wherever one looked, more of these

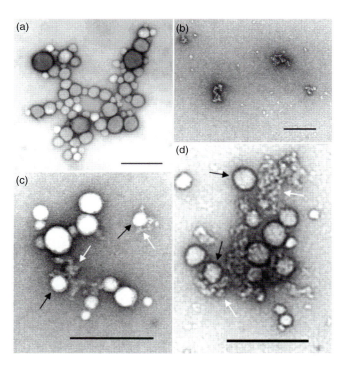

**Figure 6.12 ▶** Visualization of 14PTH (tetraiodophenolphthalein) aggregates and β-galactosidase molecules by TEM: (a) 100 μM 14PTH alone; (b) 0.1 mg/ml β-galactosidase alone; (c,d) 100 μM 14PTH with 0.1 mg/ml β-galactosidase. Representative 14PTH aggregates are marked with black arrows, and β-galactosidase molecules are marked with white arrows. The bar represents 200 nm. (Reprinted with permission from *J. Med. Chem.* **2003**, *46*, 4265–4272, copyright 2003, American Chemical Society.

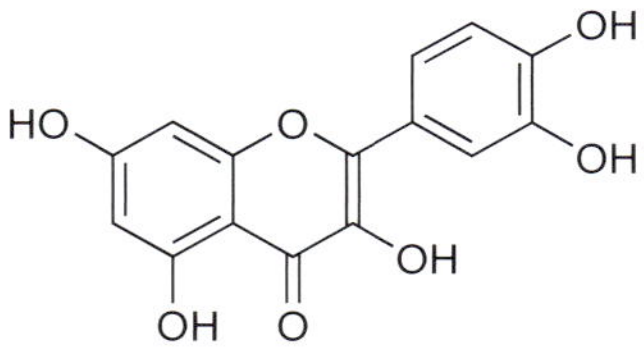

**Quercetin** 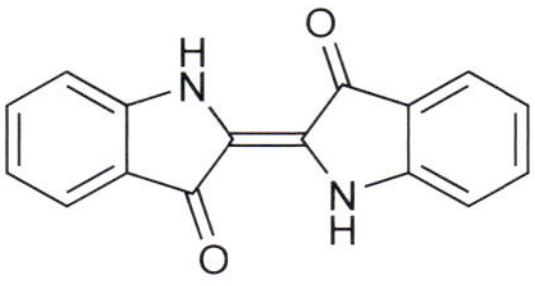   **Indigo** 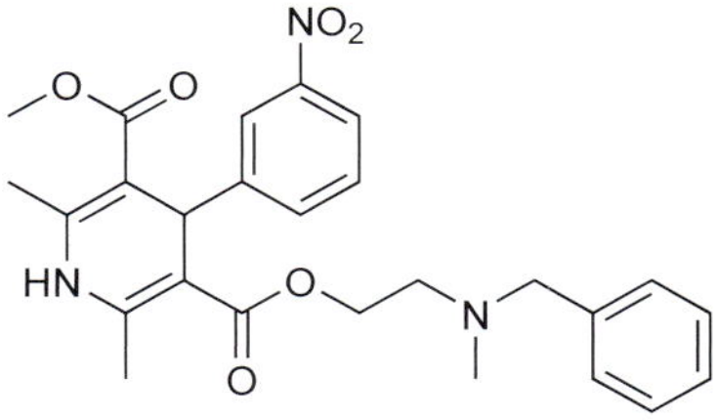   **Nicardipine**

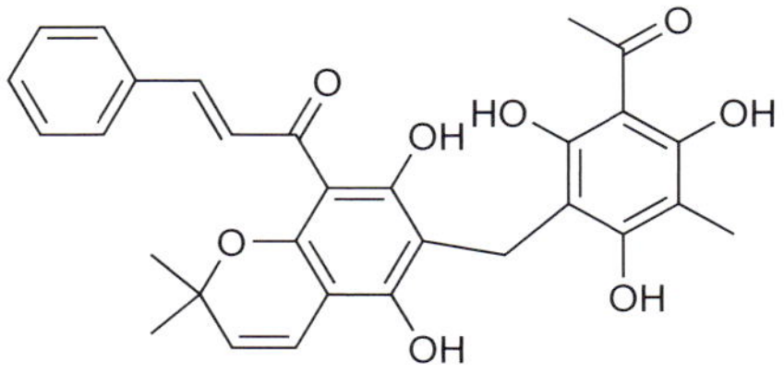

**Rottlerin** 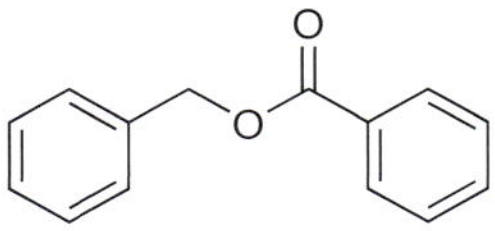   **Benzyl benzoate** 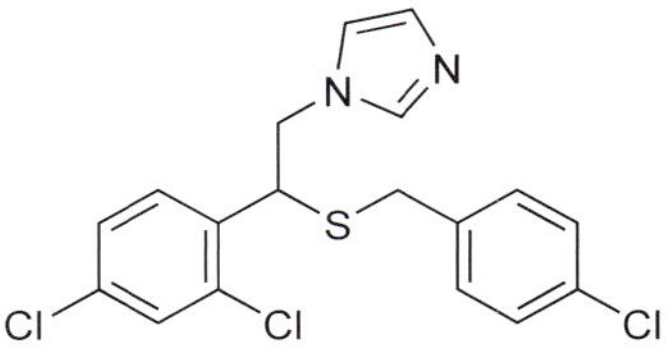   **Sulconazole**

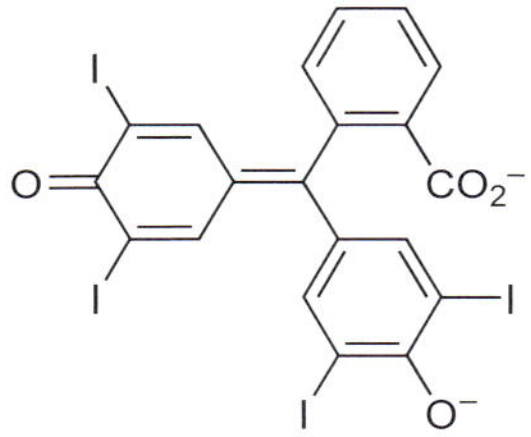

**Tetraiodophenolphthalein** 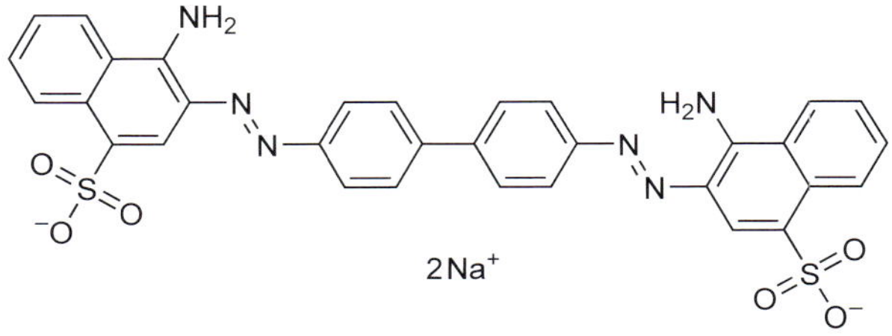   **Congo Red**

**Figure 6.13** ► Some promiscuous inhibitors acting through aggregation. (Seidler, J. et al. Identification and prediction of promiscuous aggregating inhibitors among known drugs. *J. Med. Chem.* **2003**, *46*, 4477–4486; McGovern, S.L., Shoichet, B.K. Kinase inhibitors: Not just for kinases anymore. *J. Med. Chem.* **2003**, *46*, 1478–1483.)

"gators" could be found, leading to the suspicion that HTS might be picking up far more aggregators than real hits,[33] with the troubling possibility that some of these might, like quercetin for kinases, be both. Figure 6.13 shows just a few of the promiscuous inhibitors identified that act through aggregation.

Aggregators turn out to be even more of a problem in screening *mixtures* of compounds, a popular strategy to save on time and reagent costs. Here either artifactual synergy or antagonism might be observed.[34] To date little information on what exactly aggregators might do in cellular assays has been offered in the literature, though this will no doubt change. Regarding in vivo effects, however, recent research by a group including the legendary drug hunter Dr Paul Janssen demonstrated the presence of aggregates formed by some non-nucleoside reverse transcriptase inhibitors (NNRTIs) in simulated gastric fluid and proposed that this accounts for their unusually favorable absorption into the lymphatic system.[35] Conceivably, one day pharma might consider aggregation a desirable property!

An understanding of how these promiscuous inhibitors work has made it possible to devise ways to identify them. Dr Shoichet and his group found that adding a little non-ionic detergent can have a major effect on them. "Low concentrations of Triton X-100 prevent the aggregate–enzyme interaction, although many aggregate particles remain in solution. At higher concentrations of detergent, aggregate particles themselves disintegrate."[36] Although other methods like *dynamic light scattering* (*DLS*) can be used, they suggest that one way to identify promiscuous inhibitors is to counterscreen the library against an enzyme like β-lactamase in the presence and absence of 0.01% Triton X-100. Compounds acting through aggregation will tend to lose their activity in any assay with detergent, and ambiguous results can be followed up in other ways.[37] And, fortunately, many HTS assays incorporate similar amounts of non-ionic detergents to begin with, which should minimize the number of false positives due to this phenomenon.

Given the many problems with both false positives and false negatives presented above and reiterated in Box 6.2, how should a medicinal chemist approach HTS screening?

"I would say two things," Dr Shoichet advises. "One is put away some of the biases that you go into it with, like 'this is druglike' or 'that is not druglike'. It's easy to actually do unambiguous experiments that will tell you whether it's behaving well. You can figure that out. You can just do the experiments. Put aside this, 'Oh, I don't like this functional group' stuff. You go into high throughput and virtual screening looking for something *new*.

"The converse of that is that you can insist that your biochemist colleagues bring you good measurements. 'Let's see an $IC_{50}$ curve with a Hill slope of 1, thank you very much.' Incubate it: is it reactive? These are all controls that you can do. You can't expect them to be done on every molecule, but you don't have to. You can focus it down pretty quickly to a small number of compounds. So I would be open and critical."

---

**Box 6.2   Some Common Sources of HTS Artifacts**

▶ Compound fluorescence/fluorescence quenching

▶ Light scattering due to insolubility

▶ Assay-specific artifacts

▶ Cell membrane permeabilization

▶ Cytotoxicity (in cellular assays)

▶ Reactive compounds

▶ Mechanical problems (plumbing, evaporation, etc.)

▶ Edge effects

▶ Compound aggregation

***Truth #3: Artifacts can far outnumber confirmable hits in initial HTS data.***

*Myth #4: All compounds screened should follow the Rule of Five.*

The publication of Lipinski's 'Rule of Five'[38] (ROF) without a doubt improved screening libraries in a major way. Most libraries nowadays are designed from the outset to be "rule of five compliant". Gone are the days when a combi compound might have two *n*-octyl groups and a MW of 700 Da. Without the emphasis Lipinski's rule placed on small, reasonable molecules, the screening of combi libraries would have resulted in many fewer clinical candidates than it has to date.

One needs to keep in mind, though, that there are cases where the famous rule admittedly doesn't apply, is too strict, or isn't strict enough. Natural products, as we'll see, have always been given a free-pass based on the empirical observation that some of them violate the rule but are still orally bioavailable. It wouldn't be appropriate to deselect compounds from natural product libraries because they violate the rule of five. Compounds can be actively transported into or out of cells, and the rule doesn't apply to these. Intramolecular H-bonding can increase cell permeability by a factor of 10, but whether or not this will happen with a given compound can't be reliably predicted.[39]

Antibacterials tend to be addressed by drugs based on natural products that frequently fall outside the norm for rule-of-five compliance. Figure 6.14 illustrates LogP and MW values for marketed drugs used for CNS indications and for antibacterials.[5] The former generally meet even stricter criteria than ROF.[40] The latter, though, tend to be larger, much more

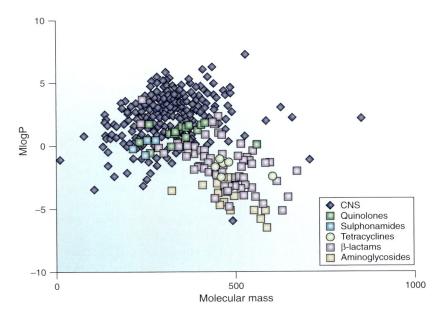

**Figure 6.14** ▶ Properties of marketed antibacterials differ from those of other drugs, such as CNS therapeutics. Note the negative LogP values and higher molecular weights (units in Daltons) for the former. (Reprinted with permission from Payne, D.J. et al. Drugs for bad bugs: confronting the challenge of antibacterial discovery. *Nat. Rev. Drug Discov.* **2007**, *6*, 29–40, copyright 2007, Macmillan Pubishers Ltd.)

hydrophilic compounds that have LogP values closer to *negative* 5 than to 5. For this special case, it might make more sense to design these properties into antibacterial libraries than to look at compounds clustered in traditional ROF space.

The rule, of course, is normally used to maximize the likelihood of oral bioavailability. Although oral administration is favored for most drugs, it isn't an absolute requirement for them all, otherwise, for example, no biologics for cancer or inflammatory diseases would exist. These latter drugs, however, have the distinct advantage of acting on cell surface receptors or extracellular targets, so they don't need cell permeability, a property that usually goes along with oral bioavailability and ROF compliance. So, although the rule is generally a good one to follow in compound selection, there are a few cases (like those of IV drugs and extracellular targets) where its violation shouldn't result in any jail time for the researcher.[41]

One final factor that needs to be mentioned has to do with the optimization of hits and leads. As we'll see in a later section, it's almost always easier to gain activity by adding mass rather than taking it away, which means the final drug candidate will tend to be heavier than the original hit or lead. For most programs (non-natural product, non-antibacterial, non-IV drug) this means that starting points with barely acceptable rule-of-five compliance will be morphed into compounds unlikely to have oral bioavailability. Applying criteria even stricter than Lipinski's rule to the original hits is probably more appropriate in most cases, and this is one of the strengths of *fragment screening*, discussed later.

The take-home message from all of this is that although the rule of five is generally a great help and has improved the quality of hits obtained from HTS screening, it's best considered within the specific context of each project and library, not just blindly applied to them all. Overly strict adherence to the rule "can have the adverse effect of restricting diversity . . . and hence also reducing similarity to natural products."[42]

**Truth #4: In some cases it's inappropriate to apply Lipinski's rule of five to screening libraries. Different or even stricter criteria may be more useful.**

*Myth #5: Screening diversity libraries is always the best way to find good hits.*

Diversity libraries have always been of great interest. The hope was that they'd enable new pharmacophores or molecular motifs to be found that might lead to drugs very different from those already known, especially for targets that have been hard to address with other methods. At the same time, the possibility of making directed libraries with hundreds of close analogs of active compounds, the aim being improvement in some particular property like selectivity, solubility, or CYP inhibition, hasn't gone unnoticed either. Over the years, the directed approach has proven to be more successful and more popular than diversity libraries. Although leads arising from diversity libraries are still occasionally reported, in 2001 it was already noted (partly because such libraries had become commercially available) that "Large pharmaceutical companies are shifting internal resources away from massive, diversity-driven, library production and focusing combinatorial chemistry/parallel synthesis on project-specific or target-specific work."[43] Today, it's parallel synthesis and small, directed libraries that the chemist is likely to deal with on a regular basis.

The problem is that diversity space tends to be so huge that precious little of it is actually sampled by even the best "diversity" library. And, as pointed out by Dr Lipinski, practical limitations like chemistry success, selection bias, and reagent availability bias mean that true diversity just does not exist experimentally.[2] High-throughput screening of diversity libraries is still routinely used and can produce hits, but no one should assume that it's more likely to do so than other methods.

Referring back to the GSK work cited earlier, few hits and even fewer leads arose from a massive campaign of 67 different target-based high-throughput screens against new bacterial targets. A shift in strategy followed. The focus changed to structural modifications of compounds with known antibacterial activity, and to finding new leads "the old-fashioned way: screening a small, discrete library of compounds for antibacterial activity." This proved more successful than the previous HTS approach, with multiple compounds moving into clinical development.[5]

**Truth #5: Large-scale screening of diversity libraries isn't necessarily the best or the only way to find new leads.**

*Myth #6: When it comes to compound libraries quantity matters more than quality.*

Relying on ever-increasing numbers of compounds to find one good one has always seemed a little paradoxical. When trying to find a needle in a haystack, it's been noted, "the best strategy might not be to increase the size of the haystack."[44] Again, early combi libraries suffered from poor physical properties, limited purification (often by *solid phase extraction* or *SPE* only), and an underemphasis on QC (only a fraction of the wells normally being analyzed), but they were nothing if not large. Some had over one million members. Many were *compressed*, with multiple compounds per well.[45] These particular haystacks eventually yielded so many useless needles, in the form of hits that couldn't be validated, that most companies have long since discarded them in favor or more modern libraries with higher standards, though not numbers. Smaller, cleaner libraries have proven to be far more likely to provide validated hits in the shortest time. According to Dr Sam Danishefsky of Columbia University, a small collection of "smart" compounds "may be more valuable than a much larger hodgepodge collection mindlessly assembled."[46]

"The *quality* of combinatorial libraries determines the success of biological screening in drug discovery programs,"[47] [italics added] and these days that lesson has been learned: high-throughput HPLC purification of every compound often follows high-throughput synthesis.[48]

**Truth #6: In the absence of good QC, increasing library size only makes things worse.**

*Myth #7: Running an HTS assay once at a single concentration is sufficient.*

To minimize the time and money involved, HTS campaigns have often been limited to one measurement per compound at a single concentration. But putting all your eggs in one basket and relying on a single data point per compound turns out to have its disadvantages. Although assay developers do everything they can to minimize the errors involved, running an assay in high-throughput mode just turns out to be "noisier" than doing a more careful experiment by hand. And since actives are often picked out by defining some activity threshold (like 50% inhibition at 20 μM), or by falling within, say, the top 0.5% of the most active compounds, having a larger standard-deviation means a given data point might arbitrarily end up on the wrong side of the bar. False positives and false negatives, as we've seen, do abound, and errors in measurement are among the many reasons for this.

Having two data points helps, and these can be obtained by running the assay more than once: *replication*. In theory, taking two measurements reduces imprecision (standard error of a mean) by 29% so that "replicates make minimally and moderately active compounds easier to detect."[30] An experiment at Novartis found that running the same high-throughput assay for nuclear receptor recruitment twice resulted in up to 35% of the hits not being confirmed, with the percentage depending on where the activity bar was set.[49] So, although replication can be expensive, its cost needs to be weighed against the time and effort saved by not pursuing as many false positives and by missing fewer actives that wouldn't have been picked up otherwise. Replication is increasingly being used in HTS screens these days to increase the quality of the data from that obtained with "single shot" screens.

Screening at two different compound concentrations is popular as well. Finding that the activity at the low concentration exceeds that at the high concentration can flag the wells as potentially problematic; having data at the lower concentration can help rank order the actives; and seeing activity coincide with concentration inspires confidence that they're real. So when it comes to high-throughput screening, many are finding that, to quote Jacqueline Susann, "once is not enough."

**Truth #7: Better data can be obtained through assay replication and multi-concentration measurements.**

*Myth #8: False positives are worse than false negatives.*

In keeping with Sir James Black's rule, all of the factors that contribute to false positives could not, of course, have been predicted in the early days of HTS, so the high ratio of false positives to genuine hits it turned up came as a surprise. As we've seen, eliminating many of the false positives, by, say, replicating an assay or screening them in the presence of a detergent to look for aggregation effects, takes some work that originally wasn't foreseen as being necessary. As a result, false positives were, and often still are, considered as the major drawback of high-throughput screening. False negatives, on the other hand, are more like "the little man who wasn't there." They're easier to ignore, but all the more insidious.

"If 0.1% of a million compounds to be screened are truly active, a low false-negative rate of 2% represents 20 potential candidates lost."[30] Realistically, though, how big a problem this is depends on the nature of the library and the compounds: "With synthetic compound collections, the potential loss may be lessened because they are made from a set number of basic scaffolds. Thus, in practice, missing an active compound may not matter if related compounds are detected. When screening natural products or extracts, however, truly unique chemical entities will go undetected."[30] Not detecting a single active that might lead to a drug would be tragic, so the problem of false negatives is an important one.

A new variation on HTS called *quantitative High-Throughput Screening* (*qHTS*) was recently reported by researchers at the NIH Chemical Genomics Center.[50] Using this new high-throughput screen for pyruvate kinase activators and inhibitors, they were able to test more than 60,000 compounds at no fewer than seven different concentrations each, essentially generating dose–response curves for every one. Advantages included a significant reduction in the numbers of both false positives and false negatives relative to a "single shot" assay, with an incredibly small 2% of the actives identified being false positives, while 40% of the actives qHTS identified wouldn't have been detected by screening at a single concentration (false negatives). The method also provided something that many chemists would truly appreciate: a dataset complete with $IC_{50}$s that enables SAR analysis directly from a primary screen.

The quality of data was such that even differences in activity between *identical compounds supplied by different vendors* sufficient to cause them to be misclassified using less rigorous assays could be noted. "For example, resveratrol would be identified as active at 2.3 μM in one sample but inactive in the other, when a 50% threshold is used." The authors suggest that qHTS "shifts the analysis of screening data from a statistical to a pharmacological process ... In single-concentration screening, the identification of compounds depends highly on assay design. In contrast, qHTS is resilient to variations in assay sensitivity."[50]

Although whether false positives or false negatives are the bigger problem is subject to debate, the desirability of having a method to cut down on both of them, be it qHTS or something else, is not.

*Truth #8: Both false positives and false negatives are problematic. The former can usually be subtracted out in follow-up, but it's hard to add the latter back in.*

*Myth #9: Compound libraries are forever.*

Despite the film title, even crystalline carbon (diamonds) can be destroyed, as Lavoisier demonstrated by burning some long ago. Compound libraries turn out to be much more thermally sensitive than that. The preferred method for long-term storage of any compound is neat; at low temperature, typically below –20°C; and under an inert atmosphere. Combinatorial libraries, though, are usually kept as DMSO stock solutions in deep well-plates from which *daughter plates* can be made when it's time to screen them. Plates containing the DMSO solutions are, of course, stored cold. But there are a couple of problems.

A study done at Evotec compared 778 diverse compounds stored as DMSO solutions under three different storage conditions: under argon at 15°C, under argon at –20°C, and under air at −20°C. Based on analysis data spanning 6 months, they calculated that in 4 years the average percentage degradation would be 58%, 12%, and 21%, respectively. Compounds that were less pure to begin with tended to degrade more quickly.[51]

Freeze–thaw cycles are rough on library plates too. If the seal on every well isn't perfect and if the plates aren't immediately transferred to a dessicator upon removal from the freezer, water from condensation can get into the DMSO solutions, changing concentrations and perhaps accelerating hydrolysis. What isn't always appreciated is that even though pure DMSO freezes at 18°C, mixtures of DMSO and water can remain liquid even below −70°C.[52] Putting thawed-out plates containing such mixtures back into a freezer at −20°C will not cause them to re-freeze, and compound degradation will probably be accelerated.

Finally, long-term storage and freeze–thaw cycles may cause precipitation in many wells, which highlights one of the fundamental problems of compound libraries: the concentration of a given compound in a given well isn't really known. The nominal value, say 10 mM, will hopefully hold for many of them, but unless careful eyeball observations of every well or fairly extensive experiments are done and repeated at intervals to ensure it, the concentration of many compounds may actually differ from that value. For weakly active compounds, this could result in false negatives. Alternately, precipitation can cause false positives as we've seen.

*Truth #9: Compounds dissolved in DMSO have a limited shelf life even under the best storage conditions.*

What should be obvious by now is that, as is so often the case, with HTS the devil is in the details. The fact that hits and leads and ultimately drugs can come out of it, making it a useful tool for drug discovery, is a tribute to persistence and creativity of the scientists who work in the field. Understanding the issues they face shouldn't discourage the chemist, but should instead encourage him or her to learn more about the process screening compounds go through as a kind of reality check in interpreting the data that comes out of it.

# 6.5   Approaches to Hit Generation

Even a short survey of all the different approaches that have been used to discover a small molecule starting point for a project would be longer than this book. This early stage in the drug discovery process is vast, fascinating, and constantly changing. In a few instances, the whole process can even be bypassed: as Dr Richard Silverman points out, a few drugs like Penicillin V were discovered intact and with no need for hits and leads.[53] Exposure of industrial workers to disulfiram caused them to swear off drinking and led to its use as the drug Antabuse. Such serendipitous cases, though, are increasingly rare. A much longer process with a lot more early-stage work is normally involved. And just as one can't make every possible small molecule, one can't screen every available compound in every possible assay. Choices need to be made, and some sort of overall strategy needs to be followed. These tend to fall into two basic categories, as summarized in Box 6.3 and described below.

## 6.5.1   Random or Non-Directed Methods

Given infinite resources, it would be hard to argue against throwing every available compound into a new target-directed or phenotype screen as it comes along. In the real world, though, practical limitations intervene. Even high-throughput screening, discussed earlier, can be

---

**Box 6.3   Some Common Approaches to Hit or Lead Discovery**

*Random Screening of*

- ▶ in-house or purchased compound collections
- ▶ known drugs and drug candidates (SOSA)
- ▶ combinatorial diversity libraries
- ▶ fragment libraries
- ▶ DOS libraries
- ▶ natural products

*Directed Methods Based on*

- ▶ endogenous ligands or substrates
- ▶ known compounds
- ▶ active metabolites
- ▶ target families (chemogenomics)
- ▶ virtual screening
- ▶ *de novo* design

limited by protein availability, cost of reagents, or available time. Factors like the work involved in making new plates; depletion of compounds that aren't likely to be active in a given screen; and the random use of compounds that aren't likely to be soluble, may have reactive functional groups, or wouldn't be good starting points for drugs are all likely to come into play. Just using everything would involve spending too much time deselecting (after the process) troublesome compounds that shouldn't have been there in the first place. Because of this, some sort of preselection filter will be used in almost every case. This might consist of a computer program used to filter out reactive groups, or a group of chemists deciding which structural types should or shouldn't be screened, or perhaps both. The goal is to minimize the problems and maximize the likelihood of finding a worthwhile hit. But as a result, even "random" screening isn't entirely random. Some bias is always built in.

Existing knowledge about the target structure and/or compounds interacting with it can definitely improve the odds of finding hits. In these cases other approaches, discussed later, will probably be more fruitful, and random screening will likely be a secondary consideration, to be used if the expected active compounds aren't found or if others are needed. But in target-directed projects where no such information exists, random screening will be the *de facto* first choice. It's a good match for phenotype screening as well, where multiple mechanisms and therefore a plethora of structural types may be involved. Let's now consider the types of compounds used for random screening and where they might come from.

## 6.5.2  Screening of Synthetic Compound Collections

One advantage that established big pharma companies have enjoyed can be found in their compound archives, which contain many thousands of hand-crafted samples representing a large number of chemotypes and going back decades. Many of these were made before the dawn of combinatorial chemistry, when elemental analyses, crystalline compounds with sharp melting points, and neat storage were the norm. The level of purity that they represent is hard to match. Available quantities may be higher as well, since in the days of tissue or whole animal primary screening, milligram quantities often weren't sufficient to test. In addition to in-house compounds, most pharma companies have always been eager to add to their collections by acquiring intermediates and final products from academic groups they worked with, often paying by the compound. Put these all together and what you end up with isn't a library; it's a resource.

Without a doubt each company's compound collection will be biased in favor of structural classes they've invested in most heavily over the years. But the preponderance of target-directed drug discovery projects being a relatively recent phenomenon, many of the "medchem" compounds in such collections arose from more chemistry-driven considerations. Consequently, they tend to be relevant to a range of different biological targets, which may or may not have been appreciated when they were made.

The thinking that led to the discovery and exploitation of the benzodiazepines by Dr Leo Sternbach and his associates illustrates the way this was done. Any molecular scaffold they decided to work on "would be expected to fulfill the following criteria: (1) be relatively unexplored, (2) be readily accessible, (3) give the possibility of a multitude of variations and transformations, (4) offer some challenging chemical problems, and (5) 'look' as if it could lead to biologically active products."[54] While the rate of compound synthesis in those days

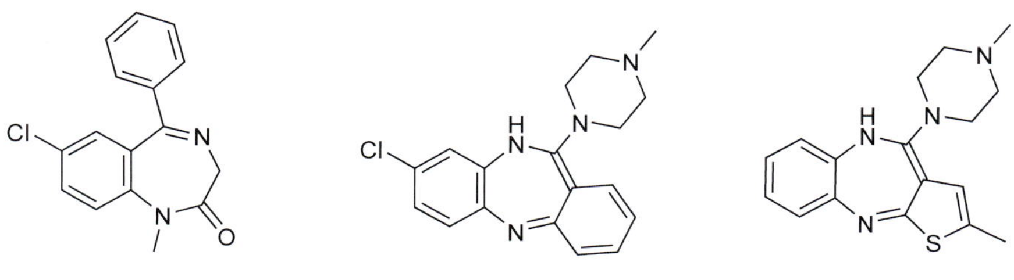

**Diazepam**                    **Clozapine**                    **Olanzapine**

**Figure 6.15** ▶ Three CNS drugs, examples of activity at multiple receptors. Left to right: Valium (diazepam), Clozaril (clozapine), and Zyprexa (olanzapine).

pales beside what even a mediocre combinatorial chemist can achieve now, the success rate was much greater, and such "historical" collections remain a rich source of leads.

This is largely because pharmacologically active small molecules often act on more than just a single target, or even a single family of targets. Whether that's viewed as a regrettable lack of selectivity or a welcome, diverse sampling of biological target space depends on whether it's wanted or not. Benzodiazepines are a good example of this. Besides high-affinity binding to benzodiazepine receptors, diazepam (Figure 6.15) is a micromolar phosphodiesterase inhibitor.[55] Other drugs like the antipsychotic agents clozapine and olanzapine inhibit at least ten different types or subtypes of neuronal receptors at nanomolar levels.[56]

The strategy of screening drugs and drug candidates for off-target effects to uncover new leads is known as *Selective Optimization of Side Activities* (*SOSA*), which was named by its leading proponent, Dr Camille Wermuth.[57] Pharma compound collections are likely to already be enriched for such compounds, but any company lacking such a legacy can still buy such collections, which, numbering from a few hundred to just over a thousand molecules, are quicker and cheaper to screen than large combi libraries.

The idea, of course, is to identify a drug with weak side-activity against the desired target and to use it as a starting point for optimizing that activity while minimizing the activity for which it was originally developed.[58] It differs from repurposing in that the drug itself probably won't end up being redirected into a new indication, but instead will only become the starting point for chemical modification, and also in that the redirection originates in the biochemistry lab, not the clinic. Starting with a compound that has at least made it to the clinic without a lot of observed toxicity seems like a good approach to arriving at another relatively safe compound, though that might depend on the nature of the modifications made. And as always, optimized compounds will tend to be heavier than the original structures, so it's probably more likely to provide an oral drug if the original drug was relatively "skinny" as in the case shown below.

Many examples of this approach have been cited,[59] including one at BMS in which the old bacteriostatic agent sulfathiazole was found to have double-digit micromolar binding to the endothelin A receptor, a target of interest for cardiovascular drugs. Replacement of the thiazole ring with an isoxazole (Figure 6.16) led to almost a 100-fold improvement in activity, and subsequent modifications led to a series of highly potent, orally bioavailable $ET_A$ antagonists that have gone on to clinical testing.[60]

9 sulfathiazole
ET$_A$ IC$_{50}$ = 69 μM

10 sulfisoxazole
ET$_A$ IC$_{50}$ = 0.78 μM

11 BMS-182874
ET$_A$ IC$_{50}$ = 0.15 μM

12 BMS-193884
ET$_A$ $K_I$ = 1.4 nM

13 BMS-207940
ET$_A$ $K_I$ = 0.010 nM

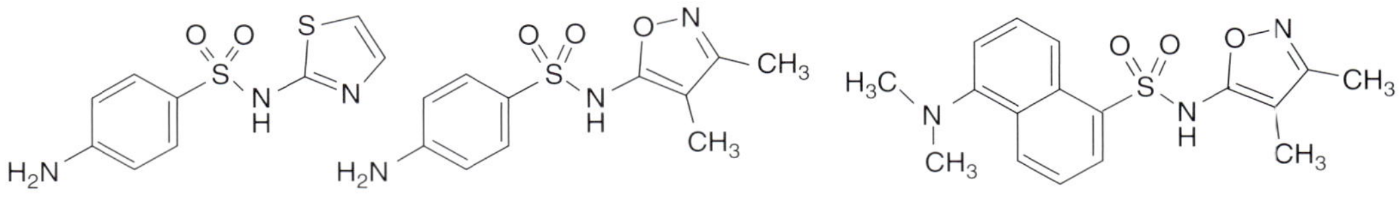

**Figure 6.16 ▸** An example of the SOSA approach. (Reprinted with permission from Wermuth, C.G. Selective optimization of side activities: the SOSA approach. *Drug Discov. Today* **2006**, *11*, 160–164, copyright 2006, Elsevier).

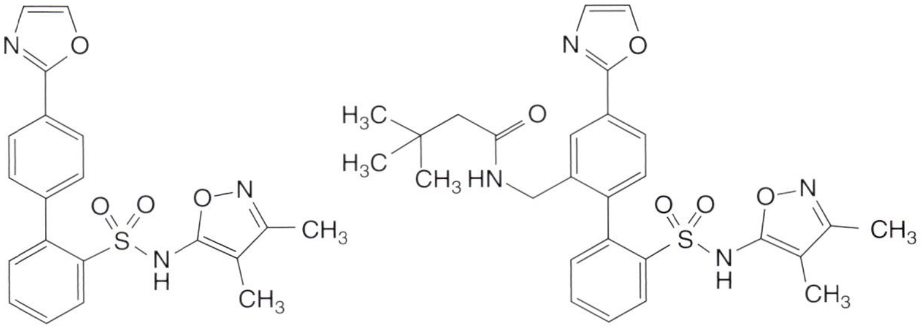

Even without a legacy compound collection or the purchase of SOSA libraries, it's still possible to put together an interesting collection of compounds that "look like they might be active." Searching a database of commercially available materials like the *Available Chemicals Directory* (*ACD*), which lists well over a million compounds, using software like IsisBase® or Daylight® followed by applying other selection criteria allows anyone who can pay for it to assemble a reasonable screening library. Many interesting compounds have recently become available from former Eastern Bloc countries. And certain vendors like ASDI BioSciences not only sell compounds but will even weigh out desired amounts (e.g. 0.2 mM) of them into vials, which turns out to be tedious and time-consuming work when putting hundreds or thousands of them together.

Several practical limitations apply to store-bought compounds, though. A few vendors, especially among those with the most compounds listed, tend not to have a very large percentage of their inventory in stock, and will backorder or sometimes may never be able to deliver the rest. Looking at the literature one finds that usually only about 70–80% of selected commercial compounds end up making it into screening: most of the rest probably couldn't be delivered within a few weeks or months. Some foreign suppliers' compounds are only available via importers who usually double or triple the prices. A small minority of vendors sometimes supply very impure or incorrect compounds. All of this goes with the territory, as does the potentially bigger problem that the same compound you've ordered may have been ordered by a competitor who's also screening it against the same targets.

"Historic" compound collections no doubt have the advantage when it comes to quality, availability, and comprising chemically interesting and sometimes still proprietary structures, but by definition newer companies won't have them. And even the companies that do are prodigious purchasers of commercial screening compounds.

The shift of academia to translational research has brought with it a need for such libraries in universities as well, and this need is being filled by compounds from a variety of academic, government, and corporate sources. For example, Dr Jim McKerrow's group at UCSF, which investigates the roles of proteases in parasitic diseases, works with compounds made available by a group at the University of Capetown, South Africa, Walter Reed Army Institute of Research, Washington, DC, and GlaxoSmithKline, among many others.[61]

Recently as part of the NIH Roadmap Initiative[62] the government has gotten into the act too, via the formation of the Molecular Libraries Small Molecule Repository (MLSMR) which currently stores about 100,000 small molecules, mostly diversity compounds, and plans to increase its inventory to one million. Given the exorbitant costs involved in developing hits into drugs, these are mainly intended for finding useful chemical tools for elucidating biological pathways. Plates are distributed for screening at the Molecular Libraries Screening Center Network (MLSCN).[63]

### 6.5.3  Screening of Combinatorial Diversity Libraries

In a sense, drug discovery approaches have been somewhat bipolar in the last few decades. In the early 1980s, computer-aided drug design (CADD) efforts seemed to proceed on the assumption that so much about what it takes to design a drug was already understood that simply bringing it all together through more and better crystal structures (or homology models or even pharmacophore maps), improved software, smart design, and faster computers would lead to a flood of new drugs, often designed from scratch.[64] Taking an opposite tack to this "knowledge-heavy" approach, combinatorial chemistry in the early 1990s assumed that so little was known about what's really needed that making and screening a very large number of pseudorandom compounds was a better way to go.

Expectations of the new HTS/combi paradigm were that "so many potential lead molecules would be found that the bottleneck would be in deciding which lead series to optimize, and then optimizing several of these series in parallel to reduce the risk of failure in the development phase." In this modern, high-throughput environment, "the 'rational' approach to drug discovery, taken to the extreme by *de novo* structure-based design but really an all-embracing term for quantitative medicinal chemistry, would be unable to compete with the speed and productivity of the new technology."[65]

So in the early days of combichem, the goal was to make as many molecules as possible as quickly as possible by combining as many available reagents as possible. Solid phase syntheses using split–mix protocols predominated, and compounds were often made as *compressed* (more than one-per-well) libraries. Tea bags, PINS, and other odd-sounding techniques made their way into organic synthesis labs. Purification was often considered unnecessary because in solid-phase syntheses impurities "can just be washed away." Even with solution phase, one-per-well libraries, running agonizingly slow HPLC purifications on each and every well seemed a sure way to crush any prospects of rapid compound production, and was consequently shunned.

The result was an accumulation of library compounds which often looked and acted more like detergents than drugs. Solubilities were minimal, hits were almost always artifacts due to detergent effects, insolubility, fluorescence, or, probably, aggregation. Even if genuine hits were identified they weren't likely to be transmutable into drugs. In the words of Dr Chris Lipinski, "the combinatorial libraries in the early years were so flawed that if you took the libraries across Pharma from 1992 to 1997 and stored them in giant dumpsters you would have improved productivity."[66] Reportedly, most companies have done exactly that, using appropriate waste disposal procedures, of course.

The only thing larger than the number of compounds produced in those days was the number of dollars invested in the new technology. With little to show for it, subsequent disappointment even spilled over into the hallowed pages of the *Wall Street Journal*, with claims that the industry was spending billions of dollars on it and "getting nothing."[67] Management's attitude was appropriately summed up by "We bought the robots. Now where are the drugs?"[68]

The truth is, though, that like computer-aided drug design (now SBDD), combinatorial chemistry (now often under the aegis of DOS libraries or parallel synthesis) has evolved and matured, the painful lessons of its youth having finally been learned. As we've seen, limits on molecular weights and LogP values are now considered from the start and compounds now meet high standards of purity. Libraries are smaller than they used to be, but more likely to find confirmed hits that represent reasonable starting points for drugs. Overall, the emphasis has shifted from diversity libraries meant to find hits to focused libraries for optimizing existing hits or leads. But screening of high-quality combinatorial diversity libraries, now largely purchased from vendors or outsourced, is still done routinely at most companies, and the technique has had some successes. A recent report concludes that, based on data provided by 54 HTS directors, at least 104 current clinical candidates and 4 marketed drugs are based on hits originally obtained through HTS.[69] To keep things in perspective, Dr Barry Bunin, author of *The Combinatorial Index*,[70] notes that "A general truth about new technologies is that their short-term impact is often over-estimated ... and their long-term impact is often under-estimated."[71]

Industrywide experience to date has turned up two basic problems with the use of high-throughput screening of combi libraries for hit discovery. The first is that, as we've seen, the method frequently fails to produce confirmed hits for a given target despite the vast number of compounds tested. This is particularly true for "difficult" targets that it was hoped this technology would best address. The second troubling observation is that even when confirmed hits are found, they're often difficult to develop into real leads and even more of a challenge to turn into drugs. Taken together, these observations account for much of the disappointment with the method (Table 6.1).

Hit rates for any screen can be improved by artificial methods like lowering the bar (say, to 10% inhibition instead of the original criterion of 50%) or using a looser definition of what constitutes a validated hit (say, not excluding compounds with funny dose–response curves), but doing this would only waste everyone's time since such "hits" are likely to be assay noise or artifacts. After maintaining reasonable standards and taking the time and effort required to rule out artifacts like aggregators, fluorescent compounds, redox compounds, etc., few hits might remain, especially in screening combi diversity libraries against new and different targets. But why should that be?

**TABLE 6.1** ▶  **Combinatorial Chemistry Then and Now**

| Combi libraries circa 1995 | Combi libraries circa 2006 |
| --- | --- |
| Large libraries: $\sim 10^5$ members | Smaller libraries, mostly $< 10^3$ members |
| Lots of rule of five violations | Mostly rule of five compliant |
| Many solid phase syntheses | Many solution phase syntheses |
| Often multiple compounds per well | Usually one compound per well |
| Minimal purification the norm | Extensive purification common |
| Primary use: diversity screening | Primary use: property optimization |
| Mostly called combinatorial libraries | Mostly called parallel synthesis libraries |

One of the rationales for making such libraries in the first place was that they would allow a quicker, more diverse sampling of all possible "chemical space"[72] than traditional medchem methods. Though that may be true, it seems that chemical space is actually so vast that even a comprehensive collection of every existing diversity library would cover only a tiny slice of it. For example, with only 30 common, non-hydrogen atoms, no fewer than $10^{60}$ compounds are conceivable.[73] To put this in perspective, only about $10^{17}$ seconds have now elapsed since the Big Bang. And, of course, overall combi library diversity isn't increased much by the many directed libraries that are being built around known active scaffolds, now a more popular approach to their use.

Another reason for the low validated hit rate observed in HTS screening of combi libraries is that the molecules represented tend to be too big and often are too complex. Big molecules just have more "things that go bump in the receptor" via unfavorable steric or electronic interactions that prevent productive binding. A certain amount of complexity seems to go along with high affinity: one wouldn't expect to see much good activity with, say, chlorobenzene.[74] But beyond a certain point, having a lot of complexity begins to favor the detection of non-specific binding and to actually *decrease* the chances for finding specific, optimizable interactions. "As a molecule becomes more complex the chance of observing a useful interaction for a randomly chosen ligand falls dramatically, and the probability of mismatches increases exponentially!"[75]

This effect was demonstrated by a computational study done at GlaxoSmithKline based on a simplified model, and is shown in Figure 6.17. The authors point out that beyond a point "as molecular complexity increases, there is an increasingly small chance of an individual molecule being a hit in an assay. But only complex molecules allow us to make the numbers we want (e.g. through several points of diversity). This is a classic catch-22 situation whereby we need more molecules to increase the odds, but these large numbers of molecules . . . tend to be more complex. This actually decreases the chance of any individual compound being a hit in the first place."[76]

Observed difficulties in *optimizing* validated hits may reflect at least two different phenomena, one theoretical and one very much empirical. A real-world analysis of 3000 hits from about one million HTS screening results found that 83% had $IC_{50}$ values above $1\,\mu M$, the average molecular weight was about 400 Da, and the average cLogP was about 4. The problem is that optimization usually involves adding more atoms and increasing MW.

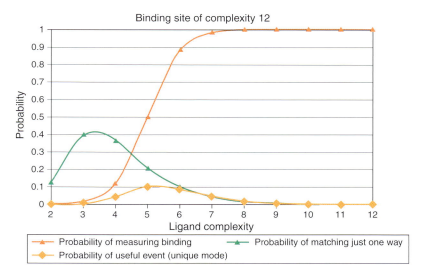

**Figure 6.17** ▶   The success landscape. Increasing the complexity of a ligand increases its odds of binding but, beyond a point, decreases the probability of finding useful interactions. (Reprinted with permission from *J. Chem. Inf. Comput. Sci.* **2001**, *41*, 856–864, copyright 2001, American Chemical Society).

An analysis of prior lead optimization data suggest that increasing the potency of these compounds into the normally acceptable range for drugs would involve adding enough molecular weight and increasing their LogP values "out from under the envelope of druglike property distributions into regions where poor bioavailability would be predicted."[77]

Starting with too big a hit will give you too big a drug candidate. For this reason, it's been pointed out that different definitions apply to what's "leadlike" versus what is "druglike"[78]. Rather than apply Lipinski's Rule of Five to hits, Congreve and his colleagues have suggested that the more rigorous *Rule of Three* might be more appropriate: MW < 300 Da, H-bond donors ≤ 3, H-bond acceptors ≤ 3, and cLogP ≤ 3. In addition, they suggested that the number of rotatable bonds should be < 3 and the polar surface area (PSA) ≤ 60.[79] Many validated hits from combi diversity libraries would not satisfy these criteria and they, in fact, imply a different approach, that of *fragment screening*, described in the next section.

A theoretical rationale for the difficulties involved in optimizing validated hits from combi libraries is shown in Figure 6.18. If most of the available binding energy is from deep pocket *a*, starting out with a combi hit that presents *A* in a less-than-optimal fashion while picking up other, less favorable interactions at *b*, *c*, and *d*, results in optimization becoming problematic. Attempting to change *B* to present *A* better will cause the loss of favorable interactions, even perhaps "bumping" at *C* and *D*, and the resulting SAR could be totally bewildering. Even structure-based drug design might be of limited utility, it being impossible to get crystal structures on inactive molecules that don't bind. More than one part of the molecule would need to be changed simultaneously for efficient optimization.[80] It might be better to deconstruct such a combi hit, a process known as *hit fragmentation*, and to start optimizing the pieces from scratch.[81] But if that's a viable strategy, why not just screen small pieces to begin with?

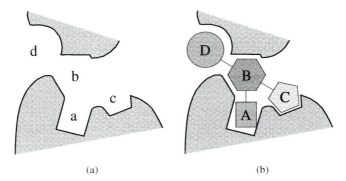

(a)                                                        (b)

**Figure 6.18** ▶  Schematic representing a compound *ABCD* interacting with an enzyme with four pockets *a*, *b*, *c*, and *d*. (a) The pockets, where most binding affinity could be obtained from deep pocket *a*. (b) A poor-quality screening hit in which the *A* moiety is poorly presented to pocket *a* but interactions at the other pockets prevent easy optimization by varying one part at a time. (With kind permission from *J. Comput Aided Mol. Des.* **2002**, *16*,741–753, "The consequences of translational and rotational entropy lost by small molecules on binding to proteins", Murray, C.W.; Verdonk, M.L., Figure 4, copyright 2002, Springer Science and Business Media).

### 6.5.4  Fragment Screening

In the last few years, *fragment screening* has become an increasingly popular method to circumvent these problems. Rather than trying to stuff a relatively large, ornate molecule into a receptor site ill-suited to accept it in order to find a micromolar hit, this sort of minimalist method looks for a starting point consisting of a very small molecule (usually with a molecular weight between 100 and about 300 Da) which binds at a subsite of interest albeit with lesser, sometimes even millimolar, affinity. Although at first glance uncovering a millimolar hit instead of a micromolar one seems more than a bit counterintuitive, there's been a lot of support for this approach based on both theoretical and experimental success.

Since low MW fragments have many fewer atoms than the average combi hit, fewer compounds are theoretically possible, so that a bigger slice of chemical space can be sampled with smaller libraries. These typically run from several hundred to several thousand compounds for fragments, versus tens-to-hundreds of thousands for combi screening libraries. A good fragment library offers at least as much in this sense as the Borgesian-sounding combi "library of libraries" in terms of diversity. Its general utility means that it can be used over and over as new targets come along. Or, as with combi libraries, "directed" or "targeted" fragment libraries containing a pharmacophore or structural motif likely to be active for a given target can be assembled, like the small libraries of fragments meant for kinase and phosphatase screening that have been described.[82] Either way, the greater commercial availability of low-MW compounds means that large parts of fragment libraries don't need to be synthesized: they can be bought. Keeping in mind that the average chemist costs over a thousand dollars a day to employ (including overhead and benefits), this is a very economical alternative.

With fragments, binding at a given site on a protein isn't likely to be accidentally blocked by unnecessary substituents, so falling into an optimal binding mode is more likely than it is for larger compounds. In fact, since such a relatively large proportion of a fragment needs to be involved in good interactions for any observable binding to occur, they are by nature more efficient, and this *ligand efficiency*[83] (the binding energy per non-hydrogen atom) may make them better starting points for leads and drugs: They carry no unnecessary baggage that might make for ADME liabilities later on. Hence the suggestion that fragment screening be used to "deconstruct" bigger combi library hits to identify minimum binding motifs that represent better starting points for optimization. And almost by definition, most fragments satisfy the 'Rule of Three'. Elaborating them by adding 100–200 Da to pick up more interactions can still give a molecule in the desirable MW and LogP ranges for drug candidates.

The consensus to date is that higher hit rates—by definition with low affinity—are, in fact, observed with fragment libraries, and some methods of detection can even make fragment screening less prone to false positives than traditional HTS bioassays. Having good methods of detecting the low affinity interactions that fragments provide is key to their utility, so the more popular detection methods are elaborated on below.

### 6.5.4.1   Detecting Fragment Binding

There's absolutely no *a priori* reason why fragment libraries can't be screened at higher concentrations in traditional bioassays with a fairly good (even high) throughput. But to do that, the fragments need to be sufficiently soluble and the bioassay needs to tolerate the conditions involved, which may, for example, require higher DMSO concentrations than are usually used. Some assay development work will be necessary. Not all fragments or all bioassays are suitable, but with sufficient effort and attention, many can be made to work as high-concentration assays. Some companies, like Plexxikon of Berkeley, California, seem to run such bioassays routinely, while others don't.[84] So a fragment screening campaign might begin with a bioassay that is capable of detecting $K_i$s of up to a few millimolar, then proceed in the usual way: artifact deselection, dose–response curves, small libraries centered around the best hits, crystallization attempts (if possible) on the best of those, etc. Better still, the bioassay might be run in parallel with other screening techniques that provide quicker information on the binding mode.

Biophysical methods to detect fragment-binding are often used. An important early driver for this sort of methodology was the *SAR by NMR* technique pioneered by Dr Stephen Fesik's group at Abbott. They took $^{15}$N-labeled protein targets, assigned the amide resonances, and then ran spectra in the presence of potential ligands. A molecule binding to the protein could be detected by chemical shift changes in the amide proton or $^{15}$N signals of residues at the binding site as determined by $^{15}$N-heteronuclear single quantum correlation ($^{15}$N-HSQC) NMR. This *protein-detected* technique provided immediate information on the location of binding and offered an opportunity to link fragments binding to adjacent sites as described later.[85] The method has found extensive use since then, for example in finding "warheads" that bind to the active site zinc in matrix metalloproteases as shown in Figure 6.19.[86]

The technique as originally described only works for soluble proteins with MWs less than about 30 kDa and requires fairly large amounts (~200 mg) of protein as well as $^{15}$N-labeling. Recent improvements such as *target immobilized NMR screening (TINS)*, which allows

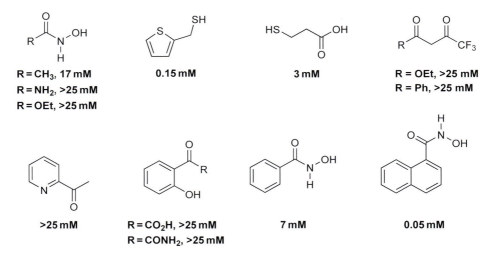

**Figure 6.19** ▶   $K_D$ values of fragments tested for their ability to chelate the active site zinc in stromelysin as determined by NMR. (Hajduk, P.J. et al. NMR-based modification of matrix metalloproteinase inhibitors with improved bioavailability. *J. Med. Chem.* **2002**, *45*, 5628–5639.)

proteins bound to a solid support to be used and reused, may remove some of these constraints.[87] Still, these techniques call for dedicated NMR time, specialized expertise, for example, in assigning protein resonances, and other resources that not every company is likely to have.

Another approach to using NMR to detect binding interactions looks not at the target but at the potential binders themselves. Researchers at Vertex Pharmaceuticals developed such a *ligand-detected* technique they call SHAPES. Here proton NMRs are run on samples (about four at a time) from a very small screening library based upon the scaffolds most frequently observed in drugs[88] in the presence of the target protein. Differential line broadening or transfer NOE experiments are used to identify binders, and pulsed-field-gradient NMR diffusion data are used to calculate binding affinities. Unlike SAR-by-NMR, the method has no MW limit for proteins (since protein resonances don't need to be observed) and doesn't require labeled protein or peak assignments. On the negative side of the ledger, it provides no information about where the hits might be binding, although competition experiments with ligands known to bind at defined sites could still provide some structural information. But the authors observe that "hit rates for follow-up compounds chosen on the basis of an initial SHAPES screen are up to tenfold higher than hit rates for compounds chosen randomly."[89] So the primary value of the method probably lies in prescreening and enriching subsequent libraries to be screened by some other method. Other ligand-detected as well as protein-detected NMR screening techniques have been reported as well.[90]

X-ray crystallography, of course, provides the ultimate in structural information and is a favored method of detection for fragment screening. Not so long ago X-ray crystallography was considered a low-throughput art, not a high-throughput screening method. But advances in automation for crystallization and sample-handling as well as the development of software that semi-automates the structure-solving process have changed this picture quite a bit. Astex

Therapeutics of Cambridge, UK was founded on the idea of high-throughput crystallography for fragment screening, and many other companies, including many big pharmas, now use the approach either in-house or via collaborations.

Fragments at double-digit millimolar concentrations are usually soaked into pre-formed crystals, though sometimes co-crystallization is used. Robotic systems handle the crystals and the data are collected using a standard X-ray source or the quicker and better (but less easily accessible) synchrotron radiation. Software like Astex's AutoSolve® is used to automate data analysis and interpretation. The end result can be tremendously useful: high-resolution pictures of fragment-binding that can jumpstart subsequent SBDD optimization efforts.

Limitations include the need for tens of milligrams of a soluble protein that can produce X-ray quality crystals. Many targets, like GPCRs, won't work. Solubility of the fragment is another concern, as it is with any detection method. In common with high-concentration bioassays, adding more than a tiny bit of DMSO to solubilize the fragment can have a bad effect here: crystals can actually be damaged. And fragments initially dissolved at their kinetic-solubility limit can precipitate and give a mess. One strength of the method is that unlike any other method discussed here, it tends to resists false positives, assuming crystal–packing artifacts aren't occurring. But false negatives can happen. Even nanomolar inhibitors sometimes mysteriously refuse to appear at the crystal sites we know they bind to. And X-ray crystallography, like the more elaborate NMR methods, requires expensive equipment and expertise. The routine crystallographic screening of fragment libraries, at least so far, might be best left to companies already experienced in it.

Any organic chemist considering the above-mentioned fragment-binding detection methods of NMR and X-ray crystallography can tell what's coming next: mass spectrometry (MS). If a fragment can form a covalent linkage to the target, say via disulfide formation with an active site or adjacent site cysteine residue, MS can quickly identify which fragment it was that bound. This has been used, for example, with Sunesis's tethering technology.[91] Another, less familiar, detection technique that can be used is surface plasmon resonance (SPR). Here either the target protein or a library of linked fragments needs to be immobilized on a gold-coated solid support. The other component then passes over it in solution and tiny changes in refractive index that are induced by binding can be detected. Binding affinities can be calculated, but no structural information is obtained.[92] This is sometimes also used as a secondary screen to confirm fragment hits turned up using one of the other detection techniques.

Table 6.2 gives a summary of the strengths and weaknesses of the above mentioned methods, but the reader should keep in mind that this isn't a comprehensive list, and new methods may well appear.

## 6.5.4.2  Optimizing Fragment Hits

Regardless of the detection method used to find it, a fragment that binds with an affinity in the hundreds of micromolar range, which is the usual outcome for a successful fragment screen, seems a far cry from a drug which will probably need to be nanomolar instead. But lots of drugs have started out with high micromolar leads, and skeptics need only look at Figure 6.20 to see an example of just how much it might be possible to improve the binding of a fragment.

**TABLE 6.2 ▶  Some Detection Methods Used for Fragment Screening**

| Detection method | Advantages | Disadvantages |
|---|---|---|
| **High concentration bioassay** | Provides real-world activity data | Not always possible to adopt assays to higher solute & DMSO concentrations. Doesn't give structural information |
| **Protein-detected NMR** | Provides detailed structural information | Resource intensive. Usually requires >200 mg of protein and $^{15}$N labeling. MW limit on protein |
| **Ligand-detected NMR** | Can be done quickly with limited resources. Can be used for larger proteins | Provides limited structural information |
| **X-ray crystallography** | Exquisite structural information provided. Few false positives | Limited to crystallizable proteins in moderate amounts. Expensive instrumentation & expertise required |
| **MS for covalent binders** | Unambiguously identifies the best binder in a fragment mixture | Limited to proteins with an appropriate reactive group (e.g. cysteine) |
| **Surface plasmon resonance** | Very direct technique | Requires protein or fragment linkage to solid support. No structural information provided |

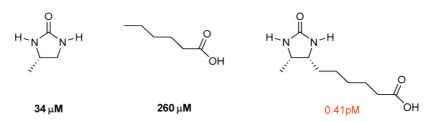

**Figure 6.20 ▶**   $K_d$ values for the binding of dethiobiotin (right) and two of its fragments to avidin. (Green, N.M. Avidin. *Adv. Protein Chem.* **1975**, *29*, 85–133.)[93]

The remarkable affinity of dethiobiotin for avidin, which would not have been expected based on its component fragments, deserves some thought. A very naive view of fragment-binding at two adjacent subsites of a protein would be that the two binding energies should just be additive so that $\Delta G_{AB} = \Delta G_A + \Delta G_B$. This would mean that multiplying the $K_i$s for the inhibitors $A$ and $B$ would give the $K_i$ for the inhibitor $AB$ where the two are perfectly linked together. On that basis alone fragment screening would be a good idea, since linking two 1 mM inhibitors could in theory produce a 1 μM inhibitor.

But there's another factor at work that makes it even better. In order for any molecule to bind to a receptor it has to pay a price to overcome lost rotational and translational entropy.[94] That price has been estimated to be fairly steep: about three orders of magnitude of binding affinity.[80] But it's essentially independent of the size of the molecule. Link two fragments together and you pay only one penalty, not two, thereby freeing up much more binding affinity and allowing for *superadditivity*. Linking two different millimolar inhibitors in the right way might give you a *nanomolar* inhibitor. Thus, linking fragments binding to adjacent subsites is one popular strategy for following up optimized fragment screening hits, as illustrated in panels (b) and (c) of Figure 6.21.

The second fragment might even be selected by re-assaying a fragment library in the presence of the first binder, solubility permitting, so that only fragments binding at a different site are detected. And although linking fragments can be attempted in the absence of any structural information through combinatorial methods,[95] using a method like SAR-by-NMR or high-throughput crystallography, which does give structural information, provides a considerable advantage in suggesting the best way of linking the two. In practice, the linking strategy, which looks seductive on paper, often proves elusive in the lab, so it's best to stack the deck as much as possible.

On the basis of models built from SAR-by-NMR data, researchers at Abbott, for example, linked the 50 µM naphthylhydroxamic acid fragment shown in Figure 6.19 with a 20 µM biphenyl fragment identified in some earlier work, to produce a 62 nM inhibitor of the metalloprotease MMP-3. Although superadditivity wasn't observed in this case, the

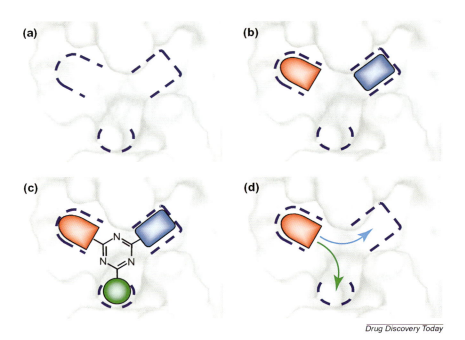

*Drug Discovery Today*

**Figure 6.21 ▶**  Structure-based fragment screening. (a) A protein with three different binding pockets. (b) Structure-based screening can identify molecular fragments that bind into one, two (shown) or all three pockets. (c) A lead compound can then be designed by arranging the fragments around a core template; or (d) growing out using iterative structure-based design from a single fragment. (Reprinted with permission from Carr, R.; Jhoti, H. Structure-based screening of low-affinity compounds. *Drug Discov. Today*, **2002**, *7*, 522–527, copyright 2002, Elsevier).

compound obtained proved to have vastly improved bioavailability compared to the earlier one, demonstrating the potential of fragment-screening to switch out troublesome parts in a lead series and improve properties.[86] Many other examples of fragment-linking can be found in a recent review.[96]

A good example of scaffold "growing", the use of structural information not to link binders at adjacent positions but to build out into them from a single fragment-binding subsite, as shown in panel (d) of Figure 6.21, is some work that was done at Plexxikon in discovering a new chemotype of PDE4 inhibitors. A 20,000-member diversity library of commercially available compounds was initially screened against five different PDE subtypes in a high-concentration bioassay. The 316 compounds showing more than 30% inhibition at 200 µM were co-crystallized with two of the subtypes, and 107 structures were solved. Simple, low-MW pyrazole ethyl esters consistently bound in the same favorable mode, so this scaffold was chosen for optimization. A very small set of analogs revealed that N-phenyl substitution greatly improved activity, and further modification of the phenyl ring based on modeling and crystal structures led to the double-digit nanomolar inhibitors shown in Figure 6.22.[97]

These optimized hits might not be considered lead compounds in themselves if the ethyl group of the ester, which proved essential for activity, ends up being hydrolyzed in plasma.[98] Still, several things are worth noting, particularly the fact that this method allowed for a 4000-fold increase in activity to be achieved through the synthesis of only 21 compounds. Second, nanomolar inhibitors with molecular weights below 300 Da were obtained, something that doesn't tend to happen with combi library hits. Third, bioassays were used to decide which compounds to crystallize. As one experienced enzymologist not associated with Plexxikon put it, "An activity assay without a structure can lead to a drug; a structure without potency information is, well, *nice to look at*." Assay data tether structural information to the real world of biology. Having both is best. And finally, great care was taken on the project to ensure that small changes in substituents didn't result in a change in binding mode. Why was that?

Structure-guided optimizing of fragment-binding, either by linking or growing, depends on the assumption that the binding mode of the fragment remains constant as modifications

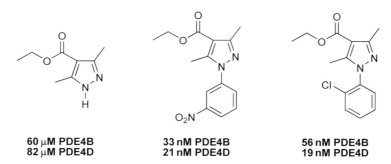

| 60 µM PDE4B | 33 nM PDE4B | 56 nM PDE4B |
| 82 µM PDE4D | 21 nM PDE4D | 19 nM PDE4D |

**Figure 6.22 ▶** An initial hit and two of its optimized analogs. (Card, G.L. et al. A family of phosphodiesterase inhibitors discovered by cocrystallography and scaffold-based drug design. *Nat. Biotechnol.* **2005**, *23*, 201–207.)

are made. If the crystal structure of a fragment bound to an enzyme shows, say, binding that's stabilized by a series of hydrogen bonds with the protein at *A* and reveals a small hydrophobic pocket at *B* that looks ripe for filling, one expects an analog with a methyl group at *B* to bind in the same way, only better. But suppose we make the methyl analog and find that it now binds at the site in a totally different way or binds at a different site altogether! It's certainly interesting and maybe even useful: the new binding mode could conceivably turn out to give better inhibitors than the old one. But one thing is for sure—it will complicate structure-based optimization attempts and make the SAR harder to predict.

Experience has shown that analogs of fragment hits are more likely to adopt alternate binding modes than analogs of tighter binders, although the same thing can sometimes happen with the latter too. Most of the time the assumption has been made that the binding mode won't change during fragment optimization. Evidence for this has been found, for example, by "deconstructing" deoxyuridine monophosphate (dUMP), the natural substrate for thymidylate synthetase, into component fragments that were shown to bind the enzyme in the same general way, albeit with reduced affinities.[99]

But this isn't always the case. A recent paper from Dr Brian Shoichet's lab showed that none of the three separate fragments of a 1 μM AmpC inhibitor bound the way they did in the intact inhibitor. In fact, two of them bound to a new pocket not previously observed for the enzyme, formed through the protein's unexpected conformational flexibility. Only a fairly elaborate "fragment" containing most of the important binding interactions found in the complete 1 μM inhibitor bound the same way that it did. The authors conclude that "fragments of a larger inhibitor need not recapitulate its binding".[100] So, making a few analogs of fragment hits early on to check for alternate binding modes, then using the results to select the best hits to optimize, as in the Plexxikon work, is a very reasonable idea.

One other caveat applies to fragments. Many molecules just aren't soluble enough to screen at high concentrations, regardless of the detection method. Testing a compound with a MW of 300 Da at 5 mM in, say, a bioassay tolerant of 2.5% DMSO means that 60 mg/ml of the compound will have to dissolve in DMSO to make the stock solution and that 1.5 mg/ml needs to dissolve in 97.5% buffer. At least twice that solubility might be needed for crystallography. Many compounds just won't go in to that extent, leaving the scientist to either leave them out or test them at some much lower maximum solubility limit. Worst of all, solubility is hard to predict. Computational predictions of it, as we'll see, just can't be counted on.

Box 6.4 summarizes some of the pros and cons of fragment screening that were discussed above. Rather than highlighting it as an alternative to other methods like traditional combi-library screening, many believe its role is best seen "as orthogonal to that of HTS; the screening technique, affinity, size of compound libraries and chemical strategies for the subsequent 'hit-to-lead' stage are all different. It therefore represents an alternative and complementary strategy, and also presents interesting challenges, both organizationally and culturally, for companies that already have significant infrastructure for HTS. A simplistic summary is that the fragment approach emphasizes efficiency and design, whereas HTS emphasizes affinity and numbers."[101] Right now fragment screening is a young technique, probably still in the "hype" stage of its expectations curve. At the moment it almost seems that reviews outnumber primary research papers in this field, so it's important to keep in mind that it isn't necessarily straightforward or simple and certainly is no panacea. That being said, it has managed to result in several clinical candidates in a remarkably short amount of time.

---

**Box 6.4    Some Advantages and Disadvantages of Fragment Screening**

*Advantages*

- ▶ Better sampling of chemical space means more diversity with smaller libraries

- ▶ More likely to find hits than combi libraries

- ▶ Smaller, more efficient hits mean leads with better ADME properties

- ▶ Hit-to-lead optimization requires fewer compounds

- ▶ Structural information from detection methods jumpstarts SBDD efforts

*Disadvantages*

- ▶ Compound solubility can be limiting

- ▶ Expensive equipment and specialized expertise are usually required

- ▶ All detection methods have limitations (amount of protein required, etc.)

- ▶ Alternate binding modes can obfuscate results and frustrate efforts

---

Figure 6.23 shows what's probably the most advanced of the current fragment success stories, actually more a case of fragment growing by virtual screening than anything else. It began with benzamidine, a known 200 μM inhibitor of the serine protease Factor Xa, and used a program called PRO_SELECT to design optimal substituents.[102] Subsequent efforts were able to overcome the lack of oral bioavailability that amidines bring by replacing it with an indole, and the end result was Lilly's LY517717, which recently proved successful in Phase II trials in preventing venous thromboembolism after hip and knee replacement surgery.[103]

Astex, the pioneer of high-throughput crystallography for fragment screening, currently has two compounds in clinical trials. The first is a cyclin-dependent kinase (CDK) inhibitor which began with a fragment hit with >100 μM activity and ended up as AT7519, "with a CDK2 $IC_{50}$

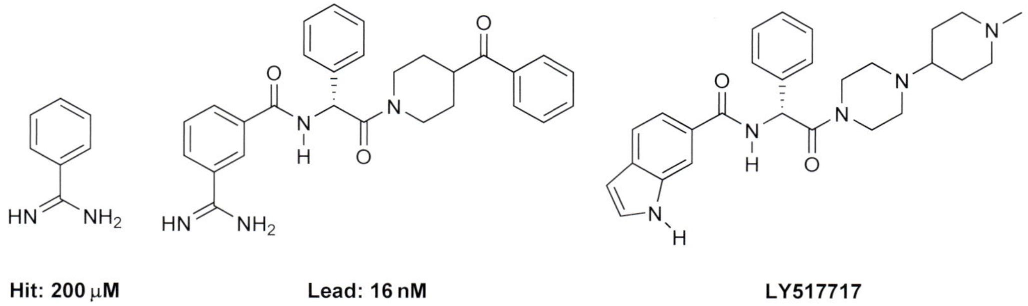

| Hit: 200 μM | Lead: 16 nM | LY517717 |

**Figure 6.23 ▶** A fragment hit to lead to clinical candidate success story.

<50 nM, good druglike properties and a molecular weight below 400."[104] This is currently in Phase I studies for patients with cancer. Another compound to come out of their Pyramid® technology is AT9283, an Aurora kinase inhibitor, also in Phase I studies for cancer. As is the norm for clinical drugs, structures of the compounds have not been published yet.

## 6.5.5  Screening of Natural Products and DOS Libraries

Up until recently the biggest source of new drugs had been natural products.[105] These are normally *secondary metabolites*, structurally complex and interesting compounds not absolutely essential for their source's growth and development. The biological activities associated with plants like opium poppies, *cannabis*, and willow trees have been known for centuries. The drug discovery paradigm symbolized by Sir Alexander Fleming and his Petri dish is firmly ensconced in the minds of scientists as well as the general public. Compounds like Taxol, Lovastatin, and Rapamycin were discovered through natural product screening. According to one analysis, 6% of all NCEs between 1981 and 2002 were natural products, while another 27% were natural product derivatives.[106] And up until about 20 years ago, the screening of plant and microbial extracts was done routinely in pharma. Some old-timers might still remember the days when their employers encouraged them to bring back soil samples when going on vacation to exotic places so that broth extracts might be brewed up in the search for new drugs.[107]

The use of natural-products screening began its long decline sometime in the 1980s and probably reached its nadir a couple of years ago. Its decline paralleled those of phenotype screening and infectious disease research, with which it was usually associated, and it was dealt a near-fatal blow by the rise of combinatorial chemistry and HTS. These seemed to offer a "faster, cleaner, easier" paradigm for how drug discovery should be done. But as we've seen, combi/HTS has largely failed to live up to expectations, and this at least partly explains the recent resurgence in interest in natural-products screening, the drug hunter's long-neglected old friend.

No one should start out with the assumption that natural-products screening is particularly easy or undemanding. Getting access to even small amounts of exotic plants and marine organisms can be challenging for both practical and legal reasons.[108] Most countries in the tropical world aren't eager for biodiversity prospectors taking what they want in return for little or nothing. Access to larger amounts once an interesting source is identified, aka the "supply issue,"[109] is much more difficult except for fermentation broths, where more can just be re-brewed on a larger scale, which is quite an advantage. In some cases, with hard work and luck, plant cell cultures might be used.

As the reader is no doubt aware of already, many natural products are incredibly complex and difficult to synthesize, representing the summit of the synthetic chemist's art and making them the natural targets for top- notch university groups worldwide.[110] While these kinds of syntheses may represent a kind of macho exercise for academic synthesis groups, industry prefers routes that are shorter, faster, and more reliable, hence the previous bias in favor of combinatorial libraries. Even just isolating active natural products already present in extracts is no small undertaking, while the same goes for elucidating their structures. In fact, isolation and identification are generally the realm of specialized in-house groups, and the process can take a lot of time.

Screening itself is tricky. Whole extracts used to be screened, but nowadays fractionation is often done up front to some extent so that mixtures of a few compounds or even pure, isolated natural products can be screened. Libraries of such compounds can now be purchased from companies like BioFocus DPI and Sequoia Sciences, obviating the need to deal with the collection and initial fractionation of samples. Mixtures of many compounds are more prone to give false positives and false negatives than discrete compounds are, especially when they contain compounds like tannins or fatty acids that can interfere with the assay.[111] With extracts, not only the structures of the components, but even their concentrations are initially unknown, so some may be present at very low concentrations. Unless pure, identified compounds are used, natural-product screening requires a step called *dereplication*, meaning that one needs to be sure, before putting the effort involved into isolating and identifying an active component, that it isn't something that's already known. Usually this is done by searching UV and MS data against a database of known natural products.

If all of this sounds expensive and time-consuming, it is. In light of that, why should anyone be interested in them at all? Figure 6.24 gives us a clue: in a word, *diversity*. Looking

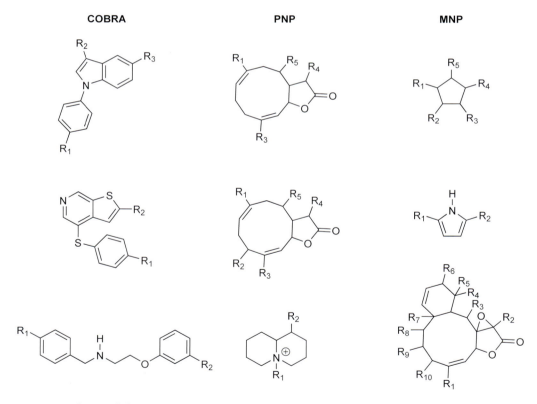

**Figure 6.24 ▶**  Some of the most common atom-based scaffolds exclusively found in a database search of druglike compounds (COBRA), pure natural products (PNP), and marine natural products (MNP). (Grabowski, K., Schneider, G. Properties and architecture of drugs and natural products revisited. *Curr. Opin Chem. Biol.* **2007**, *1*, 115–127.)

at these structures, some of the most common scaffolds found in one recent study comparing available druglike molecules (COBRA), pure natural products (PNPs), and marine natural products (MNPs),[112] shows that Mother Nature has taken a very different approach to designing in biological activity than medicinal chemists have. It's conceivable that many chemists might suggest a project based on an *N*-phenylindole scaffold, but how many would have suggested making pentasubstituted oxabicyclo[8.3.0]tridecadienes instead?

Keep in mind that these scaffolds depend on the particular database that was analyzed, but some observable trends are broadly applicable. Another study compared the properties of drugs, natural products, and combichem molecules.[113] Some of its conclusions are graphically summarized in Figure 6.25. Compared with drugs and compounds considered druglike (e.g. rule-of-five compliant), natural products tend to have more chiral centers, more rigidity, fewer nitrogen atoms, and more oxygen atoms, this last factor probably being the legacy of using $CO_2$ in biosynthesis. Fewer of the rings they contain are aromatic, making them less flat and stackable. They also tend to be heavier than drugs, but keep in mind that some are cell permeable and orally bioavailable nonetheless. Their properties differ even more from those of combichem compounds, which, for example, tend to have few chiral centers, are more lipophilic, and have more nitrogen atoms, which were probably put in as handles to allow the molecule to be further modified.

Figure 6.26, from a recent publication, shows one way of visualizing the differences between drugs, natural products, and other types of compounds.[114] The method used, principal component analysis, was briefly mentioned in Chapter 2, but the bottom line is that the further apart any two data points are the more dissimilar are the structures they

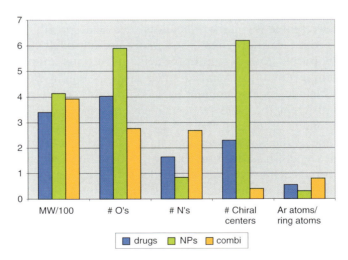

**Figure 6.25 ►** A comparison of mean molecular weight, number of oxygen atoms, number of nitrogen atoms, number of chiral centers, and the ratio of aromatic atoms to total ring atoms for drugs, natural products, and combinatorial chemistry compounds. Note that in some cases median values were significantly different from the mean, resulting in broad property distributions. (Feher, M., Schmidt, J.M. Property distributions: Differences between drugs, natural products, and molecules from combinatorial chemistry. *J. Chem. Inf. Comput. Sci.* **2003**, *43*, 218–227.)

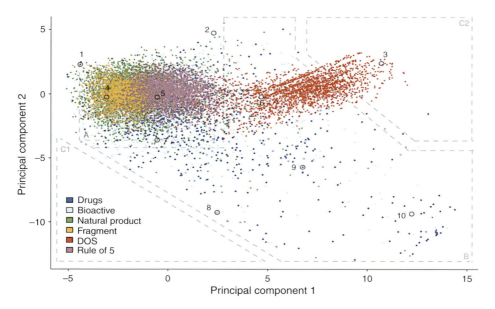

**Figure 6.26** ▶  Principal component analysis of chemical space properties for groups of compounds categorized as drugs, bioactives, natural products, fragments, diversity oriented synthesis (DOS) and Rule-of-Five. Numbered compounds representative of the various categories are given in the reference. (Reprinted with permission from Shelat, A.A., Guy, R.K. The interdependence between screening methods and screening libraries. *Curr. Opin. Chem. Biol.* **2007**, *11*, 244–251, copyright 2007, Elsevier.)

represent. Drugs, represented by dark blue dots, are widespread (which points to the difficulty in defining the term "druglike") but many are clustered near the "rule-of-five" space, represented by purple dots. Natural products (green dots) are more widely dispersed than the latter, reaching out into chemical space such as region *B*, where "rule-of-five" compounds, which include most combinatorial compounds, can't go. Another startling observation is how different the *DOS* (*Diversity Oriented Synthesis*) compounds are from all the rest, almost alone in occupying region *C2*.

DOS represents one of the ways researchers have tried to join together the advantage of diversity that's often seen with natural products with the known structures and compound purity that combinatorial chemistry affords. The idea is to generate combinatorial libraries of complex, natural-product-like compounds in 3–5 steps, ideally through a reaction or series of reactions that allows multiple core scaffolds to be generated from a given type of intermediate. A paper by Dr Robert Armstrong's group in 1997 suggesting the use of squaric acid to generate a *multiple core structure library* (*MCSL*) illustrates the concept, as shown in Figure 6.27.[115] Dr Stuart Schreiber at Harvard University has been involved in much of theory as well as practice of DOS, which he named, since his original publication on the subject.[116] His group has used solid-phase split-pool synthesis to generate such libraries (see Figure 6.28 for an example), and he points out that one of the biggest issues is a shift in thinking from the organic chemist's usual target-oriented planning by retrosynthetic analysis to a completely different, forward-looking approach.

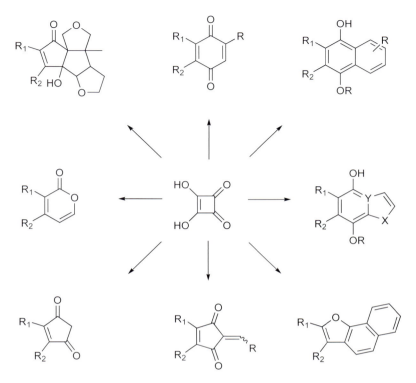

**Figure 6.27** ▶   A scheme for generating a multiple core structure library (MCSL). (From Tempest, P.A.; Armstrong, R.W. Cyclobutendione derivatives on solid support: Toward multiple core structure libraries. *J. Am. Chem. Soc.* **1997**, *119*, 7607–7608.)

The challenges involved in generating good DOS libraries shouldn't be underestimated. By Dr Schreiber's own admission in 2003, "the field of DOS has not yet come close to reaching its goals. Although short and efficient DOS pathways have yielded complex and diverse small molecules with novel biological properties, the structural diversity is misleading. Each small molecule depicted derives from its own DOS pathway. Even a qualitative analysis of members emanating from a given pathway reveals that they are disappointingly similar."[117]

But progress continues to be made, and more and more biologically active compounds continue to come out of DOS libraries. Another approach to the generation of synthetic compounds more closely resembling natural products, "evolutionarily selected chemical structures encoding the properties required for binding to proteins," called *Biology-Oriented Synthesis* (*BIOS*) was developed by Dr Herbert Waldmann at the Max Planck Institute.[118]

Although natural products are diverse, the real interest in them stems from their activity, and it's this which is now leading to a resurgence of attention by pharma,[66] usually via collaborations with companies specializing in it. Why natural selection should have acted on a plant or sea slug to produce a metabolite that's therapeutic for human

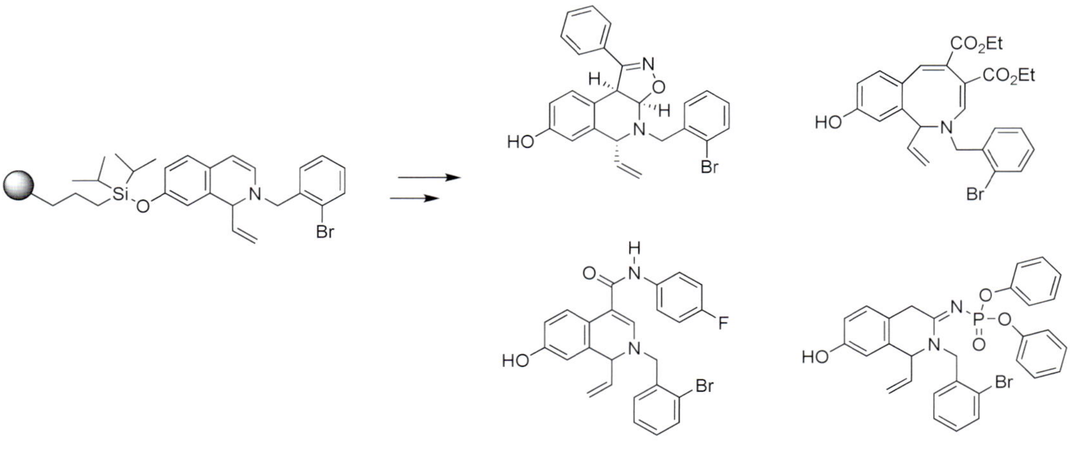

**Figure 6.28** ▶ Some different scaffolds prepared through diversity oriented synthesis (DOS). (Taylor, S.J. et al. Synthetic strategy toward skeletal diversity via solid-supported, otherwise unstable reactive intermediates. *Angew. Chem. Int. Ed. Engl.* **2004**, *43*, 1681–1685.)[119]

diseases has always seemed enigmatic. But the answer probably lies in genetic similarities and conserved protein folds.

"We share our gene families with other organisms," points out Dr Lynn Caporale, a consultant and author of *Darwin in the Genome* (New York, McGraw-Hill, 2003). "Whether inside bacteria or inside us, there is a limited number of ways that the structural components of proteins, such as α-helices or β-sheets, can arrange themselves in space.... Domain families of many human proteins were already on Earth back when we were relatives of fungus. When you have no idea where to begin in a drug discovery program, nature is a good starting point. It would be unlikely that nature had not yet seen a structure similar to what you need, even if the compound isn't curing schizophrenia in a plant."[120]

### 6.5.6  Directed or Knowledge-Based Methods

Empirical observations over many years have shown that, all other things being equal, there's a direct correlation between how much is known about a given target or disease state and how easy it is to find a molecule capable of interacting with it. So random methods like high-throughput screening of combi diversity libraries will normally be used where little is known about a target or as a supplement to directed approaches where problems (e.g. lack of structural novelty) exist with the compounds these turn up.

The type of knowledge used for directed approaches encompasses everything from detailed X-ray crystallographic structures of the protein to its similarity to another well-characterized protein or even just an understanding of what biochemical pathway it's involved with. All of these can improve the odds of finding a lead by narrowing down the sheer volume of chemical space that needs to be searched. What follows is a brief description of the types of knowledge-driven approaches that can be used along with a few examples of how they've been used.

## 6.5.6.1  Methods Based on Endogenous Ligands or Substrates

One of the richest sources of lead compounds for new drugs has been Mother Nature, not through what are normally thought of as "natural products," but instead through substances found in vivo that interact with their cognate receptors to regulate a particular biological response. Compounds like adenosine, dopamine, and serotonin (5-hydroxytryptamine) are natural starting points in the quest for drugs that act on their receptors (A, D, and 5-HT, respectively). Compounds mimicking natural ligands in occupying binding sites on nuclear hormone receptors, prostanoid receptors, and others have occupied several generations of medicinal chemists. A multiplicity of receptor subtypes allows for specific responses in specific tissues, if sufficient subtype-specificity can be built into the drug.

"We always started with a biological lead," notes Sir James Black. "I have never ever started without such a lead!"[121] And those leads led directly to some important new classes of drugs. Back in the days when the division of adrenergic receptors into two subtypes, α and β, was new and controversial, Black proposed that "annulling the actions of the sympathetic hormones, noradrenaline and adrenaline, on the heart" would be a good strategy for anti-anginal agents.[122] So a β-adrenergic receptor antagonist was sought.

The endogenous ligand adrenaline (epinephrine) had been the starting point for the synthesis of a compound called isoproterenol (Figure 6.29), which is just racemic adrenaline with an N-isopropyl group in place of its N-methyl group. Although isoproterenol is still a receptor agonist, it was found to be selective for the β-receptor subtype. Furthermore, work at Lilly had produced dichloroisoprenaline (DCI), the analog of isoproterenol where the phenols had been replaced by chlorines, and this proved to be a selective partial agonist. Subsequent series modifications at ICI, in both the aromatic group and its linker, were able to abolish all agonism and ultimately result in the drug Inderal (propranolol, Figure 6.29), which is still in use today for a number of indications. Although the path from hormone to drug here was neither simple nor straightforward,[123] in the absence of the natural ligand adrenaline as the starting point for synthetic efforts, success would have been impossible.

Sir James Black's next experience, at SmithKline & French, bore an uncanny resemblance to the propranolol story, and again began with an endogenous ligand, in this case histamine. Antihistamines that were known at that time failed to have an effect on histamine-induced gastric acid secretion, an important factor in ulcers. Proceeding on the assumption, later of course confirmed, that two different subtypes of histamine receptors existed, the SK&F group looked for a way to block the binding of histamine to its newly proposed receptor. Two different types of guinea-pig tissue were used to assay the compounds: ileal muscle for

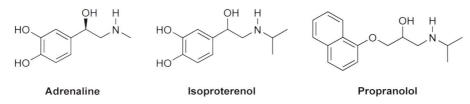

**Adrenaline**          **Isoproterenol**          **Propranolol**

**Figure 6.29 ▶**  Compounds binding to β-adrenergic receptors.

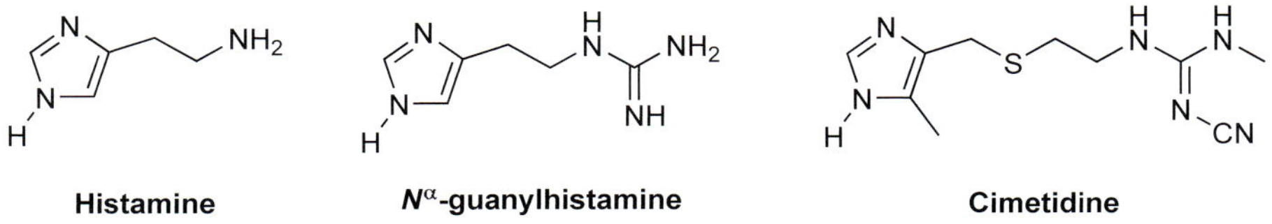

**Figure 6.30** ▶  Compounds binding to histamine receptors.

H1 binding and atrial tissue for H2 binding. $N^\alpha$-Guanylhistamine (Figure 6.30) turned out to be a partial agonist but selective for H2 receptors, the role played by DCI in the β-adrenergic receptor story. Once again, synthetic modifications eventually completed the evolution to a selective, pure receptor antagonist, Tagamet (cimetidine, Figure 6.30), which many of us have since taken before, during, and/or after histamine-inducing meals.[124]

These are two classic examples of the use of endogenous ligands as starting points for drug design; scores of others exist. Sir James Black observes, "I have seen the tremendous success that the pharmaceutical industry has achieved by basing its new drug strategy around the naturally occurring molecules, hormones, and substrates etc. These native molecules were the leads. Close analogues and derivatives were then designed around these leads. Classical bioassay and biochemistry was able to select-in those compounds that competed with the native molecule for the same active site. Compounds with a high degree of selectivity were regularly produced."[125]

For enzymes, endogenous substrates can play a role analogous to that of receptor ligands. The aspartyl protease renin is involved in blood pressure regulation by cleaving its only known endogenous substrate, the peptide angiotensinogen, to produce the decapeptide angiotensin-1, an important player in the renin–angiotensin pathway. Knowing the structure of angiotensinogen enabled the design of small peptides, where the amide bond which renin targets had been replaced with other groups, with these substrate-based analogs being the first step on the road to more powerful and less peptidic transition state analogs.[126] The optimum subsite-binding elements might differ between native substrates and synthetic inhibitors—for one thing, nature has only 20 common building blocks to work with while a chemist has many more—but the substrates do, empirically, seem to provide good starting points and are far more likely to lead to inhibitors of the enzyme than randomly chosen screening compounds.

### 6.5.6.2  Methods Based on Other Leads

Often by the time a particular project is launched at a given company, bioactive compounds for the target are already known. For established targets, meaning those already shown to have the desired physiological effect in late-stage clinical trials, papers and patents showing earlier generations of molecules will probably exist. The structure of a clinical candidate is valuable intellectual property, so it's often not revealed until Phase II or Phase III, but it's frequently speculated on based on publications, meeting notes, conjecture, the phase of the moon, etc. If a marketed drug already exists, plenty of real information on it and other analogs will be available by that time.

Even for novel targets (those so far lacking clinical validation) the structures of bioactive compounds, which are probably less-than-perfect at that point, may well be known. A simple search of PubMed can often lead to such structures. All of these represent potential starting points probably equivalent to confirmed hits in the HTS paradigm, because patentability, a usual prerequisite for a compound to be classified as a "lead," will be lacking.

Variations and structural modifications would then concentrate on turning such a confirmed hit into a proprietary lead. A close reading of existing patents and patent applications, although not exactly fun for most chemists, might turn up substituents or minor scaffold variations that aren't covered (yet). Another route to proprietary IP is through the use of a computer-aided technique called *scaffold hopping*, which relies on "the availability of a template—a chemical structure displaying the desired biological activity, and it is based on the assumption that the same biological activity can be exerted by other compounds that maintain some essential features of the template but are structurally different otherwise."[127] Scaffold hopping is further discussed in the next chapter.

Alternately, good old-fashioned medchem intuition guided by structural information or pharmacophore hypotheses can light the way to major scaffold modifications that maintain or even improve activity while remaining well beyond the scope of existing patents. Many chemists will find this the preferred route to new leads.

Active metabolites of clinical drugs or known bioactive compounds can be another source of potential leads or even new drugs in themselves.[128] Two famous examples of the latter, both antihistamines, are shown in Figure 6.31. Claritin (loratidine) loses its ethyl carbamate group in vivo to form the active metabolite, Clarinex (desloratidine). The latter is about 10-fold more potent than the former, has better PK properties, and consequently can be used at half the dose. Seldane (terfenadine) was found to block the hERG potassium ion channel, which is not a good thing, as it could lead to cardiac toxicity as discussed in Chapter 10. In the body, CYP3A4 converts terfenadine to its major metabolite, fexofenadine, a compound that lacks hERG affinity. Because CYP inhibition by other drugs like the antifungal agent ketoconazole would cut off this detoxifying mechanism and possibly lead to toxicity, Seldane was withdrawn from the market in 1998, but its active metabolite, Allegra (fexofenadine) was able to take its place.

In these examples the active metabolites themselves became drugs. In other cases, active metabolites might only provide leads that need further optimization. In this way Schering-Plough's development of the active metabolite of its lead candidate SCH48461 led to the discovery of Zetia (ezetimibe), a cholesterol absorption inhibitor, as illustrated in Figure 6.32.[129] And active metabolites might even display enhanced off-target effects, becoming starting points in an SOSA-like approach to find a drug for a different indication.

Active metabolites have many potential advantages. They may be more likely to have good safety profiles via predictable Phase II (conjugation) reactions, will probably be more soluble than the parent compound, and might even be patentable in their own right. That being the case, one might wonder why more discovery work doesn't focus on them. Instead, they're seen fairly infrequently and when they are, it's usually late in the development process.

Today's emphasis on early cellular ADME and rodent PK data means that many active metabolites are probably generated. Unfortunately, they'll almost always go unnoticed. The reason is that these tests focus on the stability and plasma concentration of the parent

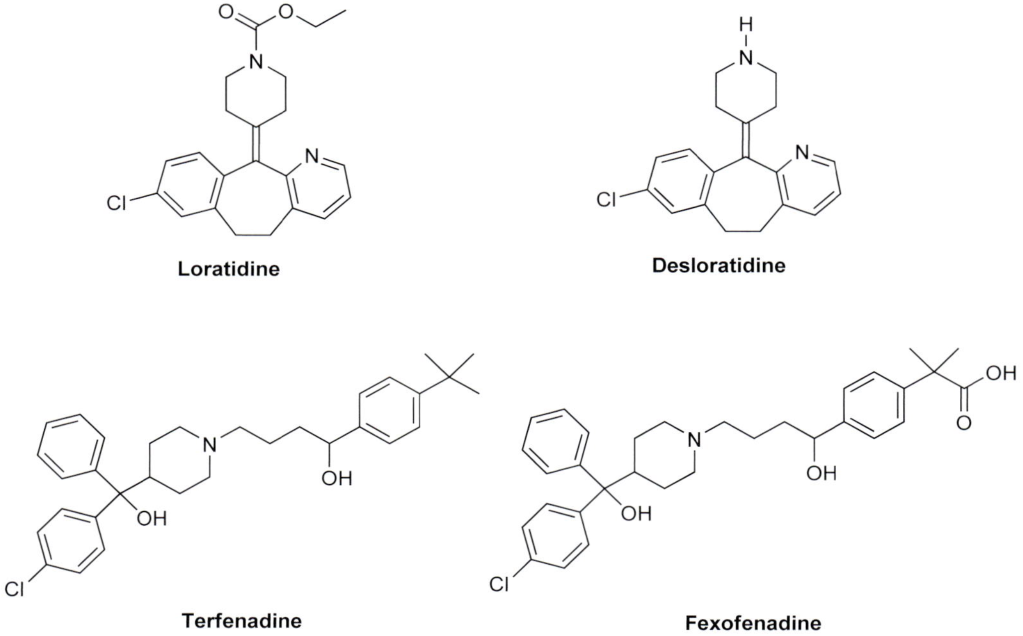

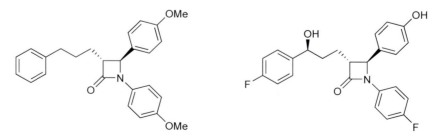

**Loratidine**

**Desloratidine**

**Terfenadine**

**Fexofenadine**

**Figure 6.31 ▶**   Top: Claritin (loratidine) and its active metabolite Clarinex (desloratidine). Bottom: Seldane (terfenadine) and its active metabolite Allegra (fexofenadine).

compound only. Compounds producing interesting active metabolites will be de-selected if those metabolites are produced quickly, giving the parent compound a short half-life. The fact that metabolism *usually* deactivates compounds provides some justification for this. It's conceivable that things could be done otherwise. Treating collections of compounds with microsomes, hepatocytes, or tissues and assaying the resulting mixtures versus their parent compounds could be one way of improving activity and generating diversity at the same time, but the mixtures obtained would face the same issues of interference and deconvolution that

**Figure 6.32 ▶**   The pursuit of an active metabolite of SCH48461 (left) led, after optimization, to the discovery of the cholesterol absorption inhibitor Zetia (ezetimibe), shown on the right.

natural product extracts or compressed combi libraries face. A new MS method that may be able to routinely pick out and characterize metabolites during PK studies has been described.[130] But that's only the first step in the process. To look for active metabolites, resynthesis, structure confirmation, and activity measurements would need to follow, all of which are resource-intensive.

So active metabolites, when they are found, generally come out of in vivo testing of a relatively small number of compounds. Here odd effects like an apparent disconnect between in vitro and in vivo activity or between observed pharmacokinetics (PK) and pharmacodynamics (PD) can give them away.[131] For the present, then, using active metabolites as new leads remains an intriguing possibility that in the future might be optimized for, but which currently is an infrequent occurrence.

The fact that certain structural classes of small molecules tend to interact with multiple members of a target family is the basis for *chemogenomics*, the systematic screening of many such molecules against many members of the cognate target family.[132] Nothing ever being simple when it comes to an "omics," a different meaning for this word is sometimes used by scientists, namely, the perturbation caused by a chemical on an organism's genome, perhaps even in a pharmacogenomic sense; so one again needs to note the context it's used in.

Numerous examples of "family favorite" structures exist, such as steroids for nuclear hormone receptors, hydroxamic acids for metalloproteases, and hydroxyl-containing transition state analogs for aspartyl proteases. The principle behind chemogenomics would not have surprised medicinal chemists half a century ago. As we've seen, it's long been realized that small molecules resembling adrenaline, for example, can interact with different adrenergic receptors and in different ways, whether they were intended to or not. But several factors involved in modern chemogenomics probably *would* have surprised them. Our knowledge of the protein sequences of all the potential targets and the ability to express them recombinantly, peek into them frequently using crystallography, and use them for isolated in vitro screens are among these, as is the sheer number of analogs that can now be made through parallel synthesis and focused libraries. So the aim of chemogenomics is not to replace medicinal chemistry "wisdom of the ages," but rather to implement it more systematically and extensively than was formerly possible. The basic idea is that "organizing early-stage drug discovery research by gene family can significantly enhance the efficiency of drug design."[133]

An example of the kind of efficiency that can be gotten was mentioned by Dr Mark Murcko and quoted in Chapter 5 in the context of kinase inhibitor work. Having a war chest of a few hundred to a few thousand compounds designed to inhibit this important family of enzymes, any proposed new kinase target can be probed by first testing these compounds against it. The odds of finding active compounds are much improved over random methods, and the SAR obtained in the process can aid in subsequent optimization efforts. In such cases, biological validation of the target is more likely to stand in the way of getting a drug than lack of a good starting point, but even here the hits afforded by chemogenomics can come to the rescue in the form of tool compounds to probe disease biology.[134]

The chief strength of chemogenomics, improving the odds of getting a hit by limiting the chemical space searched to a very small region where it's known to overlap with a limited region of biological space (a given group of targets), however, can also be viewed as its main weakness. Hits found through this approach will tend to look alike, offering little diversity and presenting the challenge of designing in selectivity versus other family members. This is often an acceptable tradeoff for finding a starting point when other methods fall short, but it does point out that chemogenomics, like every other method of hit and lead generation, isn't the only approach that should be considered.

### 6.5.6.3  Computational Methods

Dr William Jorgensen, a professor at Yale with many years of experience in computational drug discovery, reflects on a question often posed to such researchers, "Is there really a case where a drug that's on the market was designed by computer?" "When asked this," he notes, "I invoke the professorial mantra ('All questions are good questions.'), while sensing that the desired answer is 'no'. Then, the inquisitor could go back to the lab with the reassurance that his or her choice to avoid learning about computational chemistry remains wise."[135]

Actually, not learning about computational chemistry or considering it as a possible aid to finding hits is anything but wise. Computational methods are highly integrated into today's drug discovery environment. They're not found at one point (i.e. early discovery); they're found at almost every turn. They do not aim to replace the in-depth thinking and creativity that have always made drug design a stimulating and challenging pursuit; instead they're meant to assist them.

This assistance can begin at the hit discovery stage with a logical way of narrowing down the list of compounds to screen. This selection process can be as simple as running a list of, say, all commercially available secondary amides through a filter to remove compounds that violate the ROF, have molecular weights greater than 300 Da, contain heavy metals, cost too much, or only come from unreliable vendors. This is just an easy way of filtering out undesirables and shouldn't be too controversial. As we'll see later, computer programs which can flag a given compound for potential toxicity can be useful as well.

More sophisticated approaches to compound selection aim to use knowledge of ligand and/or receptor to improve the odds of finding hits, thereby leading to a good *enrichment factor*, "the increase in the proportion of hits found in any given sample of compounds compared with the proportion expected for a random sample."[136] For example, if screening a full library of 100,000 compounds turns up 20 sub-100 $\mu$M hits (a 0.02% hit rate) whereas just 4 compounds from a 100-member subset of those compounds predicted to be active by *virtual screening* turned out to be that active (a 4% hit rate), the enrichment factor would be 200.[137]

One way of looking for such enrichment would be to search a large database for compounds that might resemble a pharmacophore that's already known for a given target. Known inhibitors of the serine protease thrombin, for example, have a cationic center about 10 Å away from a defined liphophilic center.[136] Building a 3D model of such a pharmacophore and then running structures from a database through it to see if any can fit would be a *ligand-based* use of virtual screening.

Here too the devil will be in the details. Requiring compounds to fit too highly-restrictive a model will only turn up ones looking like clones of known inhibitors, whereas loosening up the criteria might select lots of diverse, but dead, compounds. Plenty of other problems can crop up too. Looking at the tautomeric forms of the imaginary screening compound shown in Figure 6.33, it's easy to see that a computer program might view them as two separate entities, only one of which has a potential H-bond donor and planar sidechain. One can't blame computers too much for this. After all, a widely held, but incorrect, view of the

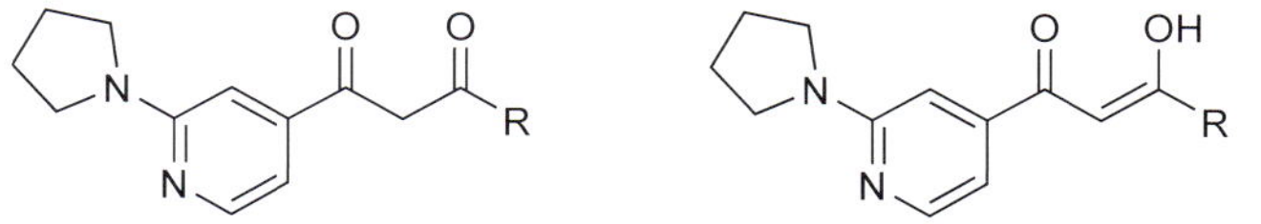

**Figure 6.33 ►** Tautomers can be a real problem in virtual screening.

predominant tautomers for guanine and thymine may have held up Watson and Crick's elucidation of the structure of DNA![138] Incorrect tautomers, pentavalent carbons, impossible aromatic rings, and other curiosities populate many databases. "The databases are filled with junk," points out Dr. Brian Shoichet of UCSF, who's been involved in virtual screening for many years, "and removing that junk will help you."

In *target-based virtual screening* structural information from crystallography, homology modeling, or some other method provides the coordinates for the receptor site so that *in silico* docking can be done. Ideally, a variety of *holo* (ligand-bound) crystal structures will be available to provide the best information on the conformation of the binding site.[139] Interrogated compounds will then be docked in using various poses and scored as to how well they fit. Hits will therefore tend to be more structurally diverse than those uncovered using a ligand-based approach.

Lots of problems, major and minor, conceptual and practical exist, of course. Docking and scoring receptor–ligand interactions are especially problematic. Although there once was a time when some computational chemists wouldn't hesitate to present lists of predicted binding affinities ("Let's see, the thiazole should be about 100 nM, the furan 260 nM, and the substituted pyridine 78 nM..."), hard experience has shown that, in the words of Dr Shoichet, "docking cannot predict absolute binding affinity and cannot even reliably rank molecules."

"The thing that docking has a decent shot at," he explains, "is categorizing molecules as unlikely-to-bind and likely-to-bind. That's where you increase your chances. Among the 'likely' what I can tell you is that a thoughtful and experienced medicinal chemist will go through that list and he or she will not see too many absurd things. And if 10% of those reasonable-looking molecules actually inhibit at micromolar concentrations, you'll be pretty pleased. But forget about rank ordering them."

If virtual screening can't produce lists of numbers that eliminate the need to run bioassays, what it *can* do is provide enrichment. That enrichment can sometimes be based on predicted binding modes that are not, in fact, the ones found. It might even be based on "screening out compounds that are wrong for the target rather than selecting those that are right."[140] It wrongly predicts that many compounds will bind which really don't: virtual screening's own set of false positives.

But however many bloopers, wild pitches, hit-by-pitches, and unearned runs this particular hitter benefits from, he manages to get on base. What's more, he works cheap. One analysis puts the cost required to screen an average compound in HTS at $15 in a campaign taking 5–7 days, while virtual screening of a compound costs about 1 cent and takes 1 second.[141] All of this begins to address a question that may have already appeared in the reader's mind: in light of all the hoopla about modern high-throughput screening capacity, why not just screen everything? Wouldn't that give the most solid, comprehensive results?

Aside from the potentially high cost of HTS, some companies and especially university groups may not have ready access to either large compound collections or robotic equipment. Virtual screening allows them to begin in a few weeks with a few hundred compounds at a reasonable price. The enrichment it affords means that a small number of compounds can be carefully tested in lower throughput assays, which tend to give more reliable results. It can also be used where other methods have failed to uncover good hits. Importantly, it often uncovers *different* compounds compared with HTS even for the same compound set, a reflection of the fact that each method finds different false negatives and false positives, a vice that can be tolerated for each as long as it identifies a few genuine and interesting hits. This observed complementarity is probably the greatest argument for the use of virtual screening by pharma.[142]

Although one wouldn't expect a method that works best with crystallographic data to be readily applicable to the field of GPCRs, where X-ray structures are as rare as hens' teeth, two excellent examples of virtual screening's successes are in exactly that area. In one recent case, virtual screening was used to find compounds interfering with protein–protein interactions, those between G$\beta\gamma$ subunits and proteins involved in downstream signaling. The crystal structure of the G$\beta_1\gamma_2$ heterodimer complexed with a short peptide was solved and seemed to provide a 'hotspot' for protein–protein interactions. Virtual screening was used to examine a library of almost 2000 compounds at this site. Consensus scoring (a way to combine results from various available algorithms to produce results better than any single one of them)[143] was employed and 85 compounds were selected for testing in a competition assay. Nine of them were found to inhibit peptide binding at concentrations below 60 $\mu$M including compound M119, as shown in Figure 6.34.[144] A small, focused follow-up library for M119 proved the bottom hydroxyl groups to be essential for activity. Both compound M119 and M201, a virtual-screening hit that did not show up as active in the competition assay, were found to differentially modulate G$\beta\gamma$ interactions with effectors. M119 could increase the analgesic potency of morphine, the classic $\mu$-opioid agonist, by 11-fold when coadministered to mice by intracerebroventricular injection. In this way virtual screening played an important role in demonstrating that small molecules interfering with

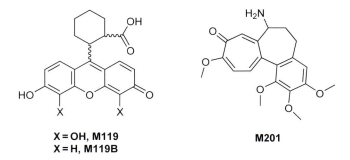

**X = OH, M119**
**X = H, M119B**

**M201**

**Figure 6.34 ▶**  Compounds interfering with the binding of G$\beta_1\gamma_2$ to SIGK identified by virtual screening. (Bonacci, T.M. et al. Differential targeting of G$\beta\gamma$-subunit signaling with small molecules. *Science* **2006**, *312*, 443–446.)

protein–protein binding in GPCR downstream signaling pathways might be useful as new therapeutic agents.

The discovery of a potentially useful new therapeutic agent currently in Phase III clinical trials has been attributed to virtual screening as well. A company called Predix Pharmaceuticals (now part of Epix Pharmaceuticals) developed a non-homology, non-rhodopsin-based, *de novo* GPCR model of the 5-HT$_{1A}$ receptor based on the sequence's most stable predicted structure in the membrane environment.[145] Virtual screening of 40,000 compounds led to 78 virtual hits being selected and tested in a radioligand displacement assay. Sixteen of the compounds were hits with $K_i$ values <5 µM. The best of them, the 1 nM compound shown in the center of Figure 6.35, was selected for optimization. Using *in silico* 3D modeling as an aid, only 31 compounds needed to be synthesized for hERG-binding and lack of selectivity, problems for the original hit, to be overcome. A candidate, PRX-00023, was selected and put into the clinic less than 2 years after project inception.[146] It's currently in Phase III trials for generalized anxiety disorder. Although one could argue that most chemists would have earmarked the 1 nm hit for screening without the benefit of computational assistance, there's no doubt that structure-based optimization played a role in the speed with which PRX-00023 got to the clinic.

Virtual screening typically processes a set of compounds that are either at hand (in-house compound archives or libraries) or that are available for purchase. Several million such compounds may exist, and if structures described in the *Chemical Abstracts* database that are probably not commercially available but can theoretically be made are considered, this number might expand to about 25 or 30 million.[147] But remember that something on the

**Figure 6.35 ▶** Buspirone (top), an approved 5 HT$_{1A}$ partial agonist, Predix's 1 nM hit (center), and PRX-00023 (bottom), currently in Phase III trials for GAD. (Becker, O.M. et al. An integrated in silico 3D model-driven discovery of a novel, potent, and selective amidosulfonamide 5-HT$_{1A}$ agonist (PRX-00023) for the treatment of anxiety and depression. *J. Med. Chem.* **2006,** *49,* 3116–3135.)

order of $10^{60}$ small molecules are conceivable. What if instead of relying on this tiny, highly patented sampling of chemical space, good structures for binding to a target that might not exist yet could be put together *in silico*, docked, evaluated and, if warranted, synthesized? This extension of virtual screening to virtual molecules is virtually the definition of *de novo design*.

As with "traditional" virtual screening, *de novo* design can be based upon the conformations of known ligands or of the targets themselves, with the latter probably allowing for more diverse resulting molecules. Commercially available programs like LUDI, CAVEAT, GrowMol, and SPROUT are often used in *de novo* design. Although many find the method promising, so far its use seems to be less widespread than that of virtual screening.

Three problems have been noted. Simply combining functional groups, fragments, and sidechains to make something that fills a receptor site properly can afford molecules that may be very impractical to make. Even the most favorably-disposed chemist might end up avoiding them after weighing the challenge they represent against the odds that at the end of a long and difficult synthesis they might really be active. A simple solution to this problem involving bringing synthetic chemists into the initial design and selection process, so that synthetically tractable molecules are generated, has been proposed.[148]

A second problem is that successful attempts at *de novo* design are unlikely to succeed without a lot of scaffold hopping and structural modifications. But in the process "molecules will be synthesized and tested that have only marginal activity; for example, binding constants in the medium- or even high-micromolar range. Such candidates would usually not be followed up or might not even be recognized at all in a drug discovery project."[149] It's one thing to get such a hit with a fragment ordered from Aldrich for $20, but quite another to find it after weeks or months of traversing "blank, lightless galaxies" of chemical space "into which good ideas at their peril wander."[150] Few chemists and fewer managers have the patience.

Finally, lots of proposed binders will probably be uncovered in *de novo* design. A way to triage them is needed. But since the method relies on docking and scoring, the same problem that confronts virtual screening applies here too: not even reliable rank ordering can be expected. Chemical tractability, not predicted binding affinity, will probably determine which compounds get made first or at all.

With all of its problems, though, *de novo* design can still be a worthwhile addition to the lead discovery armamentarium. An example to how useful it may be involved the *de novo* synthesis of inhibitors of lanosterol 14α-demethylase (CYP51). This interesting member of the cytochrome superfamily is involved in the biosynthesis of ergosterol, a steroid important for fungal cell membrane integrity, and it's consequently the target of many antifungal agents like ketoconazole, which has the unfortunate effect of also inhibiting a CYP more important to mammals, CYP3A4. In an effort to find CYP51 inhibitors that don't rely on imidazole binding to the enzyme's heme iron atom, the expected cause of this CYP promiscuity, *de novo* design was used. A series of oximinochromanes was produced, one of which was only slightly less active than fluconazole for several strains of fungus.[151]

But perhaps the biggest impact of *de novo* design will ultimately not reside in the compounds produced but in the minds of the chemists introduced to it. "An important goal of *de novo* design is to inspire medicinal chemists through the chemical motifs that are identified. Ultimately, the aim is to offer support for hit and lead identification and widen the chemical horizon."[150] Wide horizons are in fact a prerequisite for hit discovery. Neither rational nor irrational approaches, *de novo* design, substrate-based design, or any of the other methods mentioned here, can single-handedly bear the burden of supplying enough good starting points. Successful hit generation requires as diverse a community of approaches as it does of scientific disciplines.

# Notes

1. Shoichet, B.K. Interpreting steep dose-response curves in early inhibitor discovery. *J. Med. Chem.* **2006**, *49*, 7274–7277.

2. Lipinski, C. Academic drug discovery: The chemistry challenges of target choice and screening library selection. *Collaborative Drug Discovery Community Meeting*, University of California San Francisco, March 5, 2007.

3. To use Senator Hillary Clinton's quote. For more on the interdisciplinary nature of HTS, see Gordon, E.J. Small-molecule screening: It takes a village.... *ACS Chem. Biol.* **2007**, *2*, 9–16.

4. See, for example, PhRMA's Profile2007, Available at www.phrma.org/files/Profile%202007.pdf, which credits the 5–10K number to a study done at Tufts University in conjunction with PhRMA.

5. Payne, D.J. et al. Drugs for bad bugs: Confronting the challenge of antibacterial discovery. *Nature Rev. Drug Discov.* **2007**, *6*, 29–40. This paper contains numerous insights on the "many a slip 'twixt the cup and the lip" that go on in drug discovery and should be required reading for anyone going into the field.

6. Laverty, P. et al. Simple absorbance-based assays for ultra-high throughput screening. *J. Biomol. Screen.* **2001**, *6*, 3–9.

7. Palmer, J.T. et al. Design and synthesis of tri-ring P$_3$ benzamide-containing aminonitriles as potent, selective, orally effective inhibitors of cathepsin K. *J. Med. Chem.* **2005**, *48*, 7520–7534.

8. Grabowska, U.B. et al. Recent developments in cathepsin K inhibitor design. *Curr. Opin. Drug Discov. Devel.* **2005**, *8*, 619–630.

9. Kemnitzer, W. et al. Discovery of 4-Aryl-4*H*-chromenes as a new series of apoptosis inducers using a cell- and caspase-based high-throughput screening assay. 1. Structure–activity relationships of the 4-aryl group. *J. Med. Chem.* **2004**, *47*, 6299–6310.

10. Schroeder, K.S. FLIPR: A new instrument for accurate, high throughput optical screening. *J. Biomol. Screen.* **1996**, *1*, 75–80.

11. Johnston, P.A., Johnston, P.A. Cellular platforms for HTS: Three case studies. *Drug Discov. Today* **2002**, *7*, 353–363.

12. For further information on fluorescent probes or labeling in general, an invaluable resource is Invitrogen's *The Handbook—A Guide to Fluorescent Probes and Labeling Technology*. Available on the web at http://probes.invitrogen.com/handbook/.

13. See, for example, Baxter, D.F. et al. A novel membrane potential-sensitive fluorescent dye improves cell-based assays for ion channels. *J. Biomol. Screen* **2002**, *7*, 79–85.

14. See Owicki, J.C. Fluorescence polarization and anisotropy in high throughput screening: perspectives and primer. *J. Biomol. Screen.* **2000**, *5*, 297–306.

15. Andrews, B.J.; Turchi, J.J. Development of a high-throughput screen for inhibitors of replication protein A and its role in nucleotide excision repair. *Mol. Cancer Ther.* **2004**, *3*, 385–391.

16. Zlokarnik, G. et al. Quantitation of transcription and clonal selection of single living cells with β-lactamase as reporter. *Science*, **1998**, *279*, 84–88.

17. Pope, A.J. et al. Homogeneous fluorescence readouts for miniaturized high-throughput screening: Theory and practice. *Drug Discov. Today* **1999**, *4*, 350–362.

18. Wu, S., Liu, B. Application of scintillation proximity assay in drug discovery. *Bio Drugs* **2005**, *19*, 383–392.

19. Mundt, S. et al. Development and application of a scintillation proximity assay (SPA) for identification of selective inhibitors of 11β-hydroxysteroid dehydrogenase type 1. *Assay Drug Dev. Technol.* **2005**, *3*, 367–375.

20. See PerkinElmer's product brochure at http://las.perkinelmer.com/Content/RelatedMaterials/Brochures/BRO_AlphaScreen2004.pdf.

21. Nicholson, R.L. et al. Small-molecule screening: Advances in microarraying and cell-imaging technologies. *ACS Chem. Biol.* **2007**, *2*, 24–30.

22. Zhang, J.-H. et al. A simple statistical parameter for use in evaluation and validation of high throughput screening assays. *J. Biomol. Screen.* **1999**, *4*, 67–73.

23. Sills, M.A. et al. Comparison of assay technologies for a tyrosine kinase assay generates different results in high throughput screening. *J. Biomol. Screen.* **2002**, *7*, 191–214.

24. For example, see Ferrand, S. et al. Statistical evaluation of a self-deconvoluting matrix strategy for high-throughput screening of the CXCR3 receptor. *Assay Drug Dev. Technol.* **2005**, *3*, 413–424.

25. Kashem, M. et al. Three mechanistically distinct kinase assays compared: Measurement of intrinsic ATPase activity identified the most comprehensive set of ITK inhibitors. *J. Biomol. Screen.* **2007**, *12*, 70–83.

26. Wu, X. et al. Comparison of assay technologies for a nuclear receptor assay screen reveals differences in the sets of identified functional antagonists. *J. Biomol. Screen.* **2003**, *8*, 381–392.

27. Gribbon, P., Sewing, A. Fluorescence readouts in HTS: No gain without pain? *Drug Discov. Today* **2003**, *8*, 1035–1043.

28. Walters, W.P. et al. Virtual screening—an overview. *Drug Discov. Today* **1998**, *3*, 160–178.

29. Walters, W.P.; Namchuk, M. Designing screens: How to make your hits a hit. *Nat. Rev. Drug Discov.* **2003**, *2*, 259–266.

30. Malo, N. et al. Statistical practice in high-throughput screening data analysis. *Nat. Biotechnol.* **2006**, *24*, 167–175. The authors suggest that edge effects would be less of a problem if positive and negative controls were alternated along the edges instead of clumped together on the left and right, respectively.

31. Seidler, J. et al. Identification and prediction of promiscuous aggregating inhibitors among known drugs. *J. Med. Chem.* **2003**, *46*, 4477–4486.

32. McGovern, S.L., Shoichet, B.K. Kinase inhibitors: Not just for kinases anymore. *J. Med. Chem.* **2003**, *46*, 1478–1483.

33. A recent analysis of over 70,000 compounds in the NIH Chemical Genomics Center Library showed that 95% of the actives identified in a β-lactamase screen were aggregate-based inhibitors. See Feng, B.Y. et al. A high-throughput screen for aggregation-based inhibition in a large compound library. *J. Med. Chem.* **2007**, *50*, 2385–2390.

34. Feng, B.Y., Shoichet, B.K. Synergy and antagonism of promiscuous inhibition in multiple-compound mixtures. *J. Med. Chem.* **2006**, *49*, 2151–2154.

35. Frenkel, Y.V. et al. Concentration and pH dependent aggregation of hydrophobic drug molecules and relevance to oral bioavailability. *J. Med. Chem.* **2005**, *48*, 1974–1983.

36. McGovern, S.L. et al. A specific mechanism of nonspecific inhibition. *J. Med. Chem.* **2003**, *46*, 4265–4272.

37. Feng, B.Y., Shoichet, B.K. A detergent-based assay for the detection of promiscuous inhibitors. *Nat. Protoc.* **2006**, *1*, 550–553.

38. Lipinski, C. et al. Experimental and computational approaches to estimate solubility and permeability in drug discovery and development settings. *Adv. Drug Deliv. Rev.* **1997**, *23*, 3–25.

39. Owens, J. Chris Lipinski discusses life and chemistry after the Rule of Five. *Drug Discov. Today* **2002**, *8*, 12–16.

40. Dr Lipinski's rules for druglike CNS compounds were presented by him at the Drew University Medicinal Chemistry Special Topics Course in July 1999 and have been published in Pajouhesh, H., Lenz, G.R. Medicinal properties of successful central nervous system drugs. *NeuroRx Exp. NeuroTher.* **2005**, *2*, 541–553.

41. Or at least he might get out on pyrrole (Figure 6.24)!

42. Feher, M., Schmidt, J.M. Property distributions: Differences between drugs, natural products, and molecules from combinatorial chemistry. *J. Chem. Inf. Comput. Sci.* **2003**, *43*, 218–227.

43. Golebiowski, A. Lead compounds discovered from libraries. *Curr. Opin. Chem. Biol.* **2001**, *5*, 273–284.

44. Lahana, R. How many leads from HTS? *Drug Discov. Today* **1999**, *4*, 447–448.

45. Which brings to mind an old joke. A chemist tells his boss, "The good news is that I've made 60 compounds today. The bad news is they're all in the same well."

46. Quoted in Borman, S. Organic lab sparks drug discovery. *Chem. Eng. News* January 14, **2002**, 23–24.

47. Yan, B. et al. Quality control in combinatorial chemistry: Determination of the quantity, purity, and quantitative purity of compounds in combinatorial libraries. *J. Comb. Chem.* **2003**, *5*, 547–559.

48. Yan, B. et al. High-throughput purification of combinatorial libraries I: A high-throughput purification system using an accelerated retention window approach. *J. Comb. Chem.* **2004**, *6*, 255–261.

49. Zhang, J.H. et al. Probing the primary screening efficiency by multiple replicate testing: A quantitative analysis of hit confirmation and false screening results of a biochemical assay. *J. Biomol. Screen.* **2005**, *10*, 695–704.

50. Inglese, J. et al. Quantitative high-throughput screening: A titration-based approach that efficiently identifies biological activities in large chemical libraries. *Proc. Natl. Acad. Sci. USA* **2006**, *103*, 11473–11478.

51. Ilouga, P.E. et al. Investigation of 3 industry-wide applied storage conditions for compound libraries. *J. Biomol. Screen.* **2007**, *12*, 21–32.

52. Havemeyer, R.N. Freezing point curve of dimethyl sulfoxide-water solutions. *J. Pharm. Sci.* **1966**, *55*, 851–853.

53. Silverman, R.B. *The Organic Chemistry of Drug Design and Drug Action* (San Diego, Academic Press, 2004). Anyone who hasn't already read this outstanding guide to medicinal chemistry is urged to do so at once!

54. Sternbach, L.H. The benzodiazepine story. *J. Med. Chem.* **1979**, *22*, 1–7.

55. Beer, B. et al. Cyclic adenosine monophosphate phosphodiesterase in brain: Effect on anxiety. *Science* **1972**, *176*, 428–430.

56. Schaus, J.M.; Bymaster, F.P. Dopaminergic approaches to antipsychotic agents. *Annu. Rep. Med. Chem.* **1998**, *33*, 1–10.

57. Dr Wermuth is also the editor of *The Practice of Medicinal Chemistry* (San Diego, Academic Press, 2003). Like Dr Silverman's book, this is an extremely useful volume for the medicinal chemist to have.

58. See Wermuth, C.G. Selective optimization of side activities: The SOSA approach. *Drug Discov. Today* **2006**, *11*, 160–164.

59. Wermuth, C.B. Selective optimization of side activities: Another way for drug discovery. *J. Med. Chem.* **2004**, *47*, 1–12.

60. Murugesan, N. et al. Biphenylsulfonamide endothelin receptor antagonists. 4. Discovery of *N*-[[2'-[[{4,5-Dimethyl-3-isoxazolyl}amino]sulfonyl]-4-(2-oxazolyl)[1,1'-biphenyl]-2-yl]methyl]-*N*,3,3-trimethylbutanamide (BMS-207940), a highly potent and orally active ET$_A$ selective antagonist. *J. Med. Chem.* **2003**, *46*, 125–137.

61. See www.collaborativedrug.com/media/CDD_General_Information.pdf.

62. See http://nihroadmap.nih.gov/molecularlibraries/.

63. For more about this, see Livingston, D., Lease, T. Behind the scenes at the NIH molecular libraries small molecule repository. *Soc. Biomol. Sci. News*, #19, December 2005. Available at http://mlsmr.glpg.com/MLSMR_HomePage/.

64. A few of us fondly remember the prediction that someday, perhaps in 1990 or 1995, companies would have large departments full of computer modelers and only a few humble bench chemists required to put together the structures (= drugs) that came out of their efforts.

65. Valler, M.J.; Green, D. Diversity screening versus focused screening in drug discovery. *Drug Discov. Today* **2000**, *5*, 286–293.

66. Quoted in Rouhi, A.M. Rediscovering natural products. *Chem. Eng. News* October 13, **2003**, 77–91.

67. Landers, P. Drug industry's big push into technology falls short. *Wall St. J.* February 24, **2004**.

68. Livingston, D. A report on high-throughput technologies. *Drug Discov./Tech. News* **2000**, *3*, 1.

69. Fox, S. et al. High-throughput screening: Update on practices and success. *J. Biomol. Screen.* **2006**, *11*, 864–869.

70. Bunin, B.A. *The Combinatorial Index* (San Diego, Academic Press, 1998).

71. Bunin, B. Increasing the efficiency of small-molecule drug discovery. *Drug Discov. Today* **2003**, *8*, 823–826.

72. See Lipinski, C., Hopkins, A. Navigating chemical space for biology and medicine. *Nature* **2004**, *432*, 855–861.

73. Bohacek, R.S. et al. The art and practice of structure-based drug design. *Med. Res. Rev.* **1996**, *16*, 3–50.

74. Although once in a while surprisingly simple molecules will show activity. For example *N*-cyanopyrrolidine is a micromolar, somewhat selective inhibitor of some cysteine proteases. See Falgueyret, J.-P. et al. Novel, nonpeptidic cyanamides as potent and reversible inhibitors of human cathepsins K and L. *J. Med. Chem.* **2001**, *44*, 94–104.

75. Fattori, D. Molecular recognition: The fragment approach in lead generation. *Drug Discov. Today* **2004**, *9*, 229–238.

76. Hann, M.M. et al. Molecular omplexity and its impact on the probability of finding leads for drug discovery. *J. Chem. Inf. Comput. Sci.* **2001**, *41*, 856–864.

77. Teague, S.J. The design of leadlike combinatorial libraries. *Angew. Chem. Int. Ed. Engl.* **1999**, *38*, 3743–3748.

78. Rishton, G.M. Nonleadlikeness and leadlikeness in biochemical screening. *Drug Discov. Today* **2003**, *8*, 292–293.

79. Congreve, M. et al. A 'rule of three' for fragment-based drug discovery? *Drug Discov. Today* **2003**, *8*, 876–877.

80. Murray, C.W., Verdonk, M.L. The consequences of translational and rotational entropy lost by small molecules on binding to proteins. *J. Comput Aided Mol. Design* **2002**, *16*, 741–753.

81. Mitchell, T., Cherry, M. Fragment-based drug design. *Innov. Pharm. Tech.* **2005**, issue 16. Available at www.sareum.co.uk/pdfs/InnPharm2005_Sareum.pdf.

82. Hartshorn, M.J. et al. Fragment-based lead discovery using x-ray crystallography. *J. Med. Chem.* **2005**, *48*, 403–413.

83. Hopkins, A.L. et al. Ligand efficiency: A useful metric for lead selection. *Drug Discov. Today* **2004**, *9*, 430–431. For more about its quantitation, see Abad-Zapatero, C., Metz, J.T. Ligand efficiency indices as guideposts for drug discovery. *Drug Discov. Today* **2005**, *10*, 464–469.

84. Mitchell, P. Fragment-based drug design delivers. *Pharma. Disc. Devel.* May/June **2006**, 32–34.

85. Shuker, S.B. et al. Discovering high-affinity ligands for proteins: SAR by NMR. *Science* **1996**, *274*, 1531–1534.

86. Hajduk, P.J. et al. NMR-based modification of matrix metalloproteinase inhibitors with improved bioavailability. *J. Med. Chem.* **2002**, *45*, 5628–5639.

87. Vanwetswinkel, S. et al. TINS, target immobilized NMR screening: An efficient and sensitive method for ligand discovery. *Chem. Biol.* **2005**, *12*, 207–216.

88. Bemis, G.W., Murcko, M.A. The properties of known drugs. 1. Molecular frameworks. *J. Med. Chem.* **1996**, *39*, 2887–2893.

89. Fejzo, J. et al. The SHAPES strategy: An NMR-based approach for lead generation in drug discovery *Chem. Biol.* **1999**, *6*, 755–769.

90. Meyer, B., Peters, T. NMR spectroscopy techniques for screening and identifying ligand binding to protein receptors. *Angew. Chem. Int. Ed. Engl.* **2003**, *42*, 864–890.

91. Erlanson, D.A. et al. Site-directed ligand discovery. *Proc. Natl. Acad. Sci. USA* **2000**, *97*, 9367–9372.

92. Vetter, D. Chemical microarrays, fragment diversity, label-free imaging by plasmon resonance—a chemical genomics approach. *J. Cell. Biochem.* **2003**, *87*, 79–84. Also see Graffinity's website at www.graffinity.com.

93. Green, N.M. Avidin. *Adv. Protein Chem.* **1975**, *29*, 85–133.

94. Jencks, W.P. On the attribution and additivity of binding energies. *Proc. Natl. Acad. Sci. USA* **1981**, *78*, 4046–4050.

95. See, for example, Maly, D.J. et al. Combinatorial target-guided ligand assembly: Identification of potent subtype-selective c-Src inhibitors. *Proc. Natl. Acad. Sci. USA* **2000**, *97*, 2419–2424.

96. Erlanson, D.S. Fragment-based drug discovery. *J. Med. Chem.* **2004**, *47*, 3463–3482.

97. Card, G.L. et al. A family of phosphodiesterase inhibitors discovered by cocrystallography and scaffold-based drug design. *Nat. Biotechnol.* **2005**, *23*, 201–207.

98. Most companies would probably include plasma stability as a necessary criterion for a lead compound.

99. Stout, T.J. et al. The additivity of substrate fragments in enzyme-ligand binding. *Structure* **1998**, *6*, 839–848.

100. Babaoglu, K., Shoichet, K. Deconstructing fragment-based inhibitor discovery. *Nature Chem. Biol.* **2006**, *2*, 720–722.

101. Rees, D.C. et al. Fragment-based lead discovery. *Nat. Rev. Drug Discov.* **2004**, *3*, 660–672.

102. Liebeschuetz, J.W. et al. PRO_SELECT: Combining structure-based drug design and array-based chemistry for rapid lead discovery. 2. The development of a series of highly potent and selective factor Xa inhibitors. *J. Med. Chem.* **2002**, *45*, 1221–1232.

103. Agnelli, G. et al. A phase II study of the oral factor Xa inhibitor LY517717 for the prevention of venous thromboembolism after hip or knee replacement. *J. Thromb. Haemost.* **2007**, *5*, 746–753.

104. Carr, R. Fragment-based drug discovery. *Drug Discovery* **2006**. Available at www.touchbriefings.com/pdf/2287/carr.pdf.

105. For an overview, see Brown, D. Sources of drugs, Section 1.10 in *Comprehensive Medicinal Chemistry* Vol. 1 Triggle, D.J. and Taylor, J.B., Eds. (London, Elsevier, 2007). A fascinating account of the historic origins of many drugs can be found in Mann, J. *Murder, Magic, and Medicine* (Oxford, Oxford University Press, 1992).

106. Newman, D.J. Natural products as sources of new drugs over the period 1981–2002. *J. Nat. Prod.* **2003**, *66*, 1022–1037.

107. These days this practice would probably constitute a violation of a number of laws!

108. Mays, T.D., Mazan, K.D. Legal issues in sharing the benefits of biodiversity prospecting. *J. Ethnopharmacol.* **1996**, *51*, 93–109.

109. Koehn, F.E., Carter, G.T. The evolving role of natural products in drug discovery. *Nat. Rev. Drug Discov.* **2005**, *4*, 206–220.

110. For a look at a few of these targets, see Nicolaou, K.C., Snyder, S.A. The essence of total synthesis. *Proc. Natl. Acad. Sci. USA* **2004**, *101*, 11929–11936.

111. See Butler, M.S. The role of natural products chemistry in drug discovery. *J. Nat. Prod.* **2004**, *67*, 2141–2153.

112. Grabowski, K., Schneider, G. Properties and architecture of drugs and natural products revisited. *Curr. Opin Chem. Biol.* **2007**, *1*, 115–127.

113. Feher, M., Schmidt, J.M. Property distributions: Differences between drugs, natural products, and molecules from combinatorial chemistry. *J. Chem. Inf. Comput. Sci.* **2003**, *43*, 218–227.

114. Shelat, A.A., Guy, R.K. The interdependence between screening methods and screening libraries. *Curr. Opin.Chem. Biol.* **2007**, *11, 244–251.*

115. Tempest, P.A., Armstrong, R.W. Cyclobutendione derivatives on solid support: Toward multiple core structure libraries. *J. Am. Chem. Soc.* **1997**, *119*, 7607–7608.

116. Schreiber, S.L. Target-oriented and diversity-oriented organic synthesis in drug discovery. *Science* **2000**, *287*, 1964–1969.

117. Schreiber, S.L. The small-molecule approach to biology. *Chem. Eng. News* March 3, **2003**, 51–61.

118. Nören-Müller, A. et al. Discovery of protein phosphatase inhibitor classes by biology-oriented synthesis. *Proc. Natl. Acad. Sci. USA* **2006**, *103*, 10606–10611.

119. Taylor, S.J. et al. Synthetic strategy toward skeletal diversity via solid-supported, otherwise unstable reactive intermediates. *Angew. Chem. Int. Ed. Engl.* **2004**, *43*, 1681–1685.

120. Quoted in Rouhi, Reference 66.

121. Quoted in Anon. Sir James Black: Learning by doing. *Mol. Interv.* **2004**, *4*, 139–142.

122. Black, J. Drugs from emasculated hormones: The principle of syntopic antagonism. *Science* **1989**, *245*, 486–493.

123. Some of the difficulties that had to be overcome can be found in Sir James's account in the reference in Note 122.

124. Black, Note 122, also addresses this work. Another nice account of it is provided in Chapter 18, "Cimetidine—a rational approach to drug design" in *An Introduction to Medicinal Chemistry*, by Graham L. Patrick, second edition (New York, Oxford University Press, 2001).

125. Black, J. Future perspectives in pharmaceutical research. *Pharm. Policy Law* **1999**, *1*, 85–92. This is a thought-provoking paper well worth reading.

126. See Rahuel, J. et al. Structure-based drug design: The discovery of novel nonpeptide orally active inhibitors of human renin. *Chem. Biol.* **2000**, *7*, 493–504.

127. Böhm, H.-J. et al. Scaffold hopping. *Drug Discov. Today Technol.* **2004**, *1*, 217–224.

128. For an excellent review of active metabolites and their uses, see Fura, A. et al. Discovering drugs through biological transformation: Role of pharmacologically active metabolites in drug discovery. *J. Med. Chem.* **2004**, *47*, 4339–4351.

129. Rosenblum, S.B. et al. Discovery of 1-(4-Fluorophenyl)-(3*R*)-[3-(4-fluorophenyl)-(3*S*)-hydroxypropyl]-(4*S*)-(4-hydroxyphenyl)-2-azetidinone (SCH 58235): A designed, potent, orally active inhibitor of cholesterol absorption. *J. Med. Chem.* **1998**, *41*, 973–980. See also Clader, J.W. The discovery of ezetimide: A view from outside the receptor. *J. Med. Chem.* **2004**, *47*, 1–9.

130. Li, A.C. et al. Simultaneously quantifying parent drugs and screening for metabolites in plasma pharmacokinetic samples using selected reaction monitoring information-dependent acquisition on a QTrap instrument. *Rapid Comm. Mass Spec.* **2005**, *19*, 1943–1950.

131. Fura, A. Role of pharmacologically active metabolites in drug discovery and development. *Drug Discov. Today* **2006**, *11*, 133–142.

132. For a good overview of the field, see Kubinyi, H. Chemogenomics in drug discovery. *Ernst Schering Res. Found. Workshop* **2006**, *58*, 1–19. Available at www.kubinyi.de/schering58-2006.pdf.

133. Harris, C.J., Stevens, A.P. Chemogenomics: Structuring the drug discovery process to gene families. *Drug Discov. Today* **2006**, *11*, 880–888.

134. Caron, P.R. et al. Chemogenomic approaches to drug discovery. *Curr. Opin. Chem. Biol.* **2001**, *5*, 464–470.

135. Jorgensen, W.L. The many roles of computation in drug discovery. *Science* **2004**, *303*, 1813–1818.

136. Waszkowycz, B. et al. Large-scale virtual screening for discovering leads in the postgenomics era. *IBM Systems J.* **2001**, *40*, 360–376. Available at www.d2ol.com/papers/vls.pdf.

137. To be fair, though, confirmed hits (not artifacts) from the same compound collection binding to the desired site under identical assay conditions need to be compared, and this isn't always the case in literature reports of enrichment factors.

138. See Kubinyi, H. Drug research: Myths, hype and reality. *Nat. Rev. Drug Discov.* **2003**, *2*, 665–668.

139. McGovern, S.L., Shoichet, B.K. Information decay in molecular docking screens against holo, apo, and modeled conformations of enzymes. *J. Med. Chem.* **2003**, *46*, 2895–2907.

140. Leach, A.R. et al. Prediction of protein–ligand interactions. Docking and scoring: Successes and gaps. *J Med. Chem.* **2006**, *49*, 5851–5855.

141. Mattioni, B.E. In silico ADME screening for enhanced hit rates in drug discovery. *Talk Presented at the 232nd ACS Meeting*, San Francisco, California, September 12, 2006. COMP 172.

142. See Doman, T.N. et al. Molecular docking and high-throughput screening for novel inhibitors of protein tyrosine phosphatase-1B. *J. Med. Chem.* **2002**, *45*, 2213–2221.

143. Charifson, P.S. et al. Consensus scoring: A method for obtaining improved hit rates from docking databases of three-dimensional structures into proteins. *J. Med. Chem.* **1999**, *42*, 5100–5109.

144. Bonacci, T.M. et al. Differential targeting of $G\beta\gamma$-subunit signaling with small molecules. *Science* **2006**, *312*, 443–446.

145. Becker, O.M. et al. G protein-coupled receptors: *In silico* drug discovery in 3D. *Proc. Natl. Acad. Sci. USA* **2004**, *101*, 11304–11309.

146. Becker, O.M. et al. An integrated in silico 3D model-driven discovery of a novel, potent, and selective amidosulfonamide 5-$HT_{1A}$ agonist (PRX-00023) for the treatment of anxiety and depression. *J. Med. Chem.* **2006**, *49*, 3116–3135.

147. Surprisingly, this means that there currently are only enough unique small molecules described so that it wouldn't even be possible to assign one to each person in America: A dozen of us would have to share one!

148. Lameijer, E.-W. et al. Designing active template molecules by combining computational de novo design and human chemist's expertise. *J. Med. Chem.* **2007**, *50*, 1925–1932.

149. Schneider, G., Fechner, U. Computer-based *De Novo* design of drug-like molecules. *Nat. Rev. Drug Discov.* **2005**, *4*, 649–663.

150. Shoichet, B.K. Virtual screening of chemical libraries. *Nature* **2004**, *432*, 862–865.

151. Ji, H. et al. Structure-based de novo design, synthesis, and biological evaluation of non-azole inhibitors specific for lanosterol 14α-demethylase of fungi. *J. Med. Chem.* **2003**, *46*, 474–485.

# Chapter 7

# Turning Hits into Drugs

## 7.1 What Now?

When the screening campaign is over and the follow-up work to confirm activity is done, the results can range from an overabundance of potential starting points to none at all. At least in the latter case the path forward is clear: test more compounds or find another project. Having a long list of confirmed hits presents one with a more difficult decision: which of them should be followed up first? Resources at this point are often limited and some sort of triage is needed even if relatively few analogs of each are to be made. Sometimes this happens quite naturally: some screening hits are so "ugly" that few medicinal chemists would think of following them up at all. HTS is notorious for finding such compounds. Fortunately, better screening collections will turn up more reasonable-looking molecules, and here prioritizing becomes harder.

At this point, further follow-up work may also be generated by a sort of computational meta-analysis of the results. Similarity searching, overlaying the hits and generating a tentative pharmacophore model can lead to ideas for new molecules to make or buy for further screening. Medicinal chemists may be eager to try out their own ideas as well. There may be more things to try than resources to try them with.

Most chemists will prioritize compounds based on potency and structure, the latter representing a combination of synthetic tractability and structural novelty. At this early stage, potency and structure may be all that's known about the confirmed hits anyway. They're certainly important because later on in the project compounds will need to be synthetically accessible, potent, and patentable. But that's just the start of the list. Plenty of other properties will be required too, including solubility, selectivity, stability, and cell permeability. An introduction to these properties and how one might begin to optimize for them is given in the following chapters.

When prioritization has been done and analogs begin to appear, the project will have entered the stage known as hit-to-lead (HTL). Here confirmed hits, pure compounds with known identities that gave good, reproducible dose–response curves in a bioassay through interactions with the desired target, are modified to try and produce a lead series. Note that a lead *series*, which suggests a number of compounds and a consequent tolerance for diverse modifications, rather than a single lead *compound*, implying less latitude for optimization, is by far preferred.

The lead series itself represents a starting point for further optimization into clinical candidates. Hits, leads, and clinical candidates differ only in the number of hurdles in the drug discovery process that they're capable of clearing. In theory, one might assay a compound from a screening library and find that it's capable of clearing them all from the start: a drug! Similarly, in theory a random individual off the streets could win a medal in the Olympic triathlon. The reality is that this is extremely unlikely, even though a few examples of an unmodified compound from screening going on to become a drug, like the natural product taxol, do exist. Even a good hit generally requires a lot of optimization before it might ever result in a clinical drug. HTL as defined herein is the process of refining the hit structures to produce proprietary compounds having additional favorable properties including functional activity in cells.

The HTL process is a relatively recent development in drug discovery research. In the classical paradigm of several decades ago, only a few molecules already active in tissue or whole animal assays would serve as starting points for analoging. The advent of combinatorial libraries and high-throughput target screening brought about the need to triage a much larger number of often poorer starting points. The last few years have seen the increasing realization that best way to deal with this is to divide subsequent efforts into two different phases. In the first, hit-to-lead, a small group of researchers quickly assesses the prospects of each interesting confirmed hit by making and examining a limited set of analogs. As the initial strike force, HTL chemists need to be quick and aggressive: Dead ends need to be identified and promising hits improved to meet preset criteria in only a matter of months and with limited resources. Considering the diversity of chemotypes, synthetic routes, and properties involved, this is no small feat. As we've already seen, perhaps only 30% of all HTS campaigns will ever get this far. If it's accomplished, though, and a genuine lead series is identified, it's time for the bulk of chemistry efforts to begin, and formal lead optimization, discussed later, will commence.

## 7.2  Biochemical Mechanism in Hit Selection

### 7.2.1  Competition and Allostery

In following up on confirmed hits, it's important to understand how they interact with their targets. Knowing this can provide insights into how valuable they might be. Before a compound even goes into HTL, a biochemical assessment of its suitability will probably be done. Acceptable biochemical mechanisms may be listed as a criterion along with good dose-response, lack of fluorescence, etc. that needs to be met before an active can be considered a confirmed hit, or that might be considered before much work goes into following up a compound already classified as a confirmed hit. If the compound seems to be acting through what's considered an undesirable mechanism (e.g. irreversible inhibition, sometimes) it might be crossed off the list. So biochemical mechanism can play a role in hit prioritization.

Medicinal chemists have traditionally honed in on reversible, competitive compounds to inhibit enzymes, and agonists, antagonists, or inverse agonists acting at the binding site of the natural ligand (*orthosteric* binders) for receptors like GPCRs. But it's worth taking a step back to look at other possibilities as well.

*Competitive enzyme inhibitors* act by competing with endogenous substrate for binding to the enzyme. Only one or the other can bind at a given time, a sort of molecular game of musical chairs, and which of the two gets to bind depends on their relative concentrations and affinities for the enzyme. For in vivo situations, optimizing PK parameters for the inhibitor can address the concentration issue, and optimizing binding interactions using tools like SBDD is an effective way of increasing inhibitor affinity. But competitive inhibitors do have their Achilles heel: their effects can be swamped out in the presence of a buildup of substrate, where their potency might not be enough to win the competition against so many substrate molecules. "The bottom line is that when substrate concentrations exceed the $K_m$, greater concentrations of competitive drug should be required to achieve the same level of inhibition."[1] In vitro enzyme testing constitutes a so-called *closed system*, where the concentration of substrate doesn't change much, and here the substrate concentration used is typically at or near the $K_m$ value. Mother Nature, on the other hand, makes use of *open systems* and even though substrate concentrations also tend to be near the $K_m$ value in the absence of an inhibitor, once an inhibitor is present substrate concentrations can build up, causing it to lose its effectiveness. "In vivo, as in theoretical open systems, the tendency of substrates to accumulate at blockage points cannot be safely ignored."[2]

Several ways around this potential difficulty exist. Binding somewhere other than the site where the substrate or ligand goes is one possibility. For enzymes, two different ways of doing this exist. *Uncompetitive enzyme inhibitors* bind instead to the enzyme–substrate complex, or in some cases to a species formed subsequently. This means that if substrate concentrations build up, so will concentrations of the ES complex, and the apparent activity of uncompetitive inhibitors will therefore *increase*, not decrease (Figure 7.1). The immuno-suppressive agent mycophenolic acid (Figure 7.2), which inhibits inosine monophosphate dehydrogenase (IMPDH), is one example of an uncompetitive inhibitor.

*Non-competitive enzyme inhibitors* can bind to both the free enzyme and the enzyme–substrate complex (or a subsequent species). The HIV drug and non-nucleoside reverse transcriptase inhibitor (NNRTI) Viramune (nevirapine, Figure 7.2) fits this description. Because of this, drugs like these don't suffer an apparent loss in potency or efficacy due to

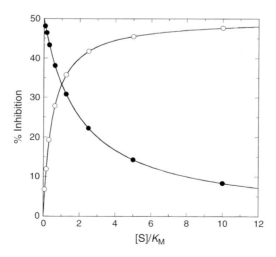

**Figure 7.1 ▶**    Effects of substrate buildup in a metabolic pathway on inhibition of an enzyme by competitive (●) and uncompetitive (○) inhibitors of equal affinity for the target enzyme. (Reprinted with permission from Copeland, R.A. *Evaluation of Enzyme Inhibitors in Drug Discovery* (Hoboken, NJ, John Wiley & Sons, 2005, copyright 2005, John Wiley & Sons.)

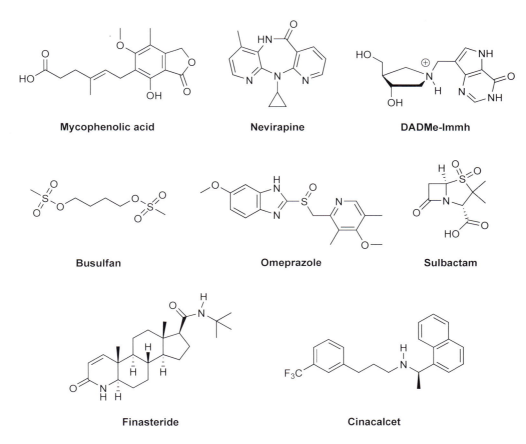

**Figure 7.2 ►** Mycophenolic acid (an uncompetitive inhibitor), nevirapine (a non-competitive inhibitor), DADMe-Immh (a slow off-rate reversible inhibitor), busulfan (a non-specific affinity label), omeprazole (prodrug of an irreversible inhibitor), sulbactam, finasteride (both mechanism-based inactivators), and cinacalcet (an allosteric activator).

buildup of substrate.[3] Uncompetitive and non-competitive inhibitors are easy to confuse since the words are so similar. One mnemonic that might help is that UNcompetitive inhibitors UNiquely bind the enzyme–substrate complex whereas NOn-competitive inhibitors are NOt so picky.

For receptors, *allosteric binding* means that a molecule needn't compete with ligand for occupancy at the orthosteric site (see Figure 7.3). Not only does this make such compounds immune to being swamped out by high ligand concentrations, but the opposite effect has been proposed as an advantage as well. The allosteric site and its consequent effects can get saturated, unlike the orthosteric site where more and more drug might increase the response to the point of toxicity (too much of a good thing). "There is a 'ceiling' to the effects of an allosteric modulator that is retained even with excessive doses."[4] Another potentially large advantage to allosteric binders lies in the area of selectivity. Orthosteric sites tend to be very similar within a given family of receptors, making discrimination between them, a

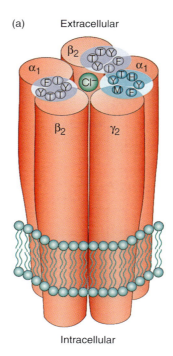

(a)                    Extracellular

Intracellular

**Figure 7.3 ►**  Schematic model of the $\alpha_1\beta_2\gamma_2$ GABA$_A$ ligand-gated chloride ion channel. The residues believed to constitute the benzodiazepine allosteric site are shown in turquoise while those of the orthosteric GABA binding sites are shown in purple. (Reprinted with permission from Christopoulos, A. Allosteric binding sites on cell-surface receptors: Novel targets for drug discovery. *Nat. Rev. Drug Discov.* **2002**, *1*, 198–210, copyright 2002, Macmillan Publishers Ltd.)

prerequisite for selectivity, difficult. The same may not hold true for allosteric receptor binding sites, where evolutionary pressure to conserve residues to accommodate an endogenous ligand hasn't necessarily been a factor.[5] The same goes for allosteric sites on enzymes where inhibitors might bind.

The subject of allosteric modulation of receptors has been well-covered in recent reviews.[6] Marketed drugs like diazepam (Figure 6.15), which acts as an allosteric modulator of GABA$_A$ receptors (chloride ion channels, see Figure 7.3) and, more recently, drugs like cinacalcet (Figure 7.2), attest to the possibilities inherent in the use of allosteric binders.[7] To date, though, such drugs are far outnumbered by drugs binding to the orthosteric sites of receptors. Why is that?

Screens for receptor binders, as we've seen, have traditionally used radioligand binding displacement assays, and compounds acting via allosteric modulation won't necessarily show up in these. Functional assays like FLIPR, which in recent years have become widespread, are better ways to find them and may explain why more of them are now turning up.

One problem, though, is that finding, characterizing, and developing allosteric receptor binders is far from trivial. Their detection depends on the probe that's used; kinetic follow-up experiments are required; and a compound acting one way in one experimental system might even act the opposite way in another, as when several allosteric binders of the cannabinoid $CB_1$ receptor which increased agonist affinity in binding assays turned out to be insurmountable antagonists of receptor function in a reporter gene assay. "These findings have highly significant implications for drug discovery based on allosteric modulators, because they highlight the possibility of misclassification of novel ligands if they are not tested in as full a spectrum of biological assays as possible."[8] Even its supporters admit that the complexities involved can sometimes mean that working with allosteric modulators turns into a "screener's nightmare."[9] But, like working with non-competitive and uncompetitive enzyme inhibitors, it can also turn up unique new drugs.

## 7.2.2 Irreversibility

A second way of avoiding the potential substrate buildup problem and improving the observed potency along the way is to "cheat" by establishing conditions under which equilibrium between enzyme, substrate, and inhibitor is not achieved, thus depriving the substrate of a fair chance to compete. The most direct way of effecting such non-equilibrium conditions is to use an inhibitor that once bound never comes off: an *irreversible inhibitor*. These generally work by forming a covalent bond between the enzyme and an electrophilic center. Using our "musical chairs" analogy, once an irreversible inhibitor wins, it never gets off the chair (target) and the game is over. Kinetically, this means that more and more active enzyme is taken out with time, so the kinetics for an irreversible inhibitor will be fundamentally different from those seen with reversible inhibitors. Because of this, $K_i$ values as determined for the latter are not appropriate for such compounds. A bit more enzymology work is necessary to determine values that are: their second-order rate constants, $K_{inact}$ or $K_I$. So inhibition constants for irreversible and reversible inhibitors can't be directly compared.

Irreversible inhibitors have been divided up into three categories: *nonspecific affinity labels*, *quiescent affinity labels*, and *mechanism-based inactivators*. Drugs exist in all three categories. Nonspecific affinity labels will react fairly promiscuously with nucleophiles, making them very unattractive compounds, but even in this category a few cancer drugs like busulfan (Figure 7.2) can be found. Quiescent affinity labels have less general reactivity and contain some structural features conferring specificity for a given enzyme's active site. Binding then provides the appropriate complementarity and effective concentration of nucleophilicity for a bond to be formed. Two of the most popular drugs of all time, aspirin and penicillin, fall into this category. The former selectively acetylates a key serine residue in COX1 and COX2, while the penicillins acylate and thereby inactivate bacterial peptidoglycan transpeptidase.

The best definition of the third type of irreversible enzyme inhibitor, mechanism-based inactivators, is provided by Dr Richard Silverman, a leading authority on the subject. "A mechanism-based inactivator (sometimes, much to my dismay, called a suicide substrate)," he writes, "is an unreactive compound that has a structural similarity to a substrate or product for an enzyme. Once at the active site of the enzyme, it is converted into a species that generally forms a covalent bond to the enzyme, producing inactivation."[10] Although both quiescent affinity labels and mechanism-based inactivators can be mistaken for substrate by the enzyme, a key difference is that the former is able to react directly to inactivate it, while in the latter case enzymatic reaction is needed to generate the inactivating species. Justice can't be done to this fascinating concept in just a few paragraphs, so those wishing to know more about it are referred to the relevant literature.[11] A number of drugs act by mechanism-based inactivation, including sulbactam (Figure 7.2), a compound for overcoming penicillin resistance, and finasteride (Figure 7.2), a drug for benign prostatic hyperplasia (BPH), a condition common in men over 50.

Although irreversible inhibitors appear advantageous in theory, one can cite many drugs acting irreversibly (whether intended to or not), and a few new candidates are made intentionally irreversible like some kinase inhibitors for cancer;[12,13] someone expecting to hear their praises sung in biotech or pharma will likely be disappointed. Representatives of several big pharma companies have publicly stated that their firms would never knowingly develop an irreversible inhibitor, a clear indication that it's passed into the realm of dogma. Most often, compounds found to be acting irreversibly in hit follow-up are likely to be dropped from further consideration on that basis alone. Why the disconnect?

The answer is fear of clinical toxicity. Conceptually, this could arise from on-target or off-target effects of such inhibitors. For many targets, it's actually undesirable to wipe out 100% of their activities, which could give rise to mechanism-based toxicity. Here activity only needs to be attenuated, not switched off altogether. Giving too high a dose of an inhibitor that once bound never comes off could amount to extending the duration, and probably severity, of toxicity for these. Off-target toxicity would be a major concern for any type of affinity label (but not necessarily for mechanism-based inactivators), as these by definition can't be entirely specific for the target. Aside from concerns about toxicity directly relatable to inhibition of a different target, there's another potentially worse problem. In theory, at least, an irreversible inhibitor might act as a *hapten*, reacting with a non-target nucleophile like the sulfhydryl group of a protein's cysteine residue to form something seen as foreign by the immune system, potentially leading to an immune response.[14]

This concern isn't entirely theoretical: penicillins, quiescent affinity label as mentioned above, cause allergies (mostly skin rashes and hives) in about 2% of the patients who take them, and even potentially fatal anaphylaxis in a few, and toxicity is believed to proceed through this sort of mechanism. With the current atmosphere about drug safety, it's no wonder that many companies will only consider working on irreversible inhibitors where the benefits clearly outweigh the risks as in the case of cancer, in the context of tool compounds, or perhaps not at all. Had penicillin been discovered today quite a few companies might choose not to develop it.

### 7.2.3  Slow Off-Rate Compounds

In practice, an inhibitor with a very slow off-rate at its target can have the advantage of "cheating" a substrate out of its fair chance for equilibrium competition while avoiding covalent bond formation that might make for toxicity. Such compounds are sometimes referred to as *functionally irreversible* or *pseudo-irreversible*. Extremely tight-binding transition state inhibitors like the purine nucleoside phosphorylase (PNP) inhibitor DADMe-ImmH (Figure 7.2) can act like irreversible inhibitors in having off-rates so low that recovery of activity can only be achieved by *de novo* resynthesis of the protein, which has been called *ultimate physiological inhibition.*[15] Such slow-off compounds can even be hard to tell apart from covalent inhibitors, sometimes requiring isolation of the enzyme–inhibitor complex, denaturation of the protein, and analysis by LCMS to see whether the protein has been covalently modified. Quite a few marketed drugs, both enzyme inhibitors and receptor binders like the H1 antagonist Clarinex (desloratidine, Figure 6.31), have fairly long off-rates, on the order of hours.[16]

As the potency of a reversible inhibitor is directly related to $k_{on}/k_{off}$, and $k_{on}$ can't be improved beyond the rate of diffusion, it isn't surprising that extremely potent inhibitors, down in the picomolar range, tend to have slow off-rates. It's probably the case that much compound work done to optimize in vitro potency really selects for longer off-rate compounds. Many advantages of slow-off inhibitors exist, including the fact that a long residence time at the enzyme can effectively shield the compound from metabolic processes that would otherwise destroy it. In fact, it may be that for drugs with very slow off-rates, plasma levels won't necessarily correlate with duration of effect. So PK requirements (plasma concentrations and $t_{1/2}$) for slow-off as well as for truly irreversible compounds may not be as critical as they are for compounds with faster off-rates, but by the same token the measurement of their pharmacodynamic (PD) effects (what the drug is doing in vivo at various time points) becomes even more important.[17]

A recent review points out many of the advantages of slow off-rate compounds, both receptor binders and enzyme inhibitors, including a potential effect on selectivity which we'll revisit later, and makes a strong argument for the measurement of $k_{off}$ in addition to $IC_{50}$ or $K_d$ in lead optimization.[18] Potential problems with this approach were not overlooked, including the possibility that such tight-binding drugs might hold a receptor too long in an altered conformation, perhaps even after eventually dissociating, thus generating epitopes that might lead to an autoimmune response. And, once again, inhibiting a target that might have mechanistic toxicity too well and too long, whether by an irreversible or a functionally irreversible inhibitor, might have adverse consequences. Still, compounds with slow-off rates have many advantages and to date no strong bias against them seems to exist in industry, where picomolar inhibitors are well-appreciated.

The subject of non-equilibrium situations such as pseudo-irreversible inhibition in drug design has been reviewed by Dr David Swinney, who concluded that "first, in the absence of mechanism-based toxicity, drug mechanisms that create transitions to a non-equilibrium state will be more efficient; second, in the presence of mechanism-based toxicity, biochemical mechanisms that maintain equilibrium to achieve a tolerable balance between efficacy and toxicity are desired."[1] "Non-equilibrium states," of course, include slow-off rate and irreversible inhibition.

In another review the same author went on to analyze biochemical mechanisms of action for 85 NMEs approved by FDA between January 2001 and November 2004.[19] He found that 80% of those drugs with known molecular targets were competitive, binding at the same site as the endogenous ligand, but for only 20% of the drugs with known targets was equilibrium competition with ligand or substrate by itself sufficient to achieve therapeutic effects (Figure 7.4). The majority of the drugs needed something else as well, coupling equilibrium binding with a second process, either a conformational change (agonism, allosteric activation, non-competitive inhibition, etc.) or non-equilibrium kinetics (irreversible inhibition, slow dissociation, etc.). A surprisingly high number, 14, of the drugs (23%) relied on outright irreversible inhibition or a slow off-rate for their efficacy and 12% hit multiple

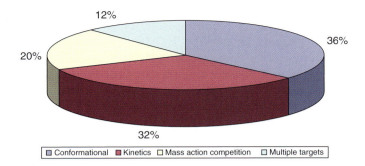

**Figure 7.4** ▶ Biochemical properties for 60 NMEs approved by US FDA between January 2001 and November 2004. Chart shows additional biochemical factors involved in drug binding. Data from Swinney. (Swinney, D.C. Biochemical mechanisms of new molecular entities (NMEs) approved by the United States FDA during 2001–2004: Mechanisms leading to optimal efficacy and safety. *Curr. Top. Med. Chem.* **2006**, *6*, 461–478.)

targets. Another study of marketed drugs that target enzymes also found a surprisingly high percentage of irreversible and non-competitive inhibitors among them.[20] All of this goes to show that focusing on selective, competitive, rapidly reversible compounds isn't the only way to get a drug.

### 7.2.4   Why Mechanism Matters

Having gone through a brief outline of some of the important biochemical factors involved in prioritizing hits, some readers may be asking themselves, "Why does it matter?" After all, medicinal chemists make compounds, turn them in for screening, and see from the results whether they're potent or not. Isn't that good enough? Why should it be necessary for you to bend your mind and try to think like a biochemist? The answer is that, while an individual chemist might sometimes be able to ignore biochemical mechanism, neither the project nor any clinical candidate it may produce can afford to do so. Prioritizing hits and optimizing activity without regard to the biochemical mechanisms involved just won't happen at most companies.

In the real world, no chemist will be allowed to ignore mechanistic factors for long. Actives or hits shown to be irreversible will often be deprioritized or dropped, while enzyme inhibitors proving to be non-competitive or uncompetitive might even find additional support on that basis. Hits that may be allosteric receptor binders will require a lot of work in biochemistry and some specialized expertise. Attempts to use structure-based drug design on uncompetitive inhibitors which, you'll recall, uniquely bind to the enzyme–substrate complex, will be futile if crystallization is carried out in the absence of substrate because it's been assumed that the inhibitor is competitive instead.[3] Mechanism matters.

This being the case, the question becomes "Is it best to select for competitive, non-competitive, or uncompetitive enzyme inhibitors, and orthosteric, or allosteric receptor binders?" The answer is a resounding "Yes!" "It is seldom clear a priori what inhibition modality will give the most desirable cellular and organismal effects . . . Compound diversity, both in terms of pharmacophore structure (i.e. the minimal structural elements of a compound that are required for inhibition) and inhibition modality, must be an important consideration in medicinal chemistry efforts. When possible, it is best to run parallel lead optimization efforts on pharmacophores that conform to different inhibition modalities to give the best chance of demonstrating maximal efficacy and safety in cells and in vivo."[21]

Just as genetic diversity is a successful strategy for avoiding species extinction, so experience has shown that starting out with multiple, diverse hits is the best way to avoid project extinction. The many problems that crop up are less likely to terminate a project that has many different starting points than a project relying on a single compound and a few close analogs with a single mode of action, which is clearly a case of putting all of one's eggs in one basket. In the real world there's sometimes no choice: there's only one egg! But having multiple lead series would be far better, and these begin with multiple hits. For this both structural and mechanistic diversity are most desirable.

## 7.3  Druglikeness

### 7.3.1  What is It?

Since the ultimate aim of all compound optimization is to transmute a hit into a drug, it makes sense at this point to stop and consider a couple of definitions. A *drug* can be defined as a human therapeutic agent approved for sale by a regulatory agency such as the Food and Drug Administration (FDA) in the United States or the European Medicines Evaluation Agency (EMEA) in Europe.

If this definition is easy and even obvious, trying to define the related and frequently used term *druglike* is anything but that. It's been defined in various, somewhat overlapping, but far from identical ways, which can make for some confusion. Here are some of the most frequently cited definitions of "druglike" or "druglikeness":

▶ "Those compounds that have sufficiently acceptable ADME properties and sufficiently acceptable toxicity properties to survive through the completion of human Phase I clinical trials."[22]

▶ "Small molecules, similar to drugs with respect to calculated descriptors that are thought to capture synthetic feasibility, protein binding functionality, and favorable ADME and Tox properties."[23]

▶ Having "physiochemical properties that improve probability of success in drug development by addressing issues of absorption and bioavailability" including the filtering out of chemically reactive functional groups and peptides, compliance with ROF, etc.[24]

▶ "A compound is drug-like if all its atoms or groups are situated in (molecular) environments similar to those of existing drugs."[25]

▶ Fulfilling the three criteria of having "sufficient functionality to potentially favorably interact with a receptor or enzyme", a "relevant calculated or experimental molecular property profile . . . similar to that of known oral drugs", and "likely to have acceptable ADME properties."[26]

▶ "Molecules which contain functional groups and/or have physical properties consistent with the majority of known drugs."[27]

These definitions may be helpful, but except for the first, which is entirely empirical, and the third, they don't get into the nitty-gritty of exactly how druglikeness should be determined. A simpler, if no more specific, definition of "druglike" in the sense that it's normally used would be this: "Orally bioavailable and relatively nontoxic in humans".

## 7.3.2　Predicting Druglikeness

But if that's what it means to be druglike, then ultimately only clinical data can show whether a particular molecule is druglike or not. Short of that, predicting whether a compound existing *in silico* or in a vial will really turn out to be druglike is a bit like trying to predict what the weather will be 2 months from Tuesday. As we'll see, even predicting absorption, one component of bioavailability, is a challenge. But medicinal chemists are often called upon to decide on druglikeness in selecting compounds to screen, selecting hits from a screen to follow up, trying to decide which molecules to make, or in the frustrating case of trying to understand disappointing in vivo results for compounds they've already made. So a more detailed look at this issue is in order.

　　The above definitions hint that two different approaches to predicting druglikeness can be considered: structural similarity to known drugs on the one hand and calculated or measured molecular properties like MW and LogP on the other. Each of these has its limitations. How can one calculate any sort of "similarity" to known drugs when these are so different from one another? Looking at Figure 7.5, one might well conclude that the only real similarity between these drugs is that they all contain carbon, hydrogen, and oxygen! Largely for this reason one generally won't find papers listing *Tanimoto similarity* coefficients—which are derived from matching up connectivity fingerprints for two compounds to find how similar they are on a scale of 0 to 1—versus known drugs as a rationale for making specific compounds.[28] But making minor modifications to an existing drug (e.g. adding a methyl group) to maximize the likelihood of having a druglike compound is consistent with this thinking, as are the incorporation of structural features most commonly found in drugs[29,30] and the use of privileged scaffolds.[31] At the very least, these molecular pieces shouldn't be "undruglike."

　　Can a good medicinal chemist just look at a list of structures and tell which ones are druglike and which ones aren't? Most would like to think so. After all, medicinal chemists specialize in SAR analysis, which depends largely on pattern recognition. Shouldn't it be possible to at least pick out the bad ones? Going through compounds available in ACD (the Available Chemicals Directory) for instance, one can find quite a few compounds with structural features like heavy metals or multiple P–N bonds whose diversity would probably not be missed by leaving them out of potential drug candidates. After years of experience in drug discovery research, shouldn't a chemist have developed an eye for what looks good and what doesn't? Perhaps she can't explain the rules she's using or program them into a

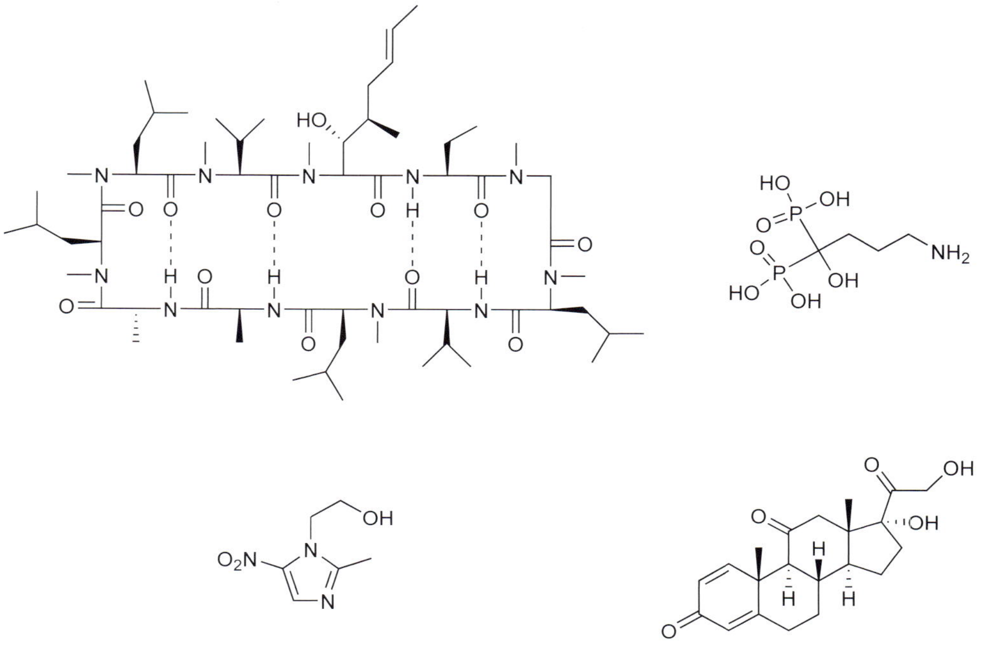

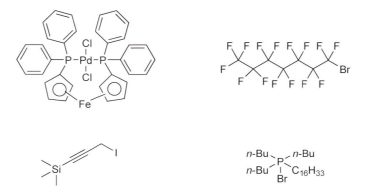

**Figure 7.5 ▶** How similar are drugs to each other? Upper left: cyclosporin A, upper right: alendronic acid, lower left: metronidazole, lower right: prednisone.

computer, but shouldn't she at least be at the level of the man at the art gallery who "doesn't know anything about art, but knows what he likes"? After all, most of us would agree that the compounds shown in Figure 7.6 are not at all druglike, although perhaps useful as synthetic reagents.

**Figure 7.6 ▶** Some HTS screening compounds that would not be progressed into HTL. (Compounds from Davis, A.M. et al. Components of successful lead generation. *Curr. Top. Med. Chem.* **2005**, *5*, 421–439.)

This issue was studied at Pharmacia by having 13 different experienced medicinal chemists review multiple lists of 2000 compounds each, all of them having already passed through a computational filter designed to remove undruglike compounds. Each chemist was asked which compounds should be rejected and the resulting sets of compounds were compared. Little overlap between the sets was found. Less than one-quarter of the compounds rejected by one chemist were rejected by another. Apparently each chemist judges by his or her own criteria, and the situation doesn't depend on how long one's worked in the field. The authors concluded that "*chemists are inconsistent in the compounds they reject*" and that "this can have serious implications in hit follow-up or lead optimization studies because the choice of which compounds to take forward depends on who is doing the review."[32] So we're still left with the need for an approach for predicting druglikeness that doesn't lie in the eye of the beholder.

Plenty of evidence that calculable parameters like molecular weight, LogP, and polar surface area do correlate with the likelihood of a given compound to become a successful oral drug in man comes from a number of sources. One of the simplest and most convincing of these arguments can be seen in Figure 7.7. As compounds progress through clinical trials the successful ones are more and more likely to have lower molecular weights: while the average MW for a Phase I candidate was 423 Da, compounds going on to gain market approval tended to have molecular weights close to the mean for all marketed drugs, which is about

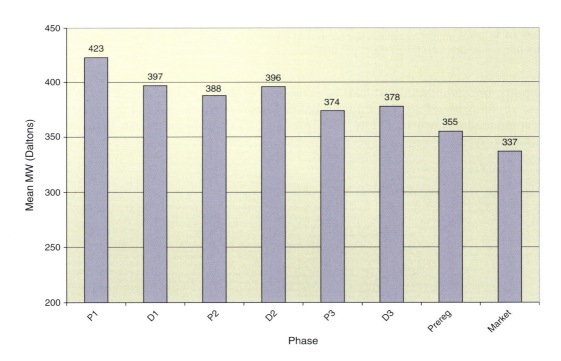

**Figure 7.7 ▶** Mean molecular weights for drugs in different development phases. P1, Phase I, D1, Discontinued Phase I, etc. (Data from Wenlock, M.C. et al. A comparison of physiochemical property profiles of development and marketed oral drugs. *J. Med. Chem.* **2003**, *46*, 1250–1256.)

337 Da. Thus chemists focus on small molecules for a very good reason: although there are exceptions like cyclosporin A (Figure 7.5), which has a MW of 1203 Da, smaller molecules are more likely to become drugs.

Our provisional definition of the term, "orally bioavailable and relatively nontoxic in humans," turns out to involve really three major issues: absorption, metabolism, and toxicity. These are the core of what's known as *ADMET* (*Absorption, Distribution, Metabolism, Excretion, Toxicity*), which was once an area that early drug discovery had little inclination to address, preferring to concentrate on potency and selectivity and leave the rest for others to worry about. Dr Lipinski's famous rule (further discussed in Chapter 9) not only made for better combi libraries but also catalyzed a revolution in thinking about exactly what's important in a drug and brought druglikeness and ADMET considerations to the fore. As a result, these properties now rank high on the list of things to select and optimize for early on using in vitro methods. This emphasis has resulted in smaller molecules being made (a good thing!) and a plethora of new techniques for predicting and assessing absorption, metabolism, and toxicity, as we'll see. These have become a great aid in designing new drugs, but unfortunately can't always predict everything that's going to happen in vivo, so animal experiments as well as clinical trials remain as important, and sometimes as surprising, as ever.

## 7.4 Multidimensional Optimization

It's no secret—in fact it's a form of job security for medicinal chemists—that hundreds or thousands of compounds might need to be made for a single project. This is largely because so many different properties need to be optimized. Consider the simple case of an enzyme inhibitor hit, $A$, which has a $K_i$ value of $10 \mu M$ against its target. Seeking to improve its activity, a chemist makes analogs $B$–$G$, which gratifyingly (and somewhat unrealistically) leads to a rapid improvement in activity as shown in Figure 7.8(a). In fact, the best analog, $G$, even meets the activity threshold specified in the target product profile, $K_i < 10 \, nM$. The one-dimensional line plot shown in Figure 7.8(b) shows the success of the approach.

Unfortunately, though, potency isn't the only thing involved. Hundred-fold selectivity versus an antitarget is also required, and when this second dimension is added, $G$ drops out as too non-selective, as represented in the now-2D plot in Figure 7.8(c). If selectivity against another antitarget or some other property like aqueous solubility is also needed, a third dimension would be added. Each required property adds yet another dimension to the graph, making it less and less likely that a given compound will meet all the necessary criteria.

Rarely will making a change in a molecule significantly affect only one property. Finding a part of the molecule where, say, target potency is fairly insensitive to modification might allow for easy optimization of another, for example, increasing solubility by adding in a piperazine group. Even then other properties like absorption could be adversely affected by the change. Each property tends to have its own rules and display an independent SPR although factors like size and lipophilicity (intimately involved in druglikeness, remember) have a sort of meta-effect, influencing many of them.

The failure of all other important properties to "stay put" while one of them is optimized means that linear optimization (first target potency, then perhaps selectivity against an antitarget, then solubility, etc.) is just not an option. It would be nice and straightforward

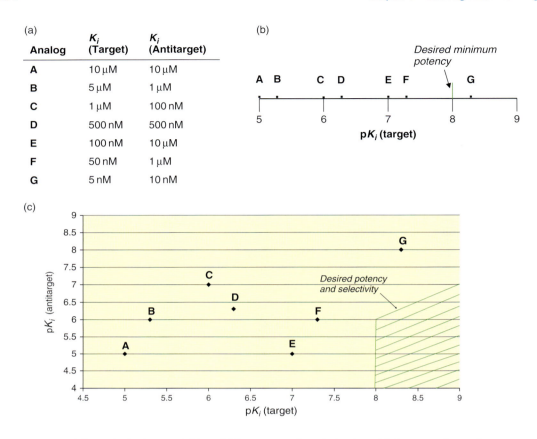

**Figure 7.8 ▶**   A hypothetical case of 1D and 2D compound optimization. (a) Seven analogs of a compound and their activities against target and antitarget. (b) A 1D logarithmic graph showing potencies against the target, with G meeting a pre-specified criterion. (b) A plot of antitarget potency versus target potency. None of the compounds are sufficiently potent and selective enough to fall into the desired (green cross-hatched) space.

if one could do this, but unfortunately this kind of serial approach can be instantly derailed when deficiencies in other properties not yet looked at eventually appear, leaving a long trail of useless compounds in their wake. Narrowing in on a series of potent, selective compounds that end up being too insoluble to formulate or too metabolically unstable to use in vivo would be a major mistake.

For this reason today's scientist will always seek as much property data as possible as early on as possible. This may require resources formerly reserved for late stage development (cell permeability assays, solubility determinations, etc.), now done in a more "quick and dirty" (high throughput) way than used to be the case. It means that compounds are more likely to drop out sooner as more properties are examined early, so that more analogs are needed sooner and parallel synthesis becomes a corollary of parallel property optimization. It also means that medicinal chemists must be able to operate on the verge of information overload and that to understand what's going on they need to have a working knowledge of factors not previously considered to be within their purview. And the parallel optimization process itself represents quite a challenge for the scientist. Still, although she or he might not be able to optimize in ten dimensions at once, quickly identifying where the most serious problems lie is the first step in solving them.

## 7.5  Lead Optimization Versus HTL

What medicinal chemists do in the lead optimization (LO) stage—modifying the structures of compounds to improve various properties—is methodologically no different from what they did starting with confirmed actives during the earlier HTL phase. Some of the properties involved will be the same but will need to meet stricter criteria. Potency and selectivity sufficient to declare a compound or series to be a lead (a successful conclusion to HTL) will probably not be sufficient for a potential clinical candidate (the aim of LO) and may need to be improved even for an in vivo POC compound.

But other properties, especially those relating to bioavailability and toxicity, will become the focus as the project progresses through lead optimization. Just as HTL is largely driven by in vitro potency data, PK data is what often determines success or failure in the LO phase. Since in vivo testing is not by nature high-throughput (although *cassette* or *n-in-1* dosing allows for testing more than one compound at a time) in vitro screens for ADMET properties like cell permeability and microsome stability are extensively used in LO to prioritize compounds for testing in animals. In order to optimize so many properties both old and new, LO calls for more chemists, often 10–15 of them as opposed to perhaps 3 or 4 that were involved in HTL (Table 7.1).

Bringing in more chemists to make more analogs necessitates more planning, better communications, and tighter coordination of their activities so that on the one hand there isn't duplication of effort (two of them, perhaps from different groups, independently trying to make the same analog), and on the other hand, useful information like synthetic tricks, when found, are disseminated to the whole team as quickly as possible. There's also likely to be more need for common intermediates on a larger scale, so that if this material isn't contracted out or provided by another group, project research chemists may need to crank out large batches of them for group use. From his or her point of view, compared to HTL all of this may translate into closer reporting of results, less latitude in the modifications made, and maybe even repeated forays into "group duties." As a result, both expectations and the "feel" of this stage of the project will be different from that for HTL.

**TABLE   7.1 ▶   Some   Differences   between   the   Hit-to-Lead   (HTL) and Lead Optimization (LO) Stages in a Drug Discovery Project**

| Factor | HTL | LO |
|---|---|---|
| *Focus* | Potency, structural novelty | PK and in vivo properties |
| *Number of chemists* | Maybe 3 or 4 | Probably 10 or more |
| *Number of compounds* | Hundreds | Thousands |
| *Number of scaffolds* | Many | Fewer |
| *Desired outcome* | Lead series | Potential clinical candidate |

As the number of analogs made for the project goes up, scaffold diversity tends to go down. In HTL, a collection of diverse confirmed actives probably provided the starting point, and a few structural changes both minor (methyl, ethyl, propyl, etc.) and major (scaffold hopping, cyclization, moving around heteroatoms, etc.) were tried out on each to see what effect they might have on potency, selectivity, etc. Unsuccessful modifications dropped out, so fewer series remain, particularly as additional properties begin to be examined. By the time a lead series has been reached, SAR has already begun to show what sorts of changes might or might not be tolerated, and acceptable scaffold horizons have narrowed quite a bit. In lead optimization, then, chemists concentrate on making deeper and narrower sets of analogs than they did at the HTL stage. Parallel syntheses, "bookshelves," and libraries are used more extensively.

This **isn't** because diversity is unwelcome in LO and it isn't because major modifications can no longer be tried out. As we've seen, maintaining maximal diversity can avoid being checkmated by problems that crop up later on. Project leaders, also motivated to push out into new IP space, would much rather face the problem of allocating resources among multiple lead series than the problem of what to do when the only one they have is failing. Ultimately, having diverse "backup" clinical candidates is something that's fully appreciated by everyone up to and including the CEO.

The problem is that it's so hard to achieve. Compounds need to meet many tough criteria and attrition rates are high. Acceptable major scaffold changes are just hard to come by, even using the knowledge gained by working with a given target for many months. Smaller variations in scaffolds known to be active can solve many of the problems that come along, and consequently at the LO stage lots of them are tried out.

Major variations can happen then too, although less frequently. A new and better lead series can come to the fore and the look of project compounds can suddenly change. Plenty of synthesis challenges await the chemist in LO as well. What they do ends up being no less exciting or creative than HTL chemistry. The emphasis is just different. The quest for potency and structural novelty that's often the primary (although, of course, not the only) driver in HTL is still a concern. And, as we've seen, parallel optimization of many properties begins early on in a project these days. But by the time a compound or series enters LO, at least some minimal "handle" on potency has been found and as the project progresses the focus largely shifts to improving in vivo properties, metabolism, and potential toxicity. To the tweaking of such properties there can sometimes be no end: thousands of analogs might be required, hence the emphasis on more chemists and more compounds more quickly.

## 7.6  Using Structure-Based Drug Design

### 7.6.1  Definition, History, and Goals

Being able to see how a ligand binds to a receptor and thus having a real basis for understanding and improving binding interactions was, up until the last few decades, only a pipedream for drug discovery scientists. The importance of shape complementarity, van der Waals interactions, electrostatic interactions, and hydrogen bonding had been known for a long time and plenty of models of how a compound *might* be interacting with its receptor had been proposed, but rarely could these interactions actually be observed. Solving the X-ray crystallographic structure of a single protein could take years.

This began to change in the late 1970s and early 1980s. As we've already seen, biotechnology made it possible to produce large enough quantities of a given protein to study, while advances in both X-ray crystallography (synchrotron sources, robotic crystallization, etc.) and NMR spectroscopy (higher-field magnets, automated assignment, etc.) led the way to much more rapid structure determinations.[33] Even today not every protein can be induced to form X-ray quality crystals—membrane bound ones like GPCRs and ion channels being particular problematic—but for those targets which are amenable to X-ray or NMR approaches, it would now be almost unthinkable not to obtain and use structural information as an aid to drug discovery efforts. Being able to visualize binding interactions might reveal new areas for hydrophobic, electrostatic, or H-bonding interactions; the best ways of filling in an existing space; the kind of conformational constraints to be built into the ligand; and many other improvements that wouldn't be obvious lacking such a picture.

Early on, any use of computers in drug discovery was usually referred to as *computer-aided drug design* (*CADD*), but more recently their use, in conjunction with X-ray or NMR protein structures, goes by the name of *rational drug design* or *structure-based drug design* (*SBDD*). It's been pointed out that since the goal here is only an improvement in binding affinity and not necessarily druglike properties, it would be more accurate to call it *structure-based ligand design*,[34] but few people do.

Regardless of what it's called, like target validation, *SBDD is an ongoing process, not a one-time event*. Improvements and even declines in potency due to modifications suggested by a first structure will need to be followed up by another to suggest what to do next. This is also important to determine whether binding was as predicted and thus prevent wrong assumptions from being made. Rarely will the structures be like two peas in a pod. Often there will be surprises, like different binding modes for close analogs or the unexpected movement of protein residues resulting in new possibilities for improving interactions. A succession of base hits (modest improvements in potency) is more likely than a home run (a quantum improvement). So the process is a cyclic one, requiring ongoing input and feedback. For this reason it was termed *iterative protein crystallographic analysis* in 1991 by researchers at Agouron, who propounded this method as graphically depicted in Figure 7.9.[35]

By definition, then, structure-based drug design is an interdisciplinary effort. A protein expression group starts the process by providing the target. At that point the cycle can begin. X-ray crystallography, or sometimes a structural NMR group, determines the structure, usually in a complex with an initial binder of known potency. Medicinal chemistry often provides that compound as well as subsequent analogs, each of which gets assayed in another group. The chemists and modelers, then having a structure of the ligand–receptor complex as well as its activity, together devise the next modification and the cycle begins again.

The interplay between crystallographers or NMR structural specialists, biochemists, medicinal chemists and modelers is vital. "Successful teams and companies need to be congratulated, whereas the search for one responsible individual or computer program is counterproductive. There is not going to be a 'voilà' moment at the computer terminal. Instead, there is systematic use of wide-ranging computational tools to facilitate and enhance the drug discovery process."[36]

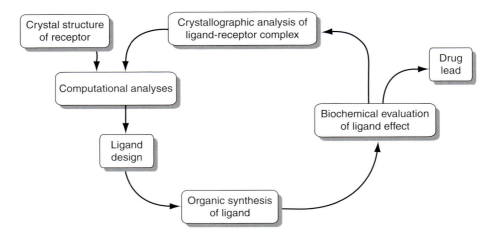

**Figure 7.9 ▶**   An iterative cycle for the discovery and elaboration of lead compounds. (Reprinted with permission from *J. Med. Chem.* **1991**, *34*, 1925–1932, copyright 1991, American Chemical Society).

Computational tools to enhance affinity can be divided into two types, those focusing on the receptor or receptor–ligand complex, and those based on the structure of the ligand alone. X-ray crystallography and NMR structural elucidation as mentioned above are examples of the former approach, which has a better track record. This isn't too surprising because including the receptor coordinates takes much more information into account. *Homology modeling*, where information from a known receptor structure is mutated to build a model of a related protein for which the coordinates are not available, is another way of doing *receptor-based* drug design. When information about the 3D structure of the receptor isn't available through any of these methods, however, a *ligand-based* approach, where the overlay of a number of ligands in a proposed active conformation is used to generate a *pharmacophore map*, is still possible.

It's reasonable to wonder exactly which thermodynamic properties are being optimized by SBDD. Those of us who weren't too traumatized by p-chem courses (and some of us who were) might remember the equation to calculate Gibbs free energy, $\Delta G = \Delta H - T\Delta S$, which, when applied to binding, shows that either minimizing enthalpy or maximizing entropy can be used to make it more negative, thereby improving affinity. Sometimes structural modifications might improve one term at the expense of the other, leaving no net difference,[37] but these two separate factors are always involved.

Consider the following observations: (1) when a ligand binds to a receptor, ordered water molecules at hydrophobic surfaces are released leading to a more positive $\Delta S$ value and (2) on the other hand, entropy is lost, to the detriment of binding affinity, when a ligand with lots of rotatable bonds gives up its conformational freedom to bind to a receptor in a fixed geometry. Taken together, these observations suggest the grand strategy of working the entropy side of the binding affinity equation by using conformationally constrained (low entropy loss on binding), hydrophobic (lots of ordered water displacement)[38] molecules as ligands. This turns out to describe most drugs.

Although it may seem strange that entropy, not enthalpy, drives the binding affinities of drugs, it's believed to be the dominant factor for many of them, like the HIV protease inhibitors indinavir, nelfinavir, and saquinavir.[39] The enthalpic factors for these three even proved to be unfavorable, and that after many rounds of structure-based optimization! Although the strategy of using hydrophobic, conformationally constrained ligands has been proven to work, it does come at a cost. Hydrophobicity, of course, can bring with it problems with solubility and other properties. And importantly, such compounds have been "tweaked" for very specific binding interactions, and the same inflexibility built into them to minimize their entropy loss and improve their binding affinities for the targeted receptor means that they may not work well against mutant forms thereof arising through microbial resistance or perhaps pharmacogenomic variants.

It's been suggested that focusing on the other side of the equation, enthalpy-dominated interactions, and loosening up conformational constraints might be a better idea, especially when facing targets like HIV protease where mutation and subsequent resistance is a major factor,[39] but this strategy is not without its own set of challenges.

### 7.6.2 Potential Limitations

#### 7.6.2.1 Conformational Flexibility

When the binding of a ligand to its receptor is considered, many of us instinctively think of the century-old *lock and key* model first proposed by Emil Fisher.[40] As shown in Figure 7.10(a), here the shape of the ligand is complementary to the binding site of a receptor so that a fixed conformation of each simply comes together to form the complex.

As a recent perspective has pointed out,[41] this was the sole model of protein–ligand binding until 1958, when Dr Daniel Koshland, Jr, then at Brookhaven National Laboratory,[42] published a paper in which he introduced the concept of *induced fit*.[43] Here binding of a ligand to its receptor can induce conformational changes in the binding site of the latter to achieve an optimized interaction as depicted in Figure 7.10(b). Together these two models could be used to explain almost any type of binding, whether the bound receptor conformation differed from that observed for the unliganded (*apo*) protein or not.

The lock and key model is simple and appealing, but the implied assumption that proteins are rigid and unyielding isn't realistic. Evidence that proteins are flexible is ubiquitous. Ion channels open and close, and GPCRs adopt different conformations in the presence or absence of agonists or inverse agonists. X-ray structures themselves reveal so much of a ligand being "buried" within a receptor that it's difficult to explain how it could have gotten in without protein flexibility. Sometimes they show two different bound conformations of receptor present in the same unit cell.[44] And not only is the receptor flexible, but so is its ligand. Since the "average organic molecule" has eight rotatable bonds, defining conformers in 30° increments of rotation means that it has about 430,000,000 different conformers![45]

The induced-fit model shown in Figure 7.10(b) allows for some flexibility, but still focuses in on a single unbound receptor conformation. In reality, "proteins are in constant motion between different conformational states with similar energies" as stated in an outstanding review of protein flexibility in drug design.[46] So the best model these days is the *conformational ensemble* model shown in Figure 7.10(c). Here a number of protein conformers exist in

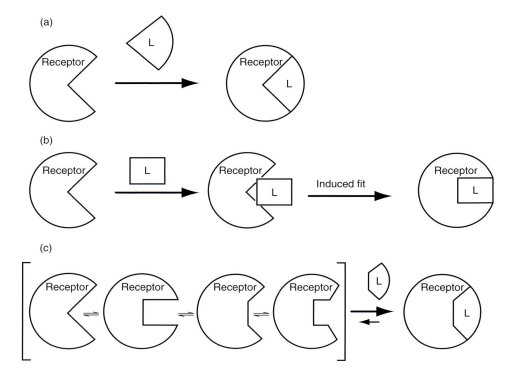

**Figure 7.10** ▶   Models of ligand (L) binding to a receptor. (a) Lock and key. (b) Induced fit. (C) Stabilization of conformational ensembles. (Reprinted with permission from *J. Med. Chem.* **2002**, *45*, 541–558, copyright 2002, American Chemical Society).

a statistical equilibrium and the ligand binds to the most complementary of these by *conformational selection*, thereby shifting the equilibrium to favor it.[47]

All of this may seem purely theoretical, but it has major implications for SBDD. Flexibility means that the protein conformation observed for a given receptor–ligand complex won't necessarily be the same as the conformation seen upon binding to a different ligand. Active site residues can be particularly mobile, an example of which is shown in Figure 7.11 for the aspartyl protease renin. Years spent studying structures where the enzyme was bound to transition state analogs like Compound 2 couldn't hint that Compound 1 might bind to this enzyme via a different conformation made possible by major changes in position for the sidechains of residues Trp39 and Tyr75.

Receptor flexibility means that not just one key, but a number of very diverse ones, might actually fit into the same lock the way that four structurally different compounds, efavirenz, nevirapine, Cl-TIBO, and UC-781 (all non-competitive inhibitors, Figure 7.12) bind to the same site on HIV reverse transcriptase. As has been pointed out, this is a nice situation from an IP standpoint,[46] allowing many different chemotypes to be patented for the same target, but it does tend to limit the predictive ability of computer modeling. These days, modeling can allow for a certain amount of receptor flexibility, but correctly anticipating the type of sidechain and/or backbone conformations to expect isn't easy.[48]

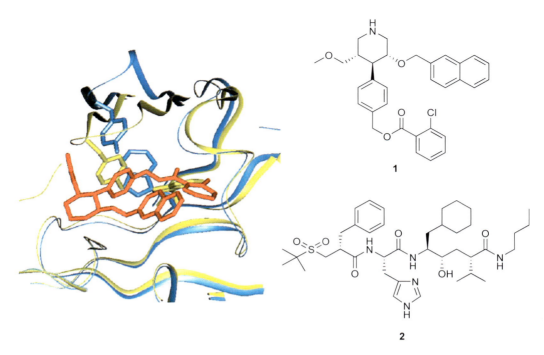

**Figure 7.11 ▶**  Two different conformations of the enzyme renin when bound to two different inhibitors. The complex between renin (blue) and Compound 1 (red) is overlaid with renin (yellow) bound to Compound 2 (not shown in complex). Note the very different positions adopted by the sidechains of residues Trp39 and Tyr75. (Reprinted by permission from Teague, S.J. Implications of protein flexibility for drug discovery. *Nat. Rev. Drug Discov.* **2003**, *2*, 527–541. copyright 2003, Macmillan Publishers Ltd.)

### 7.6.2.2  Other Limitations

Modeling a proposed change to the ligand in a ligand–receptor complex having known coordinates involves comparing the new proposed interactions with the old known ones and perhaps with those of a number of other not-yet-synthesized structures. This of course involves scoring, and as already noted in the section on virtual screening, current scoring functions leave a lot to be desired.

A recent study at GSK examined 10 docking programs and 37 different scoring functions for their ability to correctly dock and score ligand interactions with 8 different proteins. Although the docking programs often proved successful in generating the binding geometries that were close to those actually observed, the authors found "no statistically significant correlation between measured affinity and any of the scoring functions evaluated across all eight protein targets examined." They concluded that "in the area of rank-ordering or affinity prediction, reliance on a scoring function alone will not provide broadly reliable or useful information that can be applied to lead optimization."[49] Until this situation changes, observational skill, libraries of analogs, and even scientific intuition will likely remain important parts of the SBDD process.

Less profound, but often overlooked, is the fact that protein crystal structures are usually obtained with resolutions between 2 and 3 Å. This isn't enough to distinguish between a carbon

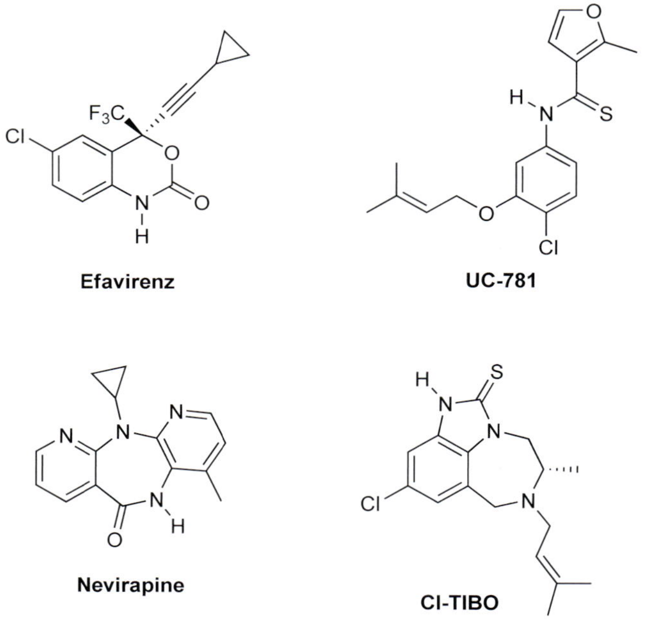

**Efavirenz**

**UC-781**

**Nevirapine**

**CI-TIBO**

**Figure 7.12 ▶**    Four structurally diverse inhibitors that all bind to the same site of HIV reverse transcriptase.

atom and a nitrogen or between nitrogen and oxygen. This can be a problem, notably with primary amides like those on the sidechains of asparagine and glutamine, where O and $NH_2$ aren't distinguishable. Assignments are usually made on the basis of which of the two possibilities would give the better hydrogen-bonding network. Other assignments can be even tougher: with an unsubstituted 2- or 3-pyridyl ring not in an H-bonding environment it might be anybody's guess which atom is nitrogen and which is carbon in a crystal structure. And if Cs, Ns, and Os can't be distinguished, one can forget about actually seeing Hs at these resolutions. These are added to the structure by modeling software, which, along with the crystallographer or modeler, must also decide on the relevant protonation states and tautomers.[50]

Finally, X-ray structures or models of ligand–receptor complexes often suggest new sites for H-bonding interactions between the two, usually by incorporation of a new −OH or −NH group into the ligand. Although this sometimes does work, the newly formed H-bond will need to be a particularly good one to pay the "desolvation penalty" required in adding such a group and still come out ahead. In practice, this can be trickier than it looks, and so such beckoning interactions don't always pan out, as experienced researchers already know and new ones are always finding out.

## 7.6.3   Examples

Structure-based drug design is now ubiquitous in drug discovery. Almost without exception, in the last 10 years SBDD has been a part of every drug discovery project for which the target is known and for which good structural information can be gotten. In some cases, SBDD was crucial to a drug's development whereas in others it played a lesser role. Just how useful it proved to a given project is often difficult to assess, as literature reports tend not to distinguish between *a priori* prediction and after-the-fact rationalization, precisely the difference between predicting the future and predicting the past.

A list of even the most recent examples of the use of SBDD would be too long to include here. Shown below instead are just a few cases that point out some of the possibilities inherent in this method and highlight some issues already mentioned.

### 7.6.3.1 HIV Protease Inhibitors

Proteases have proven to be a favored target class for SBDD. Good crystal structures can often be obtained, substrate preferences are known, and the diseases involved (hypertension, HIV, HCV, etc.) are important ones with unmet needs. HIV protease was the second target to come along in AIDS after reverse transcriptase. Comparison of the HIV-1 viral genome with those of other retroviruses suggested that it too encoded for a protease. This protein, which proved to be necessary for the production of mature, infectious viral particles, was discovered to be an aspartyl protease smaller than those found in mammals, and it was proposed that it acted as a homodimer, which several X-ray structures of the enzyme published in 1989 confirmed.[51]

HIV protease inhibitors would eventually turn out to be "among the first successful examples of receptor/structure-based designer drugs and were developed using structures of compounds bound in the active site of HIV-1 protease and with the knowledge of inhibitors of other aspartic proteases (e.g. renin)."[52] Researchers at Roche UK started out with a short peptide sequence from a known substrate and optimized a series of hydroxyethylamine isosteres to provide nanomolar and subnanomolar inhibitors. This work, done at a time when it was probably still easier to just make analogs than to do iterative crystallographic analysis, eventually resulted in the first-in-class drug saquinavir (Figure 7.13).[53] As time went on, though, SBDD began to play a larger and larger role in the design of these drugs, mainly via X-ray crystallography but also by NMR analysis.

Researchers at Abbott proposed to exploit the twofold symmetry axis seen in the enzyme's crystal structure by making $C_2$-symmetric inhibitors. The feasibility of the approach was tested by docking such an inhibitor into the active site of a closely related enzyme, RSV protease. The interactions looked favorable, leading to a series of compounds like A-74704 (Figure 7.13), a selective, 4.5 nM HIV-1 protease inhibitor whose predicted binding mode was confirmed by crystallography.[54] The $C_2$-symmetric inhibitor strategy was given a sudden jolt, though, with the publication of a paper entitled "A Symmetric Inhibitor Binds HIV-1 Protease Asymmetrically" a few years later.[55] Many monohydroxy (like A-74704) as well as dihydroxy symmetric peptides had been made, but the direction shifted, and best results eventually arose from compounds which were not in fact symmetric, like ABT-538, which came to market as ritonavir (Figure 7.13).[56]

By this point, multiple crystal structures of the enzyme had revealed something unexpected: a bound water not seen in mammalian proteases that came between and interacted with two of the carbonyl oxygen atoms of the substrate or inhibitor, and two NHs (from Ile50 and Ile50') on the flap region of the enzyme. Researchers at DuPont Merck proposed to build in selectivity and take advantage of the entropy released upon displacing this water by

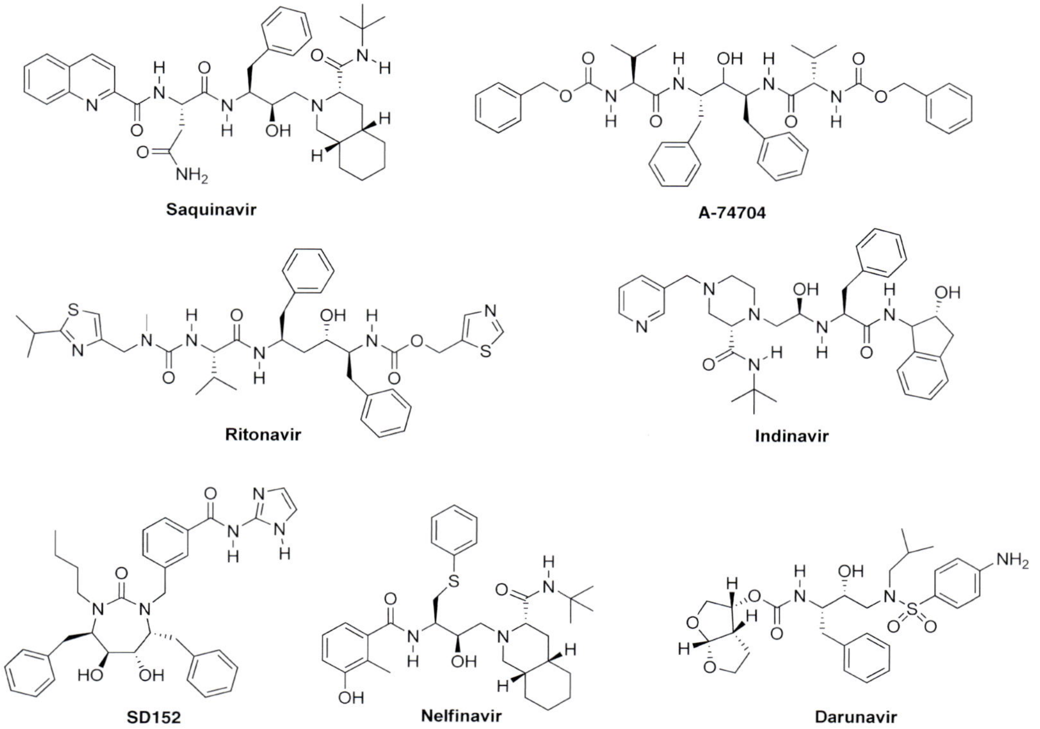

Figure 7.13 ▶  Some HIV protease inhibitors.

designing a series of cyclic—and therefore conformationally constrained—compounds where the carbonyl oxygen could occupy the same space (Figure 7.14).[57]

Use of existing crystal structures and docking programs suggested a series of 7-membered cyclic ureas, which turned out to be extremely potent, like the 25 pM SD152 shown in Figure 7.13.[58] Interestingly, $C_2$-symmetric compounds were initially synthesized here too, but turned out to have no inherent advantage over asymmetric ones. Several of the cyclic ureas reached clinical trials but, unfortunately, dropped out. The entropy gain in displacing bound waters remains an area of active interest for many targets, though, and an algorithm for calculating a given water's propensity for displacement has recently been published.[59]

More recent HIV protease inhibitors like indinavir,[60] nelfinavir,[61] and darunavir[62] (all shown in Figure 7.13) have made extensive use of SBDD. A good example of what it could accomplish can be seen in the *bis*-tetrahydrofuranylurethane (*bis*-THF) group of darunavir. Figure 7.15 shows the H-bonding interactions this unusual group makes with the enzyme. Note also that in this structure the bound water at Ile50 and Ile50′ is not displaced.

## 7.6.3.2  Other Examples

The enzyme dihydrofolate reductase (DHFR) is essential to the biosynthesis of thymidylate and other necessary biochemicals in every species from prokaryotes to man. Structural differences between host and bacterial forms of DHFR might be exploited to provide

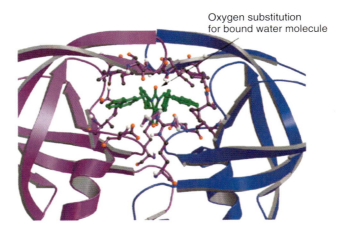

**Oxygen substitution for bound water molecule**

**Figure 7.14 ▶** Cyclic urea HIV protease inhibitor incorporating an oxygen atom in place of a bound water molecule. (Reprinted with permission from Anderson, A.C. The process of structure-based drug design. *Chem. Biol.* **2003**, *10*, 787–797. copyright 2003, Elsevier.)

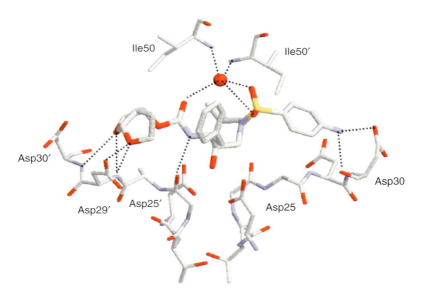

**Figure 7.15 ▶** H-bonding interactions between darunavir (UIC-94017) and HIV-1 protease. The *bis*-THF group is shown at left interacting with Asp30′ and Asp29′. (Reprinted with permission from Tie, Y. et al. High resolution crystal structure of HIV-1 protease with a potent non-peptide inhibitor (UIC-94017) active against multi-drug-resistant clinical strains. *J. Mol. Biol.* **2000**, *338*, 341–352. copyright 2000, Elsevier).

selectivity in treating diseases like tuberculosis—*if* they could be found. A crystal structure of *Micobacterium tuberculosis* DHFR complexed with the non-selective inhibitor methotrexate (MTX) revealed one such surprising difference, the presence of a bound glycerol molecule adjacent to the N-8 atom of MTX's pteridine ring.[63] The glycerol came from that used as a co-solvent to purify and crystallize the enzyme, but in human DHFR, hydrophobic residues in the same pocket precluded its binding. Researchers at the University of Bath recently incorporated the glycerol sidechain into a series of heterocyclic inhibitors like Compound 7 (Figure 7.16) and found evidence of selectivity for the *M. tuberculosis* enzyme over human or yeast versions.[64]

Interestingly, there are two sides to the glycerol incorporation coin. In the field of neuraminidase inhibitors, another area where SBDD has proven popular, researchers have been eager to get rid of the glycerol sidechain found in compounds like the inhaled influenza drug Relenza (zanamivir) owing to its adverse effects on oral bioavailability![65] Here again, structural changes that improve one property need to be balanced against adverse effects on another.

Finally, although the lack of crystallizability of membrane-bound proteins like GPCRs severely curtails the use of receptor-based drug design for these popular targets, there are exceptional cases where it still proves useful. A good example of this comes from another HIV target, CCR5, a receptor involved in viral entry into cells.[66] No crystal structure of this

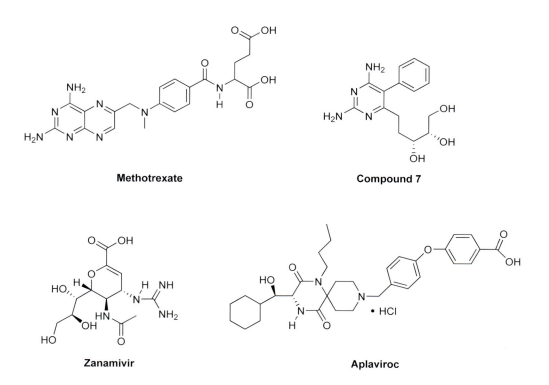

**Figure 7.16** ▶ Dihydrofolate reductase inhibitors methotrexate and Compound 7, (El-Hamamsy, M.H.R.I. et al. Structure-based design, synthesis and preliminary evaluation of selective inhibitors of dihydrofolate reductase from *Mycobacterium tuberculosis*. *Bioorg. Med. Chem.* **2007**, *15*, 4552–4576.), neuraminidase inhibitor Relenza (zanamivir), and CCR5 blocker aplaviroc.

7-transmembrane protein exists. But homology modeling, beginning with the high-resolution structure of bovine rhodopsin, with which it shares 20% sequence homology, was combined with data from site-directed mutagenesis and antibodies directed at specific parts of the protein. This afforded a working model of the binding of aplaviroc (Figure 7.16) and two other compounds that block the binding of HIV-1 gp120 to the receptor. Proper use of the site-directed mutagenesis data proved tricky: "certain amino acid substitutions reduce the binding of an inhibitor even though they are not directly interacting with the inhibitor," while "mutations at certain residues directly interacting with an inhibitor do not always cause significant loss of binding."[67] Evidence of the fact that CCR5 inhibitors can still allow the binding of their endogenous ligands, chemokines, a major concern for long-term safety was also generated. This work represents a *tour de force* and stands as an example of how modeling can still prove valuable for membrane-bound proteins, but by the same token shows just how much of a concerted effort this might require.

### 7.6.4 Working with Modelers

A few medicinal chemists in industry do their own modeling work. Some enjoy it, are proficient enough, and are willing to dedicate the time it takes. They might do this by choice or because they work for companies too small to have hired modelers yet. Most, however, do work with modelers. These valuable team members often have a background in organic synthesis so that they can tell which proposed structures are synthetically tractable and which aren't.

Dr Guy Bemis, for example, did synthesis research in the Katzenellenbogen group at the University of Illinois before going into modeling. For the last 14 years he's been a modeler working with medicinal chemists at Vertex Pharmaceuticals in Cambridge, Massachusetts. In an interview, he explained what the job involves.

"I mostly work with teams of medicinal chemists," he said. "We have a very good crystallography group. When we go after a target, typically it's been one that we can solve the crystal structure of. So I'll sit down with the med chemists and we'll have the crystal structure in front of us. We'll look at the lead molecules that they're working on, either a crystal structure or a predicted structure of how they bind to the target, and I'll suggest modifications that would improve binding. Or they'll have ideas for synthetic derivatives that they can make and they'll want to see which ones will bind and which ones won't. So that's one thing, helping to improve $K_i$s.

"Another is helping to filter out bunches of molecules closer to the end of a project, when you're looking at a lot of different molecules with very small changes. They're all likely to have very good $K_i$s or $IC_{50}$s and what you're trying to do is just tweak the molecule to improve physical properties like cell penetration, pKa, and things like that, running sets of filters to weed out bad molecules before they're made."

Practical experience has taught him that the challenges involved in SBDD, including solvation, entropy effects, and conformational effects, shouldn't be underestimated. "It's really easy to miss conformational effects: even a methyl group can change the free conformation of the ligand. When you're staring at an enzyme active site and a small molecule bound to it, it's really easy to miss things like that. It's not engineering. It *is* science."

Deviations from expected SAR shouldn't be swept under the rug, but instead call for closer scrutiny, he suggests. "That's a tip off that there's something weird going on, and it feeds into the crystallography group. Then they know 'Ah, here's an interesting molecule that we want to do a crystal structure of', as opposed to the other ten molecules with small changes that are all relatively understandable. The oddballs are the more interesting structures that they'll want to work on anyway.

"If you see something that doesn't make sense in the SAR, that's a tip off that you'd better look at the structure. And if something in the structure doesn't make sense with SAR, that's an indication you'd better look a little more closely at how you're getting your crystal structure, by soaking or co-crystallization, or whether you're looking at an inactive form. In kinases, which we look at a lot, you can have active and inactive forms."

In light of all the complications and uncertainties, how much should a medicinal chemist expect from SBDD? Dr Bemis explains, "People are realizing, more and more, that it's a tool that you use to do your project. They shouldn't expect $K_i$ predictions. But I think that dividing things into bins, where this bunch of molecules probably will be better than your starting point and this will probably be worse, is pretty doable." As a tool, it's important that medicinal chemists have access to structures, and ensuring this is also one of the functions of the modeling group. Dr Bemis continues, "Where we try to help out is in making sure they (the medicinal chemists) have the tools to quickly look at crystal structures. We make sure it's easy for them to measure distances and ask questions. This is just visualization. The more you can get chemists to look at structures, the better off you are."

### 7.6.5  Conclusions

Some early proponents of CADD believed that its use would eventually allow researchers to accurately predict $K_i$ or $IC_{50}$ values and hone in on just a few structures that when synthesized would be highly active. But note that in none of the examples cited above was a single analog based upon one crystal structure sufficient for potency optimization. Although the original dream has not come to pass, the fact that SBDD is now an integral part of almost every project in which it can be used speaks to the fact that it does play an important role in drug discovery. This often lies in suggesting which parts of the molecule to vary, where new hydrophobic or H-bonding groups might be positioned, which binding pockets to explore, or which related scaffolds might be active. The result is a kind of enrichment factor *alà* virtual screening: Projects making proper use of SBDD are more likely to reach a given level of potency (or selectivity)[68] with a given number of analogs than those not using it, allowing projects to hone in on better compounds more quickly. Its extensive utilization means that fewer analogs will probably need to be made, but not that a single "perfect" molecule will suddenly appear on a computer screen.

So, in a development that early advocates of neither method may have anticipated, knowledge-based optimization (SBDD) and the technology derived from diversity screening (combinatorial chemistry) have found their optimal real-world use—*together*. Optimizing hits uncovered in diversity screening using crystallographic and modeling methods, and the use of small, focused libraries of analogs suggested by SBDD have proven more likely to succeed than either method alone, or neither.

# Notes

1. Swinney, D.C. Biochemical mechanisms of drug action: what does it take for success? *Nature Rev. Drug Discov.* **2004**, *3*, 801–808.

2. Westley, A.M., Westley, J. Enzyme inhibition in open systems. Superiority of uncompetitive agents. *J. Biol. Chem.* **1996**, *271*, 5347–5352.

3. Those interested in more information on these and other aspects of enzyme inhibitor biochemistry are urged to consult Robert A. Copeland's excellent book *Evaluation of Enzyme Inhibitors in Drug Discovery* (Hoboken, NJ, John Wiley & Sons, 2005) which goes into far more detail than this current text and provides the reader with a fascinating look at real-world drug discovery enzymology.

4. Christopoulos, A. Allosteric binding sites on cell-surface receptors: Novel targets for drug discovery. *Nature Rev. Drug Discov.* **2002**, *1*, 198–210.

5. Some endogenous ligands for allosteric sites on receptors have, however, been identified. See Christopoulos, ibid.

6. See Christopoulos, Note 4. Also see May, L.T., Christopoulos, A. Allosteric modulators of G-protein-coupled receptors. *Curr. Opin. Pharmacol.* **2003**, *3*, 551–556.

7. See Gao, Z.-G., Jacobson, K.A. Allosterism in membrane receptors. *Drug Discov. Today* **2006**, *11*, 191–202.

8. Price, M.R. et al. Allosteric modulation of the cannabinoid $CB_1$ receptor. *Mol. Pharmacol.* **2005**, *68*, 1484–1495.

9. Christopoulos, A. Screening of Allosteric Modulators of GPCRs, presented at "New Approaches to Drug Targets", BIO2006, Chicago, Illinois, April 12, 2006.

10. Silverman, Richard B. *The Organic Chemistry of Enzyme-Catalyzed Reactions* (San Diego, CA, Academic Press, 2000), p. 67.

11. For starters, see Chapter 2 of Silverman, ibid., Chapter 8 of Copeland, Note 3, and Richard B. Silverman's *Mechanism-Based Enzyme Inactivation: Chemistry and Enzymology* (Boca Raton, FL, CRC Press, 1988).

12. Fry, D. Site-directed irreversible inhibitors of the erbB family of receptor tyrosine kinases as novel chemotherapeutic agents for cancer. *AntiCancer Drug Des.* **2000**, *15*, 3–16.

13. Kwak, E.L. et al. Irreversible inhibitors of the EGF receptor may circumvent acquired resistance to gefitinib. *Proc. Natl. Acad. Sci. USA* **2005**, *102*, 7665–7670.

14. The situation is probably more complex than that, however, requiring a second signal before an immune response can be initiated. For a good discussion of this in the context of reactive metabolites, see Uetrecht, J. Screening for the potential of a drug candidate to cause idiosyncratic drug reaction. *Drug Discov. Today* **2003**, *8*, 832–837.

15. Lewandowicz, A. et al. Achieving the ultimate physiological goal in transition state analogue inhibitors for purine nucleoside phosphorylase. *J. Biol. Chem.* **2003**, *278*, 31465–31468.

16. Anthes, J.C. et al. Biochemical characterization of desloratidine, a potent antagonist of the human histamine $H_1$ receptor. *Eur. J. Pharmacol.* **2002**, *449*, 229–237.

17. But if you use this argument to justify proceeding with a compound with poor *in vivo* properties, please don't blame it on this author!

18. Copeland, R.A. et al. Drug-target residence time and its implications for lead optimization. *Nat. Rev. Drug Discov.* **2006**, *5*, 730–739.

19. Swinney, D.C. Biochemical mechanisms of new molecular entities (NMEs) approved by the United States FDA during 2001–2004: Mechanisms leading to optimal efficacy and safety. *Curr. Top. Med. Chem.* **2006**, *6*, 461–478.

20. Robertson, J.G. Mechanistic basis of enzyme-targeted drugs. *Biochemistry* **2005**, *44*, 5561–5571.

21. Quotation from Copeland, Note 3, page 78.

22. Lipinski, C.A. Drug-like properties and the causes of poor solubility and poor permeability. *J. Pharmacol. Toxicol. Methods.* **2000**, *44*, 235–249.

23. Podlogar, B.L. et al. Computational methods to estimate drug development parameters *Curr. Opin. Drug Discov. Devl.* **2001**, *4*, 102–109.

24. Rishton, G.M. Nonleadlikeness and leadlikeness in biochemical screening. *Drug Discov. Today* **2003**, *8*, 86–96.

25. Wang, J., Ramnarayan, K. Towards designing drug-like libraries: A novel computational approach for prediction of drug feasibility of compounds. *J. Comb. Chem.* **1999**, *1*, 524–533.

26. Lajiness, M.S. et al. Molecular properties that influence oral drug-like behavior. *Curr. Opin. Drug Discov. Devel.* **2004**, *7*, 470–477.

27. Walters, W.P., Murcko, M.A. Prediction of 'drug-likeness'. *Adv. Drug Deliv. Rev.* **2002**, *54*, 255–271.

28. Such interesting publications would likely be axed by corporate legal departments which have no desire to imply that their own patented compounds aren't completely novel.

29. Bemis, G.W., Murcko, M. The properties of known drugs. 1. Molecular frameworks. *J. Med. Chem.* **1996**, *39*, 2887–2893.

30. Bemis, G.W., Murcko, M. Properties of known drugs. 2. Side chains. *J. Med. Chem.* **1999**, *42*, 5095–5099. See also Ajay et al. Designing libraries with CNS activity. *J. Med. Chem.* **1999**, *42*, 4942.

31. See Müller, G. Medicinal chemistry of target family-directed masterkeys. *Drug Discov. Today* **2003**, *8*, 681–691.

32. Lajiness, M.S. et al. Assessment of the consistency of medicinal chemists in reviewing sets of compounds. *J. Med. Chem.* **2004**, *47*, 4891–4896.

33. See Kuhn, P. et al. The genesis of high-throughput structure-based drug discovery using protein crystallography. *Curr. Opin. Chem. Biol.* **2002**, *6*, 704–710 for more about this.

34. See Smith, D.A. This is your captain speaking. *Drug Discov. Today* **2002**, *7*, 705–706.

35. Appelt, K. et al. Design of enzyme inhibitors using iterative protein crystallographic analysis. *J. Med. Chem.* **1991**, *34*, 1925–1934.

36. Jorgensen, W.L. The many roles of computation in drug discovery. *Science* **2004**, *303*, 1813–1818.

37. Lafont, V. et al. Compensating enthalpic and entropic changes hinder binding affinity optimization. *Chem. Biol. Drug Des.* **2007**, *69*, 413–422.

38. Interestingly, an alternative to displacing a lot of water molecules may be to displace just a few of the right sort. Recently the suggestion has been made that displacing a water molecule located in a hydrophobically enclosed region of a protein might give a much higher entropy increase than most. See Young, T. et al. Motifs for molecular recognition exploiting hydrophobic enclosure in protein-ligand binding. *Proc. Natl Acad. Sci. USA* **2007**, *104*, 808–813.

39. Velazquez-Campy, A. et al. The application of thermodynamic methods in drug design. *Thermochim. Acta* **2001**, *380*, 217–227.

40. Fischer, E. Einfluss der Configuration auf die Wirkung der Enzyme. *Ber. Dtsch. Chem. Ges.* **1894**, *27*, 2985.

41. Bursavich, M.G., Rich, D.H. Designing non-peptide peptidomimetics in the 21st Century: Inhibitors targeting conformational ensembles. *J. Med. Chem.* **2002**, *45*, 541–558.

42. Dr Koshland, of course, went on to do more groundbreaking biochemistry research, become Professor Emeritus at UC Berkeley, long-time editor of *Science* magazine, and the recipient of numerous awards.

43. Koshland, Jr D.E. Application of a theory of enzyme specificity to protein synthesis. *Proc. Natl Acad. Sci. USA* **1958**, *44*, 98–104.

44. Lee, A.Y. et al. Conformational switching in an aspartic proteinase. *Nat. Struct. Biol.* **1998**, *5*, 866–871.

45. See Verlinde, C.L., Hol, W.G. Structure-based drug design: progress, results and challenges. *Structure* **1994**, *2*, 577–587, and reference therein.

46. Teague, S.J. Implications of protein flexibility for drug discovery. *Nat. Rev. Drug Discov.* **2003**, *2*, 527–541.

47. Note that one shouldn't assume that the lowest energy conformation of the receptor is necessarily the one that will be bound. In fact, an energy minimum for the receptor–ligand complex, not the receptor itself, is the driving force, so higher energy receptor conformers may well be the ones that bind ligands.

48. See Wilson, E.K. Dealing with flexible receptors. *Chem. Eng. News*, May 10, **2004**, 46–47.

49. Warren, G.L. et al. A critical assessment of docking programs and scoring functions. *J. Med. Chem.* **2006**, *49*, 5912–5931.

50. For a discussion of these points as well as a description of the forces underlying binding interactions, see Ermondi, G., Caron, G. Recognition forces in ligand-protein complexes: Blending information from different sources. *Biochem. Pharmacol.* **2006**, *72*, 1633–1645.

51. See Wlodawer, A., Vondrasek, J. Inhibitors of HIV-1 protease: A major success of structure-assisted drug design. *Annu. Rev. Biophys. Biomol Struct.* **1998**, *27*, 249–284 and references therein.

52. Leung, D. et al. Protease inhibitors: Current status and future prospects. *J. Med. Chem.* **2000**, *43*, 305–341.

53. Roberts, N.A. et al. Rational design of peptide-based HIV proteinase inhibitors. *Science* **1990**, *248*, 358–361.

54. Erickson, J. et al. Design, activity, and 2.8 Å crystal structure of a $C_2$ symmetric inhibitor complexed to HIV-1 protease. *Science* **1990**, *249*, 527–533.

55. Dreyer, G.B. et al. A symmetric inhibitor binds HIV-1 protease asymmetrically. *Biochemistry* **1993**, *32*, 937–947.

56. Kempf, D.J. et al. ABT-538 is a potent inhibitor of human immunodeficiency virus protease and has high oral bioavailability in humans. *Proc. Natl. Acad. Sci. USA* **1995**, *92*, 2484–2488.

57. Lam, P.Y.S. et al. Cyclic HIV protease inhibitors: Synthesis, conformational analysis, P2/P2' structure-activity relationship, and molecular recognition of cyclic ureas. *J. Med. Chem.* **1996**, *39*, 3514–3525.

58. Jadhav, P.K. et al. Cyclic urea amides: HIV-1 protease inhibitors with low nanomolar potency against both wild type and protease inhibitor resistant mutants of HIV. *J. Med. Chem.* **1997**, *40*, 181–191.

59. Barillari, C. et al. Classification of water molecules in protein binding sites. *J. Am. Chem. Soc.* **2007**, *129*, 2577–2587.

60. Dorsey, B.D. L-735,524: The design of a potent and orally bioavailable HIV protease inhibitor. *J. Med. Chem.* **1994**, 3443–3451.

61. Kaldor, S.W. et al. Viracept (Nelfinavir Mesylate, AG1343): A potent, orally bioavailable inhibitor of HIV-1 protease. *J. Med. Chem.* **1997**, *40*, 3979–3985.

62. Tie, Y. et al. High resolution crystal structure of HIV-1 protease with a potent non-peptide inhibitor (UIC-94017) active against multi-drug-resistant clinical strains. *J. Mol. Biol.* **2004**, *338*, 341–352.

63. Li, R. et al. Three-dimensional structure of *M. tuberculosis* dihydrofolate reductase reveals opportunities for the design of novel tuberculosis drugs. *J. Mol. Biol.* **2000**, *295*, 307–323.

64. El-Hamamsy, M.H.R.I. et al. Structure-based design, synthesis and preliminary evaluation of selective inhibitors of dihydrofolate reductase from *Mycobacterium tuberculosis*. *Bioorg. Med. Chem.* **2007**, *15*, 4552–4576.

65. Sudhakar Babu, Y. et al. BCX-1812 (RWJ-270201): Discovery of a novel, highly potent, orally active, and selective influenza neuraminidase inhibitor through structure-based drug design. *J. Med. Chem.* **2000**, *43*, 3428–3486.

66. See Oppermann, M. Chemokine receptor CCR5: Insights into structure, function, and regulation. *Cell. Sig.* **2004**, *16*, 1201–1210.

67. Maeda, K. et al. Structural and molecular interactions of CCR5 inhibitors with CCR5. *J. Biol. Chem.* **2006**, *281*, 12688–12698.

68. For an example of the use of SBDD to build in selectivity against an antitarget, in this case revealing an unexpectedly different binding mode for the antitarget, see Aronov, A.M. et al. Flipped out: Structure-guided design of selective pyrazolylpyrrole ERK inhibitors. *J. Med. Chem.* **2007**, *50*, 1280–1287.

# Chapter 8

# Initial Properties

## 8.1 Why Not All At Once?

Having emphasized the importance of the role that parallel property optimization plays in the modern drug discovery process, readers might be curious as to how any of them can really be considered "initial" or how any sort of linear optimization sequence can be implied. But in the real world, you can't optimize—or even test—every compound for every property. Although a number of them are optimized at the same time in various stages of a project, trying to optimize all of them at once for everything could only be an exercise in frustration and inefficiency. There seems little point in trying to optimize the activities of compounds that are unstable, determine the protein binding of inactive compounds from a screening library, or trying to improve cell permeability for compounds that don't dissolve at all.

In practice, some properties must be "gating" for the further investigation of others. Projects need to look at these most critical properties first and create a "funnel" to reduce the numbers of compounds that move on to slower, more resource-intensive assays, not to mention the massive resources involved in analoging efforts. This chapter addresses five of the properties most often used in doing this: potency, selectivity, structural novelty, solubility, and chemical/plasma stability. Probably more than 99.9% of screening compounds and 90% of confirmed hits will be unable to run this gauntlet. The greatly reduced number of survivors and their progeny will move on to face further challenges and parallel optimization of some of these "initial" properties as well as some new ones.

Different companies have different ways of doing things, making it impossible in this book to lay out exactly what property is examined or optimized at which exact point in a project's progression. Many companies, for example, run metabolic stability screens (discussed in the next chapter) quite early on, while a few may now be prescreening compound libraries for solubility before even looking at potency. So keep in mind that the order of properties presented won't necessarily match with the order used at a given company. Only hands-on experience with a project that progresses from screening to clinical candidate selection can really provide the scientist with that information. What follows in this chapter is a "best guess" at the properties most likely to be examined at the earliest stages of a project at a typical company, beginning with potency.

## 8.2 Potency

### 8.2.1 What, Why, and How Much?

Ask any medicinal chemist what a compound needs to have in order to become a drug, and it's almost certain that potency will be #1 on the list. There's definitely a reason for this: *like good looks, great potency can make up for a wide variety of failings.* Make a compound potent enough and you won't need to get as much of it to the site of action, so problems with solubility, protein binding, PK, and toxicity due to off-target effects diminish or even disappear. Modern clinical drugs are usually dosed at just a couple of *mpk* (*milligrams per kilogram*), and at that low of a dose potency is a prerequisite, while some forms of administration like inhalation and transdermal delivery may require still higher potency.

Clinical potency, of course, isn't what's measured in biochemical assays, and the disconnect between the two can be large. Factors like cell permeability, metabolic stability, and efflux transport can stand as barriers to a potent compound having the desired effect in vivo, and optimization for these parameters is important, as we'll see. Also, the situation is a bit complicated for hits derived from phenotype screens, where the actual target might not be known, making it hard to tell whether structural changes are affecting target potency or something else. But despite all the other complicating factors, potency against the target of interest is the first requirement for having a drug with the desired effect, so in both the HTL and, later, LO stages optimizing potency will be one of the main drivers of medchem efforts.

Clinical drugs have a wide range of in vitro potencies, from millimolar to picomolar as shown in Figure 8.1.[1] Most projects aim for compounds with single-digit nanomolar $K_i$ or $IC_{50}$ values or better. Activities seen with HTS hits depend to a large extent on the type of targets screened against. Screening against aminergic GPCRs is more likely to turn up single-digit micromolar hits (excellent starting points) than screening against peptidergic GPCRs or protein–protein interactions, for instance. Ultimately, "you get what you get" and need to work with that. Often an improvement in potency of $10^3$–$10^4$ is needed, sometimes more with fragment hits, which start out with low potency but high atom efficiency.

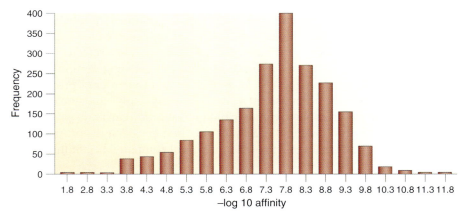

**Figure 8.1** ▶ Frequency distribution of small-molecule drug potencies (Reprinted with permission from Overington, J.P., et al. How many drug targets are there? *Nat. Rev. Drug Discov.* **2006**, *5*, 993–996, copyright 2006, Macmillan Publishers Ltd.)

## 8.2.2 Species Specificity

As clinical candidates are to be used in humans, it's the human form of the enzyme or receptor that will normally be used for screening assays. An exception to this can occur if patent considerations necessitate taking an animal version of the protein and mutating it (the protein, not the animal!) to be as close as possible to the human protein while still being technically legal, giving a sort of chimeric target. But screening against a human or near-human target comes with one potential drawback: well before the compounds are dosed in man they need to prove themselves in animal testing, often starting with mice or rats. Differences in compound potency between animals and man can be profound, and aren't necessarily predictable on the basis of the sequence homology or anything else.

A typical recent example comes from some work on tryptase inhibitors, tryptase being a serine protease of interest for the treatment of asthma. Compound 44, shown in Figure 8.2, was found to be a 2.5 nM inhibitor of human tryptase, but a 365 nM inhibitor of the mouse enzyme, and a 400 nM inhibitor of monkey tryptase.[2] Larger species specificity effects than this have been observed for some renin inhibitors, including zankiren (Figure 8.2) and the

**Compound 44**

**Aliskiren**

**Zankiren**

**Figure 8.2 ▶** Structures of the tryptase inhibitor, Compound 44 (Palmer, J.T., et al. Design and synthesis of selective keto-1,2,4-oxadiazole-based tryptase inhibitors. *Bioorg. Med. Chem. Lett.* **2006**, *16*, 3434–3439.) and the renin inhibitors aliskiren and zankiren.

**TABLE 8.1 ►   Potencies of Two Renin Inhibitors Against Renin from Different Species**

| Species | Aliskiren (nM) | Zankiren (nM) |
|---|---|---|
| Guinea pig | 63 | 9.4 |
| Rat | 80 | 1400 |
| Dog | 7 | 110 |
| Man | 0.6 | 1.1 |

Data from Wood, J.M., et al. Structure-based design of aliskiren, a novel orally effective renin inhibitor, *Biochem. Biophys. Res. Comm.* **2003**, *308*, 698–705, and Kleinert, H.D., et al. Discovery of a peptide-based renin inhibitor with oral bioavailability and efficacy *Science* **1992**, *257*, 1940–1943.

marketed drug Tekturna (aliskiren, Figure 8.2) as shown in Table 8.1. Pity the poor project team if it was counting on testing zankiren in rats!

Finding that a compound is *more* potent against an animal than a human target, which sometimes happens, isn't a bad thing, but time and again it comes as a shock to project scientists to find that their compound, which is nanomolar against a human target, is micromolar in the species they need to first test it in. This makes it perhaps the most predictable—and therefore most preventable—surprise in drug discovery. Choices then need to be made: Are there other compounds available that might work better against the animal target? Can a different species be used in an efficacy model? Should some limited SAR be done to try and optimize against the animal version of the protein? It's a far, far better thing to know that a problem exists early on in a project than to find out just before an in vivo study is due to begin. For this reason, the protein expression group may be asked to clone and express the animal target protein at the HTL stage to avoid much trouble later.

---

**Box 8.1   Key Points About Potency**

► High potency can help compensate for suboptimal

　　Solubility

　　Absorption

　　Metabolism

　　Off-target activity

► Apparent potency in cells and in vivo is a composite property

► Potencies that can be expected for screening hits depend on the target class

► Potency against a target can be highly species-dependent

---

As an alternative where this, for some reason, isn't possible, more pharmacology work can be done early to look for pharmacodynamics effects in the species of interest. Seeing a change in a relevant biomarker might lead you to infer that the compound has effectively interacted with the target and therefore must be potent against it. Or activity-based probes, if available, can be used to identify protein activity changes in treated animals. These are resource-intensive, but very informative, experiments when it comes to a preliminary assessment of the in vivo effects of a particular compound, but they're poor surrogates for simple biochemistry assay data, which is much easier to get. In short, *compounds need to be tested early on for activity against the target protein produced by the animal species to be used in efficacy studies* (Box 8.1).

## 8.3  Selectivity

Calling a compound *selective* means nothing until one specifies what exactly it's selective against. Just as all compounds have "one spot" TLCs in systems where they don't move off the origin, every compound is selective if only screened against one target. Increase the number of targets screened against and the probability of having selectivity against them all goes down exponentially.

Proteins to be selective against can be divided into two categories: those related to the target of interest and those that aren't. Most chemists dream of compounds that are highly selective against *all* targets but the one of interest, while the reality is that few, if any, such compounds exist. Toxicity owing to off-target effects is what drives the concern for selectivity, but note that even a completely "clean" drug could still have on-target, or *mechanistic toxicity*. As we've already seen, it isn't always easy to distinguish the two if something goes wrong in the clinic. But selective compounds are a safer bet, especially for diseases which require their chronic administration for months or years, hence the emphasis on selectivity beginning early in the drug discovery process.

### 8.3.1  Selectivity . . . Not!

It's easy to understand why selectivity against other target family members can be difficult to achieve. The similar active sites and identical mechanisms found in families of proteins that make the chemogenomic approach to hit generation viable also make selectivity hard to come by. Consider the human kinome, as graphically depicted by the dendrograms shown in Figure 8.3, a way of displaying the relatedness of the different enzymes. There are over 500 family members. Most kinase inhibitors, referred to as *type I inhibitors*, compete with ATP at the ATP-binding sites of the active enzymes, which tend to be very similar.[3] Selectivity problems with such compounds seem almost inevitable.

Gleevec (imatinib, Figure 1.17) is remarkably selective for a kinase inhibitor, hitting relatively few kinases at submicromolar levels (Figure 8.3). It does this through *type II inhibition*, stabilizing an inactive conformation of BCR-ABL kinase via allosteric binding near the ATP site.[4–5] Importantly, it makes use of an adjacent hydrophobic pocket, not found in many other kinases, that can be formed when the enzyme's activation loop has moved out of its normal position. Recall too that this drug actually benefits from inhibiting another kinase, c-Kit, which is believed to explain its efficacy in a different indication, gastrointestinal stromal tumors (GIST), a fairly rare type of cancer, but one for which few treatment options were available.

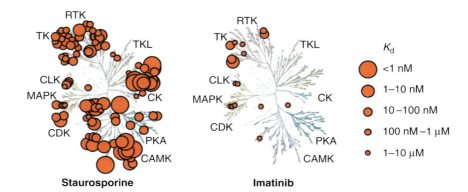

**Figure 8.3** ▶ Dendrograms illustrating potencies of two drugs, staurosporine and imatinib, against targets in the human kinome. For imatinib, potency is greatest against Abl, a member of the nonreceptor tyrosine kinase (TK) family in the NW section of the dendrograms. Potencies are proportional to red-circle size as shown at right (Adapted with permission from Nature Biotechnology, Fabian, M.A., et al. A small molecule-kinase interaction map for clinical kinase inhibitors. *Nat. Biotechnol.* **2005**, *23*, 329–336, Macmillan Publishers Ltd. The kinase dendrogram is reproduced courtesy of Cell Signaling Technology, Inc., www.cellsignal.com).

Staurosporine (Figure 8.4) is far less selective as its dendrogram shows, and one might expect it to have a much higher potential for toxicity considering the number of different targets that it hits. In fact, though, this hasn't prevented staurosporine and some of its derivatives from entering clinical trials for cancer,[6] where compounds as non-selective as alkylating agents are sometimes useful. But for less life-threatening indications or where safer therapeutic agents are available, compounds this non-selective aren't likely to be considered as clinical candidates. Even for cancer, most companies would rather have another imatinib than another staurosporine.

Clearly, hitting some targets might not matter much physiologically, or it might even be beneficial as in the case of imatinib. But inhibiting or antagonizing others, whether family members or not, might make for toxicity. So there may be some benefit to a special kind of non-selectivity, where multiple desired targets but no major toxic antitargets are affected by a

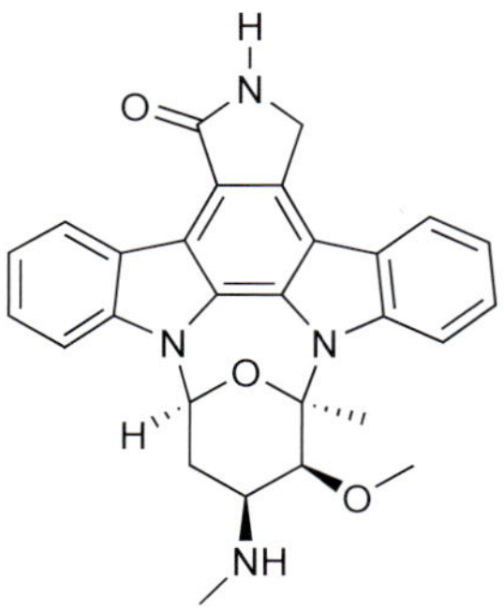

**Figure 8.4** ▶ Staurosporine.

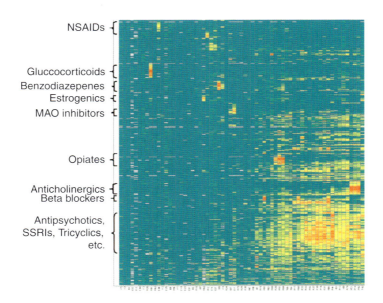

NSAIDs {

Gluccocorticoids {
Benzodiazepenes {
Estrogenics {
MAO inhibitors {

Opiates {

Anticholinergics {
Beta blockers {

Antipsychotics,
SSRIs, Tricyclics,
etc. {

**Figure 8.5 ▶** BioPrint® cluster analysis of 2000 compounds and drugs across 70 pharmacological assays (pIC$_{50}$s). Biological assays are on the *x*-axis, clustered by similarity, and compounds on the *y*-axis, clustered by their fingerprint of biological activity. Red indicates most active and blue indicates inactive (Reprinted with permission from , Hopkins, et al. Can we rationally design promiscuous drugs? *Curr. Opin. Struct. Biol.*, *16*, 127–136, copyright 2006, Elsevier.)

compound. Recalling the case of clozapine already mentioned, we know that quite a number of drugs fall into this category, particularly CNS drugs.

Figure 8.5, an example of Cerep's BioPrint® database analysis,[7] presents graphical data for 2000 compounds (*y*-axis) clustered by similarity against 70 different targets (*x*-axis) clustered by their similarity. The color-coding indicates activity, with red being most active and blue inactive. It's striking just how many different targets the CNS agents hit, and very few compounds in any category seem to be entirely selective even against this very small set of targets.

Just as the "one gene, one protein" concept dominated genetics for many years, "one drug, one target" thinking has dominated medicinal chemistry (although not clinical oncology) for several decades now. But recently a more open-minded approach seems to be emerging, largely based on the success of some non-selective cancer and CNS drugs and the advent of systems biology with its emphasis on biological network pathways. As evidence of this, non-selective compounds previously referred to as "dirty drugs" are now more commonly cited as examples of *polypharmacology*.

In theory, one of the problems with "one drug, one target" is that biological systems have evolved to be masters of compensation and redundancy. Intervention at a single point can have little effect on the whole. Effects on a single target can be compensated for in many ways. As an example, Mother Nature has another way of overcoming enzyme inhibition aside from building up substrate concentrations to overwhelm the effect of a competitive inhibitor as discussed earlier. Levels of the enzyme itself can simply be increased to have the same effect on *any* type of inhibition. A recently suggested example is that of Tekturna (aliskiren, Figure 8.2), the first-in-class renin inhibitor. Along with other types of hypotensive agents like ACE inhibitors and angiotensin receptor blockers, aliskiren may "induce reactive increases in renin secretion" by removing the feedback constraint on it so that high levels of angiotensin II (a peptide downstream in the renin–angiotensin pathway) and hypertension would normally activate.[8] Inhibit renin and the body can just make more of it. Fighting a disease by addressing a single target comes with these kinds of risks. Polypharmacology in the form of separate drugs acting on different targets in the same indication (e.g. *highly active antiretroviral therapy, HAART*, for HIV) has, many times, proven to be better. What about addressing several desired targets with the *same* drug, then?

At first it seems almost heretical that, at least in some cases, a particular lack of selectivity can not only be tolerated but might even be designed in. For the chemist who knows the challenges involved in optimizing activity for even a *single* target and who may find in this bright idea the beginnings of a migraine headache, there is some good news. At least in theory, the synergy that should accompany interference at properly selected multiple points means that much lower affinity might be needed for each of the targets. "Network models suggest that partial inhibition of a surprisingly small number of targets can be more efficient than the complete inhibition of a single target."[9]

The subject of such *designed multiple ligands* (*DML*) has recently been reviewed.[10] Challenges await both in designing and in screening such compounds. Simply linking together two ligands that act at two receptors will usually result in too big of a molecule to have reasonable druglike properties, but building up a single compound acting at both, beginning with fragment screening, has been suggested as an attractive alternative.[11] In trying to identify hits, cross-correlating hit lists from HTS for several targets might end up being a poorer approach than behavioral screening (for CNS drugs) or gene-signature-based screening.[12] And an understanding of biological pathways sufficient to predict which set of targets to hit for a given disease still lies somewhere in the future. But for the present, the loosening of the "one drug, one target" hypothesis may prove to have some value in and of itself.

### 8.3.2   Antitargets

It isn't always easy to know which targets to avoid, but the most important *antitargets* do need to be identified somehow, if only by educated guesses or literature reports of toxicity associated with them, so that compounds can be routinely screened against them as early as possible.

One example of following up on the toxicity of antitargets and what a lack of selectivity against them might do comes from some recent work with DPP-IV inhibitors at Merck. This enzyme proteolyses and thereby inactivates the endogenous peptides GLP-1 and GIP, both of which act to lower blood glucose levels, so preventing their breakdown with a DPP-IV inhibitor like Merck's recently approved Januvia (sitagliptin, Figure 8.6) has beneficial effects in Type II diabetes. Besides DPP-IV, other prolyl peptidases exist including quiescent cell proline dipeptidase (DPP7), DPP8, DPP9, and fibroblast activation protein (FAP).[13] Merck had a number of small-molecule inhibitors of these other targets and so was able to investigate how critical to drug safety avoiding them might be. Toxicity studies in rats and dogs turned up major toxic effects, including death, for a mixed inhibitor of DPP8/DPP9,

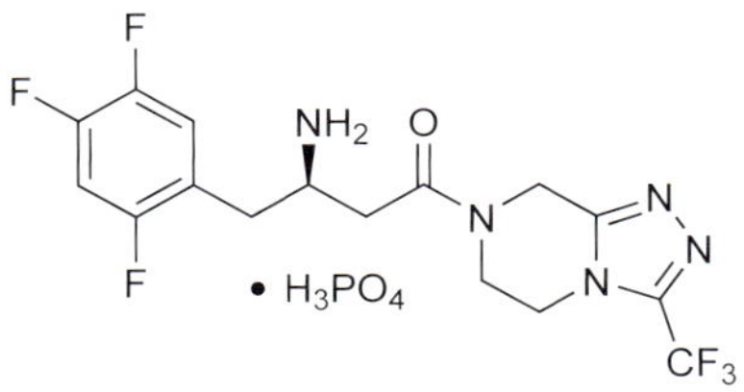

**Figure 8.6 ►** Januvia (sitagliptin).

while some toxicity (reticulocytopenia) in rats was found for the DPP7 inhibitor. A selective DPP-IV inhibitor, however, was clean.[14] It's likely that all of these other prolyl peptidases and more were considered antitargets from the beginning, and that new DPP-IV inhibitors were routinely screened against them, a good thing since even the non-lethal toxicity in rats observed for the DPP7 inhibitor would probably have been enough to derail the development of a compound insufficiently selective against it.

Selectivity is usually expressed as a ratio of $K_i$ or $IC_{50}$ values, but this isn't necessarily a reflection of what it will really be in vivo. A compound can have identical $K_i$ values for its target and an antitarget, but if its off-rates for the two differ, it will, in time, have a greater effect on the one for which it has the slower off-rate. This can make for a kind of kinetic selectivity. This important effect was recently pointed out: "The selectivity of a drug is not a static parameter, but is instead time-dependent . . . As such, the selection of 'lead' compounds based on their dissociative half-lives from primary and secondary targets might be far more relevant for the ultimate efficacy and safety of the compounds than the determination of their $IC_{50}$, $K_d$, or $K_i$ values at a fixed time point of short duration, as is the current, common practice."[15] This is likely to be an important consideration later on in lead optimization and candidate selection, but early on most companies probably still judge compounds by their one-timepoint selectivity ratios.

Up to this point, selectivity against related targets has mostly been considered, but there are plenty of other targets which need to be avoided. If chemogenomic hit generation points to the difficulties involved in achieving selectivity against kindred targets, SOSA hit generation (see Chapter 6) illustrates the inherent possibility that a compound might bind to a completely unrelated target. Some of these potential antitargets bind a wide variety of compounds and can be responsible for major toxicity. These include *hERG* (discussed in more detail later), a potassium ion channel which if inhibited can lead to life-threatening arrhythmias, and 5-HT$_{2b}$ receptors, where binding might lead to cardiac valvulopathy.[16] Affinity for many different receptor subtypes (adenosine, adrenergic, muscarinic, etc.) as well as other types of antitargets might be associated with undesirable toxicities. As a result, the screening of lead compounds against a broad variety of potential antitargets has been an accepted part of the drug discovery process for years.

Those of us who have been in the field for a while might still refer to this as "PanLabs screening," after the company (now part of MDS Pharma) that it was frequently contracted out to. These days MDS or another company like Cerep might be used, or it might be done in-house in big pharma. This kind of broad, off-target screening used to be done just before clinical trials, which is a bit like not checking to see whether your horse is still alive until the race begins. Unexpected problems would be encountered late in the game, causing some programs to be dropped and others to start all over again. Part of this was probably due to the low-throughput nature of some of the receptor pharmacology assays involved at the time, and some of it might be attributed to the disconnect between discovery and development that used to be common. Neither of these holds true anymore, and clinical success for a project is much more likely if such potential compound liabilities are identified early on.

As a recent paper by researchers at Novartis points out, with early in vitro safety pharmacology profiling "the assay panels can easily be used to aid lead optimization by prioritizing compounds for selectivity and can contribute to SAR-based design. This type of information can help chemists and biologists to rank and prioritize compounds according to pharmacological profile and to optimize their structures without losing the affinity to the primary target".[17] As an example of the power of this approach in selecting leads, consider the data summarized in Figure 8.7. Here the promiscuity of compounds based on three different scaffolds toward important antitargets can be taken in at a glance, with series 2 being more problematic and series 3 the least so. Notice, though, that the scaffold doesn't determine everything: compound 2i from the most problematic series is less so than compound 3g from the cleanest one. But having this information available in HTL or early LO would allow a project team to flag potential toxicity problems for different compounds, identifying which off-target effects need to be dealt with, and aid in prioritizing the different series (Box 8.2).

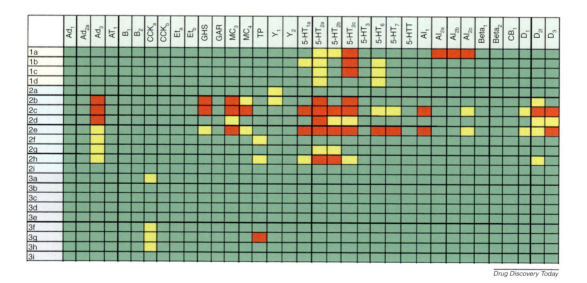

*Drug Discovery Today*

**Figure 8.7 ▶** Pharmacological activity patterns for three different classes of compounds directed at the same therapeutic target. Numbers on the left mark compounds having the same scaffold. Percent inhibition at 10 μM concentration of compounds is color coded: <50% is green, 50–75% is yellow, and >75% is red (Reprinted with permission from Whitebread, S., et al. In vitro safety pharmacology profiling: An essential tool for successful drug development. *Drug Discov. Today*, 10, 1421–1433, copyright 2005, Elsevier Publishers Ltd.)

---

**Box 8.2   Key Points About Selectivity**

▶ Achieving selectivity is crucial for avoiding off-target effects

▶ Important antitargets need to be identified early on and screened against

▶ Selectivity against closely related targets is usually hardest to achieve

▶ Selectivity against unrelated targets (safety pharmacology profiling) is important in lead selection

▶ Activity against several chosen targets (DMLs) can be a viable approach

---

# 8.4   Structural Novelty

Of the three requirements for a patent—novelty, utility, and unobviousness—novelty is often the most difficult to achieve for a typical composition-of-matter patent. Although chemical space is unimaginably vast, most projects and most drugs tend to be densely clustered in the same few tiny regions thereof, often making a clear title to intellectual property hard to come by. Just how difficult it might be depends on where the project started out. As we've seen, those beginning with a competitor's compound will probably have the most difficult time, while hits arising from an approach like fragment screening with subsequent linking or fragment "growing" are more likely to turn up novel structures. Using a chemogenomic approach and screening libraries of compounds which the company already has patent rights to can also be an advantage.

Although structural novelty isn't normally required for a compound to be considered a confirmed hit, it's important down the road in lead optimization. No manager wants to pour resources into developing a drug that might belong to someone else. Chemists will likely be most interested in structurally novel hits from the start, but limitations on the types of compounds available for high-throughput screening (commercial availability, feasible combinatorial routes, etc.) means that most of the time few hits will be novel to begin with. Instead, structural novelty may need to be engineered in while working out SAR. Whole volumes could be written just on this subject, but here only a few of the most popular methods of manipulating structures to build in novelty will be highlighted.

As was previously noted, changes made to a molecule can affect all properties. So although introduced here in the context of patentability, these same methods of structural variation are commonly used for many different purposes in HTL and LO. They'll be used to try and solve other problems like ones related to potency and stability, and improve basically any of the properties listed in this chapter or the next two. As such, they form an integral part of the medicinal chemist's art.

## 8.4.1   Bioisosteres, Group, and Atom Replacements

### 8.4.1.1   Definition and Utility

Much like the concept of druglikeness, the idea behind *bioisosteres* seems deceptively simple at first, but gets the more complex the more one tries to pin it down. The origin and evolution of the terms isostere and bioisostere are discussed in an excellent review of the subject by Wermuth.[18] Bioisosteres are "groups that can be substituted for each other in a variety of chemical classes for a variety of bioactivities."[19] The assumption is made that this substitution occurs at a part of the molecule important for interactions with the target, not in "empty space" where almost any substitution would work. The term applies even when the type of activity of the compound for the target receptor is changed, as in switching out a group of atoms in a receptor agonist and turning it into an antagonist or inverse agonist instead. Interested readers are referred to a comprehensive review of this important subject that was recently published.[20] Only a few basic points will be mentioned here.

Bioisosteres are usually divided into two types, as shown in Figure 8.8. *Classical bioisosteres* share a similar valence and geometry with the groups they replace. Going down a row of the periodic table thus produces some classical bioisosteres (Si for C, S for O, etc.). *Nonclassical bioisosteres* can vary more in size and shape, often with a property like H-bonding ability or pKa being the point of commonality with the original group. These often *won't* have very similar sterics, making the term somewhat of a misnomer.

The simplest application of the bioisostere approach to build in structural novelty would probably lie in the substitution of a single atom by one not covered in existing patents, although changes this small run the risk of being considered obvious by patent examiners,

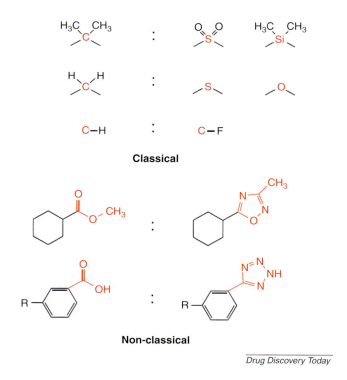

**Figure 8.8** ► Classical and non-classical bioisosteres (Reprinted with permission from , Showell, G.A., Mills, J.S. Chemistry challenges in lead optimization: Silicon isosteres in drug discovery. *Drug Discov. Today*, *8*, 551–556, copyright 2003, Elsevier.)

especially if no unexpected improvements are brought about by the change (see Chapter 3). "Oddball atoms" like silicon and phosphorus, once rare in patent applications, are now merely uncommon as a growing number of scientists try this approach.[21] But changes like this, even if they work, often come with other drawbacks: increases in lipophilicity, decreases in stability and solubility, or difficulties in synthesis which may partly explain why such substitutions weren't included in patent applications in the first place.

Simply finding a "hole" in an existing patent, like a common substituent not covered at a given position, is a related, if not-very-exciting, strategy. But with today's "bulletproof" patents with pages-long lists of substituent groups, such holes have become as rare as hens' teeth. Still, for important hits and leads, close scrutiny of the existing "prior art" with the aid of those versed in patent law is always in order, if only to see what's out there.

### 8.4.1.2 Examples

The substitution of a phenyl ring by a thiophene is probably one of the most well-known bioisosteric replacements. Figure 8.9 shows how this was applied to the non-steroidal anti-inflammatory agent (NSAID) piroxicam to produce tenoxicam, another potent NSAID.[22] The replacement of one ring system by another is one of the most familiar strategies in medicinal chemistry. Such changes tend to have major effects on properties, although not always in the desired direction. They often represent a good way of gaining an IP foothold. One of the most

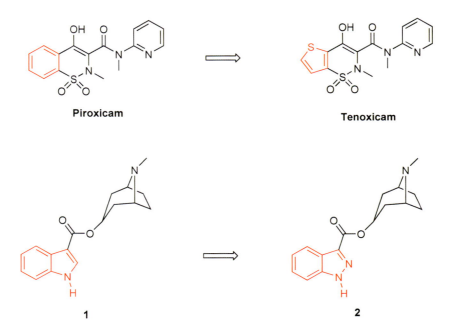

**Figure 8.9 ▶** Top row: NSAIDS piroxicam and tenoxicam (Binder, D., et al. Analogues and derivatives of tenoxicam. 1. Synthesis and antiinflammatory activity of analogues with different residues on the ring nitrogen and the amide nitrogen. *J. Med. Chem.* **1987**, *30*, 678–682.). Bottom row: Two bioisosteric 5HT$_3$ receptor antagonists (Fludzinski, P., et al. Indazoles as indole bioisosteres: Synthesis and evaluation of the tropanyl ester and amide of indazole-3-carboxylate as antagonists at the serotonin 5HT$_3$ receptor. *J. Med. Chem.* **1987**, *30*, 1535–1537.)

common one-atom replacements is that of a nitrogen for a carbon atom in an aromatic ring, for example, substituting pyridyl for phenyl. One example of this sort of bioisostere is shown in Figure 8.9, where the indole ring of the 5HT$_3$ receptor antagonist **1** was replaced with an indazole to give **2**, another potent antagonist.[23] Changing a ring or ring system more extensively than by a one-atom substitution (say, replacing a furan with an imidazole) might also be considered as an example of bioisosterism, but will be discussed below in the section on scaffold hopping.

Functional group replacement by bioisosteres has been an area of great interest, although it's frequently driven by problems like stability and cell permeability rather than structural novelty. A variety of bioisosteres for carboxylic acids have been tried, one of the most successful of which is the tetrazole group, as shown in Figure 8.10. Research directed at angiotensin II receptor antagonist uncovered **3**, for which acidity proved essential to activity but also seemed to preclude oral bioavailability. Replacement of the carboxylic acid moiety with the tetrazole bioisostere, however, proved successful, resulting in the marketed antihypertensive agent losartan.[24]

Carboxylic esters are often unstable in vivo; consequently many different bioisosteres for this functional group have been tried, most of them being small heterocycles. For a series of CDK2 inhibitors like **4** (Figure 8.10) this problem could be overcome through the use of an

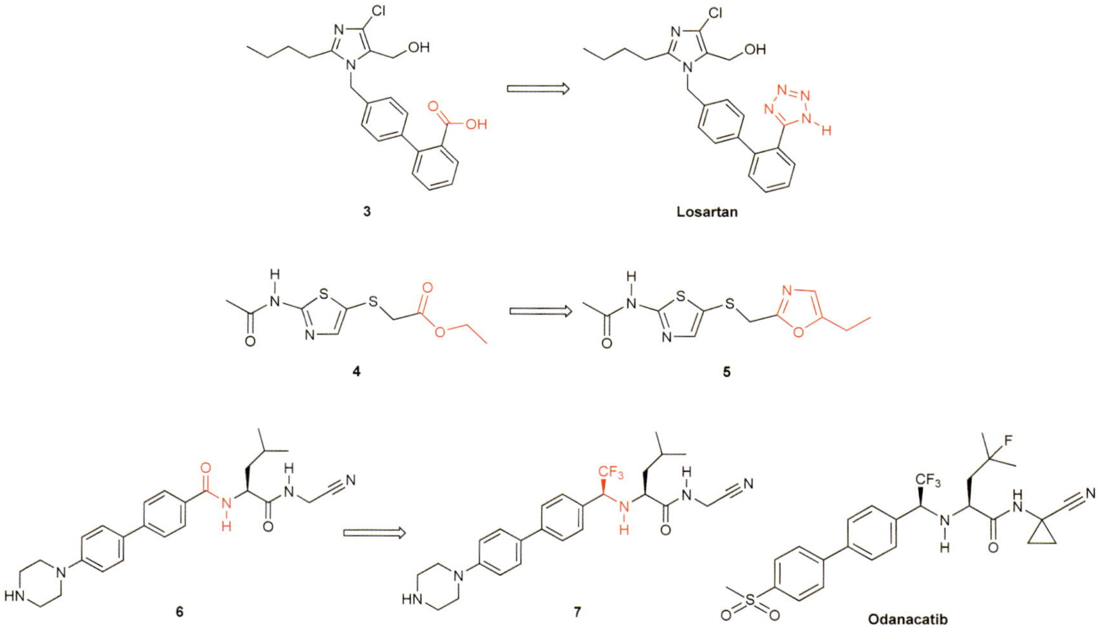

**Figure 8.10** ► Some bioisosteric replacements for carboxylic acids (top) (Carini, D.J., et al. Nonpeptide angiotensin II receptor antagonists: The discovery of a series of *N*-(Biphenylmethyl) imidazoles as potent, orally active antihypertensives. *J. Med. Chem.* **1991**, *34*, 2525–2547), esters (middle) (Kim, K.S., et al. Discovery of aminothiazole inhibitors of cyclin-dependent kinase 2: Synthesis, X-ray crystallographic analysis, and biological activities. *J. Med. Chem.* **2002**, *45*, 3905–3927.), and amides (bottom) (Black, W.C., et al. Trifluoroethylamines as amide isosteres in inhibitors of cathepsin K. *Bioorg. Med. Chem. Lett.* **2005**, *15*, 4741–4744.) The clinical candidate odanacatib (bottom right) incorporates a trifluoroethylamine amide isostere. (Gauthier, J.Y., et al. The discovery of odanacatib (MK-0822), a selective inhibitor of cathepsin K. *Bioorg. Med. Chem. Lett.* **2008**, *18*, 923–928.)

oxazole bioisostere. Compound **5** had an $IC_{50}$ of 20 nM, a significant improvement over the 170 nM for compound **4**.[25] Numerous other examples along these lines exist.

Amide bonds can be particularly problematic. Difficulties with solubility, bioavailability, and stability are common. At the same time, this group is frequently essential for activity due to the H-bonding they enable as well as their conformational and electronic effects. Research on amide bond replacements for peptidomimetic compounds has a rich history going back many decades and is still an area of active interest today. A very recent example of the successful use of a simple amide bond bioisostere can be seen in Figure 8.10.

Compound **6** is a 15 pM inhibitor of the proteolytic enzyme cathepsin K, a novel target for osteoporosis drugs as described earlier. Analysis of the crystal structures for related inhibitors revealed that although the N-H of the P2-P3 amide (the red one in the figure) participated in H-bonding to the enzyme, the oxygen did not. Simply leaving out the oxygen and using the corresponding benzylamine, however, would produce a charged compound lacking a partially polarized bond $\alpha$ to the nitrogen. Here the electron withdrawing effects of fluorine were put to good use. Trifluoroethylamine bioisostere **7** was not protonated at the relevant pH, and proved even more active (**7** had an $IC_{50}$ of 5 pM), selective, and, surprisingly, stable to metabolic cleavage at the *other* amide bond than *bis*-amides like **6**.[26] This bioisostere was used to good effect in odanacatib (Figure 8.10), Merck's anti-osteoporosis candidate currently in Phase III clinical trials.[27]

Bioisosteres can be used creatively for many other functional groups besides those containing a carbonyl moiety. Figure 8.11 shows two of these. Phenolic groups like the one in the

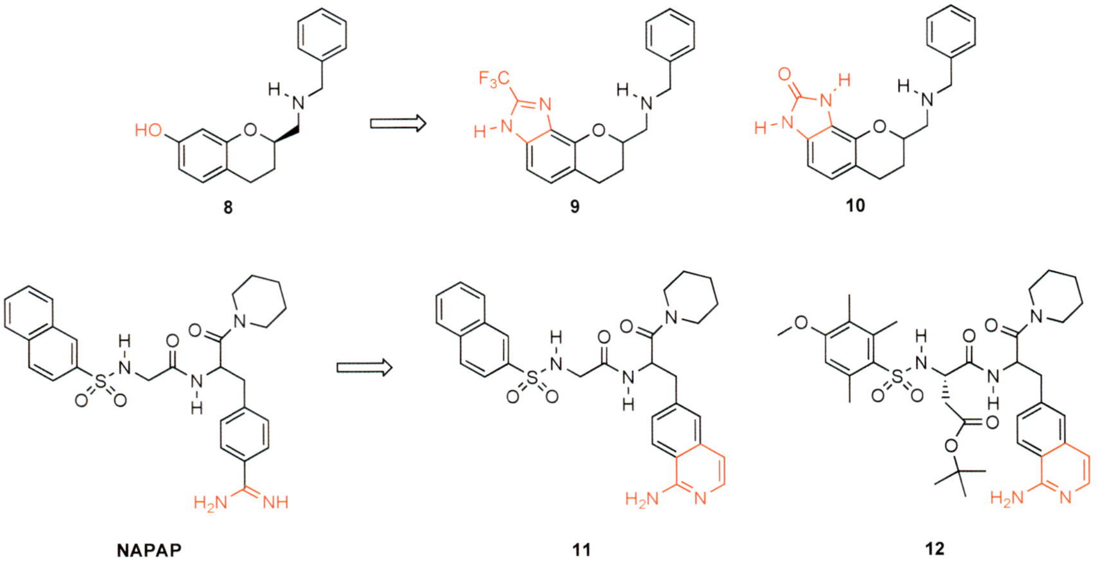

**Figure 8.11 ▶** Top: Two bioisosteres of a phenolic group (Mewshaw, R.E., et al. New generation dopaminergic agents. Part 8: Heterocyclic bioisosteres that exploit the 7-OH-2-(Aminomethyl)chroman $D_2$ template. *Bioorg. Med. Chem. Lett.* **2002**, *12*, 271–274.). Bottom: Aminoisoquinoline bioisosteres of a benzamidine (Rewinkel, J.B.M., et al. 1-Aminoisoquinoline as benzamidine isosteres in the design and synthesis of orally active thrombin inhibitors. *Bioorg. Med. Chem. Lett.* **1999**, *9*, 685–690.)

dopamine $D_2$ partial antagonist **8** can sometimes be replaced by rings substituted with electron withdrawing groups that increase the acidity of an N-H that mimics the phenolic O-H. Compounds **9** and **10** were the result of this kind of approach. Compound **9** was found to be somewhat less potent than **8**, possibly due to a propensity to exist as the other tautomer, but the (–)-isomer of **10** displayed improved potency relative to the phenol.[28]

The thrombin inhibitor NAPAP (Figure 8.11) features an amidine group that's crucial for its interaction with an active site aspartate of the enzyme. Unfortunately, benzamidines like these tend to have poor cell permeability. The 1-aminoisoquinoline bioisostere, **11**, although itself much less active than NAPAP, led to the synthesis of other aminoisoquinolines like **12**, which was almost 10-fold more potent and >10-fold more cell-permeable than NAPAP.[29]

## 8.4.2   Scaffold Hopping, Morphing, and Grafting

As chemists can't even agree on a precise definition of what constitutes a "scaffold" (although we're all sure that we know one when we see one), the term *scaffold hopping* has been criticized as "ill-defined and highly subjective."[30] The general concept, though, seems to be useful. In an admittedly fuzzy sense, it refers to making major changes in the structures of compounds while retaining affinity for a given receptor. Changes to the "scaffold," "core," "molecular framework," or "backbone" are key. These, then, must go beyond peripheral substitutions, simple functional group replacement, use of a single bioisostere, or variations that might be considered "obvious", which explains why it's so useful in the context of patentability.[31]

The methodology used to make the scaffold "hop" is important too. This is often done by taking the original hit, or template, and either running 2D or 3D similarity searches against a library of real or virtual compounds to find different ones that might also work, or by generating a *pharmacophore map* of what molecular features seem to be important to activity, then searching through such libraries to find other compounds that might fit it. Use of these types of ligand-based computational methods is often, but not always, implicit in the term "scaffold hopping". As computers get faster, as algorithms get better, and as patents proliferate, scaffold hopping, however it's defined, is becoming increasingly popular in drug discovery.

A couple of examples of scaffold hopping are shown in Figure 8.12. High-throughput screening of about 250,000 compounds for inhibitors of the interaction between the bacterial proteins ZipA and FtsZ, which is critical for bacterial cell division, turned up **13**, which had a $K_i$ of 12 μM. An X-ray structure of the ZipA-inhibitor complex was obtained. Unfortunately, though, this compound proved to be toxic as well as subject to potential patent issues. "Rather than forfeit the information contained in both the molecule and the nature of its interaction with the protein, a new shape-based computational procedure was used to identify other molecules within Wyeth's corporate database that could fill the same region of the binding pocket and make comparable interactions with the protein."[32] This turned up several compounds including **14**, which had a $K_i$ of 83 μM, but may lack the problems seen with the earlier compound, thus providing a possible starting point for this difficult protein–protein interaction and one that's certainly very different from **13**.

In a second example, researchers at Hoffmann-LaRoche in Basel applied a program they developed called CATS (Chemically Advanced Template Search) to mibefradil, a known 1.7 μM calcium channel blocker. In-house libraries were virtually screened using this pharmacophore search and the highest-ranking molecules went into a real, cellular $Ca^{2+}$ influx assay. The best of

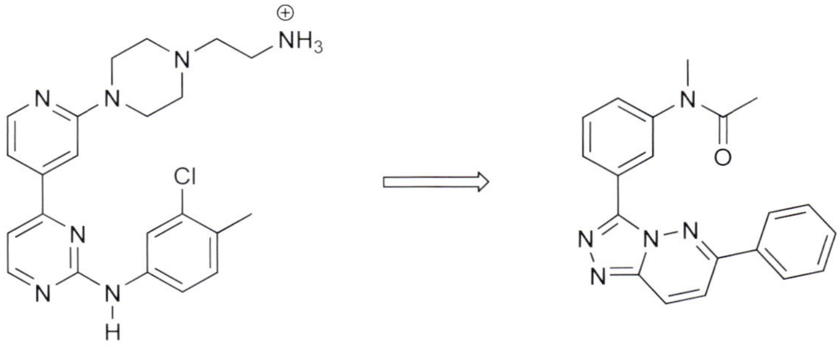

**Figure 8.12** ▶  Two examples of scaffold hopping. See text.

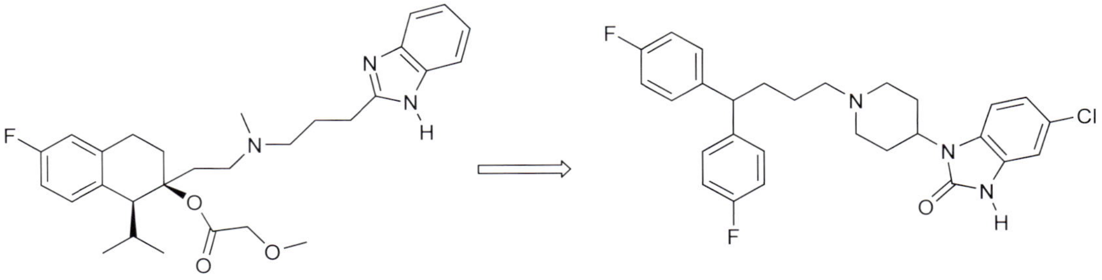

these molecules, clopimozid (Figure 8.12), had a submicromolar $IC_{50}$ and bears little resemblance to the starting compound, mibefradil.[33]

Such computational methods can obviously be useful, and in a sense they're just doing what medicinal chemists have always done: generating hypotheses based on observed activity data and using them to suggest major modifications to try. When this is done in a stepwise fashion, either with or without extensive input from ligand-based computational programs, it sometimes goes by the name of *scaffold morphing*. An early, successful application of this approach has been claimed for the angiotensin II receptor antagonist valsartan (Figure 8.13). Based on the structure of the clinical drug losartan (Figure 8.10), a postulated model of how that compound might mimic angiotensin II was developed, and an energy-minimized conformation of each was found to overlap, lending support to the hypothesis. In a reversal of the usual strategy by which peptides are made more druglike, the drug losartan was made more *peptide-like* by replacement of the imidazole ring with a valine. Although the data was best explained by modifying the initially hypothesized model, a new antihypertensive drug, valsartan (still, like losartan, bearing a tetrazole bioisostere), resulted from these efforts.[34]

Structural information about the binding of different ligands to a receptor can be useful for generating ideas on how to combine or graft parts of them together to improve properties. This increasingly popular technique is sometimes also referred to as scaffold hopping. A recently reported example is shown in Figure 8.14. Compound **2** was a subnanomolar renin inhibitor but had problems with CYP inhibition (see Chapter 10) and poor PK

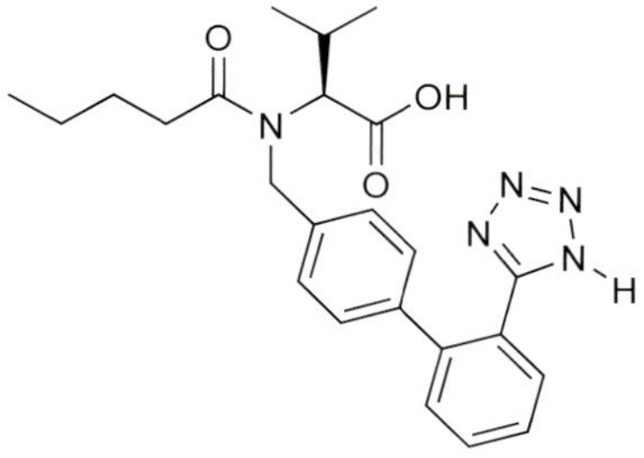

**Figure 8.13** ▶  The angiotensin II receptor antagonist Diovan (valsartan).

properties. Compound **4** was a smaller, but less potent (4 μM) renin inhibitor. Multiple H-bonding interactions of its diaminopyrimidine ring with the target were seen in the X-ray crystal structure. Overlaying a docked structure of **2** in the enzyme's active site with the crystal structure of **4** showed that the aniline phenyl ring of **4** and the aryl part of the tetrahydroquinoline in **2** nearly overlapped. This suggested the strategy of grafting the "head" of **4** onto the tail of **2**. The result, **5**, gratifyingly proved to be a potent, low-molecular-weight inhibitor with low CYP inhibition.[35]

These kinds of approaches get a lot of use these days. They also point to the fact that the boundaries between bioisosterism, scaffold hopping, structure-based drug design, and simple successive structural variations are subject to a great deal of overlap. This brings up another important point. Arguably the most frequently applied and successful strategy for transmuting a non-proprietary hit or lead compound into a patentable one is via heterocyclic ring replacement. Consider the well-known PDE5 inhibitors Viagra (sildenafil) and Levitra

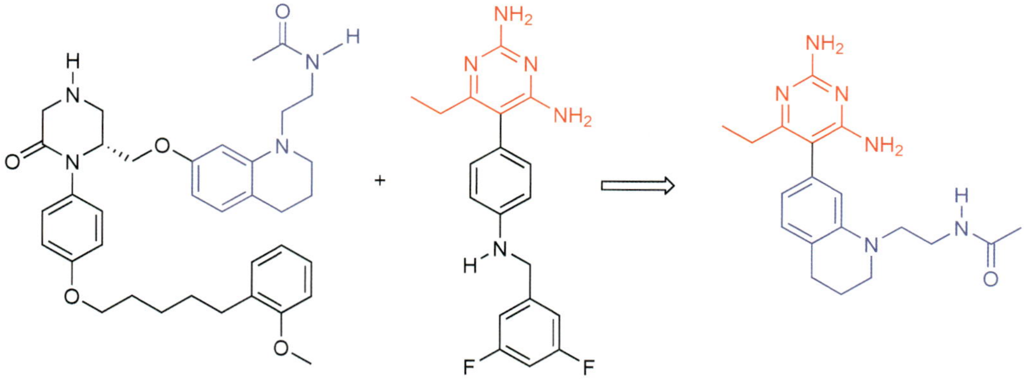

|  **2** | **4** | **5** |
| 0.50 nM | 4 μM | 178 nM |

**Figure 8.14** ▶  Combining structural features of two renin inhibitors, **2** and **4**, to produce a hybrid structure, **5**, combining the best of both. Renin inhibition IC$_{50}$ values are listed below the compound numbers. (Holsworth, D.D. Discovery of 6-ethyl-2,4-diaminopyrimidine-based small molecule renin inhibitors. *Bioorg. Med. Chem. Lett.* **2007**, *17*, 3575–3580.)

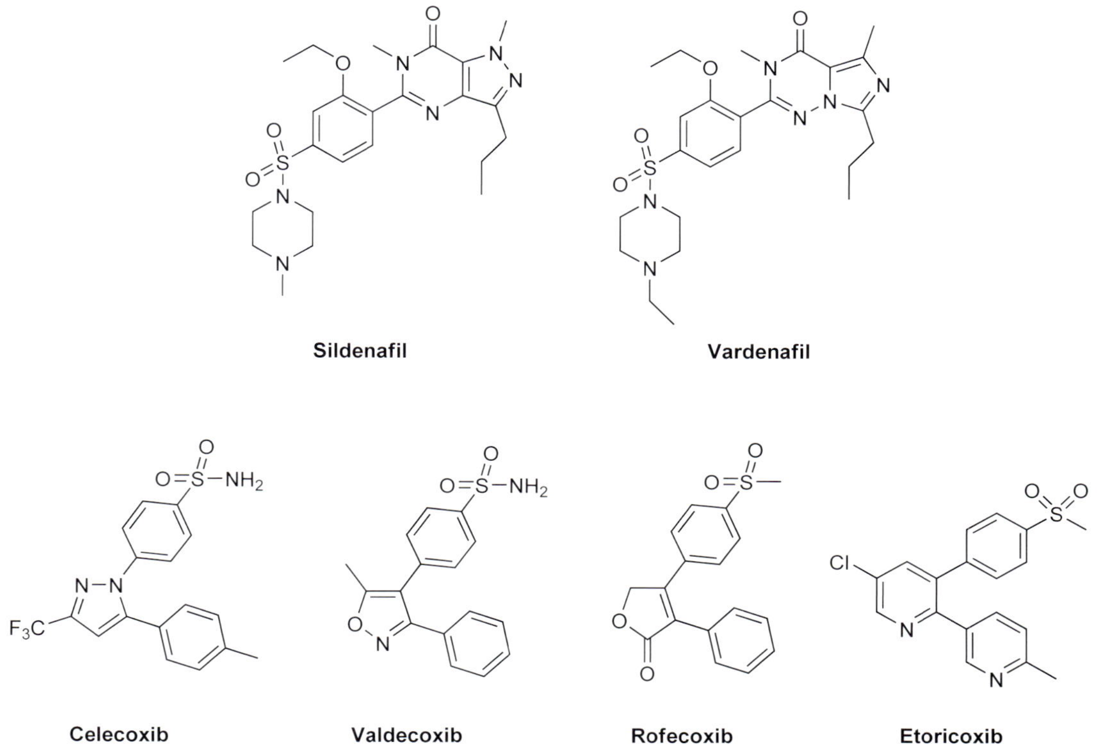

**Figure 8.15 ▶** Top: PDE5 inhibitors sildenafil and vardenafil. Bottom: COX2 inhibitors Celebrex (celecoxib), Bextra (valdecoxib), Vioxx (rofecoxib), and Arcoxia (etoricoxib).

(vardenafil), both shown in Figure 8.15. Except for an *N*-methyl versus *N*-ethyl piperazine substitution, the only difference between these two drugs lies in the fused heterocyclic core, both of which are isosteres of the purine ring. Vardenafil, which turns out to be more potent than sildenafil for reasons which aren't obvious, seems to be the result of an extensive SAR study of different heterocyclic scaffolds.[36]

Chemists will understand that the "simple" change of a heterocyclic scaffold can involve completely different and often difficult synthetic routes. Those who've tried these kinds of substitutions will also understand that most of the time such a change will result only in a massive loss of activity, and modelers will understand that it's usually difficult to say why. So no one should view any proposed heterocyclic scaffold replacement as "a sure thing". That being said, apparent examples of heterocyclic core switching can also be found in the COX2 inhibitors shown in Figure 8.15 like celecoxib, valdecoxib, rofecoxib, and etoricoxib.

## 8.4.3 Cyclization and Ring Opening

Modifying a structure by forming a new ring has several potential advantages. Most importantly, if not part of the "prior art" it can be a good way of gaining structural novelty. But it can also lead to improvements in potency, either through giving a better "fit" with the

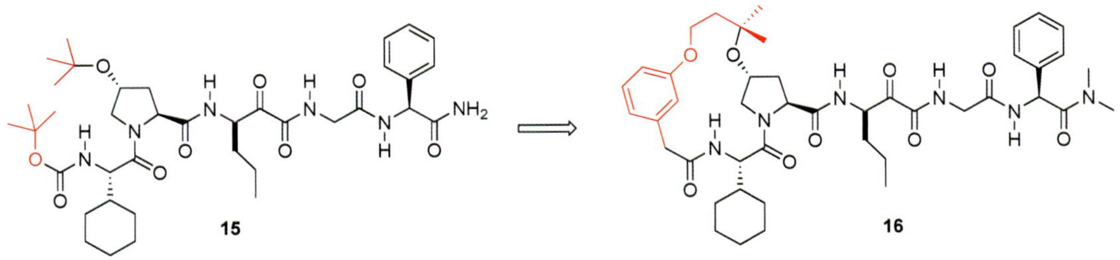

**Figure 8.16** ▶ Cyclization of an HCV NS3 protease inhibitor. (Chen, K.X., et al. Novel potent hepatitis C virus NS3 protease inhibitors derived from proline-based macrocycles. *J. Med. Chem.* **2006**, *49*, 995–1005.)

enzyme or by freezing the conformation into the necessary binding conformation, thus eliminating the entropy penalty that must be paid by floppy ligands, a factor discussed in the SBDD section. Other advantages, like improvements in PK properties and/or a reduction in toxicity (e.g. through decreased binding affinity for hERG receptors as discussed in Chapter 10) might also accrue.[37]

A good example of how cyclization can be used in drug design involves some research focusing on inhibitors of Hepatitis C NS3 Protease that was done at Schering-Plough. Like HIV, HCV codes for a polyprotein required for proper processing, in this case a serine protease with a shallow binding pocket that makes it a challenging but important target for drug discovery. The α-ketoamide pentapeptide, **15** (Figure 8.16), a 220 nM inhibitor, "lines up" with the enzyme so that the highly electrophilic central carbonyl group can interact with the active site serine residue. Modeling it into the crystal structure of the enzyme showed the unexpectedly close proximity of the two *tert*-butyl groups shown in red in Figure 8.16, which suggested the strategy of linking to form a macrocyclic ring. Compound **16**, a 16-membered macrocycle, proved to be a 6 nM inhibitor of the enzyme. An X-ray structure showed the expected complementarity.[38]

Conformational constraints can also be introduced into a molecule via intramolecular hydrogen bonding or by including substituents to hinder rotation about key bonds, but although such changes may be advantageous for other reasons, it's harder to argue that they're structurally more novel than a compound resulting from cyclization.

The opposite strategy, that of opening up a ring in a cyclic scaffold to generate acyclic analogs, is less common but also works at times. Consider the drugs acyclovir (used to treat herpes simplex virus, HSV) and ganciclovir[39] (used to treat retinitis caused by cytomegalovirus, CMV) shown in Figure 8.17. Both can be thought of as analogs of guanosine where the ribose ring has been opened up. Although obviously not the result of hits in search of structural novelty, they do illustrate the fact that in some cases ring opening can provide active compounds and even real drugs.

A recent example of ring opening comes from research into urotensin-II receptor agonists. Cell-based high-throughput screening turned up the isochromanone AC-7954 (Figure 8.17), a selective 300 nM agonist of this GPCR.[40] Opening up the heterocyclic ring provided **17**. One might anticipate major differences in conformation for these compounds, but they proved to be almost equipotent. Further optimization of the acyclic series led to the more potent amide **18**.[41]

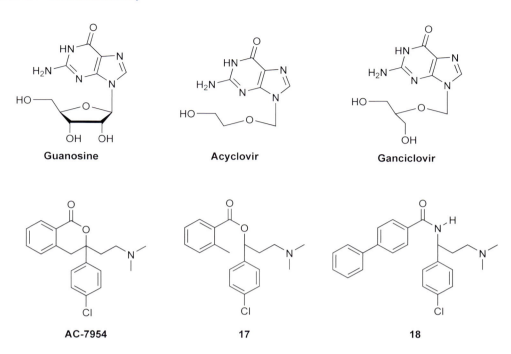

**Figure 8.17 ▶** Top: Guanosine and the acyclic nucleoside drugs Zovirax (acyclovir), and Cytovene (ganciclovir). Bottom: three urotensin-II receptor agonists. (Croston, G.E., et al. Discovery of the first nonpeptide agonist of the GPR14/Urotensin-II receptor: 3-(4-Chlorophenyl)-3-(2-(dimethylamino)ethyl)isochroman-1-one (AC-7954). *J. Med. Chem.* **2002**, *45*, 4950–4953.)

So, although they may need to fight entropy effects, ring opening might still provide potent compounds that have a major advantage in being clearly distinct from their cyclic counterparts.

## 8.4.4  Other Methods

Optimization of other properties can often lead to structural novelty as an unanticipated but much appreciated side effect. An example of this can be seen in the research that led to the HIV protease inhibitor tipranavir, a drug remarkably different from others in the class. High-throughput screening originally turned up the anticoagulant warfarin, a coumarin, as a 30 μM hit.[42] Following this up through similarity searching of the Upjohn compound collection led to the discovery of phenprocouman (Figure 8.18) as a 1 μM inhibitor. Using structure-based drug design, the coumarin ring system was replaced with a substituted pyrone, as shown for the 38 nM HIV protease inhibitor PNU-96988. An X-ray crystal structure obtained with an analog of this compound revealed a hydrophobic binding pocket near C-6 of the pyrone ring that disubstitution at that position should be able to fill.[43] Optimization of the resulting dihydropyrone series led to tipranavir, an 8 pM inhibitor of HIV-1 protease.[44] Here the progression from coumarin to pyrone to dihydropyrone, which some might consider a case of scaffold hopping or morphing, seems to have been the natural result of potency optimization guided by rational drug design.

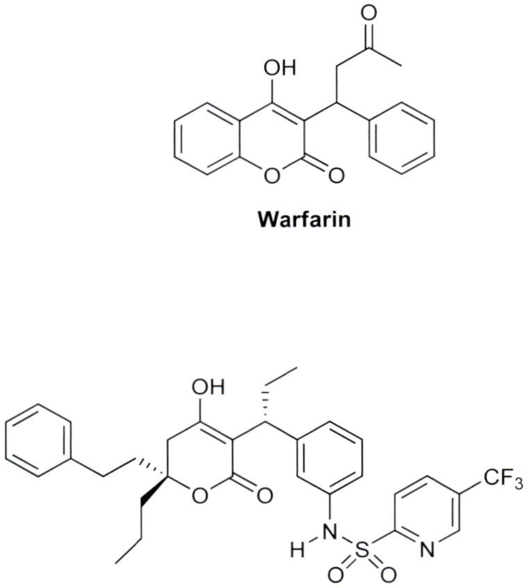

**Figure 8.18** ▶   Scaffold progression in the discovery of the HIV protease inhibitor Aptivus (tipranavir). (Turner, S.R., et al. Tipranavir (PNU-140690): A potent, orally bioavailable nonpeptidic HIV protease inhibitor of the 5,6-Dihydro-4-hydroxy-2-pyrone sulfonamide class. *J. Med. Chem.* **1998**, *41*, 3467–3476.)

In summary, we can see that just as no single method of generating hits was enough, so too a variety of approaches to building in structural novelty is needed. This can only come about through the combined efforts of medicinal chemists, computational scientists, and experts in that most relevant field of all, patent attorneys or agents (Box 8.3).

---

**Box 8.3   Key Points About Structural Novelty**

▶ Hits often lack the structural novelty needed for composition-of-matter patents

▶ Methods useful for building in structural novelty include

    Replacing parts of a molecule with bioisosteres

    Scaffold hopping and scaffold morphing

    Compound cyclization, often using SBDD

    Ring opening

▶ Medicinal chemistry, computational chemistry, and legal are key players in ensuring structural novelty

## 8.5  Solubility

### 8.5.1  Defining, Estimating, and Measuring Solubility

Solubility ranks among the most important properties a compound needs to have in order to become a drug. Most chemists have seen the contrast between compounds so soluble in water that no organic solvent can extract them out and compounds so insoluble in everything that they won't even dissolve well enough in DMSO-$d_6$ to give a decent NMR spectrum. To no one's surprise, solubility is highly solvent- and pH-dependent, and "aqueous" solubility can be very different for pure water, saline,[45] buffer, a few percent DMSO in buffer, simulated gastric fluid (SGF), or simulated intestinal fluid (SIF), all of which may be important at different times in the drug discovery process.

Importantly, two different kinds of "solubility" encountered in drug discovery need to be distinguished. *Thermodynamic* or *equilibrium solubility* is the true amount of compound that will go into solution from the solid phase given an infinite amount of time. This is approximated by shaking it in a flask with the solvent for 24–72 h,[46] filtering or centrifuging, and then quantitating the amount dissolved by using HPLC standards. Obviously this is not a high-throughput method.

*Kinetic solubility*, which might also be termed *apparent solubility*, is more commonly encountered in drug discovery, but is unfortunately *not* the same as equilibrium solubility, which can be orders of magnitude lower. Kinetic solubility is typically observed when a stock solution of a compound dissolved in DMSO is diluted into water or aqueous buffer, as is done in preparing HTS samples. Although precipitation might not be observed immediately, it often happens a few hours or days later. Here a supersaturated solution is formed upon dilution, which eventually reaches equilibrium but takes some time to do so. In contrast to equilibrium solubility, though, kinetic solubility is easy to measure quickly, making it more suitable for high-throughput determination using a method like nephelometry or turbidimetry.[47] For this reason, when a medicinal chemist gets "solubility data" on large numbers of compounds on a regular basis, which is in fact useful data to have, it's probably kinetic solubility that's been measured.

As previously stated, solubility can also be predicted by a number of different programs. The simplest method tries to correlate LogP and melting point with solubility, and things get more sophisticated from there. Reviews of this subject have recently been published.[48–49] As also mentioned previously, large errors are not uncommon for such programs, which, however, "can provide estimates to well within an order of magnitude in favourable cases."[49] Solubility programs don't get much better than that for a very good reason: the strength of crystal–packing interactions needs to be known to correctly predict it, and the type of crystal lattice that a given compound might adopt is almost impossible to foresee.

### 8.5.2  Problems Resulting from Poor Solubility

The dependence of solubility on the particular form of the solid involved has some important implications for drug discovery. An amorphous solid, often the result of lyophylization, "crashing out", or evaporating chromatography fractions, for example, will be much more soluble than the same compound when crystalline. Even when crystalline, different polymorphic forms of the same compound will have different solubilities, with the most stable polymorph often being the last to form and the least soluble. Years ago, most research compounds were recrystallized to get sharp melting points and pass elemental analysis, but this old-fashioned approach has not survived in today's age of parallel synthesis, automated analyses, and lower purity requirements. The upshot is that compounds made today are likely to possess kinetic solubilities that far exceed the true equilibrium solubilities of the most stable polymorphs of the pure materials. But if this means that compounds are more likely to be soluble enough to at least test than would otherwise be the case, why is it necessarily a bad thing? Couldn't amorphous compounds even be useful in the clinic precisely *because* they're more soluble?

Several amorphous drugs are on the market, but the danger is that in the long run thermodynamics might kill such an approach, as Warner-Lambert is said to have discovered during their clinical trials with atorvastatin, a calcium salt. Reports have it that this blockbuster compound was originally formulated as an amorphous salt, but during Phase III studies it began to crystallize, with a change in physical properties that might have proven disastrous for its development. Fortunately, additional studies were able to demonstrate the suitability of the crystalline form, which is what's used today.[50] Few companies, however, would want to gamble on being that lucky.

Why is solubility so important in drug discovery and development? For one thing, experience has shown that good solubility is important for oral bioavailability. In an analogy to the argument for low molecular weights illustrated by Figure 7.7, it's been observed that marketed drugs in general are much more soluble than research compounds. While, by one estimate, up to 50% of screening compounds may have aqueous solubilities <10 $\mu$M,[51] most oral drugs have aqueous solubilities exceeding 100 $\mu$M (see below). Poor solubility and differences between kinetic solubilities and equilibrium solubilities can cause problems at many points in the drug discovery process. In the HTS stage, precipitation of compounds after a few hours or days means re-running wells in daughter-plates to confirm whether initial actives might truly be "hit or miss"! The compound in a given well might have precipitated from solution since the first assay was run. Making up and retesting a fresh solution would give more reproducibility, but many would argue that a compound that insoluble to begin with is just better off being dropped as irreproducible.

Another problem, as Dr Lipinski has pointed out, is that poor solubility introduces uncertainty into the concentrations of compounds in the wells of HTS plates. The nominal value, say 10 $\mu$M, might really be an upper limit, with many compounds present at lower concentrations. This in turn limits the significance of the value obtained for a given well and speaks against any attempt to derive SAR or compare chemotypes that's based on primary screening data.

Solubility becomes even more of an issue farther down the road, when animal testing begins. Here less-than-ideal compounds will be tested on the road to proof of concept for either a target or a lead series. Formulations scientists are faced with finding some way of getting them into solution or at least a stable suspension. Although in vivo results obtained for a solution formulation don't depend on whether the material started out crystalline or amorphous, it's frequently observed that amorphous material is kinetically soluble enough to provide a suitable formulation that can show good bioavailability. A crystalline form of the same compound, however, won't dissolve to anywhere near that extent. At that point, rushing back to the lab to intentionally resynthesize the compound in amorphous form is probably not a good idea. It may not be a simple matter to do this again since airborne crystals may now be present to seed anything made on site, and even if new amorphous material is delivered to formulation it might crystallize out in the vehicle.

A more realistic approach to the "second batch effect" is to use a different type of formulation, often at a much lower concentration, or a suspension formulation, although results for these can be dramatically worse for the same compound. Project chemists might initially find that amorphous XYZ-9057 has an oral bioavailability of 40% (solution formulation) only to see that number go down to 2% when it's resynthesized on a bigger scale and forms beautiful crystals only suitable for micronization and suspension. For this reason, formulations people will always prefer that one large batch of compound be available for all experiments. That way a new formulation—which would give results not directly comparable with those obtained with other formulations—doesn't need to be developed each time a new batch, possibly with a different crystalline form, is synthesized.

So the different solubilities of different solid forms can make compounds hard to test and results hard to compare, which may not stop those with little appreciation for formulation "subtleties" from making such comparisons. In addition, the kinds of formulations used in early in vivo studies may be very different from those acceptable for clinical use. "Heroic measures" like excipients that aren't likely to be used clinically (DMSO, NMP, etc.), or nanoparticles dispersions may have been used just so that an experiment can be run and some kind of data obtained. The inability to produce any formulation capable of delivering reasonable plasma levels of a compound means that underlying problems in metabolism, toxicity, etc. which may be inherent to the scaffold or series will go undetected although the need to improve solubility is highlighted.[51]

Ideally, all in vivo data would be obtained for each compound's most stable polymorph, but in the real world medicinal chemists rarely make them and analytical chemists are unlikely to run slow, labor-intensive solubility experiments on the amorphous compounds and metastable polymorphs that they do make. Kinetic solubility measurements run in HT mode coupled with equilibrium solubility determinations for only a few compounds are the best that one can expect right now. The use of high-throughput crystallization and formulation techniques early on in drug discovery has been suggested, but not, so far, widely adopted.[50]

### 8.5.3  Improving Solubility

#### 8.5.3.1  Molecular Modifications

One might wonder how soluble in water a compound needs to be to become a drug. Figure 8.19, from some work done at Pfizer, shows minimum suggested equilibrium aqueous solubilities for clinical drugs.[52] Note that the solubility required is highly dependent on potency and cell permeability. Minor deficiencies in any one of this triad of properties might be made up for by good values for the other two. Clearly, getting quick feedback on all three of these properties, and not just potency, is best for compound optimization.

Methods for improving solubility include lowering hydrophobicity by reducing logP or molecular weight (e.g. excising unnecessary alkyl groups if any) and adding in polarity (hydroxy groups, pyridines in place of benzenes, etc.). Another way is to reduce crystal–packing interactions by adding substituents that might force a molecule out of its planar orientation or change the normal bond angles between rings (e.g. adding an ortho substituent

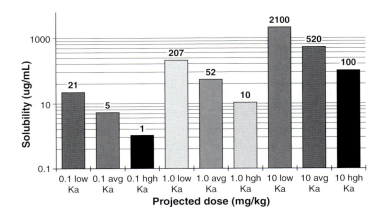

**Figure 8.19** ▶  Minimum acceptable solubilities for compounds with low, medium, and high permeability at a clinical dose. The middle three bars are for a 1 mpk dose where, with medium permeability, a solubility of 52 μg/mL is required (Reproduced with permission from Lipinski, C. "Aqueous Solubility in Discovery, Chemistry, and Assay Changes", Chapter 9 in van de Waterbeemd, H., Lennernäs, H., Artursson, P., Eds. *Drug Bioavailability*, (Wiley-VCH Verlag GmbH & Co. KGaA) p. 222, copyright 2003.).

to a biphenyl system).[53] The method most likely to have a dramatic effect, however, involves making a salt if an amine or carboxylic acid is already present in the molecule, or adding one in if it isn't. Salts are known to increase both equilibrium solubility and dissolution rate, another property which can become important down the road. Salts, like drugs, aren't all the same, and much, from solubility to hygroscopicity to FDA acceptability, depends on the counterion used. An example of this can be seen with codeine, where the aqueous solubilities (in mg/ml) are 8.3, 33, and 445 for the free base, sulfate salt, and phosphate salt, respectively.[54]

Another interesting reason for using salts instead of free bases clinically is that the presence of large amount of HCl to solubilize a free base in the patient's stomach can't always be counted on. The HIV protease inhibitor indinavir (Figure 7.13), for example, went into the clinic as a crystalline free base, but poor bioavailability was observed for patients with *achlorhydria*, a lack of HCl in gastric juice which is seen with some diseases and especially among elderly patients. Use of a sulfate salt solved this problem for the drug, Crixivan.[55]

It should be pointed out that adding an amine or carboxylic acid to a molecule, which can do wonders for solubility, can also introduce new problems with other properties. For example, carboxylic acids are often glucuronidated and thus rapidly metabolized, amines sometimes make for CYP inhibition, and either one can contribute to poor absorption. All of this points once again to the need for early monitoring of the effects a given change has on many properties—parallel optimization.

### 8.5.3.2  Prodrugs

Some arguments that are often used to justify the further development of insoluble compounds are that ultimately formulations can come to the rescue through the use of gel caps, that some other, *parenteral* (non-oral) mode of administration can be used, or that prodrugs can be made to increase solubility. Gel caps have worked at times but come with some severe

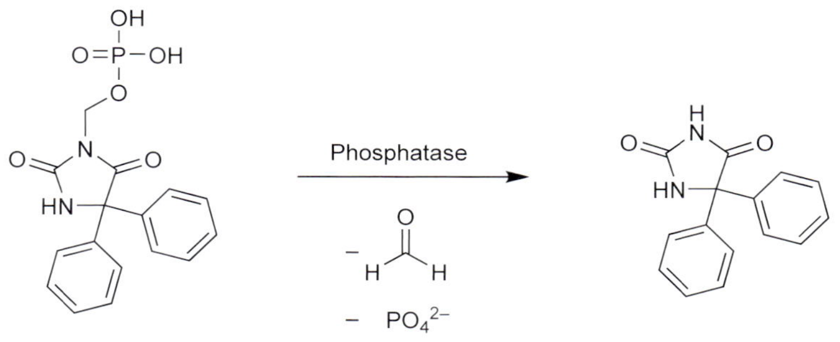

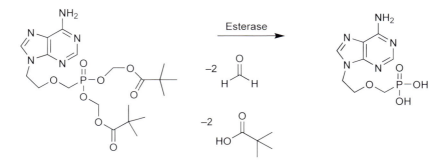

**Figure 8.20** ▶  The water soluble prodrug fosphenytoin (left) is converted by phosphatases to the active component at right, the water insoluble anticonvulsant phenytoin.

limitations.[56] Owing to the acceptable volume of a solution that can be injected, IV dosing usually requires much *higher* aqueous solubility than oral dosing.

Prodrugs made to increase solubility, however, can sometimes be a viable option. One example of this, shown in Figure 8.20, is the anticonvulsant prodrug fosphenytoin. This compound, administered IV as the disodium salt, has an aqueous solubility of 142 mg/ml and is hydrolyzed in vivo by phosphatases, decomposing to the active component phenytoin, which is "practically insoluble in water".[57] However, "one of the problems in this approach is that solubilizing groups may sometimes generate toxic effects. Thus, the phosphate prodrug fosphenytoin was designed to enhance the aqueous solubility of phenytoin for intravenous administration, but the high phosphate concentrations at the site of injection produced some hypocalcemic effects leading to mild pruritis and paresthesia," in other words, itching and burning.[58]

A chemist looking at the chemical reaction involved might expect the formaldehyde produced to be a bigger problem, but this, although far from non-toxic, is also a byproduct produced by other prodrugs such as the hepatitis B (HBV) drug adefovir dipivoxil (Figure 8.21).[59] But note the very different types of prodrugs fosphenytoin and adefovir pivoxil represent. In the former, the phosphorous-containing moiety is there only to increase solubility and is not part of the active species produced, while in the latter it

**Figure 8.21** ▶  The antiviral agent adefovir dipivoxil, a POM ester, is converted by esterase into its active component, adefovir. Two molecules of formaldehyde and two of pivalic acid are produced for each molecule of drug, the pivalic acid actually being a bigger concern for toxicity than formaldehyde.

---

**Box 8.4  Key Points About Solubility**

▶ Kinetic solubility is easy to measure in high-throughput fashion

▶ Thermodynamic solubility is harder to measure but ultimately more relevant

▶ Solubility for a given compound depends on its crystalline or amorphous form

▶ Inadequate solubility can

    Make for poor absorption

    Lead to erroneous SAR and poor compound prioritization

    Mask other problems

▶ The minimum solubility a drug will need depends on its potency and permeability

▶ Poor solubility can be improved by

    Reducing hydrophobicity

    Adding polarity

    Disrupting crystal packing interactions

    Making appropriate salts

    Using a prodrug approach

---

stays there after de-esterification, being essential to activity, an example of prodrugs designed to get around a different problem, cell permeability, discussed in the next chapter. In either case, though, the potential for toxicity inherent in the "excess baggage" that makes most prodrugs possible is something that needs to be kept in mind.

So, although prodrugs can sometimes help, it makes more sense to develop soluble candidates in the first place than to count on some sort of *deus ex machina* to come to the rescue late in the project. Using solubility data, even if only kinetic, in prioritizing compounds and designing analogs, thus investigating *QSPR* (*quantitative structure–property relationships*) in parallel with QSAR is the best way to do that (Box 8.4).

## 8.6  Chemical and Plasma Stability

### 8.6.1  Definitions and Importance

Like insolubility, compound instability can make assay results false and potentially misleading. Unlike insolubility, though, compound instability isn't necessarily visible to the naked eye, so if it isn't looked for (usually by HPLC), it won't be found. There are, however, some structural features that can often reduce stability under solution conditions at various pH values, which we'll call *chemical stability*, or stability in blood or its components, generically called *plasma stability* herein. Often, both chemical stability and plasma stability are found wanting for a given compound, the same transformation being seen in both. Another major stability issue, *metabolic stability*, is an important in vivo consideration in drug discovery as well. It's sometimes used as early as the HTL stage to prioritize and improve compounds, and is discussed in the next chapter.

Drugs need to be stable at every stage of discovery and development. Besides confounding early biochemical assay values, instability will affect everything thereafter including cellular values, in vivo results, a compound's ability to survive the pH range in the GI tract, and even shelf life. Interestingly, in some cases (prodrugs and *antedrugs* as described later), a specific type of instability in a particular compartment can actually be engineered into a drug, which makes its stability determination especially important, but in most cases unstable compounds will go nowhere. It's most important, then, to identify unstable compounds as early as possible. Some common sources of chemical and plasma instability are described below.[60]

## 8.6.2 Common Types of Instability

### 8.6.2.1 Oxidative Instability

Organic chemists are quite aware of the tendency of ether and THF to form hydroperoxides and of thiols to form disulfides given a little bit of air and a lot of time. These types of reactions usually occur in solution by free radical mechanisms and can be catalyzed by various metal ions or light. Similarly, amines can be converted to *N*-oxides, thioethers to sulfones or sulfoxides, and aldehydes to carboxylic acids. Not surprisingly, these things can happen to hits, leads, and drugs if such functional groups are present. An example of this is shown in Figure 8.22.

The ACE inhibitor captopril contains a sulfhydryl group and is therefore subject to disulfide formation both in vitro and in vivo. Although chemically stable enough to make a fine drug, when captopril was incubated with rat, dog, or human plasma for 40 min, only small amounts of unchanged material was recovered, and three products, the homo-disulfide, mixed disulfide with cysteine, and the mixed disulfide with glutathione were found (cysteine and glutathione being two of the most abundant sources of nucleophilic thiol residues in vivo). The same products were detected using [$^{14}$C]captopril in man. It was hypothesized that these products function, along with mixed disulfides formed on reaction with cysteine residues on plasma proteins, as a kind of depot for captopril, all of them being able to regenerate the drug on reaction with another thiol (cysteine or glutathione) or under enzymatic reductive conditions in vivo.[61]

Although in this case plasma instability didn't prevent the development of the drug, it did introduce complications that later ACE inhibitors which lacked a thiol group were able to avoid.

### 8.6.2.2 Chiral Instability

As seen earlier, chiral purity is important since drugs are increasingly being introduced as single stereoisomers. For drugs that are racemic to begin with, the equilibration of each epimer amounts to stability! Not so for those introduced as single stereoisomers, which can racemize or undergo chiral inversion. This often doesn't require the assistance of any protein, but as in the case of captopril, where cysteine residues on plasma proteins can be involved, the basic sidechains of arginine and lysine on albumin may be able to catalyze racemization for some drugs.[62]

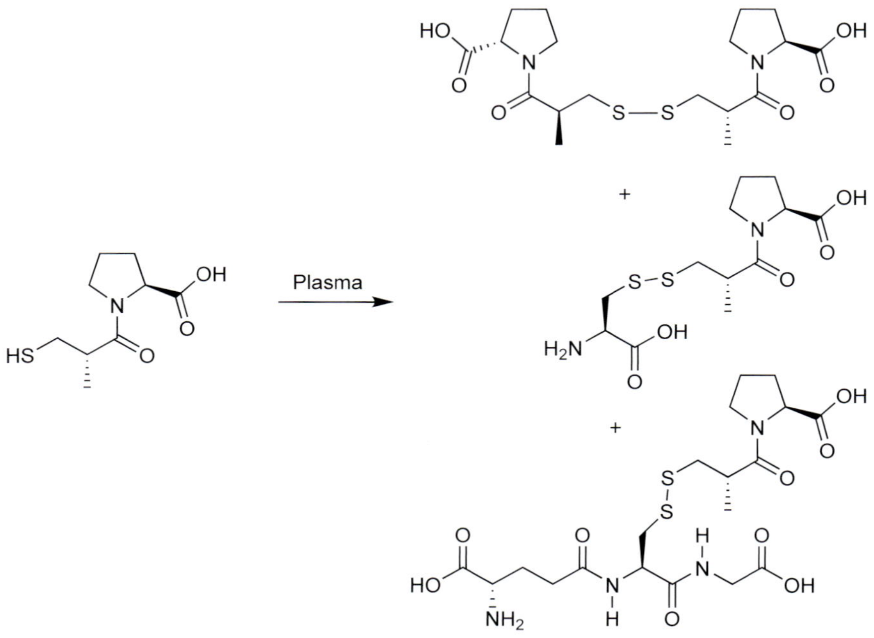

**Figure 8.22 ▶** Incubation of the ACE inhibitor captopril (left) with plasma leads to formation of the homodimer (top right), and mixed disulfides with cysteine (middle right) and glutathione (lower right).

The usual factors governing racemization rates apply: electron withdrawing groups at positions adjacent to a chiral methine can increase acidity and facilitate epimerization as in the case of moxalactam (Figure 8.23), an antimicrobial compound with a moderate ability to cross the blood–brain barrier. Moxalactam is racemic at the chiral center adjacent to the phenyl ring, but the *R*-epimer has about twice the activity of the *S*-isomer. They interconvert, however, in buffer or serum, with a half-life of about 1.5 h, making any attempt to use the pure *R*-epimer problematic.[63] For the same reason the immune modulator Actimid (an analog of thalidomide, Figure 8.23), for which the *S*-epimer has been reported to be more active than the *R*-epimer, is being developed as a racemic mixture.[64] Chiral drugs with a propensity to epimerize in vivo can be developed, but present complications that are best avoided, if possible.[65]

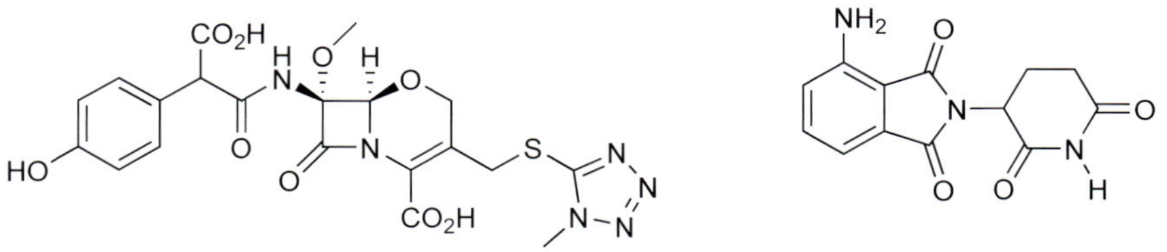

**Figure 8.23 ▶**  Left: Moxalactam. Right: Actimid.

Since epimerization can obviously afford compounds which are less active, more toxic, and/or differ in their PK properties, chiral lability is a liability and an issue that needs to be identified early on. If such instability is suspected, stability determinations should be done at several pH values and in plasma, with analysis by chiral HPLC or chiral shift NMR studies being, of course, necessary for enantiomers.

### 8.6.2.3 Hydrolytic Instability

Simple hydrolysis is probably the most commonly seen reason for compound instability. This can be a problem for certain functional groups like esters, amides, carbamates, sulfonamides, lactones, and lactams. Hydrolytic instability for such compound spans the range of simple, pH-dependent, non-enzymatic hydrolysis to hydrolytic cleavage of peptides by specific peptidases. Often, a compound with slight hydrolytic instability in buffer will be much more unstable in plasma, a rich source of esterases, lipases, and peptidases. For this reason profiling any compound of interest that contains one of the functional groups mentioned above in both assays might save the researcher later grief. High-throughput stability assays have been developed and are fairly common in drug discovery these days.[66]

Although usually shunned, hydrolytic instability, like lack of selectivity, can in some cases actually be desired, assuming it happens at the right place and at the right time. This is true for many prodrugs like adefovir dipivoxil in Figure 8.21. Here stability to esterases, enzymes necessary to unmask the active component, would make for an inactive compound, unable to generate the necessary drug. So too, though, would lack of stability to the range of pH values found in the GI tract, which would prevent oral absorption. So these kinds of prodrugs need to walk a fine line.

Another type of drug that might benefit from hydrolytic instability under certain specific conditions is called an *antedrug*. While the prefixes "pro" and "ante" aren't terribly different in meaning, there's a big difference between what's meant by "prodrug" and "antedrug". "The prodrug undergoes a predictable metabolic activation prior to exhibiting its pharmacological effects in a target tissue while the antedrug undergoes metabolic deactivation in the systemic circulation upon leaving a target tissue."[67] A good example of how the latter works can be seen in a paper describing the potential use of some matrix metalloprotease inhibitors (MMPIs) for psoriasis. Clinical MMPIs to date have tended to show systemic toxicity, so the idea was to use a topically applied one to maximize dose at the site of action and design in instability in blood to prevent systemic effects for any of the compounds that might get through to it. Figure 8.24 illustrates this concept and shows how the ester group was "tweaked" by electron withdrawing fluoro-substituents to achieve rapid breakdown in plasma. All of the compounds were found to be stable in buffer.[68]

That simple hydrolytic instability is an issue many drugs face can be proven by simply opening up a bottle of aspirin and sniffing. Here the hydrolysis of the acetyl group is slow enough so that it doesn't present too much of a problem, but the process is much faster for many hits and leads. Stability in blood is often much more of a challenge for compounds. Quite a few enzymes, most notably esterases but also peptidases and phosphatases, are present there. These can be more powerful than many synthetic reagents. For example, the relatively hindered cyclohexyl ester group used for protection in peptide synthesis normally requires treatment with HF for removal, but, at least in some cases, in plasma de-esterification happens in minutes.[69]

Stabilities can differ between whole blood (which contains cells), plasma (the fluid portion of blood), and serum (the fluid portion of blood after coagulation). All three can be run, but

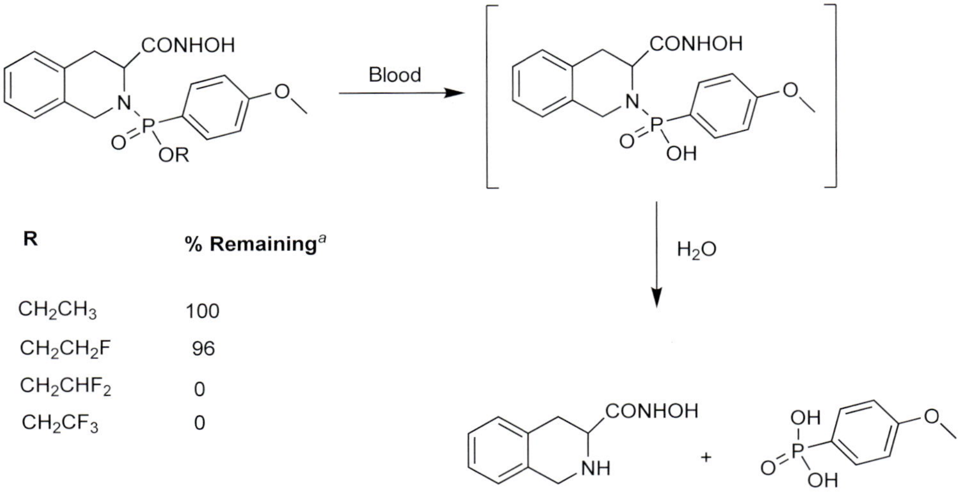

R | % Remaining[a]
--- | ---
CH$_2$CH$_3$ | 100
CH$_2$CH$_2$F | 96
CH$_2$CHF$_2$ | 0
CH$_2$CF$_3$ | 0

**Figure 8.24** ▶ An example of an antedrug. The metalloproteinase inhibitor shown at upper left is applied topically. Absorption into blood results in plasma esterase catalyzed de-esterification of the phosphoramidate leading to an unstable intermediate which spontaneously decomposes.[a] Plasma stability as percentage of parent compound remaining after 60 min in human plasma (data from Sawa, M., et al. New strategy for antedrug application: Development of metalloproteinase inhibitors as antipsoriatic drugs. *J. Med. Chem.* **2002**, *45*, 930–936.)

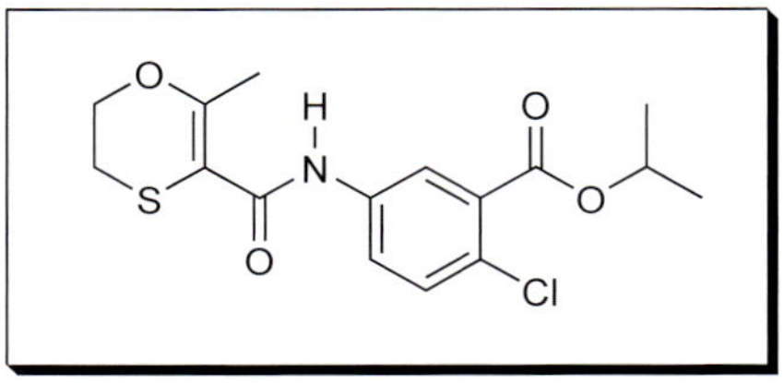

| Species | % Remaining plasma | % Remaining blood |
| --- | --- | --- |
| Mouse | 8 | 8 |
| Dog | 88 | 70 |
| Human | 98 | 69 |

**Figure 8.25** ▶ The non-nucleoside reverse transcriptase inhibitor NSC 615985 and percent remaining after incubation at 10 μg/ml for 24 h at 37°C in plasma and blood of the species listed (data from Nomeir, A.A., et al. Liquid chromatographic analysis in mouse, dog and human plasma; stability, absorption, metabolism, and pharmacokinetics of the anti-HIV agent 2-chloro-5-(2-methyl-5,6-dihydro-1,4-oxathiin-3-yl carboxamido)isopropylbenzoate (NSC 615985, UC84). *J. Pharm. Biomed. Anal.* **1998**, *17*, 27–38.)

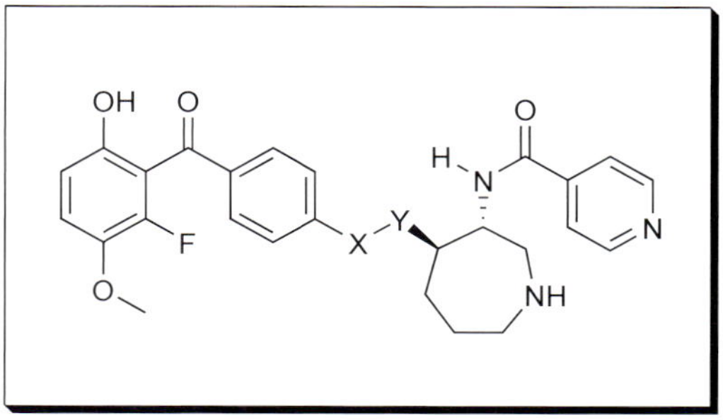

| X-Y | $t_{1/2}{}^a$ | IC$_{50}$ (PKB-$\alpha$) | IC$_{50}$ (PKA) |
|---|---|---|---|
| 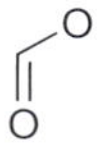 (ester) | <1 min | 5 nM | 5 nM |
| 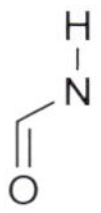 (amide) | 69 h | 4 nM | 2 nM |
| (ether) | 29 h | 355 nM | 39 nM |
|  (amine) | 161 h | 3000 nM | 800 nM |

**Figure 8.26** ▶ $^a$Activity and half-life in mouse plasma at 37°C for some PKB inhibitors (data from Breitenlechner, C.B., et al. Structure-based optimization of novel azepane derivatives as PKB inhibitors. *J. Med. Chem.* **2004**, *47*, 1375–1390.)

plasma is used the most frequently. Plasma stability is known to be species-dependent as Figure 8.25 demonstrates, so such assays should be run using plasma from the animal species to be used for in vivo studies as well as human plasma. The extent to which a compound will be sequestered by binding to proteins present in plasma, which can also have an effect on in vivo metabolism and efficacy, is similarly species-dependent as discussed in Chapter 9.

Much of the time the rule of thumb that says the smaller the animal, the faster it can metabolize holds true both in plasma and in the whole animal, but not always. Stability in plasma can even vary depending on the enzyme expression levels of the individual animal used, so it may be necessary to run controls on different batches. To address the importance of plasma stability to drug discovery as well as how low stability can be improved, some examples are in order. Three of these are given below:

*Example 1*: A series of azepane-containing inhibitors of protein kinase B (PKB) were made as potential anti-cancer compounds. A compound containing an ester linkage was cleaved almost instantly ($t_{1/2}$ <1 min) in mouse plasma, but an amide analog proved stable for days, and fortunately retained activity for the target (Figure 8.26).[70] Here the plasma stability of amide versus ester obviously parallels their reactivities toward aqueous hydrolysis.

*Example 2*: The natural product camptothecin proved to be a starting point for several important anti-cancer topoisomerase-1 inhibitors like irinotecan and topotecan (Figure 8.27). The E-ring lactone was shown to be critical for the activity of such compounds. In aqueous solution hydrolytic ring opening to the inactive carboxylic acid, although reversible, reduces the concentration of the active component considerably. The half-life of camptothecin in pH 7.4 buffer is only 17 min, with an 87:13 ratio of acid:lactone being formed. Worse still, physiological concentrations of human serum albumin (HSA) shift the equilibrium even farther toward the acid, with <1% of the lactone left intact, apparently due to preferential binding of the acid to HSA.[71] Interestingly, in mouse plasma no such effect is seen, and this may help explain camptothecin's better observed efficacy in mouse than in man.[72] Still more interestingly, considerably *more* of the lactone (5.3% versus 0.2%) was found at equilibrium in whole human blood rather than plasma, which was explained by hypothesizing increased partitioning into the hydrophobic lipid bilayer of erythrocytes where the compound was not subject to high concentrations of water.[73]

Taking this ball and running with it, researchers decided to *increase* the lipophilicity of camptothecin, even adding a *t*-butyldimethylsilyl (TBDMS) group to the final candidate as shown in Figure 8.27. For this compound, silatecan, increased lipophilicity to protect against hydrolysis by partitioning into bilayers combined with reduced affinity of the ring-opened form for HSA to give superior blood stability, with 30% of the lactone remaining at equilibrium after a long $t_{1/2}$ of 130 min.[74]

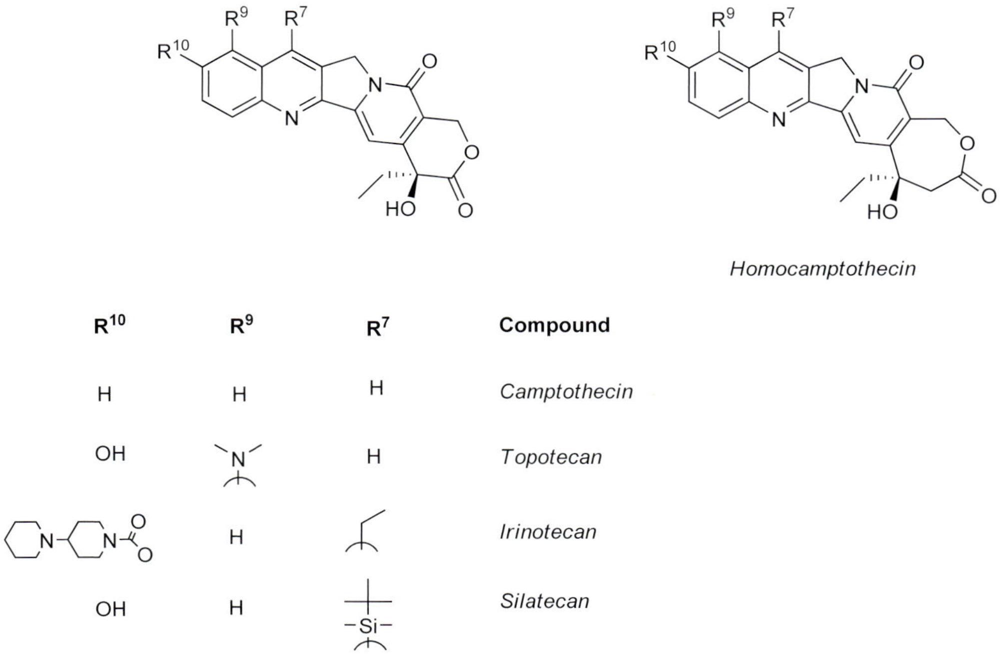

Homocamptothecin

| R$^{10}$ | R$^9$ | R$^7$ | Compound |
|---|---|---|---|
| H | H | H | Camptothecin |
| OH | (N) | H | Topotecan |
| (piperidine-piperidine-O) | H | (isopropyl) | Irinotecan |
| OH | H | (Si group) | Silatecan |

**Figure 8.27** ▶  Camptothecin and some analogous topoisomerase I inhibitors.

A second approach to deal with the instability of the camptothecin E-ring lactone was taken by researchers who decided to homologate its lactone to a 7-membered ring by inserting a methylene group between the carbonyl carbon and the alcohol-bearing carbon. By doing this it was hoped that electron withdrawing effects of the OH group, which might facilitate cleavage of the lactone, would be minimized. The compound they made, homocamptothecin (hCPT) proved to have a half-life of nearly 3 h in human plasma at 37°C, versus about 5 min for camptothecin itself, and it retained good activity against topoisomerase I.[75] Although opening of the 7-membered ring was of course irreversible in a practical sense, the concentration of the active component in blood remained far higher and for a longer time than it did for camptothecin. Thus from the same starting point, two different approaches, one based on a biological observation and the other based on organic chemistry theory, were able to address the stability problem. Both, of course, involved major synthetic efforts.

*Example 3*: Plasma instability can be a major barrier to the therapeutic use of small peptides. Even with IV administration, peptidases stand ready to degrade these compounds through amide hydrolysis. This made in vivo use of an interesting series of antiarrhythmic peptides (AAPs) impractical.[76] The hexapeptide AAP10, shown in Figure 8.28, had a half-life of only 11.8 min in human plasma.[77] But since peptides have been of interest as possible therapeutics for decades, strategies to address this issue via the use of peptidomimetics have been developed.[78]

One approach is simply to use D-amino acids, which peptidases may not recognize. Unfortunately, neither does the intended receptor much of the time. Another involves reversing the peptide sequence so that, for example, H-Phe-Leu-Ser-NH$_2$ becomes H-Ser-Leu-Phe-NH$_2$. These are the so-called *retro* peptides. Combining both of these, called the *retro-inverso* approach, produces molecules where the amide bond has been inverted (HNCO in place of CONH) while the sidechains are presented in a similar way as they were

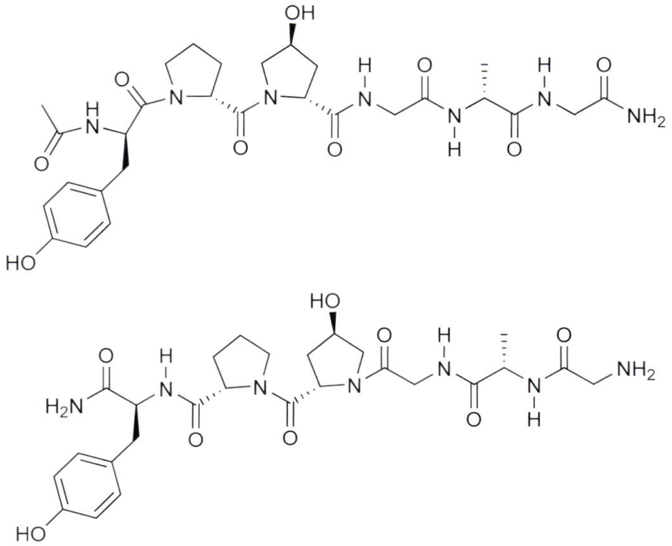

**Figure 8.28 ▶** Top: Rotigaptide (ZP123). Bottom: AAP10 (H-Gly-Ala-Gly-4Hyp-Pro-Tyr-NH$_2$).

in the original peptide.[79] Such a version of AAP10, called ZP123 or Rotigaptide (Figure 8.28), proved to be equipotent to it but had the remarkably long half-life of 14 days instead of 11.8 min. The compound has made it at least as far as Phase II studies. So overcoming a plasma stability problem via a retro-inverso approach was able to produce a compound that is, at least by one definition, druglike.

The three cases above provide examples of different ways of dealing with stability problems: by isosteric replacement and structural modifications, both major and minor. Most of the utility of plasma- and pH-stability testing in HTL, however, will probably come from its ability to raise warning flags about problems associated with particular structural motifs.

An example of how this might work was provided by researchers at Wyeth, where two scaffolds, *A* and *B* in Figure 8.29, were being pursued by a discovery team. Plasma stability studies showed good stability for *A* but rapid degradation for *B*. This information allowed the *B* effort to be dropped and increased resources to be put on *A*. Not having this information at an early stage could have led to wasted time and effort as well as unnecessary frustration. As the authors point out, such testing is also useful later on in selecting compounds for in vivo studies, examining prodrugs and antedrugs, and identifying degradation products (Box 8.5).[66]

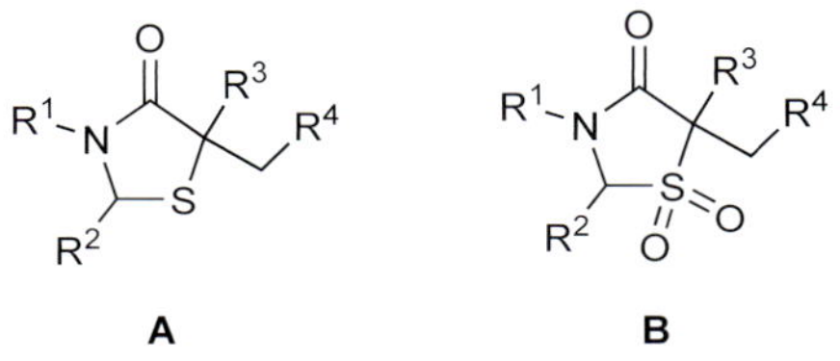

**Figure 8.29** ▶   Two scaffolds for plasma stability triage. (Di, L., et al. Development and application of high throughput plasma stability assay for drug discovery. *Int. J. Pharm.* **2005**, *297*, 110–119.)

---

**Box 8.5    Key Points About Chemical and Plasma Stability**

▶ Chemical and plasma stability screening can help to

    Identify problem compounds early

    Select compounds for in vivo studies

    Study prodrugs and antedrugs

▶ Three kinds of instability often encountered are

    Oxidative instability

    Chiral instability

    Hydrolytic instability

▶ Stability in plasma, serum, or blood is species dependent

▶ Stability can be enhanced by isosteric replacement or through wide variety of modifications

# Notes

1.  Overington, J.P., et al. How many drug targets are there? *Nat. Rev. Drug Discovery* **2006**, *5*, 993–996.

2.  Palmer, J.T., et al. Design and synthesis of selective keto-1,2,4-oxadiazole-based tryptase inhibitors. *Bioorg. Med. Chem. Lett.* **2006**, *16*, 3434–3439.

3.  Traxler, P., Furet, P. Strategies toward the design of novel and selective protein tyrosine kinase inhibitors. *Pharmacol. Ther.* **1999**, *82*, 195–206.

4.  See Bogoyevitch, M.A., Fairlie, D.P. A new paradigm for protein kinase inhibition: Blocking phosphorylation without directly targeting ATP binding. *Drug Discovery Today* **2007**, *12*, 622–633.

5.  Okram, B., et al. A general strategy for creating "inactive-conformation" Abl inhibitors. *Chem. Biol.* **2006**, *13*, 779–786.

6.  Gescher, A. Analogs of staurosporine: Potential anticancer drugs? *Gen. Pharmacol.* **1998**, *31*, 721–728.

7.  See www.cerep.fr/Cerep/Users/pages/ProductsServices/BioPrintServices.asp.

8.  Sealey, J.E., Laragh, J.H. Aliskiren, the first renin inhibitor for treating hypertension: Reactive renin secretion may limit its effectiveness. *Am. J. Hypertens.* **2007**, *20*, 587–597.

9.  Csermely, P. The efficiency of multi-target drugs: The network approach might help drug design. *Trends Pharmacol. Sci.* **2005**, *26*, 178–182.

10. Morphy, R., Rankovic, Z. Designed multiple ligands. An emerging drug discovery paradigm. *J. Med. Chem.* **2005**, *48*, 6523–6543.

11. Morphy, R., Rankovic, Z. Fragments, network biology and designing multiple ligands. *Drug Discovery Today* **2007**, *12*, 156–160.

12. Roth, B.L., et al. Magic shotguns versus magic bullets: Selectively non-selective drugs for mood disorders and schizophrenia. *Nat. Rev. Drug Discovery* **2004**, *3*, 353–359.

13. For more on this family tree, see Rosenblum, J.S., Kozarich, J.W. Prolyl peptidases: A serine protease subfamily with high potential for drug discovery. *Curr. Opin. Chem. Biol.* **2003**, *7*, 496–504.

14. Lankas, G.R., et al. Dipeptidyl peptidase IV inhibition for the treatment of Type 2 diabetes. *Diabetes* **2005**, *54*, 2988–2994.

15. Copeland, R.A., et al. Drug-target residence time and its implication for lead optimization. *Nat. Rev. Drug Discovery* **2006**, *5*, 730–739.

16. Rothman, R.B., et al. Evidence for the possible involvement of 5-HT$_{2b}$ receptors in the cardiac valvulopathy associated with fenfluramine and other serotonergic medications. *Circulation* **2000**, *102*, 2836–2841.

17. Whitebread, S., et al. *In vitro* safety pharmacology profiling: An essential tool for successful drug development. *Drug Discovery Today* **2005**, *10*, 1421–1433.

18. Wermuth, C.G. "Molecular Variations Based on Isosteric Replacements", Chapter 13 in *The Practice of Medicinal Chemistry*, Wermuth, C.G., Ed. (London, Academic Press, 1996).

19. Sheridan, R.P. The most common chemical replacements in drug-like compounds. *J. Chem. Inf. Comput. Sci.* **2002**, *42*, 103–108.

20. Wermuth, C.G., et al. "Bioisosteres", Chapter 2.16 in *Comprehensive Medicinal Chemistry* Triggle, D.J., Taylor, J.B., Eds. (London, Elsevier, 2007).

21. For a review on the isosteric use of Si, see Showell, G.A., Mills, J.S. Chemistry challenges in lead optimization: Silicon isosteres in drug discovery. *Drug Discovery Today* **2003**, *8*, 551–556.

22. Binder, D., et al. Analogues and derivatives of tenoxicam. 1. Synthesis and antiinflammatory activity of analogues with different residues on the ring nitrogen and the amide nitrogen. *J. Med. Chem.* **1987**, *30*, 678–682.

23. Fludzinski, P., et al. Indazoles as indole bioisosteres: Synthesis and evaluation of the tropanyl ester and amide of indazole-3-carboxylate as antagonists at the serotonin 5HT$_3$ receptor. *J. Med. Chem.* **1987**, *30*, 1535–1537.

24. Carini, D.J., et al. Nonpeptide angiotensin II receptor antagonists: The discovery of a series of *N*-(Biphenylmethyl)imidazoles as potent, orally active antihypertensives. *J. Med. Chem.* **1991**, *34*, 2525–2547.

25. Kim, K.S., et al. Discovery of aminothiazole inhibitors of cyclin-dependent kinase 2: Synthesis, X-ray crystallographic analysis, and biological activities. *J. Med. Chem.* **2002**, *45*, 3905–3927.

26. Black, W.C., et al. Trifluoroethylamines as amide isosteres in inhibitors of cathepsin K. *Bioorg. Med. Chem. Lett.* **2005**, *15*, 4741–4744.

27. Gauthier, J.Y., et al. The discovery of odanacatib (MK-0822), a selective inhibitor of cathepsin K. *Bioorg. Med. Chem. Lett.* **2008**, *18*, 923–928.

28. Mewshaw, R.E., et al. New generation dopaminergic agents. Part 8: Heterocyclic bioisosteres that exploit the 7-OH-2-(Aminomethyl)chroman D$_2$ template. *Bioorg. Med. Chem. Lett.* **2002**, *12*, 271–274.

29. Rewinkel, J.B.M., et al. 1-Aminoisoquinoline as benzamidine isosteres in the design and synthesis of orally active thrombin inhibitors. *Bioorg. Med. Chem. Lett.* **1999**, *9*, 685–690.

30. Brown, N., Jacoby, E. On scaffolds and hopping in medicinal chemistry. *Mini Rev. Med. Chem.* **2006**, *6*, 1217–1229.

31. For a practical review of scaffold hopping, see Zhao, H. Scaffold selection and scaffold hopping in lead generation: A medicinal chemistry perspective. *Drug Discovery Today* **2007**, *12*, 149–155.

32. Rush III, T.S., et al. A shape-based 3D scaffold hopping method and its application to a bacterial protein-protein interaction. *J. Med. Chem.* **2005**, *48*, 1489–1495.

33. Schneider, G., et al. "Scaffold-Hopping" by Topological pharmacophore search: A contribution to virtual screening. *Angew. Chem. Int. Ed. Engl.* **1999**, *38*, 2894–2896.

34. Bühlmayer, P., et al. Valsartan, a potent, orally active angiotensin II antagonist developed from the structurally new amino acid series. *Bioorg. Med. Chem. Lett.* **1994**, *4*, 29–34.

35. Holsworth, D.D. Discovery of 6-ethyl-2,4-diaminopyrimidine-based small molecule renin inhibitors. *Bioorg. Med. Chem. Lett.* **2007**, *17*, 3575–3580.

36. Haning, H., et al. Comparison of different heterocyclic scaffolds as substrate analog PDE5 inhibitors. *Bioorg. Med. Chem. Lett.* **2005**, *15*, 3900–3907.

37. For an example of cyclization used to reduce hERG binding in a series of farnesyltransferase inhibitors, see Bell, I.M., et al., 3-aminopyrrolidine farnesyltransferase inhibitors: Design of macrocyclic compounds with improved pharmacokinetics and excellent cell potency. *J. Med. Chem.* **2002**, *45*, 2388–2409.

38. Chen, K.X., et al. Novel potent hepatitis C virus NS3 protease inhibitors derived from proline-based macrocycles. *J. Med. Chem.* **2006**, *49*, 995–1005.

39. For an account of the discovery of ganciclovir, see Verheyden, J.P.H. "Ganciclovir", in *Chronicles of Drug Discovery, Volume 3*, Lednicer, D., Ed. (Washington, D.C., ACS Books, 1993).

40. Croston, G.E., et al. Discovery of the first nonpeptide agonist of the GPR14/Urotensin-II receptor: 3-(4-Chlorophenyl)-3-(2-(dimethylamino)ethyl)isochroman-1-one (AC-7954). *J. Med. Chem.* **2002**, *45*, 4950–4953.

41. Lehmann, F., et al. Novel potent and efficacious nonpeptidic urotensin II receptor agonists. *J. Med. Chem.* **2006**, *49*, 2232–2240.

42. Thaisrivongs, S., et al. Structure-based design of HIV protease inhibitors: 4-hydroxycoumarins and 4-hydroxy-2-pyrones as non-peptidic inhibitors. *J. Med. Chem.* **1994**, *37*, 3200–3204.

43. Thaisrivongs, S., et al. Structure-Based design of HIV protease inhibitors: 5,6-Dihydro-4-hydroxy-2-pyrones as effective, nonpeptidic inhibitors. *J. Med. Chem.* **1996**, *39*, 4630–4642.

44. Turner, S.R., et al. Tipranavir (PNU-140690): A potent, orally bioavailable nonpeptidic HIV protease inhibitor of the 5,6-Dihydro-4-hydroxy-2-pyrone sulfonamide class. *J. Med. Chem.* **1998**, *41*, 3467–3476.

45. Beware of making hydrochloride salts of bases for administration with saline: The common ion effect can make them very insoluble.

46. But consider what happens with a compound not entirely stable to hydrolysis for this long.

47. Kerns, E.H. High throughput physicochemical profiling for drug discovery. *J. Pharmacol. Sci.* **2001**, *90*, 1838–1858.

48. Delaney, J.S. Predicting aqueous solubility from structure. *Drug Discovery Today* **2005**, *10*, 289–295.

49. Dearden, J.C. *In silico* prediction of aqueous solubility. *Expert Opin. Drug Discovery.* **2006**, *1*, 31–52.

50. Gardner, C.R., et al. Drugs as materials: Valuing physical form in drug discovery. *Nat. Rev. Drug Discovery* **2004**, *3*, 926–934.

51. Lipinski, C.A. Avoiding investment in doomed drugs. *Curr. Drug Discovery* **2001**, *1*, 17–19.

52. This is taken from an excellent short review of the role of solubility in drug discovery, Lipinski's "Aqueous Solubility in Discovery, Chemistry, and Assay Changes", Chapter 9 in *Drug Bioavailability. Estimation of Solubility, Permeability, Absorption and Bioavailability* van de Waterbeemd, H., Lennernäs, H., Artursson, P., Eds. (Weinheim, Wiley-VCH, 2003).

53. For a brief analysis of what adding a 2-Cl substituent to a biphenyl system can do to bond angles and solubility, see Leach, A.G. Matched molecular pairs as a guide in the optimization of pharmaceutical properties; A Study of aqueous solubility, plasma protein binding, and oral exposure. *J. Med. Chem.* **2006**, *49*, 6672–6682, and supplementary information thereto.

54. Garad, S.D. How to improve the bioavailability of poorly soluble drugs. *Am. Pharmacol. Rev.* **2004**, *7*, 80–85.

55. Lin, J.H. Role of pharmacokinetics in the discovery and development of indinavir. *Adv. Drug Delivery Rev.* **1999**, *39*, 33–49.

56. Curatolo, W. Physical chemical properties of oral drug candidates in the discovery and exploratory development settings. *Pharm. Sci. Technol. Today* **1998**, *1*, 387–393.

57. According to the *Merck Index*, 13th edition (Whitehouse Station, New Jersey, Merck & Co., 2001).

58. Ettmayer, P., et al. Lessons learned from marketed and investigational prodrugs. *J. Med. Chem.* **2004**, *47*, 2393–2404.

59. For more about the toxicity of pivalic acid, another common prodrug byproduct, see Brass, E.P. Pivalate-generating prodrugs and carnitine homeostasis in man. *Pharmacol. Rev.* **2002**, *54*, 589–598.

60. For a review of chemical stability, see Kerns, E.H., Di, L. "Chemical Stability", Chapter 20 in *Volume 5* of *Comprehensive Medicinal Chemistry* Triggle, D.J., Taylor, J.B., Eds. (London, Elsevier, 2007).

61. See Wong, K.K., et al. *In vitro* biotransformations of [$C^{14}$]captopril in the blood of rats, dogs, and humans. *Biochem. Pharmacol.* **1981**, *30*, 2643–2650 and references therein.

62. Reist, M., et al. 'Drug Racemization and Its Significance in Pharmaceutical Research" in *Handbook of Experimental Pharmacology, Volume 153, Stereochemical Aspects of Drug Action and Disposition*, Eichelbaum, M., Testa, B., Somogyi, A., Eds. (Berlin, Springer-Verlag, 2003).

63. See Hutt, A.J., O'Grady, J. Drug chirality: A consideration of the significance of the stereochemistry of antimicrobial agents. *J. Antimicrob. Chemother.* **1996**, *37*, 7–32 and references therein.

64. Teo, S.K., et al. Chiral inversion of the second generation IMiD CC-4047 (ACTIMID) in human plasma and phosphate-buffered saline. *Chirality* **2003**, *15*, 348–351.

65. See www.fda.gov/cder/guidance/stereo.htm.

66. Di, L., et al. Development and application of high throughput plasma stability assay for drug discovery. *Int. J. Pharm.* **2005**, *297*, 110–119.

67. Lee, H.J., et al. Prodrug and antedrug: Two diametrical approaches in designing safer drugs. *Arch. Pharm. Res.* **2002**, *25*, 111–136.

68. Sawa, M., et al. New strategy for antedrug application: Development of metalloproteinase inhibitors as antipsoriatic drugs. *J. Med. Chem.* **2002**, *45*, 930–936.

69. See Rydzewski, R.M., et al. Optimization of subsite binding to the β5 subunit of the human 20S proteasome using vinyl sulfones and 2-Keto-1,3,4-oxadiazoles: Synthesis and cellular properties of potent, selective proteasome inhibitors. *J. Med. Chem.* **2006**, *49*, 2953–2968.

70. Breitenlechner, C.B., et al. Structure-based optimization of novel azepane derivatives as PKB inhibitors. *J. Med. Chem.* **2004**, *47*, 1375–1390.

71. Burke, T.G., Mi, Z. Preferential binding of the carboxylate form of camptothecin by Human serum albumin. *Anal. Biochem.* **1993**, 285–287.

72. Mi, Z., Burke, T.G. Marked insterspecies variations concerning the interactions of camptothecin with serum albumins: A frequency-domain fluorescence spectroscopy study. *Biochemistry* **1994**, *33*, 12540–12545.

73. Mi, Z., Burke, T.G. Differential interactions of camptothecin lactone and carboxylate forms with human blood components. *Biochemistry* **1994**, *33*, 10325–10336.

74. Bom, D., et al. The novel silatecan 7-*tert*-butyldimethylsilyl-10-hydroxycamptothecin displays high lipophilicity, improved blood stability, and potential anticancer activity. *J. Med. Chem.* **2000**, *43*, 3970–3980.

75. Lesueur-Ginot, L., et al. Homocamptothecin, an E-ring modified camptothecin with enhanced lactone stability, retains topoisomerase I-targeted activity and antitumor properties. *Cancer Res.* **1999**, *59*, 2939–2943.

76. Haugan, K., et al. Gap junction-modifying antiarrhythmic peptides: Therapeutic potential in atrial fibrillation. *Drugs Future* **2007**, *32*, 245–260.

77. Kjølbye, A.L., et al. Pharmacological characterization of the new stable antiarrhythmic peptide analog Ac-*D*-Tyr-*D*-Pro-*D*-Hyp-Gly-*D*-Ala-Gly-NH$_2$ (ZP123): *In vivo* and *in vitro* studies. *J. Pharmacol. Exp. Ther.* **2003**, *306*, 1191–1199.

78. For a glimpse into peptidomimetics and their chemistry, see Kazmierski, W.M., Ed. *Peptidomimetic Protocols* (Totowa, New Jersey, Humana Press, 1999).

79. See Chorev, M., Goodman, M. Recent developments in retro peptides and proteins – An ongoing topochemical exploration. *Trends Biotechnol.* **1995**, *13*, 438–445.

# Chapter 9

# ADME and PK Properties

## 9.1 Cell Permeability and Absorption

### 9.1.1 Definitions

What happens to a drug in vivo depends on a whole series of factors including its ability or inability to cross cell membranes. Most drugs are delivered orally and need to cross a number of lipid bilayers before ever reaching the place where they can do some good. Failure to do so will render all other considerations (potency, selectivity, solubility, etc.) moot. An understanding of the factors involved in passing through these barriers is critical to the drug discovery process. Projects now take this into account even before screening begins and it remains an important consideration all the way through to drug approval.

None of the factors involved in a drug leaving its site of administration and getting to its site of action are particularly hard to understand, but the number of variables involved makes it seem like a somewhat imposing equation. And unfortunately, definitions of some frequently used terms vary, meaning that different people—even experts—can use the same word to mean different things. Chemists and pharmacologists, especially, are often at odds in their usage of some of these terms. Many researchers have a regrettable tendency to use the terms "cell permeability," "absorption," and "bioavailability" almost interchangeably sometimes. To avoid confusion, our first order of business must be to define those terms as they'll be used in this book. Be warned that discrepancies with other books and papers are bound to arise if their definitions differ from the ones used here.

*Passive transmembrane diffusion* is the ability of a compound to *cross a membrane bilayer in a way solely driven by concentration gradient*, as described by *Fick's law of diffusion*. No special molecules assist in this transport, nor is any type of energy input required, just nature's tendency to equalize concentrations on both sides of a semipermeable barrier. This simple process turns out to be the major mechanism by which most drugs can enter and exit cells.

*Cell permeability*, which is measured by $P_{app}$, as used herein refers to the ability of a compound to *traverse an intact cell monolayer unchanged in an* in vitro *system*. Note that this definition differs from what many chemists think of as cell permeability, the ability to get inside but not necessarily pass through a cell. This too is a useful definition in the right context, but when discussing results in cell monolayer models like the Caco-2 one described later, the "through a cell monolayer" definition needs to be used.

*Oral absorption* is the ability of a swallowed drug to leave the intestinal lumen to which it's carried and *pass the first biological barrier to entry into the body either by entering the apical*

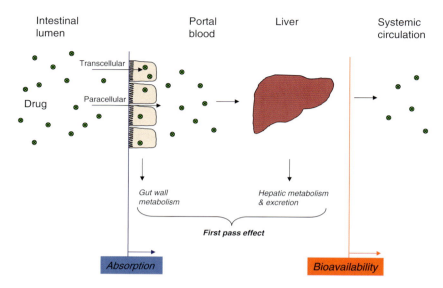

**Figure 9.1** ▶   An oral drug leaving the intestinal lumen and passing the first barrier to entry into the body by either entering the apical membrane of enterocytes or by passing paracellularly between them is considered absorbed. Administered drug that gets absorbed and is able to survive metabolism in the gut wall and excretion by the liver (first pass effect), finally reaching its site of action is considered bioavailable in the strict sense. But in practice, bioavailability is usually determined using drug concentrations in systemic circulation.[3]

*membrane of enterocytes, the cells that line the small intestine, or by passing through the narrow pores between them.* This is illustrated in Figure 9.1. Note that the absorption does not include compound initially taken into cells, which is then sent back out into the lumen by efflux transporters like *P-glycoprotein* (*P-gp*) (described later) and therefore not absorbed. Also, be aware that there are clear fault lines in the literature as to how this term is used, with many people defining it in terms of the amount of drug reaching the portal blood. This depends not only on absorption as defined here but also on how much was metabolized within the enterocytes, which, for example, represents a significant clearance mechanism for the HIV drug saquinavir.[1] The definition shown here considers that as a separate, second barrier instead.[2] Absorption of drugs delivered by other routes, like *intraperitoneal* (*i.p.*) or *subcutaeneous* (*s.c.*) injection, depends on the compounds' abilities to cross different barriers than the intestinal epithelium, like the cells surrounding blood vessels.

Not all of the drug that gets absorbed by any route ends up where it needs to be. *Oral bioavailability* refers to the *fraction of orally administered drug that ultimately reaches its site of action.* This is the strict definition, consistent with the FDA definition,[4] but in practice the concentration of the compound at the site of action is usually difficult to determine, unlike concentrations in blood. So *experimental protocols for determining bioavailability normally base it on the amount found in systemic circulation,* and this is what's usually meant when the term is used.[5] Using this definition, *intravenously* (*i.v.*) bioavailability, then, the fraction of injected compound that reaches blood, is always one.

Note that plasma concentrations don't always correlate well with the amount available at the site of action. Nowhere is this more evident than in *central nervous system* (*CNS*) indications, where plasma concentrations can be high (good bioavailability by the usual definition) but the compound might never cross the blood–brain barrier and reach its target (poor bioavailability using the strict definition). So for CNS projects, brain penetration is often even more of a concern than oral bioavailability.

Importantly, as Figure 9.1 illustrates, metabolism and biliary excretion by the liver (collectively known as *hepatic extraction*) stands as a formidable barrier between the drug delivered to portal blood and the amount that gets into systemic circulation. Gut wall metabolism plus hepatic extraction is referred to as the *first pass effect* and this is often responsible for low oral bioavailability, the bane of many a compound. In some cases, compound absorption might approach 100%, but first pass metabolism could "chew up" so much absorbed material that almost none of it reaches systemic circulation, not to mention its target. Thus, oral bioavailability, $F$, is one of the most important parameters one gets from *pharmacokinetic* (*PK*) experiments, and something that project chemists become obsessed with optimizing. But always keep in mind that bioavailability is not an immutable property of the compound like molecular weight, but depends on dose, formulation, species, fed state, co-administration of other drugs, etc.

## 9.1.2 A Closer Look at Intestinal Absorption

Intestinal absorption requires solubility: solids obviously can't pass through. Thus, poor solubility in itself can make for poor absorption. As the time a compound will spend in the gut is limited, slow dissolution of a solid formulation like a tablet can have the same effect. Once in solution of the intestinal lumen, which has a bulk pH starting out around 5.0 in the duodenum and increasing through the jejunum, reaching 7.5 at the end of the ileum (the last of the three segments of the small intestine),[6] the drug needs to penetrate a layer of mucus surrounding the epithelial cells. After that it needs to cross a monolayer of cells. As previously mentioned, a compound can get past this layer either by passing through the cells *transcellularly* (Figure 9.2a and b) or by sneaking through the *tight junctions* between the cells, which is called *paracellular absorption* (Figure 9.2c), if it's a very small molecule.

Transcellular passage can either be *passive* (Figure 9.2a) or it can occur via *carrier-mediated active transport* (Figure 9.2b).[7] For most oral drugs, simple passive diffusion through the intestinal epithelial cells is the major route of absorption. Notice that the layer of intestinal epithelial cells is asymmetric and is different on the *apical* side, which the drug initially faces, and on the *basolateral* side, which eventually leads to blood. The intestinal epithelium also contains other types of cells, like goblet cells, which secret mucus, and isn't flat but folded and convoluted into *crypts* and *villi* so that the actual surface area available for absorption in the human intestine is much greater than would otherwise be the case. In fact, it's said to be the size of a tennis court!

As its main purpose is to allow the absorption of nutrients, not drugs, the intake of certain substances like amino acids, nucleosides, and glucose is facilitated by the presence of a large variety of special transporters. These are divided up into two types; *facilitated diffusion* transporters like the non-energy dependent GLUT1 transporter for glucose, and the energy-dependent *active transporters*.[8] These vary in how picky they are about substrates and how much they're able to transport. Certain drugs do make use of active transporters,

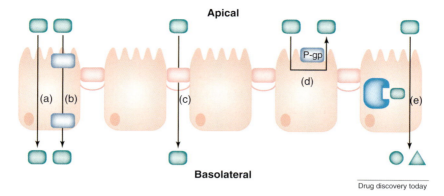

**Figure 9.2 ▶**   Possible pathways for intestinal absorption of a compound: (a) transcellular passive diffusion; (b) carrier-mediated transport; (c) paracellular absorption; (d) entry limited by P-gp, an efflux transporter; (e) metabolic enzymes within the cells might metabolize the compound. (Reprinted with permission from Balimane, P.V., Chong, S. Cell culture-based models for intestinal permeability: A critique. *Drug Discov. Today* **2005**, *10*, 335–343, copyright 2005, Elsevier.)

for example angiotensin-converting enzyme (ACE) inhibitors like captopril, which, besides being absorbed by passive diffusion, benefits from being mistaken for a "normal" dipeptide by the di/tri-peptide transporter hPEPT1.[9] Note that, however, unlike passive diffusion, which can in theory transport unlimited amounts of a drug as long as it's being somehow drained from the basolateral side (e.g. by blood flow), such carrier-mediated processes have a limited capacity that can be saturated by a high local concentration of drug or the natural substrate (e.g. after meals), making this potentially a lower capacity and less reliable route of absorption.

Intestinal epithelial cells are held together in a monolayer by a network of proteins, resulting in tight junctions with narrow pore openings between cells. The traditional view was that only very small, hydrophilic molecules (MW < 200 Da) could penetrate this paracellular space and thus not be limited to transcellular passage. Furthermore, compared to the huge area the intestinal mucosa presents for transcellular passage, these pores constitute only a tiny fraction of the available surface area. In recent years, however, a number of researchers have taken the view that paracellular absorption is something that can't be entirely neglected for some more moderately sized drugs like cimetidine (MW = 252 Da)[10] and atenolol (MW = 266 Da).[11] The tight junctions can apparently be loosened up (pore size increased) or tightened (pore size decreased) by various permeability modulators and this has resulted in a lot of research on agents that do the former, since they represent a possible way of getting otherwise non-absorbable drugs into systemic circulation.[12] The danger, however, is that prying open this particular door (or pore) will also allow in a host of molecular party-crashers, toxic agents that are normally excluded from systemic circulation for good reason.

Figure 9.2d points out an important exception to our definition of absorption that's already been mentioned. A number of different proteins exist, most notably P-gp which, in an ATP-driven process, can shoot a compound back out of enterocytes once they've gotten in. Enterocytes are also rich in CYP3A4 and some other enzymes so that significant meta-bolism can often occur within these cells (Figure 9.2e), as it does for saquinavir.

If all of this seems a bit complex you've probably read it correctly. Table 9.1 summarizes the terms. The most important things for the reader to remember are that passive transcel-lular diffusion is usually an important component of absorption, cell permeability, *and* bioavailability, and that first pass metabolism stands squarely between absorbed compound and compound in systemic circulation, presenting a different problem that the medicinal chemist needs to address in a different way than absorption difficulties.

Bioavailability is a composite property that requires PK studies to assess, but just as the routine measurements of simple in vitro potency is the first step in building clinical potency, routine computational estimates or measurements in model systems can play a key role in optimizing compounds for bioavailability in man. Once again, such parameters used to be examined only late in the drug discovery process, but nowadays the essential triad potency, solubility, and cell permeability is taken into account from the very beginning of a project. Absorption and cell permeability can be modeled in a number of in silico and in vitro systems, as discussed in the following sections. The still more important property of bioavail-ability, where metabolism can play a major role, is discussed in more detail later in the chapter.

**TABLE 9.1 ▶  A Comparison of Passive Diffusion, Oral Absorption, Cell Permeability, and Oral Bioavailability, and Some Methods Used to Predict and Assess Them**

| Property | Describes | Comment | Models |
|---|---|---|---|
| *Passive transmembrane diffusion* | Ability to cross a membrane bilayer in a concentration gradient-driven fashion | Usually the major component of cell permeability and drug absorption | IAM, PAMPA |
| *Cell permeability* | Ability to traverse an intact cell monolayer unchanged in an in vitro system | For compounds stable to first pass effect and with no solubility or dissolution limitations, often correlates with oral bioavailability | Caco-2, MDCK, etc. |
| *Absorption* | Ability to pass the first biological barrier to entry into the body (see text) | Not the same as the ability to reach portal blood by this definition | ROF, computational models |
| *Bioavailability* | Fraction of administered drug that ultimately reaches the site of action, although in practice concentrations in systemic circulation are usually measured | Includes absorption and first pass effect | In vivo PK studies, computational models |

## 9.1.3   Models of Cell Permeability and Absorption

### 9.1.3.1   Property-Based Predictions of Oral Absorption

Lipinski's *rule of five* (*ROF*) predicts that a compound is more likely to have "poor absorption or permeation" when

▶ "There are more than 5 H-bond donors (expressed as the sum of OHs and NHs)"

▶ "The MWT is over 500"

▶ "The Log*P* is over 5 (or MLog*P* is over 4.15)"

▶ "There are more than 10 H-bond acceptors (expressed as the sum of Ns and Os)"[13]

An important caveat in the present context is that *these rules are meant to apply to drugs, not leads or hits, which, as we've already seen, may need to be considerably smaller* as MW will likely be added on to them later. Many will use the "rule of three" at these early stages, as we saw in the section on fragment screening. Rule of five was never meant to apply to natural products or actively transported compounds. Whether it works by predicting "pure" passive membrane diffusion or some combination of properties including solubility to some extent and maybe even metabolic stability or efflux (smaller, less greasy molecules being less readily metabolized and less likely to be substrates for P-gp) isn't crystal clear. Many reports that ROF compliance correlates with each of the non-interchangeable properties of cell permeability, absorption, and bioavailability (all defined in various ways) can be found in the literature.

Rule of five is useful, easy to remember, and it works most of the time. Its simplicity lends itself to widespread use in combinatorial library design and in deciding on which screening compounds to buy.[14] It represents the simplest and most widely accepted way of flagging compounds unlikely to be well absorbed. It isn't used to assign predicted values to compounds or even to rank order them, but has proven its worth as a simple way of dividing them up into those likely to have problems and those that aren't. Rule of five might work largely by flagging impermeable compounds that require extensive desolvation (those having a large number of *hydrogen bond donors* (*HBD*s), and *hydrogen bond acceptors* (*HBA*s)) before they can passively diffuse through membranes, and large molecules (MW > 500 Da) with slow membrane diffusion rates.

On applying ROF, be aware that the calculated Log*P* values can differ somewhat from the better, experimentally determined numbers in the same way that a real proton *nuclear magnetic resonance* (*NMR*) might differ from the ones predicted for a compound by a program like ChemDraw® (which estimates Log*P* too, of course). For this reason, the real, experimental numbers should be used whenever possible. But, for comparing compounds in a given series calculated Log*P* values are often acceptable, and for compounds not actually made yet they're the only way to go.

So things can get a bit fuzzy, and as we've seen, the permeability or absorption that ROF seems to predict is not the only component of bioavailability. And keep in mind that ROF compliance is just one factor that often goes into making a drug, not the whole process.[15] "Lipinski's rule is not infallible, and is a far cry from having a magic machine that takes in the three-dimensional structure of a target and spits out the ideal drug candidate. Some structures that fall outside the parameters can make good drugs, and there is no guarantee that compounds within the parameters will succeed."[16]

But in the end, these criticisms don't matter much. Although it can't provide accurate estimates of observed cell permeability or absorption for a given compound, as a way of steering chemists away from bloated, overly hydrophobic or overly hydrated compounds that are unlikely to become drugs and toward smaller, more reasonable compounds that have a better chance of passing through intestinal cells and being absorbed, the use of ROF has been an overwhelming success.

Another set of rules has been proposed by Veber et al.[17] In analyzing data, for over 1100 compounds for which oral bioavailability data in rat had been determined, they found that good bioavailability was likely if the compounds had 10 or fewer rotatable bonds and either a polar surface area of less than 140 $Å^2$ or a total of 12 or fewer HBDs and HBAs. *Polar surface area (PSA)* is an easily computed property and its good correlation with fractional absorption in humans for passively absorbed drugs can be seen in Figure 9.3. For this reason, it's become a major consideration in drug design, with many preferring compounds with PSA values below 100 $Å^2$, if possible. The interchangeability of PSA with the number of HBDs and HBAs in Veber's rule suggests that it may also relate to the amount of desolvation required to cross cell membranes.

The physical significance of the number of rotatable bonds is a bit more enigmatic. Often, but not always, it correlates with MW, and the compounds with high molecular weights have much slower membrane diffusion rates that can reduce permeability. It's easy to see that a larger number of rotatable bonds mean greater entropy, for which a penalty needs to be paid by binding to a receptor, but it isn't obvious how that should adversely affect a property like oral bioavailability. Like ROF, the main utility of Veber's rule lies in warning chemists away from molecules that aren't drug-like, but accurate quantitative predictions of bioavailability

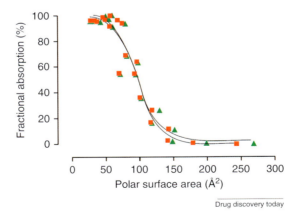

Drug discovery today

**Figure 9.3 ▶** Observed correlation between human fractional absorption and PSA calculated using "dynamic"[18] (red) and single conformer[19] (green) methods for 20 drugs utilizing passive absorption. (Reprinted with permission from Clark, D.E., Pickett, S.D. Computational methods for the prediction of "drug likeness". *Drug Discov. Today* **2000**, *5*, 49–58, copyright 2000, Elsevier).

are just too much to ask of it. Lipinski's and Veber's rules don't necessarily represent an either/or situation, and both are frequently used to flag potentially problematic compounds.

Other, more computationally intensive methods for predicting absorption as well as various PK parameters have been developed.[20] Many biotech and pharma companies now provide access to a smorgasbord of such programs. Algorithms much more complex than ROF calculations[21] which use neural networks[22–23] or 3D molecular field maps (like the program VolSurf)[24] have been developed to assist in predicting the drug likeness of sets of compounds or in optimizing the PK parameters for leads. Such programs may still not be able to predict the best compound in a given series but, at minimum, are very good at eliminating the worst ones from further consideration. Some *ex silico* methods that are widely used in predicting and measuring cell permeability and absorption are discussed below.

### 9.1.3.2  Immobilized Artificial Membranes

Liposomes, which are vesicles made of phospholipid bilayers, have been of interest for a long time as drug carriers, among other things. For example, the cancer drug Doxil is doxorubicin contained within liposomes that are specially modified to resist rapid degradation. Liposomes mimic cell membranes in having hydrophilic (phosphatidylcholine or phosphatidylcholine-like) head groups on the exterior and interior with hydrophobic tails sandwiched in between. This alone suggests the possibility of their use as models for passive transmembrane diffusion.

Such partitioning of compounds into liposomes has been done,[25] but a more convenient, higher throughput way of using the same concept would be to somehow bond a monolayer of phospholipids to the solid support of a reversed-phase *high performance liquid chromatography* (*HPLC*) column and measure the retention factor of a compound as a surrogate for membrane partitioning, in turn a surrogate for passive diffusion through cell membranes, in turn a surrogate for absorption—no shortage of surrogates here![26] This sort of column-based approach is somewhat analogous to the use of HPLC methods for measuring Log$P$. Passive diffusion through a membrane is directly proportional to the equilibrium constant $K_m$, which is reflected in $k'$-values, as shown in Figure 9.4.

The chemical structure of one of the more popular bonded phases that are used for this purpose is shown in Figure 9.5. Retention factors obtained from simple hydrophobic-bonded phases, like typical C-18 HPLC columns, are not predictive of absorption, but the *immobilized artificial membrane* (*IAM*) method has been shown to correlate well with other models and with in vivo absorption data. For example, one study found that it gave a correlation coefficient of 0.858 (taking into account MW) when plotted against rat's intestinal absorption data, for a set of 12 compounds.[27]

Immobilized artificial membrane chromatography has been used in pharma for a long time, but it does have a few drawbacks. It requires a separate HPLC run for each compound (unless more exotic *N*-in-1 techniques with MS detection are used) which makes it less suitable for high-throughput than one would like. In common with most chiral chromatography columns, IAM columns can be sensitive to abuse by solvents, pH, nucleophilic species, and gunk that doesn't wash off. Finally, the results aren't always as predictive as hoped. Immobilized artificial membrane chromatography can certainly give useful data, but it's probably not as widely used these days as some of the other methods described below.

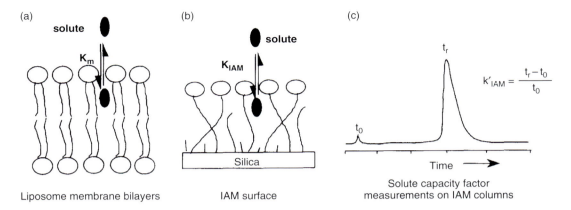

**(a)**   solute   $K_m$   Liposome membrane bilayers

**(b)**   solute   $K_{IAM}$   Silica   IAM surface

**(c)**   $t_r$   $t_0$   Time   $k'_{IAM} = \dfrac{t_r - t_0}{t_0}$   Solute capacity factor measurements on IAM columns

**Figure 9.4 ▶** *Immobilized artificial membrane (IAM)* chromatography. Binding of solute to fluid membranes in (A) can be modeled by solute binding to an immobilized membrane as in (B) through measurement of $k'_{IAM}$ as shown in (C). (Reprinted with permission from Yang, C.Y., et al. Immobilized artificial membranes—Screens for drug membrane interactions. *Adv. Drug Deliv. Rev.* **1996**, *23* 229–256, copyright 1997, Elsevier).

**Silica**

**Figure 9.5 ▶** Structure of IAM.PC.DD2, a bonded-phase used to predict permeability. See www.registech.com/iam/index.html.

### 9.1.3.3   Parallel Artificial Membrane Permeability Assay

In 1998, Dr Manfred Kansy and co-workers at Hoffmann-La Roche published a high-throughput method for determining how well compounds can passively diffuse through a model of a membrane bilayer. They called it *parallel artificial membrane permeability assay (PAMPA)*.[28]

In this experiment, carried out in 96-well plates, a buffered donor solution containing the compound is separated from a buffered acceptor solution using a filter support, which has been treated with a solution of egg lecithin (mostly phosphatidylcholine) dissolved in *n*-dodecane, thus making a kind of sandwich. The wells are usually stirred to minimize the thickness of the *unstirred water layer* (or *aqueous boundary layer*), the water-filled area immediately adjacent to the membrane that can sometimes constitute another formidable barrier to the permeation of compounds. The concentration of the compound in the acceptor well is measured, usually by UV detection, over the course of some hours and the permeability, $P_{app}$, can be calculated.[29]

The simplicity of this technique, its high-throughput capability, and its ability to often predict passive intestinal absorption (see below) have combined to make it a very popular assay. Commercial instruments are usually employed.[30] At some companies, these days, the assay may be run on most compounds made during drug discovery, providing rapid feedback on transcellular passive diffusion, which is, as we've seen, an important component of intestinal absorption.

As mentioned earlier, the pH range involved in absorption from the gut varies from acidic to slightly basic, and it's generally been assumed that charged compounds don't passively diffuse through membranes very well (the *pH partition hypothesis*), although recent research has questioned the validity of this assumption in at least a few cases.[31] But in general, acidic compounds like warfarin (Figure 8.18) are best absorbed under lower pH conditions, when a smaller fraction is ionized, while basic compounds like propranolol (Figure 9.6) reach their maximal absorption at higher pH values for the same reason. Because of this, it's important to run passive absorption models at the right pH value. Figure 9.7 shows just how much of a difference this can make. Parallel artificial membrane permeability assays can be run at several different pH values and the highest permeability result used to predict absorption.

When this was done, with PAMPA permeabilities measured at pH 5.5 and pH 7.4 for a series of 93 commercial drugs, a rough correlation between $P_{app}$ and human absorption, shown in Figure 9.8, was observed. One can see that despite the scatter, compounds with $P_{app} > 3 \times 10^6$ cm/sec are likely to be at least moderately well-absorbed. Several other things are worth noting, though, including the steep slope of the initial curve that "limits the prediction accuracy for some medium to low absorption drugs," which are likely to be those most frequently encountered early on in drug discovery.[32] Distinct subsets of compounds with low PAMPA values but high observed absorption and the opposite, those with fairly high PAMPA numbers but less than 100% absorption, were also encountered. These represented a minority of compounds, and, again, no one would seriously expect this type of simple assay, which looks only at passive transmembrane diffusion, to accurately reflect

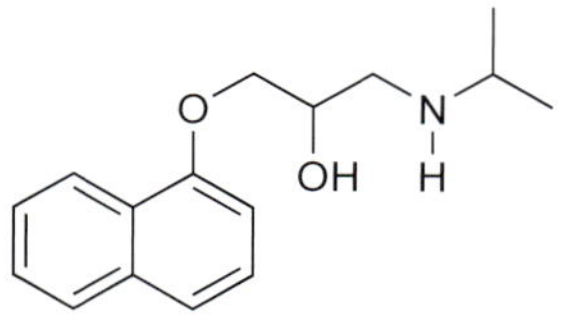

**Figure 9.6** ▶   Inderal (propranolol).

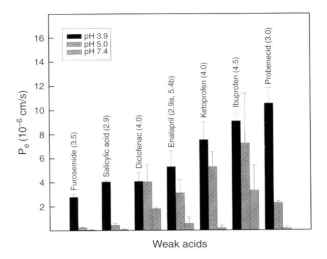

**Figure 9.7** ▶ Effective permeabilities for some weak acids, as measured by PAMPAs at three different pH values. p$K_a$ values for the acids are shown in parentheses. (Reproduced with permission from Avdeef, A., "High-throughput Measurement of Permeability Profiles", Chapter 3 in *Drug Bioavailability*, van de Waterbeemd, H., Lennernäs, H., Artursson, P., Eds. (Wiley-VCH Verlag GmbH & Co. KGaA) p. 53, copyright 2003).

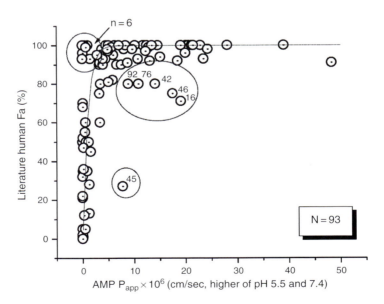

**Figure 9.8** ▶ Parallel artificial membrane permeability assay $P_{app}$ values versus literature values, for human oral absorption, for a set of 93 drugs. Note circled areas, where PAMPA data were not predictive, including #45, griseofulvin, for which bioavailability is known to be very formulation-dependent. (Reprinted with permission from Zhu, C., et al. A comparative study of artificial membrane permeability assay for high throughput profiling of drug absorption potential. *Eur. J. Med. Chem.* **2002**, *37*, 399–407, copyright 2002, Elsevier).

in vivo absorption in every case. To emphasize this caveat, cases where PAMPA $P_{app}$ values are very low but oral bioavailability is high have been noted before.[33] More often, however, the predictive value of PAMPA, in binning compounds into those that are likely to have low, medium, and high passive permeabilities, is borne out by the kind of relationship Figure 9.8 demonstrates.

An additional advantage of this method is that it tolerates solubilizers and co-solvents. As we've seen, poor solubility can make the measurement of all other properties inaccurate or even impossible. Cellular permeability assays may tolerate only a few percent dimethyl sulfoxide (DMSO), but PAMPA assays can be run in the presence of solubilizing agents like glycocholic acid[34] and even 20% *acetonitrile (ACN)*.[35] Other modifications are possible, many of them being highlighted in a recent review.[34] Albumin can be added to the acceptor side to simulate physiological absorption more realistically, with the added benefit of reducing the assay time.[36] Different membrane compositions can be used, for example, porcine brain tissue extract can be added to more closely simulate the blood–brain barrier.[37] The ease and versatility of the PAMPA methods are among its strongest points.

Aside from not reflecting paracellular, active transport, or efflux mechanisms—either a weakness or a strength, depending on how one looks at it—PAMPA suffers from a number of limitations. A compound lacking a good UV absorbance will be unsuitable for PAMPA unless *liquid chromatography–mass spectrometry (LCMS)* analysis is used, where good ionization will be required instead. As previously noted, it's a less reliable predictor of low or medium permeability than it is of high permeability, and its error bars will be larger for poorly permeable compounds due to larger errors in the measurement of the resulting dilute solutions as well as the need for a longer time-course to be accurate. Operationally, difficulties can be encountered in preparing the artificial membrane reproducibly.

Researchers at Novartis have pointed out that the 120 µM size of the PAMPA membrane solid support "renders the formation of well-characterized bilayer membranes rather unlikely."[38] Their method, referred to as *hexadecane method (HDM)-PAMPA*, substitutes a thin (10 µM) immobilized layer of hexadecane instead. Here, ionic membrane interactions are removed from the model altogether. Whereas conventional octanol–water Log$P$ or Log$D$ does not correlate well with passive permeability, HDM-PAMPA values obtained at optimal pH do, as shown for a set of 32 drugs in Figure 9.9. Again, binning compounds into those that are likely to be highly, moderately, or poorly absorbed through passive transcellular diffusion is probably the best way to use such data, which should ideally be as readily available in early stage drug discovery as activity numbers.

The popularity of the PAMPA approach in all its various incarnations has even led to some speculation that it might entirely supplant cell-based methods of permeability determination, which are more complex and variable. But, it seems more likely that the different strengths and weaknesses of these two models mean that they're best used together when possible.

### 9.1.3.4   Caco-2 and Other Cellular Assays

Caco-2 is a human intestinal cancer cell line that can form a well-differentiated monolayer on a solid support that mimics the human intestinal epithelium, where most drug absorption occurs. This makes it useful in a cellular equivalent of a PAMPA "sandwich" that's more

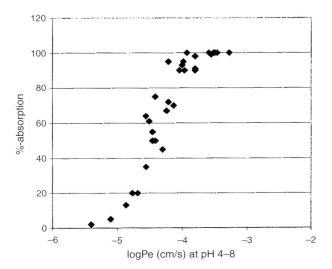

**Figure 9.9** ▶  Human GI tract absorption of a set of 32 drugs versus HDM log $P_e$ values for the best permeability in the range of pH 4–8. (Reprinted with permission from, *J. Med. Chem.* **2001**, *44*, 923–930, copyright 2001, American Chemical Society).

complex and probably more realistic than the previously described methods. For more than a decade now, Caco-2 has seen extensive use in in vitro determination of intestinal permeability. This is evidenced by the fact that the FDA recognizes Caco-2 permeability data as a way of demonstrating good absorption in applying for a waiver of in vivo bioequivalence studies when seeking abbreviated new drug application (ANDA) approval for rapidly dissolving, highly soluble, and highly permeable drugs.[39]

Caco-2 monolayers, like the intestinal epithelium diagrammed in Figure 9.2, present an apical side and a basolateral side, with tight junctions between cells. Permeability coefficients are normally determined by measuring the compound concentrations on the basolateral (receiver) side after the compound exposure on the apical (donor) side, at various time points. Unlike PAMPAs, however, more than just passive diffusion can be involved in transport through Caco-2 cells. Active transport through the cell (Figure 9.2b) or efflux back out of (Figure 9.2d) the cells, a certain amount of metabolism by the cells themselves (Figure 9.2e), and a small amount of paracellular transport (Figure 9.2c) can all take place.

The presence of active transport mechanisms means that the compounds that were initially placed on the basolateral (normally the acceptor) side can have different $P_{app}$ values on crossing the monolayer and reaching the other side than the compounds loaded onto the apical (normally the donor) side. In fact, conducting the experiment in both the directions and taking the ratio of the two values is one indication of active transport. If the ratio of $P_{app}a \rightarrow b / P_{app}b \rightarrow a$ is much above or below one, either active transport into the cell or efflux out of the cells, the latter often due to P-gp, is indicated. Needless to say, this kind of information can be invaluable in drug discovery and can't be obtained by PAMPA.

These days, many researchers believe that P-gp efflux combined with cytochrome P450 (CYP) metabolism in the intestinal epithelium, and not just problems with passive diffusion through membranes per se, can be a major factor for the poor absorption observed in many compounds.[40] Thus P-gp and CYPs, may be the molecular "bouncers" meant to bar the entry of unwanted xenobiotics into the body. Caco-2 cells contain both of these to some extent. Significantly, however, CYP3A4 levels are far lower in Caco-2 cells than those found in human intestinal epithelial cells, while Caco-2 results tend to overpredict efflux due to P-gp.[41] The latter is at least partly because at the low compound concentrations normally used in the assay this active efflux mechanism is not saturated the way it might be in vivo. To compensate for low CYP activity, using a Caco-2 subclone, TC7, which expresses higher levels of CYP3A[42] or exposing normal Caco-2 cells to vitamin $D_3$, which can have the same effect,[43] has been proposed. Inhibitors of P-gp like verapamil can attenuate an unrealistically high level of efflux. So Caco-2 and related systems may still represent tools to study the roles of intestinal CYP3A4 and P-gp in drug absorption.

Besides CYPs, other metabolic enzymes like hydrolases, phosphatases, peptidases, and even some Phase II conjugating enzymes are also found in Caco-2 cells, so careful LCMS analysis of Caco-2 wells in a detailed experiment can provide not only information about the permeability but also some information on the gut wall metabolism, raising early flags for metabolically unstable compounds. Also, unlike PAMPA, Caco-2 results can reflect a paracellular transport component, at least to a limited extent. Caco-2 monolayers have an average pore size of about 4 Å, compared to an estimated 8–13 Å found in the intestinal epithelium, but withholding $Ca^{2+}$ from the medium or chelating it with EDTA can open these tight junctions somewhat to better reflect in vivo pore size.[44]

So Caco-2 assays constitute more of a composite system than the relatively "pure" passive transcellular diffusion that PAMPA and IAM reflect. In light of this additional complexity, the trend shown in Figure 9.10 becomes quite understandable. The two methods tend to correlate with each other fairly well, except when active uptake (diamonds) or active efflux (triangles) mechanisms play a role. As most of the drugs are passively absorbed, however, this kind of analysis explains why initial screening of discovery compounds is often done by the somewhat higher throughput and cheaper PAMPA method, with Caco-2 reserved as a secondary assay for compounds of interest.[45]

A more important question, though, has to do with how well Caco-2 permeability coefficients correlated with the actual drug absorption in humans. Figure 9.11 shows the results of such a study, for a set of 25 model drugs that were carefully evaluated. Note that although a good sigmoidal correlation was observed there was still some scatter, especially for compounds with less-than-optimal absorption, again, the most typical kind one is likely to find early on in a drug discovery program.

Keep in mind that these data result from the careful measurements of drugs and drug-like compounds using a standardized protocol in a single lab experienced in such experiments. In practice, automated data on compounds not fully characterized as to solubility and stability will not be this good. Caco-2 and related cellular assays are very sensitive to the exact conditions used, particularly passage number, and $P_{app}$ values have been shown to vary as much as 100-fold between labs, making comparisons based on numbers from different labs perilous.[46] Standards need to be run along with experimental compounds for quality control (QC) purposes, and some potential problems discussed later need to be ruled out. But when that is done, the data should be sufficient to bin compounds into low, moderate, and high absorptions, and perhaps sometimes to rank order them.

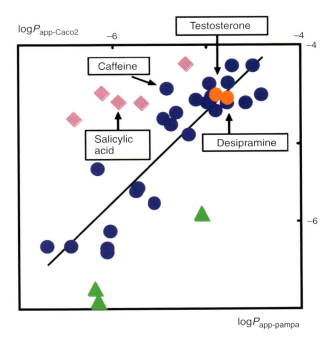

**Figure 9.10 ▶** Relationship between permeability coefficients obtained from Caco-2 and PAMPA screenings. Circles represent passively transported compounds, diamonds are actively transported, and triangles are compounds subject to efflux. (Reprinted from with permission from Fujikawa, M., et al. Relationship between structure and high-throughput screening permeability of diverse drugs with artificial membranes: Application to prediction of Caco-2 cell permeability. *Bioorg. Med. Chem.*, 2005, *13*, 4721–4732, copyright 2005, Elsevier).

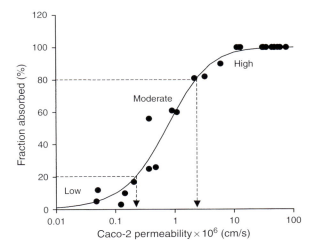

**Figure 9.11 ▶** Oral fraction absorbed in humans versus Caco-2 $P_{app}$, for a series of 25 compounds. (Reproduced with permission from Ungell, A.-L., Karlsson, J. "Cell Cultures in Drug Discovery: An Industrial Perspective", Chapter 5 in van de Waterbeemd, H., Lennernäs, H., Artursson, P., Eds. *Drug Bioavailability* **2003**, (Wiley-VCH Verlag GmbH & Co. KGaA) p. 107, copyright 2003).

The actual cutoff values for absorption range will vary by conditions and lab, but often compounds having values of $<1 \times 10^{-6}$ cm/sec are predicted to be poorly absorbed, those between 1 and $10 \times 10^{-6}$ cm/sec are predicted to be moderately absorbed, and those $>10^{-5}$ cm/sec well absorbed.

No model can predict human absorption 100% of the time, but Caco-2 has a relatively good track record for at least binning compounds. Along with the early information advanced Caco-2 studies can provide information on transport mechanism, gut wall metabolism and toxicity, and its utility in establishing structure property relationships, this makes it a valuable addition to the armamentarium of drug discovery.

Potential problems with the method do exist, many of them being shared in common with PAMPA assays. Assay pH matters in both the cases, with Caco-2 cells not being amenable to quite as broad a range as PAMPA. In both the cases, compounds first need to be soluble, and as a cellular assay Caco-2 is somewhat less tolerant of co-solvents than PAMPA, with no more than 1–2% DMSO usually being acceptable. Compound retention on the "meat" of the sandwich, either the artificial membrane for PAMPA or the cells themselves for Caco-2 (so-called *cacophilicity*[45]) or even the plastic used in the wells, is a big problem. Sometimes most of the compounds will end up sticking to the Caco-2 cells, and in such cases data obtained will be meaningless. A good practice is to analyze the amount of compound remaining on the donor side as well as on the acceptor side to either correct for this effect (recovery) or, if it's too large, to flag the data. As with PAMPA, adding *BSA* (the biologist's *bovine serum albumin*, not the chemist's benzenesulfonic acid!) can help minimize this effect. If *mass spectrometry* (*MS*) detection is not used, UV absorption probably will be, and in that case the compound will need a good $\varepsilon_{max}$ value.

Adding up all the insoluble compounds, lipophilic compounds that are likely to extensively stick somewhere, compounds lacking good ionizability or UV $\varepsilon_{max}$ values, and compounds that may not be stable in water over the course of hours (another complication) one can see that quite a few discovery stage compounds, particularly screening compounds, may not be amenable to either PAMPA or Caco-2 assays. Plenty of toxic or detergent-like compounds can kill or maim Caco-2 cells as well, and some compounds can appear to be permeable by opening up the tight junctions between them, so some additional work is needed to validate results.

One drawback in using the Caco-2 assay is that it takes three weeks to prepare differentiated cells. At a large organization routinely doing these assays, this just means some planning is required. But to speed things up, other cellular assay systems have been developed, including a Caco-2 system that can be used in as little as three days.[47] Other cell lines, including *Madin-Darby canine kidney* (*MDCK*) cells, which have lower levels of transporters and metabolic activity,[48] and 2/4/A1 cells, from rat, have the advantage of shorter differentiation times. The latter cell line also produces a monolayer with looser tight junctions that allow it to mimic paracellular absorption more closely than Caco-2 does, and at least one report showing that it gave a better correlation with human absorption than Caco-2 did for a collection of 30 compounds has been published.[49]

Cellular models of blood–brain barrier permeability are also important, both for drugs meant to penetrate the CNS and those meant to stay out of it. Various Caco-2-like assays have been used for this, such as one using MDCK cells engineered to overexpress P-gp, a major barrier to CNS drug penetration. Other cell lines such as brain microvessel endothelial

TABLE 9.2 ▶   One Possible Staging Scheme for the *In Silico*
and In Vitro Prediction of Absorption in Early Stage Drug Discovery[52]

| Stage | Method | How many? |
| --- | --- | --- |
| Library selection & screening | Computational | All |
| Hit-to-lead | Computational | All |
| | PAMPA (2 pH) | All soluble |
| | Caco-2 (a → b) | Some |
| Lead optimization | Computational | All |
| | PAMPA (2 pH) | All soluble |
| | Caco-2 (a → b and b → a) | Some |

cells[50] and human umbilical vein endothelial cells (ECV304) co-cultured with rat glioma cells (C6) have also been used.[51]

Cellular assays, of course, are by no means the last word in predicting drug absorption in man. More realistic results can come from assays using excised intestinal tissue in an Ussing chamber,[53] or portal vein cannulated animals, which allow direct measurement of "absorption" by a different definition than the one used here.[54] Ultimately, the absorption of a drug in man can even be measured using various intestinal perfusion techniques.[55] Needless to say, none of these methods are suitable for the routine screening of discovery compounds, which must therefore rely on the less predictive but much more convenient computational and in vitro techniques discussed above. This being the case, Table 9.2 shows one possible way that such techniques might be used through a discovery stage project, but in practice each company will have its own optimized way of doing this. Resource limitations and price considerations may also mean that routine screening in such assays is likely to be more common in big pharma than at smaller biotech companies. Although, as we've seen, everything can be contracted out these days, that doesn't mean it can always be afforded.

### 9.1.4  Improving Cell Permeability and Absorption

#### 9.1.4.1  Molecular Modifications

As we've already seen, there are several factors that clearly influence passive transcellular diffusion, usually the most important component of intestinal absorption. High molecular weight compounds have slow diffusion rates. Charged compounds have problems crossing cell membranes. Compounds generally need to be desolvated in order to cross membranes, so those with a lot of HBDs and HBAs will tend to be impermeable. An illustration of this last point can be seen by perusing the data in Figure 9.12, where the loss of a single HBD seems to make the difference between a poorly–moderately absorbed compound and a well-absorbed one.[56] This effect has been used to partly explain the observation that, in general, peptides, even relatively small ones, are poorly absorbed by the passive transcellular route with the possible exception of some cyclic peptides, like cyclosporin A (Figure 7.5), which are capable of forming intramolecular H-bonds.[57]

Some logical ways, then, to improve cell permeability of compounds will obviously include cutting down on the number of HBDs and HBAs. Increasing lipophilicity, a natural

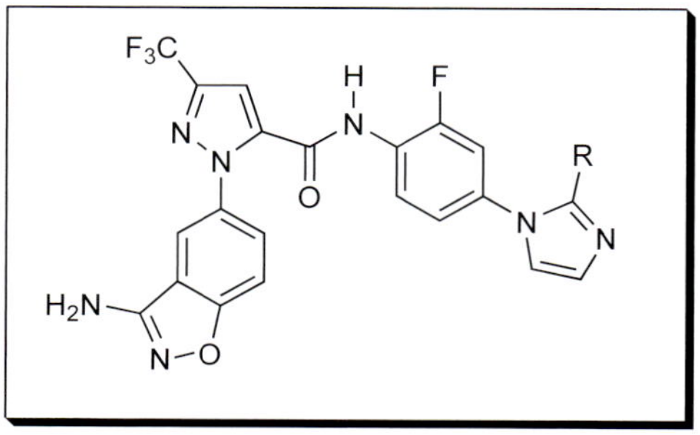

| R | $fX_a$ $K_i{}^a$ | Caco-2 $P_{app}{}^b$ | % $F^c$ |
|---|---|---|---|
| Me | 0.70 | 7.41 | nd |
| CH$_2$OH | 1.00 | 0.10 | nd |
| CH$_2$NHMe | 0.17 | 0.20 | 27 |
| CH$_2$NMe$_2$ | 0.19 | 5.56 | 84 |
| 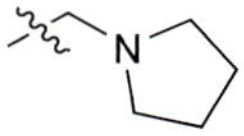 (pyrrolidine structure) | 0.37 | 23.00 | nd |

**Figure 9.12** ▶ Activity, cell permeability, and oral bioavailability data for a series of inhibitors of the serine protease Factor Xa.[56] (a) $K_i$ in nanomolar. (b) Permeability in cm/sec $\times 10^{-6}$. (c) Oral bioavailability in dog.

consequence of cutting down on HBAs and HBDs, will probably also help up to a point (Box 9.1). Balancing out an ionic charge, e.g. by adding an amino group to reduce the net charge, an approach used for the ACE inhibitor lisinopril (Figure 9.13, which also partly owes its success to active transport), can also improve absorption.

---

**Box 9.1   Key Points About Cell Permeability and Absorption**

▶ Cell permeability, absorption, and bioavailability have distinct meanings, but passive diffusion across lipid bilayers is usually important to all of them

▶ Computed properties can often flag molecules that are likely to be poorly absorbed

▶ PAMPA can provide a moderate-to-high throughput assessment of passive diffusion

▶ Caco-2 and other cellular assays can provide information about passive diffusion and active transport

▶ Absorption can often be improved be by removing charges, HBAs and HBDs and sometimes by increasing lipophilicity

▶ Absorption problems can be overcome through prodrugs designed to improve passive diffusion or active transport

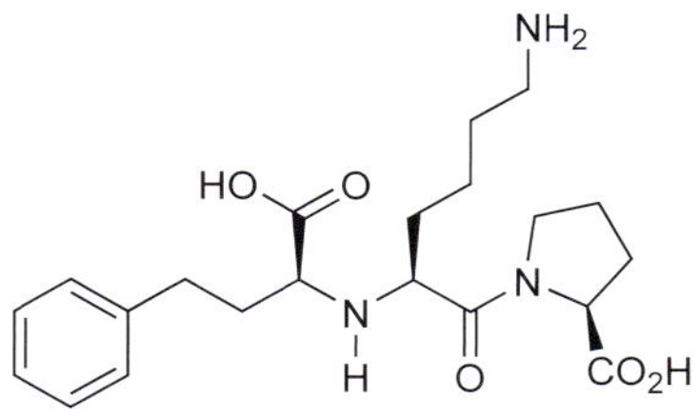

**Figure 9.13 ▶**   Zestril (lisinopril).

Note that the most powerful method of improving solubility, adding in carboxylic acids or amines in order to form soluble salts, should have quite an adverse effect on permeability, unless it reduces overall charge as for lisinopril or coincidentally produces a substrate for carrier-mediated transport (e.g. via an amino acid or dipeptide transporter). In short, structure property relationships for solubility and cell permeability, like many properties that need to be optimized in going from a hit to a clinical candidate, operate by different rules, and often go in opposite directions.

### 9.1.4.2   Prodrugs

The barrier presented by intestinal absorption represents the greatest opportunity for the use of prodrugs and is what they're most frequently used to overcome. As stated previously, prodrugs—with a few exceptions like simvastatin, a lactone opened up to the corresponding hydroxy acid drug—generate other molecules as byproducts that present their own potential toxicities. Although these can sometimes be fairly innocuous, like ethanol or $CO_2$, prodrugs still tend to be used in "last resort" mode to rescue compounds that normal medchem modifications have been unable to fix. But in that role, they've proven to be quite valuable.[58]

Prodrugs can help overcome absorption problems in two ways. The first approach makes use of passive transcellular diffusion. An example would be adefovir dipivoxil, which was already shown back in Figure 8.21. The drug, a phosphonic diacid, is unlikely to either be orally absorbed or cross the necessary cell membranes to get to its site of action. Simple phosphonate esters (e.g. dimethyl) could eliminate the anionic charges and make it more permeable, but these wouldn't function as prodrugs since enzymes that cleave phosph*ate* esters won't cleave them to produce the drug. Instead, a *pivaloyloxymethyl* (*POM*) diester was prepared. Esterase-catalyzed attack at each carbonyl carbon results in hydrolysis with loss of formaldehyde to generate the phosphonic acid. Simple chemical hydrolysis of such acyloxymethyl esters can happen too, but with the bulky pivalic ester this non-enzymatic process, which can make these esters unstable, is acceptably slow. Again, the balance between chemical stability, typically to hydrolysis at the range of pH values found in the GI tract, and ease of enzyme-catalyzed unmasking is an issue many prodrugs must face.[59]

Phosphonic and carboxylic acids aren't the only functional groups presenting potential barriers to absorption and thus natural targets for prodrugs. The thrombin inhibitor mela-gatran, shown in Figure 9.14, is poorly permeable due to the presence of not only a carboxylic acid but also of a highly basic amidine, both being required for good activity against its target. Its Caco-2 $P_{app}$ was found to be only $3.0 \times 10^{-8}$, giving an oral bioavail-ability between 3% and 7%.[60] The acid was masked as the ethyl ester while a

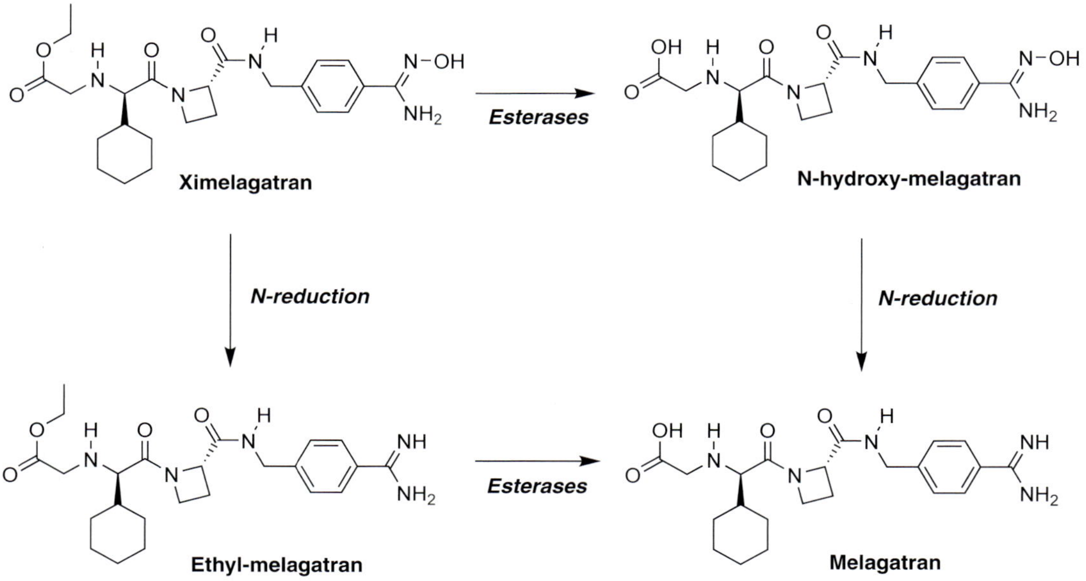

**Figure 9.14 ►**    An example of the use of a prodrug to improve passive transport. Ximelagatran, a double prodrug used to overcome the poor absorption characteristics of the thrombin inhibitor melagatran, requires desterification and hydroxyamidine reduction to generate the active species.[61]

hydroxyamidine replaced the amidine, lowering its basicity by orders of magnitude, to provide the double prodrug, ximelagatran.

In this case, a hepatic enzyme system proved to be capable of reducing the hydroxyamidine to the active amidine.[61] Ximelagatran's Caco-2 $P_{app}$ was found to be $2.4 \times 10^{-6}$, 80-fold better than melagatran's, and its oral bioavailability was between 18% and 24%, about four times better. Although such double prodrugs involve more complicated in vivo analyses (and regulatory submissions!) as Figure 9.14 points out, ximelagatran made it through clinical trials but never received drug approval due to concerns about liver toxicity,[62] showing once again that although the prodrug approach can provide viable clinical candidates, poor absorption isn't the only problem a compound can have.

The second way prodrugs can be used to improve cell permeability and absorption is by taking advantage of active transporters. Recall that these exist in the intestinal epithelium for the purpose of bringing in necessary compounds like vitamins, nucleosides, peptides, and amino acids. Modifying a drug to make it a substrate for such transport can increase absorption as it does for the antiviral compound valacyclovir, a prodrug of acyclovir that shows 3–5-fold better oral bioavailability than the parent compound.[63] As shown in Figure 9.15, valacyclovir is simply the L-valine ester of acyclovir, which allows it to be transported across intestinal epithelial cell membranes by the *human oligopeptide transporter* (*HPT*1) and the oligopeptide transporter PEPT1.[64] The corresponding D-valine ester is not well-transported.[65] Afterward, valacyclovir is hydrolyzed by an enzyme dubbed, appropriately, valacyclovir hydrolase to generate the antiviral agent acyclovir.

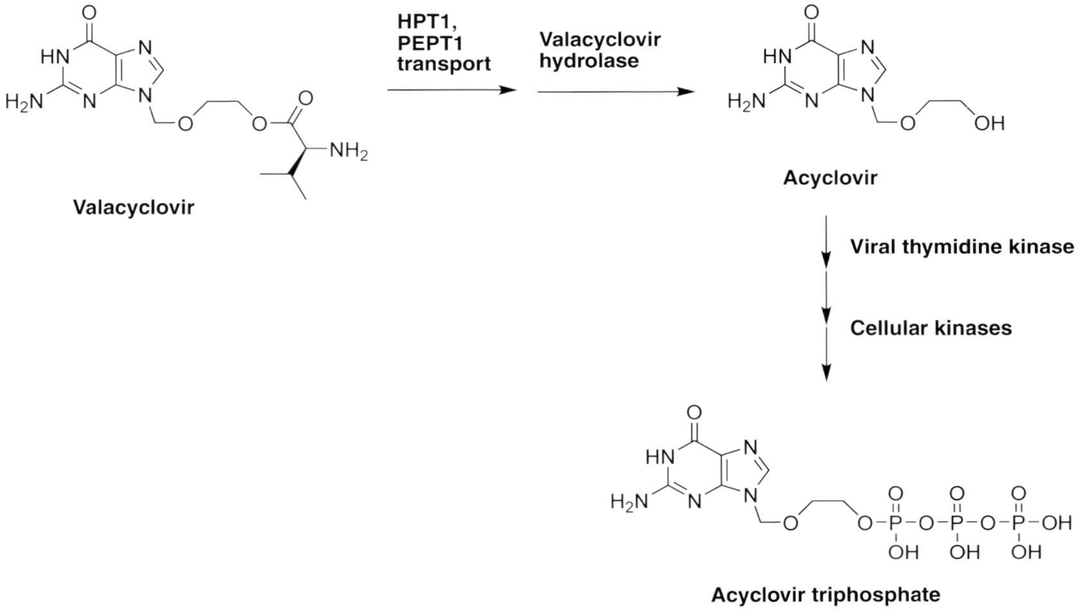

**Figure 9.15 ▶** An example of the use of a prodrug to improve carrier-mediated transport. Valacyclovir, an L-valine ester prodrug, is transported across intestinal epithelial cell membranes by the HPT1 and PEPT1 transporters, and is then enzymatically hydrolyzed to the antiviral agent acyclovir before being converted to the true active component, acyclovir triphosphate, in cells.

Note, however that, although this is the "drug" the prodrug delivers, it isn't really the active species. Only conversion of acyclovir to its triphosphate in virally infected cells gives a compound capable of inhibiting DNA polymerase and terminating DNA chains, the MOA for this drug. One implication is that, in a sense, valacyclovir is a prodrug of a prodrug. Another is that the structure–activity relationship (SAR) for antiviral compounds like acyclovir, AZT (incidentally, another actively transported compound), and many other nucleoside and nucleoside-like compounds is particularly hard to deconvolute, since it represents a composite of absorption via passive and/or active transport, varying affinities for subsequent kinases, and the activity of the triphosphate eventually produced.

Other drugs including the ACE inhibitor enalapril (Figure 9.16) have taken advantage of active transporters as well. Although a glance at the structures of the active component, enalaprilat, a diacid, and enalapril, a monoacid monoester, might lead one to believe that absorption has been improved by making the compound less polar and thus better passively transported, the evidence is that enalapril is actually a better substrate for carrier-mediated transport than the diacid.[66] Results like these have piqued interest in increasing absorption via prodrugs designed to be actively transported,[67] and ways of screening compounds for their affinities for transporters have appeared.[68]

As with the use of prodrugs to overcome solubility problems, reliance on the use of suitable prodrugs to increase absorption is really contraindicated in the early stages of

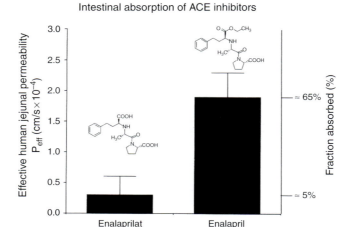

**Figure 9.16 ▶** Effective permeabilities and fraction absorbed for the ACE inhibitor enalaprilat and its prodrug, enalapril. (Reproduced with permission from Petri, N., Lennernäs, H. "In vivo Permeability Studies in the Gastrointestinal Tract of Humans", Chapter 7 in van de Waterbeemd, J., Lennernäs, H., Arturrson, P., Eds. *Drug Bioavailability*, Wiley-VCH Verlag GmbH & Co. KGaA, p. 171, copyright 2003).

drug discovery. Although prodrugs remain a sometimes valuable option, compounds which can be well-absorbed without their assistance are much preferred . . . *if* they can be found!

## 9.2 Metabolic Stability

Metabolically unstable compounds will tend to have poor PK properties like bioavailability and in vivo half-life due to rapid degradation. Metabolic instability can also make for other problems like toxicity or drug–drug interactions caused by metabolites. As with other important drug properties like absorption, metabolic stability wasn't emphasized until late in the game under the old system, and careful, time-consuming studies sometimes revealed metabolic flaws in compounds that chemists had spent many months optimizing for activity. That lesson having been learned, these days it's a property optimized in parallel with others from the early stages of a project.

The exact point in a project where the metabolic stability screening begins and its relative weighting against other properties like cell permeability and potency is done will vary somewhat by project and by company. Using metabolic stability early on, as a strict criterion for making analogs of confirmed hits risks dropping of promising hits that might easily have been "fixed," say by converting a labile ester to an amide. But looking at it too late could mean spending all your time making compounds that have little chance of working in vivo. So, somewhere in the HTL or LO stage it will become an important factor in every project. The information it provides at the discovery stage can alert chemists to potentially problematic compounds unlikely to work in vivo, to establish SPRs, which can guide them to better ones, and to aid in selecting compounds to use in animal studies.

## 9.2.1 Common Metabolic Transformations

Xenobiotics are transformed and cleared from the body through a highly evolved system that effectively prevents the accumulation of lipophilic and potentially toxic substances. The process used often involves two stages. In the first, *Phase I metabolism*, the drug's polarity is slightly increased through what's usually an oxidative or hydrolytic process like hydroxylation or ester hydrolysis. This can be followed by *Phase II* metabolism, where the polarity is further increased through an enzyme-catalyzed conjugation of the resulting compound with glucuronic acid, sulfate, etc. Having turned a lipophilic compound into a hydrophilic one, the now water-soluble substance can be excreted. Alternately, if a conjugatable functional group like a phenol or carboxylic acid is already present in the parent drug, a Phase II process can occur without Phase I preceding it, making these terms a little fuzzy. Note that not all drugs need to be metabolized—much less metabolized by a hepatic oxidative process—in order to be cleared. Some, like valsartan (Figure 8.13) are mostly excreted unchanged.

The study of drug metabolism is fascinating, but vast and complex. For those wishing to learn more, two excellent sources are recommended.[69–70] Texts on drug metabolism tend to focus on the various enzymes involved, like CYPs (Figure 9.17), *flavin monooxygenases* (*FMOs*), etc., their structures when known, and their reaction mechanisms. But, from the point of view of a medicinal chemist trying to design compounds which avoid too-rapid metabolism or, in rare cases, attempting to hasten the demise of a too-stable molecule,[71] what really matters is the correlation between the structure and metabolism. Which sites on a given molecule are likely to be metabolized? How can it be stopped or slowed down? Are there particular metabolic enzymes that need to be avoided? Since, every medchem compound can't be tested in the clinic or even in animals, how will the metabolic stabilities of project

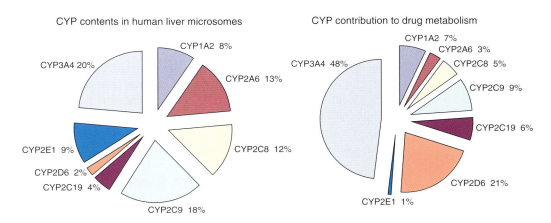

**Figure 9.17 ▶** Left: Percentage of total CYP protein content (pmol/mg) in one sample of pooled HLMs.[90] Numbers don't add up to 100, since not all CYPs are shown. Right: Contribution of individual CYPs to the metabolism of 403 marketed drugs.[72] Note how a disproportionately large amount of this pie belongs to only two enzymes, CYP3A4 and CYP2D6.

compounds be determined? It's this practical point of view that this section addresses, beginning with a very brief rundown on the types of metabolic transformations the chemist is most likely to deal with.

### 9.2.1.1   Hydrolysis of Esters and Amides

As reactions of carboxyesterases and peptidases were already discussed in the plasma stability section, only a few more comments will be made here. Carboxyesterases aren't confined to blood, but are expressed ubiquitously and can be found at high levels in liver and kidney. Their roles in unmasking free carboxyl groups often pave the way for subsequent Phase II (conjugation) reactions. Also of note, other types of enzymes are sometimes involved in hydrolysis, like CYP3A4, which hydrolyzes the drug fluticasone (Figure 9.21) by an oxidative mechanism.

### 9.2.1.2   Oxidation of Arenes, Alkenes, and Alkynes

It's been called "nature's most versatile biological catalyst"[73] and compared to a molecular blowtorch for its ability to catalyze oxidations that would otherwise require extremely high temperatures.[74] Altogether, it represents a most extraordinary series of in vivo synthetic reagents, one that will react with almost anything a medicinal chemist can throw at it. When it comes to Phase I metabolism, CYP-catalyzed oxidative reactions usually predominate.[75] The most versatile of them all, CYP3A4,[76] has been estimated to be involved in the metabolism of about half of all marketed drugs (Figure 9.17). In the presence of reduced *nicotinamide adenine dinucleotide phosphate* (*NADPH*)-*CYP reductase*, these heme-containing enzymes can transfer one oxygen atom of $O_2$ to a substrate and convert the other to water, making them *mixed function oxidases.*

Drugs containing aromatic rings are often suitable substrates for CYP-catalyzed hydroxylation, especially if electron donating substituents are present, and the regiochemistry one ends up with tends to be that predicted for electrophilic aromatic substitution. The reaction usually involves the formation of an epoxide intermediate (arene oxide) for *ortho-* and *para*-hydroxylations, but can also proceed via hydrogen abstraction and direct insertion. The intermediacy of an electrophilic arene oxide, which normally rearranges to a phenol, means that other products might be obtained. Arene oxides can be trapped by a nucleophilic species like glutathione, or (much worse!) a cellular protein that might make for toxicity, as discussed later. Alternately, arene oxides can, with the help of another metabolic enzyme, *epoxide hydrolase*, be opened up to diols, or a substituent might migrate via the famous *National Institutes of Health* (*NIH*) *shift.*[77] In most cases, however, the simple phenol is the predominant product and often the major observed metabolite.

CYPs are good at epoxidizing alkenes and alkynes as well, which is probably one reason why more drugs don't contain these groups. Then again, just because a reaction *can* happen doesn't mean that it will. For carbamazepine (Figure 9.18), epoxidation followed by diol formation is the major metabolic route, but for tamoxifen, metabolic epoxidation of the central double bond hardly happens, and other CYP-catalyzed reactions, including arene epoxidation and N-demethylation (discussed in a later section) are much more in evidence.

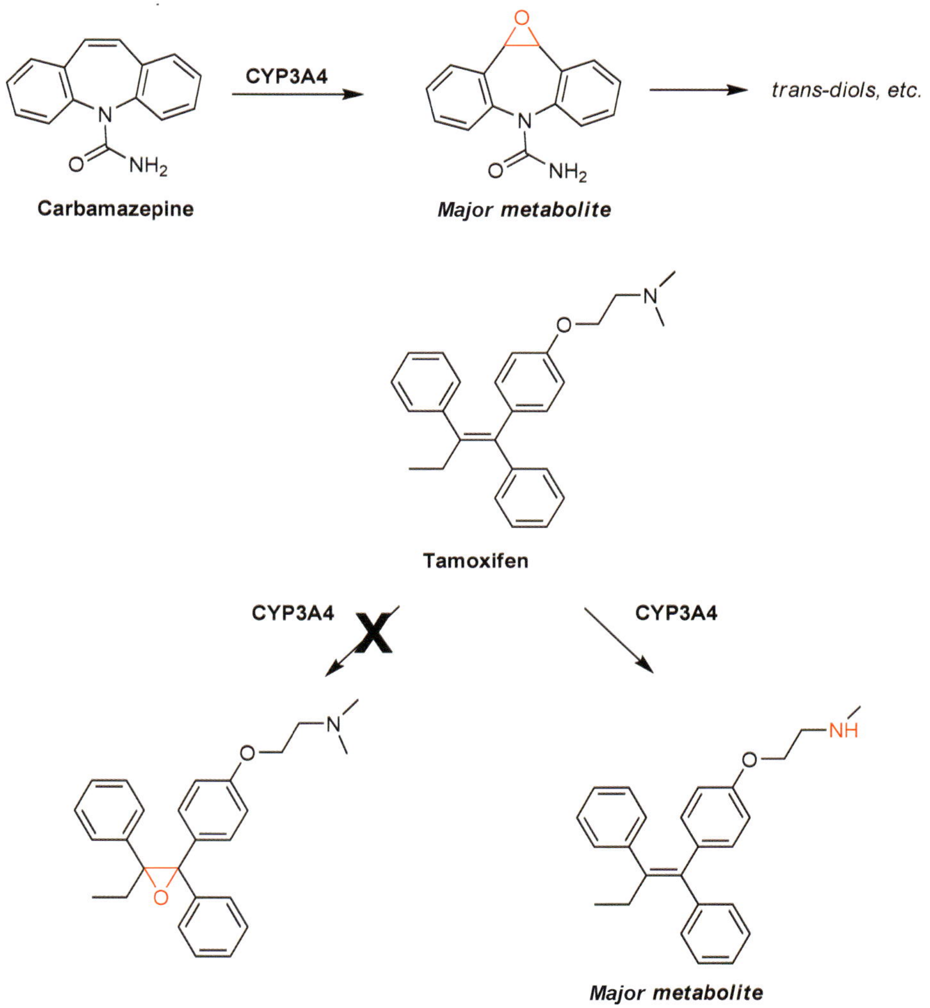

**Figure 9.18** ▶  Relative contribution of CYP3A4-catalyzed double bond epoxidation to drug metabolism. For carbamazepine, this represents the major metabolic route, while for tamoxifen CYP3A4-catalyzed epoxidation of the alkene is negligible and the enzyme instead *N*-demethylates it.

## 9.2.1.3  Aliphatic Hydroxylation

CYP-catalyzed hydroxylations of aliphatic carbons are one of the most common metabolic transformations observed for drugs. As in CYP-catalyzed direct insertion into arenes, hydrogen abstraction to produce the radical is the rate-limiting step, so it isn't too surprising that hydroxylation at allylic or benzylic positions is most often observed. Calculations of H-radical abstraction energies have thus played a major role in computational programs designed to predict sites of CYP metabolism.[78] In a testament to the versatility of CYPs, however, even unactivated primary methyl groups can be substrates. Hydroxylation of the terminal methyl group in an aliphatic side chain is known as *ω oxidation*, while the reaction at the next position is referred to as *ω-1-oxidation*. Simple *n*-hexane can be hydroxylated at all three possible positions.[79]

### 9.2.1.4   Oxidations At or Adjacent to Heteroatoms

Amines, ethers, and thioethers can sometimes seem like magnets for oxidative metabolism, being subject to a variety of reactions catalyzed by CYPs or FMOs. An attempt to simplify, but hopefully not oversimplify, the possible metabolic fates of amines in the presence of various CYPs is shown in Figure 9.19. A radical cation is initially formed at nitrogen. This sometimes goes on to yield an amine *N*-oxide for tertiary amines or a hydroxylamine for primary or secondary ones. More often, however, an *α*-hydrogen is extracted and a carbinolamine, the net product of *α*-hydroxylation, is formed.

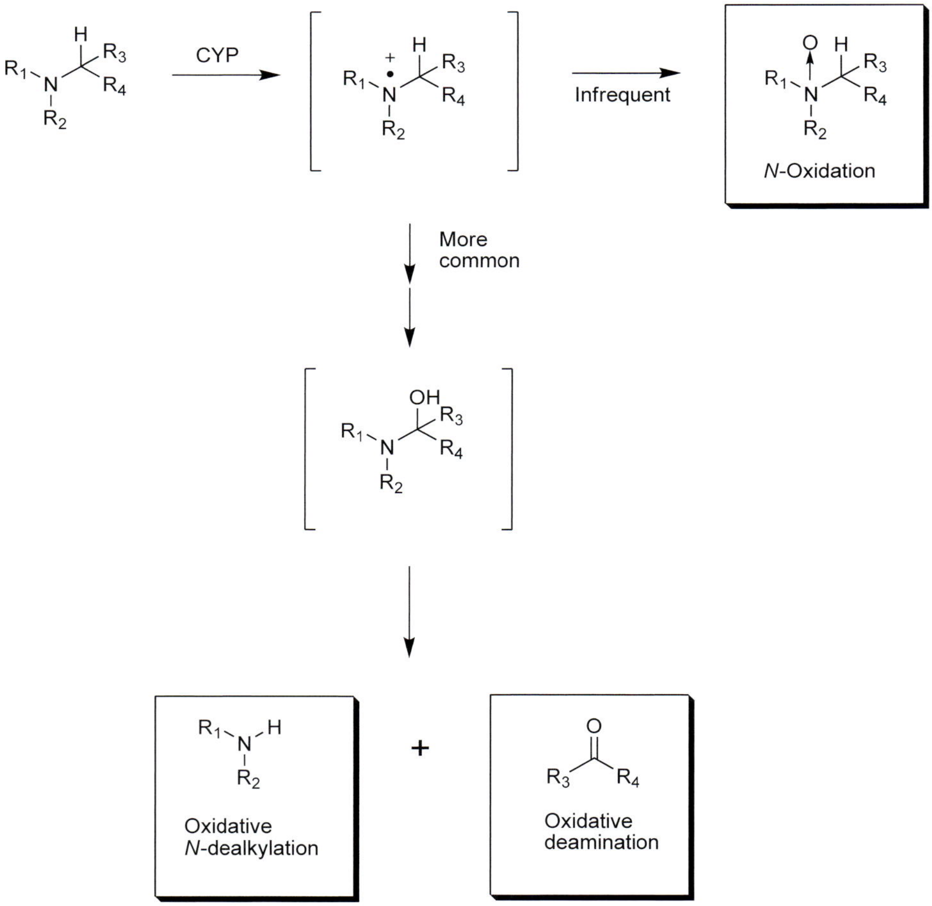

**Figure 9.19** ▶ Simplified scheme illustrating the pathways involved in the CYP-catalyzed oxidative metabolism of amines.

Decomposition of this unstable intermediate produces an amine plus either an aldehyde or a ketone. The term used to describe this process depends on which of the two cleavage products one is interested in. If most of the original molecule remains with the amine (e.g. the N-demethylation of olanzapine shown in Figure 9.21), it's considered *oxidative N-dealkylation*, but if it ends up in the carbonyl-containing fragment (e.g. conversion of a primary phenethylamine to a phenylacetaldehyde and ammonia) it would be instead referred to as *oxidative deamination*. Notably, FMOs can also carry out N-oxidations of secondary and tertiary amines, for example the pyrrolidine nitrogen of nicotine, but not α-hydroxylations to produce the cleavage products that CYPs can provide.[80]

Both CYPs and FMOs can oxidize the sulfur atoms in sulfides and sulfoxides, making these frequent metabolic sites, and the latter enzymes (there being five known families of FMOs) can oxidize trivalent phosphorus compounds as well. In a process analogous to amine α-hydroxylation, CYPs can also oxidize carbons alpha to oxygen or sulfur as in $ArXCH_3 \rightarrow ArXCH_2OH \rightarrow ArXH + H_2C = O$.[81] Like carbinolamines, the intermediate hemiacetals and hemithioacetals decompose to give net dealkylation. The same type of mechanism involving an α-hydroxylated intermediate explains the ability of CYPs to sometimes cleave esters and carbamates (e.g. conversion of loratidine to desloratidine, both shown in Figure 6.31, by CYP3A4).

Substrate specificities can be broad and overlapping both among CYP family members and between CYPs and FMOs, and since metabolic transformations are often sequential (e.g. aliphatic hydroxylation being followed by oxidation by alcohol dehydrogenase, further oxidation to the acid, etc.), many enzymes and many metabolites can be involved in processing a single drug. Only a few of the most important enzymes involved in Phase I transformations have been mentioned here. For these and many others (monoamine oxidase, xanthine oxidase, etc.), further information can be found in the previously cited reviews.[69–70] Bear in mind too that not all Phase I reactions are oxidative: enzymes like carbonyl reductases are important in metabolism as well.

### 9.2.1.5 Glucuronidation

Drug metabolism seems to aim at three things: reducing biological activity, increasing water solubility, and facilitating excretion. Although Phase I processes can sometimes do the trick,

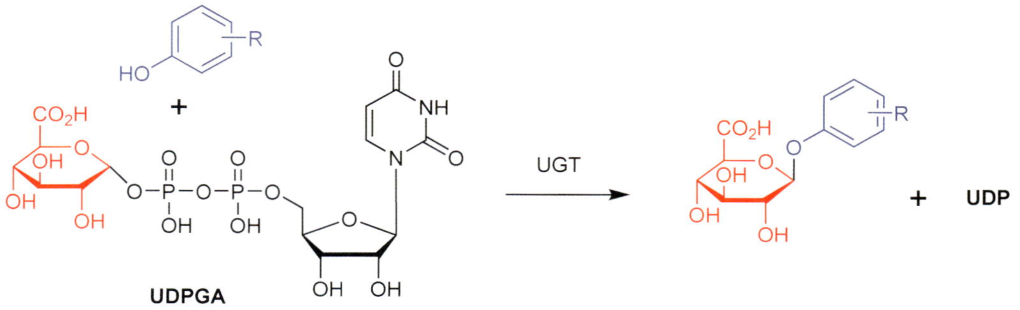

**Figure 9.20 ▶** Glucuronidation of a phenol by UDPGA, catalyzed by a UGT. Note the β-configuration of the product.

they often require a little help from further Phase II reactions. The most common of these is glucuronidation, the process by which a compound with an *N*, *O*, *S*, or sometimes even a *C* nucleophile is tagged with the large and polar glucuronic acid group of *UDP-α-glucuronic acid* (*UDPGA*) in a reaction catalyzed by a member of a family of enzymes called *UDP-glucuronosyl transferases* (*UGT*s), as illustrated in Figure 9.20.

Special organic anion transporters then ensure rapid excretion of the conjugate in bile or urine. Still, things aren't always that simple. In brain, UGT2B7 catalyzes the formation of a morphine (mono)glucuronide at each of the two possible sites, the 3-phenol and the 6-alcohol. While the 3-*O*-glucuronide has no analgesic activity, the 6-*O*-glucuronide is a more potent analgesic than the parent drug, and its charge may prevent it from easily crossing the blood–brain barrier, giving it a longer residence time in the CNS than morphine itself.[82]

Another interesting complication stems from the fact that the β-linkage of glucuronides makes them (unlike UDPGA) substrates for β-glucuronidase, which is present in intestinal microflora. So a drug can be glucuronidated and secreted into bile, which passes into the intestines. But there, instead of the glucuronide being excreted, it might be cleaved by microbial enzymes, releasing the deconjugated compound for absorption back into the bloodstream. If this is the parent drug, say, formed from glucuronidation of a phenol originally present in the molecule (e.g. mycophenolic acid, Figure 7.3)[83] plasma levels of the drug, which had been falling, will begin to rise again hours after administration and a second peak in the PK graph will be observed. In yet another affront to standardized nomenclature, this effect is variously known as *enterohepatic circulation*, *enterohepatic recirculation*, *enterohepatic cycling*, or *enterohepatic recycling*, and it can happen with other conjugates not discussed here, like sulfates, as well.[84]

More often, though, glucuronidation does a straightforward job of taking the drugs out of circulation, thus frustrating the medicinal chemist. Ways for her to fight back are discussed later.

### 9.2.1.6   Overview

The best way to get a quick overview of the real-world metabolism is to see how popular prescription drugs are metabolized. Figure 9.21 outlines the major metabolic sites for the world's 10 top-selling small molecule prescription drugs of 2006.[85] The appearance of 11 structures in this top 10 list is no mistake, since both fluticasone propionate and salmeterol xinafoate (counterion not shown) are components of the #2 drug, Advair, used for the treatment of asthma. Only the one or two most predominant primary metabolic processes are shown.

It's apparent that CYPs are involved in the metabolism of all but one, Plavix (clopidogrel), and even for that compound they're important downstream, after initial hydrolysis of the ester. What's not apparent from the figure is that metabolism isn't always something that limits drug effectiveness. Some drugs like valsartan are mostly excreted unchanged, while for others, like the top-seller Lipitor (atorvastatin) metabolism affords compounds that are still active. In Lipitor's case, active metabolites are said to account for 70% of the pharmacodynamic (PD) effect the drug produces. Much more worrisome would be reactive (electrophilic) metabolites that make for toxicity or metabolites that affect CYP expression or function.

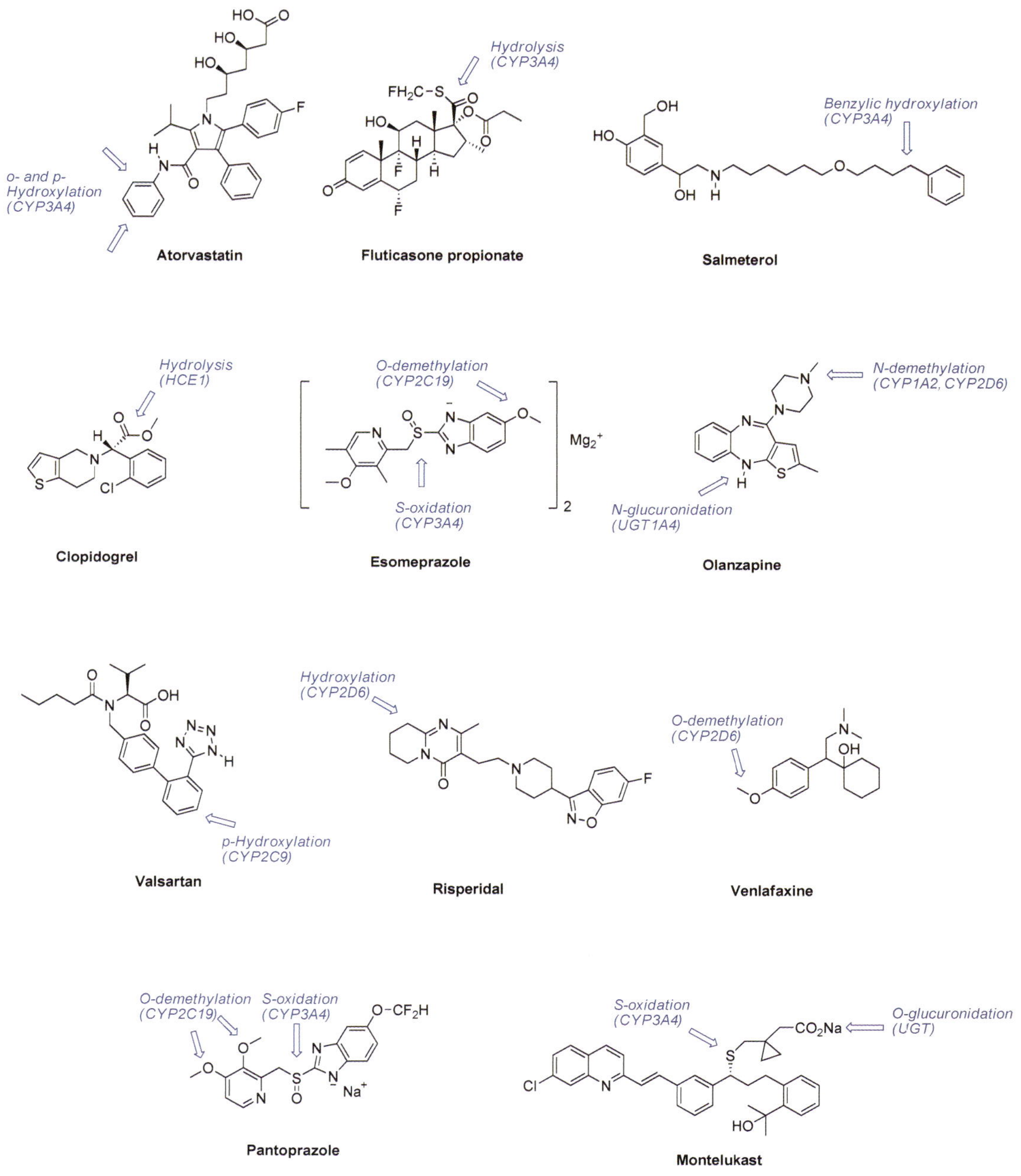

**Atorvastatin**

*o- and p-Hydroxylation (CYP3A4)*

**Fluticasone propionate**

*Hydrolysis (CYP3A4)*

**Salmeterol**

*Benzylic hydroxylation (CYP3A4)*

**Clopidogrel**

*Hydrolysis (HCE1)*

**Esomeprazole**

*O-demethylation (CYP2C19)*

*S-oxidation (CYP3A4)*

$Mg_2^+$

**Olanzapine**

*N-demethylation (CYP1A2, CYP2D6)*

*N-glucuronidation (UGT1A4)*

**Valsartan**

*p-Hydroxylation (CYP2C9)*

**Risperidal**

*Hydroxylation (CYP2D6)*

**Venlafaxine**

*O-demethylation (CYP2D6)*

**Pantoprazole**

*O-demethylation (CYP2C19)*  *S-oxidation (CYP3A4)*

**Montelukast**

*S-oxidation (CYP3A4)*

*O-glucuronidation (UGT)*

**Figure 9.21** ▶ The top 10 small molecule prescription drugs by worldwide sales for 2006,[85] their major metabolic sites, and the enzymes principally involved in their metabolism.[86] Counterions for atorvastatin and salmeterol xinafoate are omitted. Updates to this figure will be posted to www.realworlddrugdiscovery.com.

These will be discussed in later chapters, but first a look at how metabolic stability for project compounds is assessed is in order.

## 9.2.2  Assessing Metabolic Stability

Metabolic stability assays lack a convenient "universal" endpoint like the fluorescence used in many HTS activity assays. Instead, an HPLC peak for each individual compound needs to be measured. A gradient capable of separating sample from metabolites and contaminants is required, and this takes a few minutes, unlike a fluorescence read. As a result, metabolic stability assays, although now reportedly capable of running 1200 samples/week with one dedicated robot, one dedicated LCMS, and one very dedicated scientist,[87] aren't as high-throughput as many activity screens. The term higher throughput is often used. Still, the capacity to run even a few hundred samples per week makes it possible for medicinal chemists to get timely data on many of their compounds which can feed back into the next round of synthesis. Listed below are the most common ways of assessing stability to hepatic metabolism, which is usually the most important contributor to drug biotransformation. Those interested in computational models are referred to a recent review.[88]

### 9.2.2.1  Recombinant Drug Metabolizing Enzymes

Individual *drug metabolizing enzymes* (*DMEs*) like CYPs, FMOs, and UGTs are now routinely expressed, usually in insect cells using a baculovirus. Ultracentrifugation provides vesicles called *microsomes*, which contain these membrane-bound DMEs. One popular type of recombinant CYP preparation called Supersomes™ comes with CYP reductase, a necessary cofactor, co-expressed.[89] Not only individual CYPs (2B6, 3A4, etc.) but also allelic variants like CYP2C9*2 are available for the study of possible pharmacogenomic effects on metabolism.

The stability of a given compound could be assayed against a full collection of these individual DMEs or a cocktail thereof as a primary screen for stability to Phase I metabolism; running each individually would obviously be the most straightforward route to SPR. But in the real world, this turns out to be more expensive and less practical than simply using liver microsomes instead, as described in the following section. Because of this, *recombinant cytochrome P450s* (*rCYPs*) normally play a role later on in *reaction phenotyping*, the determination of which DME is involved in a particular metabolic transformation, for a few compounds of interest instead.[90]

### 9.2.2.2  Liver Microsomes

Microsomes from liver, which are vesicles containing hepatic *endoplasmic reticulum* (*ER*), contain membrane-bound DMEs including CYPs, FMOs, and UGTs. Screening for metabolic stability using microsomes is the most popular way of looking at a compound's stability to most, but not all important metabolic transformations. Microsomes from many different species (mouse, rat, monkey, etc.) including man are commercially available and are stable for months when kept at $-80°C$, although their activities drop off at room temperature after an hour or two, making them less useful for long time-course experiments. Vendors provide specifications for the measured DME activity levels of each batch. Microsomes aren't "complete" metabolizing systems, so other things like NADPH-regenerating system for CYPs and UDPGA for UGTs need to be added to ensure each type of activity. Notably, microsomes from liver, although the type most frequently used for testing, aren't the only kind available. Microsomes from small intestines and microsomes derived from kidney and brain, for example, also have their uses.

The metabolic stability data that are available to scientists in a drug discovery project at the HTL or LO stage will most likely come from liver microsome stability testing, often run in parallel using both rodent and human microsomes due to species differences in metabolism. A brief (10–30 min) incubation of the test compound with microsomes and co-factors will be followed by LCMS analysis to determine what percentage of the parent compound remains. For primary metabolic stability screening, only a single time point will probably be used. Results will be expressed either as percentage (%) remaining at, say, 30 min, or perhaps a $t_{1/2}$ estimated from the single time point.[91] Sometimes, these resulting half-lives will be converted into a value known as *hepatic intrinsic clearance*, $Cl_{int}$.[92] Obviously, microsome stability experiments could provide a lot more information than this, including multiple time point curves, reaction kinetics, and even metabolite identification, but this implies more effort and lower throughput, so these types of experiments are reserved for compounds later found to be of particular interest. For primary screening purposes, a simple estimate of stability to microsomes is enough.

As usual, though, the devil is in the details. The conditions used to run a particular microsome stability screen can radically affect the results. Microsomes are very sensitive to the presence of co-solvents, so DMSO concentrations normally need to be kept at no more than 0.2% and ACN at no more than 1%.[87] Insoluble compounds, then, can be particularly problematic. Compound precipitation can mean that most of the material is unavailable for metabolism, but may dissolve when later diluted with ACN for analysis, resulting in a falsely high stability estimate. Ways of optimizing assays to minimize such effects have recently appeared in the literature.[93] The concentration of compound used in the assay can make for more than a 70-fold difference in the percent remaining at the measured time point, while the concentration of the microsomes is also important.[87]

However, the biggest variable of all is probably the metabolic activity of a given batch of microsomes, which varies an amazing amount from vendor to vendor, batch to batch, animal to animal, and donor to donor. As previously mentioned, these are determined and specified by the vendors, but they do vary a lot. For *human liver microsomes* (*HLMs*), much depends on sample preparation, the pharmacogenomics of the donor, and the enzyme-inhibiting or enzyme-inducing substances they may have been exposed to. Because of all these variabilities, microsomes are often pooled from a group of 10 or more donors, although microsomes from individual donors are also available. Correlating the stability of a compound with the CYP activities of different individual microsome batches can, like the use of rCYPs, be useful for reaction phenotyping.

But for the routine stability screening, even with pooled microsomes, variability in enzyme activity can make it difficult to compare results obtained for different compounds using different batches of microsomes. For this reason, standards of known drugs will be run as controls and incubation times may be varied for the different batches of microsomes to try and normalize the results.[87] Including a particular project compound as a benchmark in each run is also a good practice.

Keeping in mind that liver microsome stability testing doesn't represent all possible metabolic processes and that microsome batch variability sometimes makes comparisons difficult, the data obtained from these assays can still be some of the most useful information a medicinal chemist can get.

### 9.2.2.3   Liver Cytosol and S9 Subcellular Fractions

Hepatic cytosol contains three important soluble DMEs that aren't found in microsomes, *N-acetyl transferase* (*NAC*), *sulfotransferase* (*SULT*, sometimes abbreviated *ST*), and *glutathione S-transferase* (*GST*). When the appropriate co-factors, like Acetyl CoA for NAC, are added stability to these three conjugating enzymes can be studied.

Microsomes contain membrane-bound DMEs and cytosol has soluble ones, but another available product, liver subcellular fraction S9, the supernatant obtained from centrifugation at $9000\,g$, contains both. When all of the necessary co-factors are added, a more complete representation of the hepatic metabolism of a compound can be obtained. However, an even better one could come from the use of whole cells, hepatocytes, as discussed below. Consequently, S9 most often finds its use in conjunction with the Ames test for mutagenicity (Chapter 10). A metabolite rather than the parent compound can often be the genotoxic species, and incorporating S9 incubation allows both to be evaluated.[94]

### 9.2.2.4   Hepatocytes

Freshly isolated hepatocytes lose much of their metabolic activity after just a few hours. For rodents, this might still be acceptable if sacrifice and isolation are done in the lab just prior to the assay being run. For obvious reasons this isn't done for human hepatocytes. Fresh primary human hepatocytes are the gold standard for metabolism research, but are difficult to get (often coming from liver resection surgery) and very expensive, preventing their more widespread use. Fortunately, advances in cryopreservation techniques have recently made it possible to store frozen human hepatocytes for up to a year, then thaw them out and use them with only a slight loss of activity. As a result, human hepatocyte stability screening has become widespread in drug discovery recently. Although microsomes are still cheaper and more popular for primary metabolic stability screening, cryopreserved human hepatocytes can now be used in 96-well plate format as an aid to early stage drug discovery efforts.[95]

Hepatocytes contain both membrane-bound and soluble DMEs and they represent a more realistic model of hepatic metabolism than one could get from subcellular fractions. Even using combined microsomes and cytosol or S9, for example, couldn't model effects due to transport across cell membranes that are inherent in the stability data obtained using hepatocytes. Being fairly self-reliant, hepatocytes don't require the addition of enzyme co-factors for metabolic stability testing. As with hepatic microsomes, metabolizing abilities can vary from batch to batch and donor to donor, so cryopreserved hepatocytes are usually pooled except for reaction phenotyping or pharmacogenomic purposes. Many other factors that apply to microsome stability screening (the need for low co-solvent concentrations and soluble compounds, lack of universal detection methods) also apply to hepatocyte stability screens.

Hepatocyte stability data, being more complete, are preferred to microsome data when doing in vitro in vivo *extrapolation* (*IVIVE*),[96] that is, using in vitro ADME data to predict in vivo properties like hepatic clearance and the risk of *drug–drug interactions* (*DDIs*) that we'll see later. Being living cells, cultured hepatocytes can be used to study the changes in gene transcription upon exposure to drugs[97] and have proven to be useful to model enzyme induction in man.[98]

Other ways of looking at hepatic metabolism (e.g. transformed cell lines and liver slices) have a place in drug metabolism studies too, but are less frequently seen early on in drug discovery and won't be discussed here. Instead it's time to tackle the critical issue of how poor metabolic stability, once determined, might be improved.

### 9.2.3   Improving Metabolic Stability

#### 9.2.3.1   Metabolite Identification

Having metabolic stability data alone will suffice for rank ordering a series of compounds. Sometimes that's all a chemist wants to do, for example in choosing a compound to go into a PK study. If the goal is to improve stability, though, knowing the structure of the principal metabolite(s) can help. Later on, when the project produces a clinical candidate, major metabolites will need to be unambiguously identified and characterized, which involves a fair amount of effort from metabolism, radiochemical synthesis, and toxicology groups.[99] In the early stage drug discovery, though, fewer resources and fewer rigors are involved in "metabolite ID". Just how far metabolite identification will be pursued early on depends on the prevailing *mores* and its importance to a given project. Ways it can be looked at include educated guesses based on the kinds of known metabolic patterns described earlier that fit with observed SPR, computer programs like MetaSite[100] that are designed to predict the most prominent metabolic sites, and the hard data produced by LC/MS/MS analysis after microsome or hepatocyte incubation.

Figure 9.22 demonstrates the type of information the last of these can give. The chromatogram shows that after 1 h incubation with rat liver microsomes, more *O*-debenzylated material is present than the parent compound, and that hydroxylation of the leucine side chain is occurring as well. Knowing what the compound's metabolic hotspots are, structural modifications to prevent these reactions can be carried out if the aim is to prolong in vivo exposure to the parent compound, which is normally the case except for prodrugs and antedrugs.

#### 9.2.3.2   Caveats

The benefits of improved metabolic stability have already been mentioned, but keep in mind that not every compound will need to have its stability toward metabolism improved. Some may be stable enough to begin with. For some, excretion as the unchanged parent drug will be the major route of elimination. A few may even be *too* stable (for days or weeks), which has the downside of making it impossible to quickly terminate exposure to the drug when desired. In this case "soft" groups capable of being metabolically transformed, like the *para*-methyl group of celecoxib, which is oxidized by CYP2C9,[101] might intentionally be introduced into the molecule to *reduce* stability.

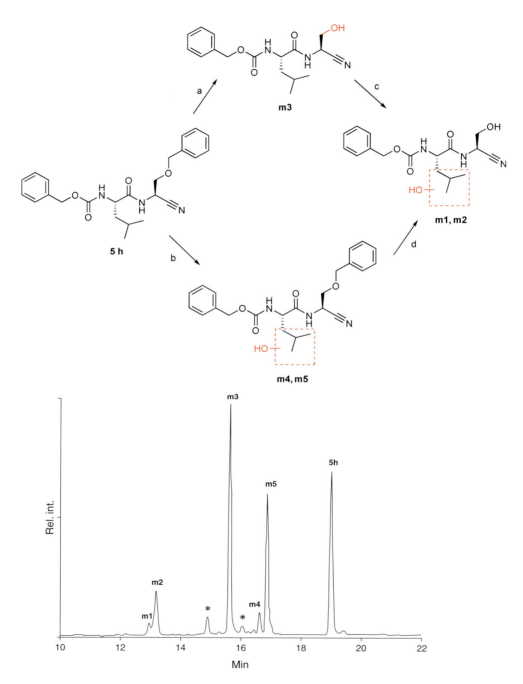

**Figure 9.22** ▶    Proposed metabolic pathways of the Cathepsin K inhibitor **5h** and selected ion chromatogram 1 h after its incubation with rat liver microsomes. (Reprinted with permission from Altmann, E., et al. Dipeptide nitrile inhibitors of cathepsin K. *Bioorg. Med. Chem. Lett.* **2006**, *16*, 2549–2554, copyright 2006, Elsevier).

For the many that do require modifications to increase stability, the advantages that come with these improvements can come at a price. As an outstanding review on the subject of metabolic stability in drug design points out,[102] the following caveats need to be kept in mind:

1. Fixing a metabolic stability problem at one position can lead to a new problem somewhere else on the molecule, a second meaning (the other being a PGx-induced changes in metabolic pathways) of the term *metabolic switching*.

2. Fixing a metabolic stability problem often leads to problems with absorption since, as we'll see, the reduced lipophilicity that often helps improve the former can have the opposite effect on the latter. Since both absorption and stability to first pass metabolism are important components of bioavailability, increasing one while decreasing the other can leave one with either no net gain or even a net loss— something like winning a battle but losing the war.

3. Fixing a metabolic stability problem won't necessarily result in better in vivo exposure for another reason too: the compound might just be excreted faster instead.

4. Fixing a metabolic stability problem might be accomplished by unintentionally turning the compound into a CYP inhibitor, which, as discussed later, is probably a bigger problem for a potential drug to have than insufficient exposure, since it affects other drugs as well. For this reason CYP inhibition assays should be run alongside metabolic stability assays.

5. Fixing a metabolic stability problem in animals won't necessarily help in man and vice versa, since DMEs and metabolites can vary with species. Stability testing therefore should be carried out in the relevant systems for both.

All of these things should be kept in mind, but shouldn't discourage the researcher or keep him from trying to improve the metabolic stability of compounds requiring such optimization. None of these obstacles are unavoidable and, as shown below, plenty of successful examples of improved metabolic stability exist.

### 9.2.3.3  Structural Modifications to Improve Metabolic Stability

As metabolic stability implies resistance to reactions catalyzed by many enzymes, some of which allow for multiple binding modes and resulting attack at more than one site in a given molecule, improving this property can be much less straightforward than one would like. There may not be any royal road to improved metabolic stability, but there are a number of methods that sometime work. These are summarized in Box 9.2.

The most direct way to prevent metabolic degradation is to somehow alter the molecule at or very near the site of metabolism. The favorite of these *site-directed* methods involves blocking metabolism by replacing the labile group with a more stable one, which often means replacing an aryl C–H bond which can be hydroxylated with C–F bond which can't.[103] Figure 9.23 shows several examples of this strategy. The anti-anxiolytic drug buspirone is oxidatively metabolized by CYP3A4, with one of the major sites of hydroxylation being the

---

**Box 9.2    Some Methods for Improving Metabolic Stability**

*Site-directed methods*

▶ Replace atoms (i.e., with a bioisostere) at the site of metabolism

▶ Alter electronic or steric effects to disfavor metabolism

*Lipophilicity-directed method*

▶ Reduce LogP (or LogD) by adding in polarity

*"If you can't beat 'em…" method*

▶ Pre-form a stable active metabolite

---

5-position of the pyrimidine ring. Incubation with the recombinant enzyme showed a half-life of only 4.6 min. Replacement of the 5-H with a fluorine ("Compound **5**") resulted in slightly more than a 2-fold loss of activity, but an increase in half-life to 52.3 min.[104]

The taxane analog "Compound **3**" showed promise in antitumor models, but even though it was fairly stable ($t_{1/2} > 30$ min) to incubation with *mouse liver microsomes* (*MLMs*), it was much more rapidly metabolized by HLMs, with only 60% of the parent drug being present after a 5 min incubation.[105] Chemical synthesis showed the major product, an M + 16 peak in the MS, to be the result of 5-hydroxylation on the pyridine ring. Blocking this with the 5-fluoro analog (Compound **35c**) increased the amount present after 5 min of incubation with HLMs to 99%. The 5-methyl analog, Compound **35a**, however, was even less stable than Compound 3, with only 29% remaining under the same conditions. Here, it's likely that although aryl hydroxylation was blocked, benzylic oxidation was proffered instead, which a DME took advantage of.[106]

Drugs containing phenolic groups offer a handle for Phase II conjugation reactions without the need for any Phase I processes and as a result these are often quickly glucuronidated. For Compound **2** in Figure 9.23, glucuronidation was determined to be the main source of its poor (0.6%) oral bioavailability in mice, although N-demethylation also occurred to a small extent. Bioisosteric replacement of the phenol, however, gave Compound **19**, only slightly less active, but with a quantum improvement in bioavailability (87%). Its *N*-desmethyl analog, Compound **20**, was less bioavailable (42%), possibly because the addition of another N–H bond adversely affected absorption, a common theme as mentioned in the caveats above.[107]

Blocking metabolism needn't always involve direct replacement of an atom at the metabolic site. Electronic effects from substituents at other positions can sometimes do the trick as well. For example, *para*-hydroxylation of a benzene ring can sometimes be curtailed by *meta,meta*-difluorination. If changing the electronics of a molecule can work, it's a cinch that chemists will try using steric effects as well. The improved rat PK properties of Compound **23**, the matrix metalloprotease inhibitor shown in Figure 9.24 (Clp 44 mL min$^{-1}$ kg$^{-1}$), versus Compound **2**, which lacks the α,α-dimethyl group (Clp 86 mL min$^{-1}$ kg$^{-1}$) were attributed in part to steric hindrance of metabolism of the hydroxamic acid.[108]

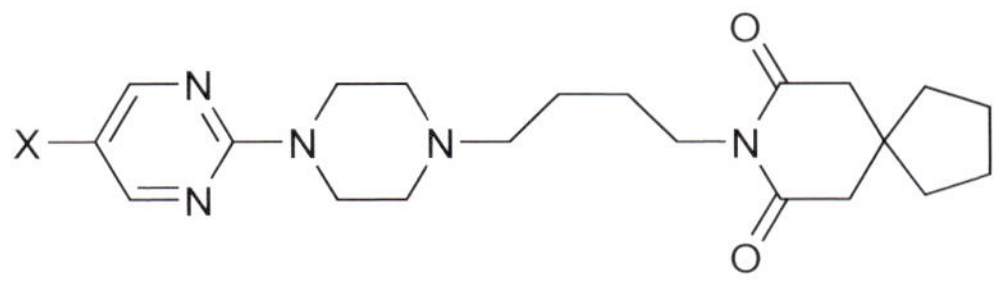


X = H, Buspirone
X = OH, major metabolite
X = F, Compound **5**

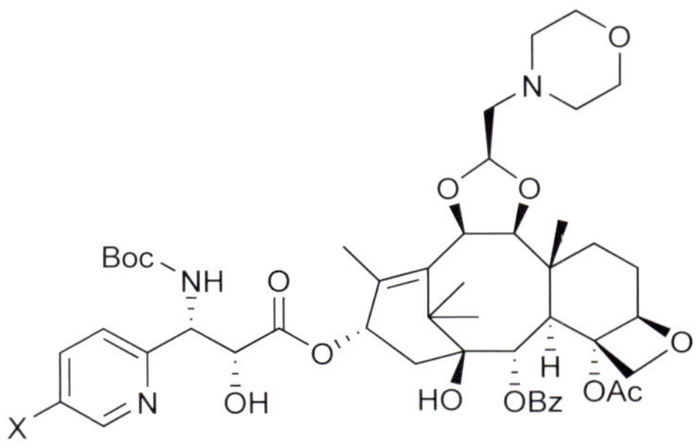


X = H, Compound **3**
X = Me, Compound **35a**
X = F, Compound **35c**

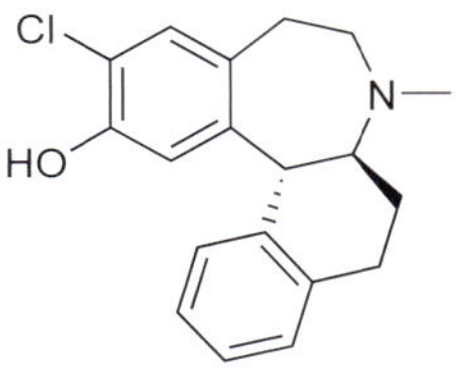


Compound **2**

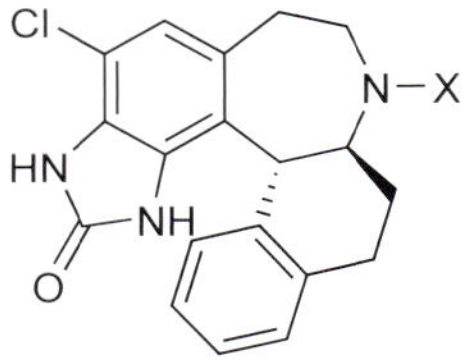


X = Me, Compound **19**
X = H, Compound **20**

**Figure 9.23 ▶** Upper left: 5-$HT_{1A}$ ligands from Tandon, M., et al.[102] Upper right: Taxoids from Takeda, Y., et al.[104] Bottom: $D_1/D_5$ antagonists from Wu, W.-L., et al.[105]

Drug metabolizing enzymes are, after all, just enzymes, even if particularly versatile and accommodating ones, so sometimes changing the structure of a molecule by favoring a different conformation or using a different stereoisomer can push it beyond what a given DME will tolerate in a substrate. Differential metabolism of epimers is common, for example, which also means that stability data obtained on racemic mixtures can be misleading.

Overall, lipophilicity is a particularly important requirement for CYP metabolism, since their *raison d'être* is to make greasy compounds more polar and thus easier to excrete. Because of this, one of the best ways of reducing metabolic instability is to reduce the Log*P* or Log*D* value of the compound. Since chemists rarely leave unnecessary hydrophobic

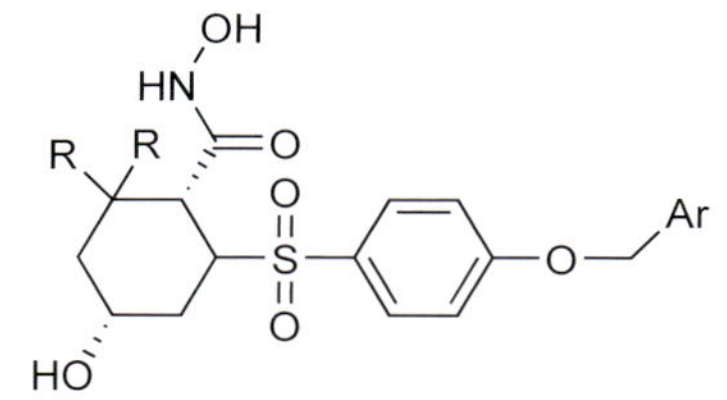

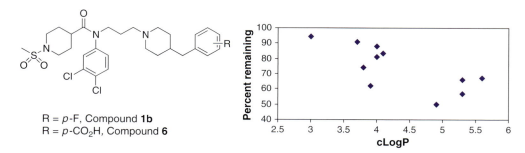

Ar = 2-Methyl-5-fluorophenyl, R = H, Compound **2**
Ar = 2,4-Dichlorophenyl, R = Me, Compound **23**

**Figure 9.24** ▶ *Matrix metalloproteinase-13 (MMP-13) inhibitors from Noe, M.C., et al.[108]*

groups "hanging around" in a molecule, although an occasional alkyl group might be pruned down a bit, in practice reducing LogP usually means adding in polarity somehow. Doing this might create other problems such as poorer absorption, increased protein binding or even the introduction of a new metabolic site, although other properties like solubility might benefit. Still, a long record of success for this method puts it high up on the list of things to try.

A series of CCR5 antagonists like the ones shown in Figure 9.25 were found to be oxidatively metabolized at the left-hand piperidine ring, with only 67% of Compound **1b** remaining after 20 min incubation with HLMs. Stability was improved by introducing polar groups at the opposite side of the molecule where the target activity was less sensitive to structural changes, thereby lowering LogP values. For the carboxylic acid, Compound **6**, 94% of the parent compound survived the 20 min incubation. As the authors noted, not every reduction in LogP led to increased metabolic stability.[109] Plotting out the percentage of parent compound remaining versus cLogP for compounds in the series shown reveals more of a trend than a tight correlation, but the tendency is there and the strategy worked.

Another example, this one involving epoxide hydrolase inhibitors, is shown in Figure 9.26. Here, only 21% of the *para*-fluoro Compound **16d** was left after 1 h incubation with HLMs,

R = *p*-F, Compound **1b**
R = *p*-CO₂H, Compound **6**

**Figure 9.25** ▶ Left: Structures of two Takeda CCR5 antagonists. Right: A plot of percent parent compound remaining after 20 min incubation with HLMs versus cLogP for 11 series compounds substituted at the phenyl ring, including the ones at left.[109]

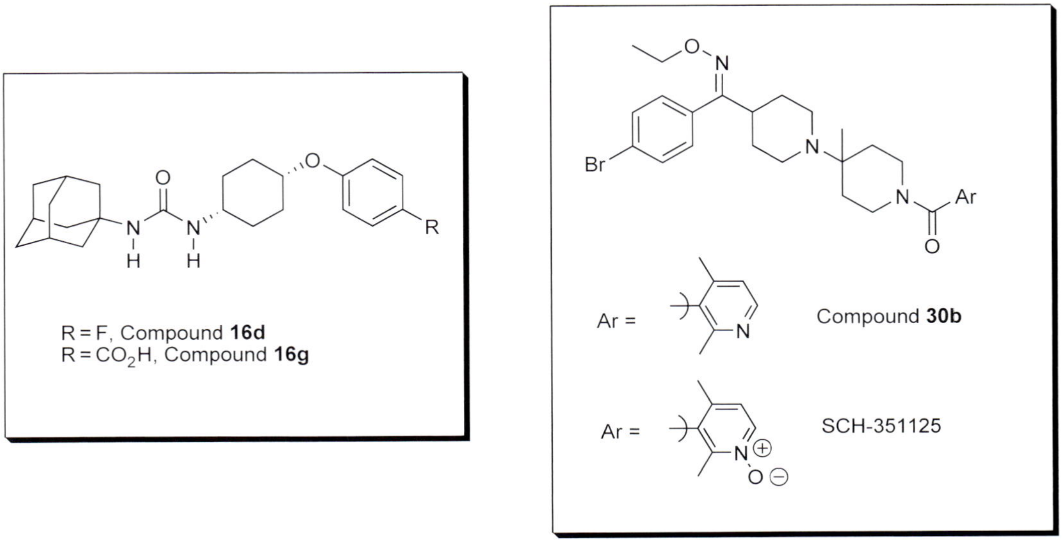

**Figure 9.26** ▶ Left: Epoxide hydrolase inhibitors.[110] Right: CCR5 receptor antagonists.[112]

but 98% of the corresponding carboxylic acid, Compound **16g**, remained under the same conditions. This difference in metabolic stability was reflected in a >100-fold difference, in the *area under the curve* (*AUC*) between the two when dosed orally in dog, a remarkable result.[110] More often, improving metabolic stability by adding in polar groups involves a struggle against decreased absorption and perhaps with other properties as well.[111]

In some cases, metabolic processes themselves can be used to improve metabolic stability. Although most metabolites are less active against the desired target than their parent compounds are, sometimes their activities can be comparable or even greater. It's already been noted that most of the observed PD effects of atorvastatin are due to active metabolites, that desloratidine, the active metabolite of loratidine (both shown in Figure 6.31), is a marketed antihistamine with some advantages over the other compound, and that active metabolites in general can be a fertile source of leads. Major metabolites tend to be fairly stable or they wouldn't be observed as major metabolites. Pursuing such an active metabolite as a drug can be a way of following the advice that "If you can't beat 'em, join 'em."

N-Oxidation of pyridines is one metabolic process that sometimes leads to stable, active metabolites. An example of this is shown on the right side of Figure 9.26. Compound **30b**, a CCR5 receptor antagonist from Schering–Plough, was one of a series of compounds that came out of an effort to stabilize an earlier series against benzylic oxidation by incorporating an oxime group at the labile position. The pyridine *N*-oxide SCH-351125 was identified as a major (M + 16) metabolite. In rat oral PK studies, this metabolite was superior to Compound **30b**, with a 3-fold higher AUC. It was orally bioavailable in rat (63%) and monkey (52%) and showed only modest protein binding.[112] Also called SCH-C, it was successfully used to establish proof-of-concept (POC) in clinical trials for CCR5, as a target for HIV.[113] Although, eventually dropped due to concerns about hERG binding and potential cardiac effects, this compound along with other active metabolites shows just how useful they can be (Box 9.3).

---

**Box 9.3  Key Points About Metabolic Stability**

▶ Metabolic stability is normally essential for good PK properties

▶ Commonly observed metabolic reactions include the following:

  Hydrolysis of esters and amides (Phase I)

  Hydroxylation of arenes and alkanes (Phase I)

  Epoxidation of alkenes and alkynes (Phase I)

  Oxidations at heteroatoms (Phase I)

  *O*-, *N*-, or *S*-glucuronidation (Phase II)

▶ Metabolic stability is typically assessed by incubation with liver microsomes or hepatocytes

▶ Fixing a metabolic stability problem might:

  Cause new instability somewhere else on the molecule

  Decrease cell permeability and absorption

  Lead to faster excretion

  Turn the compound into a CYP inhibitor

  Not improve metabolic stability in a different species

▶ Ways to increase metabolic stability include the following:

  Bioisosteric replacement at the metabolic site

  Altering electronics, sterics, or conformation to disfavor metabolism

  Reducing LogP by adding in polar groups

  Pre-forming an active metabolite

---

# 9.3  Plasma Protein Binding

## 9.3.1  Is It Important?

When exposed to blood, serum or plasma,[114] anywhere from about 1% to more than 99% of a compound will bind to the proteins found there. Lots of evidence over many years has given rise to the *free drug principle*, which states that only free (that is, not bound to non-target protein) drug concentrations available at the site of action can cause a physiological response: "*Free fraction drives drug action.*"[115] This can present a challenge to chemists, since it means that for highly protein-bound drugs, which are common, much higher in vivo concentrations can be needed in order to see an effect. For example, by the free drug principle, if 99% of a compound binds to plasma proteins, making the *fraction unbound*

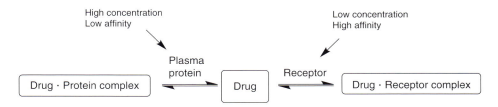

**Figure 9.27 ▶** Simplified scheme representing competition between binding to plasma proteins and binding to a drug's target receptor.

($f_u$) 0.01, 100-fold higher concentrations will be required to get the same effect as in the absence of any protein binding.[116]

As shown in Figure 9.27, from the point of view of a drug, binding to plasma proteins represents a competition between low affinity (perhaps hundreds of micromolar) binding to proteins which are, however, abundant and high affinity (probably nanomolar) binding to a target receptor that's extremely scarce by comparison.

This model is conceptually useful for thinking about protein binding, but it's too simple to describe the real-world situation for most drugs, where the target lies not in blood but elsewhere in a tissue that the drug still needs to reach for any receptor binding to occur. Here, the situation is better described by the scheme shown in Figure 9.28. Plasma protein binding acts to restrict tissue distribution of a compound by keeping it in blood. Only unbound drug can get into interstitial fluids, where another equilibrium with local proteins is established, and free drug from *that* equilibrium can then get into target cells, if the target is intracellular. Even there, binding to the target competes with relatively non-specific binding to higher abundance proteins.

Although this system seems complicated, it isn't all bad news from the point of view of the medicinal chemist. For one thing, plasma proteins are too big to get out of the bloodstream, so while bound to them compounds are protected from both the *glomerular filtration* that would cause them to end up in urine and from partitioning into hepatocytes, where they might be metabolized. Compound's half-life may thus be extended albeit at the price of lower free drug concentrations being available to "drive the action". It should be noted, however, that, as Dr. Rang points out, "the effect of protein binding on drug elimination is not straightforward" and such effects aren't always large, caveats that are mentioned only infrequently.[117]

Another thing to keep in mind is that the kind of partitioning equilibria shown in Figure 9.28 might sometimes allow distribution to the target tissue to actually be driven by a preference for non-specific binding to the kinds of proteins found there rather than the ones in plasma. This could shift the equilibria over to the right in the figure. In such cases, binding to tissue (but not plasma) proteins could be beneficial. Recent publications from a group at Pfizer[118] and another at GlaxoSmithKline[119] on work along these lines have appeared, bearing important implications for the design of CNS drugs.

Human plasma and serum contain at least two proteins that are "the usual suspects" when it comes to binding drugs. The first and most abundant of all proteins is *human serum albumin* (*HSA*). This has a MW of 66.5 kDa, an average concentration of about 640 μM, and lots of

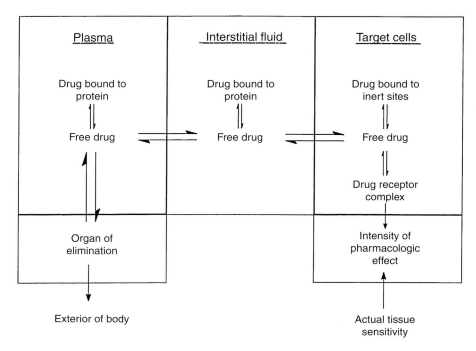

**Figure 9.28** ▶    Equilibria relating the concentration of a drug in plasma to its effects at an extravascular site of action. (Reprinted with permission from Sparreboom, A., et al. The (ir)relevance of plasma protein binding of anticancer drugs. *Neth. J. Med.* **2001**, *59*, 196–207, copyright 2001, Elsevier).

lysine and arginine residues near its binding sites, which help to explain why it tends to bind acidic drugs (although non-acidic drugs can sometimes bind to it as well). Two major binding sites for small molecules have been characterized. The first, known as *Sudlow Site I*, binds warfarin. X-ray crystallography done on the protein has localized this to a pocket in the subdomain IIA of the protein.[120] A second major site, *Sudlow Site II*, which binds diazepam and ibuprofen, is located in subdomain IIIA. Serum albumin is notoriously flexible and has a number of other binding sites, allowing it to bind many things including endogenous fatty acids. Still, knowing the structure of this protein sometimes allows excessive protein binding to be designed out of a molecule as we'll see later.

The other frequent binder is a plasma protein called $\alpha_1$-*acid glycoprotein* (*AAG*). This has a molecular weight of 40 kDa, varies in concentration from about 15 to about 35 μM, and usually prefers to bind basic compounds. For both of these proteins, blood concentrations can change due to factors like pregnancy and disease, and this might result in changes to unbound drug concentrations, making them less efficacious or more toxic if doses aren't adjusted. Pharmacogenomic factors can also be involved in protein binding, as illustrated by a recent report that imatinib was 94% bound to one variant (F1-S) of AAG, but only 82% bound to another (A), resulting in potentially a 3-fold difference in the concentration of free drug.[121]

It's long been believed that problems can arise when a drug that binds to plasma proteins and reaches relatively high concentrations (say phenylbutazone) displaces another drug that's extensively bound to the same protein at the same site, but has a low therapeutic index, like warfarin. Competitive displacement that only changed warfarin's protein binding from 99% to 98% would double the free drug's concentration, which could result in toxicity. The potential for such drug–drug interactions (DDIs) has gotten a lot of attention in the literature over the years, but current thinking holds that for a variety of reasons such interactions are, at the very least, clinically uncommon.[122] CYP inhibition, which is discussed in Chapter 10, turns out to be a far greater problem for DDIs.

Species differences in protein binding are universally acknowledged in drug discovery research. Figure 9.29 shows some not atypical results for three compounds reported in the literature. The data for the staurosporine derivative and clinical cancer candidate UCN-01 are especially worthy of comment. Here, extremely tight protein binding to human AAG must have conferred quite a lot of protection against metabolism: the compound's in vivo half-life in one Japanese clinical study was nearly *two months!*[123] This was not predicted by preclinical studies in the other species shown, where protein binding wasn't as extensive. Clearly, for compounds going into animal efficacy studies, having plasma protein binding data for the relevant species, and possibly even strain,[124] is a prerequisite and human protein binding values certainly need to be investigated before clinical candidates are selected.

But, views about the importance of getting protein binding data earlier on and about how heavily to weight it in compound optimization vary a lot. There's probably more disagreement about this than about any of the other properties (solubility, microsome stability, etc.) discussed in this book. There are those who feel that, for example, the fact that seven out of the top 10 best-selling small molecule prescription drugs of 2006 are at least 95% protein bound (see Table 9.3) means that protein binding should never be used as a criterion for choosing or discarding compounds.

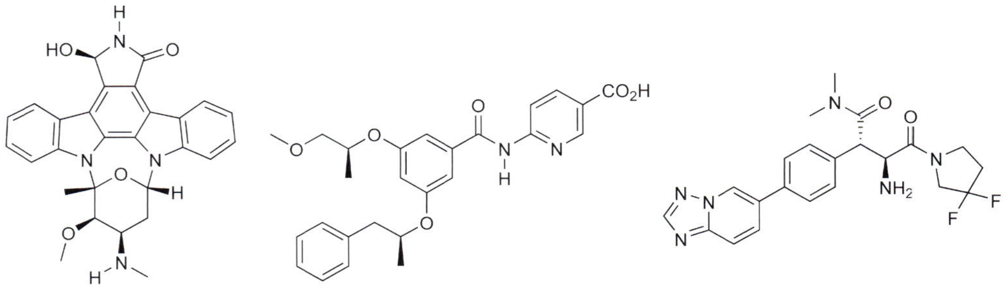

|  | UCN-01[a] | GKA50[b] | Compound 6[c] |
|---|---|---|---|
| *Mouse* | 1.17 | NA | 4 |
| *Rat* | 1.75 | 0.40 | 9 |
| *Dog* | 0.49 | 1.46 | 36 |
| *Human* | <0.02 | 0.44 | 18 |

**Figure 9.29 ▶** Unbound fractions (%) in plasma of different species for staurosporine analog UCN-01,[123] glucokinase activator GKA50,[140] and a DPP-IV inhibitor.[125] "NA" = Not available.

**TABLE 9.3 ▶   Percent Protein Binding for the Ten Best-Selling Small Molecule, Single Agent Prescription Drugs**

| Drug | % Protein bound |
|---|---|
| Atorvastatin | 98 |
| Clopidogrel | 98 |
| Esomeprazole | 97 |
| Amlodipine | 97.5 |
| Olanzapine | 93 |
| Valsartan | 95 |
| Risperidone | 90 |
| Venlafaxine | 27 |
| Pantoprazole | 98 |
| Montelukast | 99 |

Source of data on sales: Humphreys, A. MedAd News 200 world's best-selling medicines, *MedAdNews* 26, No. 7, July 2007, 14–40. Protein binding data from DrugBank, http://redpoll.pharmacy.ualberta.ca/drugbank

This "So what?" argument tends to treat protein binding out of context, though. Its importance or lack thereof depends on the concentration of free drug needed to drive physiological effects, which depends, among other things, on the potency of the drug and on how sensitive its receptor is to modulation. When using a very potent drug, extensive protein binding may not matter, since only low free drug concentrations need to be achieved. But, for less potent drugs or with receptors that need to be "knocked out" at concentrations exceeding, say, the $IC_{90}$ in order for an effect to be seen, squeezing out that last bit of free drug by decreasing protein binding might make a lot of difference.

Of course, early on in a project, especially the ones involving new targets or truly novel compounds, one usually can't predict either the ultimate in vitro potency of project compounds or the plasma levels of free drug that will be needed to see an effect. A glimpse into how important protein binding may prove to be isn't usually obtained until compounds start going into animals. By this logic, looking at protein binding in HTL, before anyone knows that it will even be an issue, may be premature. But, ignoring it in LO seems unwise, especially if reasonable concentrations of drug are getting into blood without causing much of a PD effect. Still, opinions vary and everyone seems to have his or her own philosophy on this.

## 9.3.2  Measuring Plasma Protein Binding

The plasma levels one gets from PK studies say nothing at all about protein binding. They represent *total*, that is, bound plus unbound, concentrations. Traditionally, a couple of methods have been used to determine protein binding. The first, *equilibrium dialysis*, is considered the "gold standard" method. A sample of the compound in plasma at pH 7.4 and at 37°C is dialyzed through a membrane with a 10 kDa cutoff into buffer for some hours.

After standard sample preparation, detection and quantitation of both solutions are done by LC/MS/MS. The expected problem of compounds sticking to the membrane or container can occur, but careful quantitation will at least tip one off that this is happening. Another problem that crops up is that the compounds need to be stable to plasma over an hour long time course or the results will be wrong and misleading. Equilibrium dialysis is a bit time- and labor-intensive, but it is possible to improve the throughput to some extent, for example by using pooled samples[126] or 96-well dialysis plates.

In the second method, *ultrafiltration*, centrifugal force is applied for a few minutes to a sample of the compound in plasma, pushing the unbound fraction through a permeability selective membrane. LC/MS/MS analysis of the ultrafiltrate and retentate is then done. This has the advantage of being faster than dialysis. Methods like 96-well plate have been developed here too.[127] As in HTS, edge effects have been noted with these, and one study comparing such a method with results from individually run samples for a dozen research compounds showed an $r^2$ value of 0.60.[128]

Note that for any of these methods exquisite detection sensitivity can be required to analyze highly protein bound compounds, which is why radiolabeled compounds have sometimes been used. Many other methods exist as well, including *ultracentrifugation*, the use of HPLC columns to which HSA has been bonded,[129] and surface plasmon resonance.[130] All of them have their drawbacks, and in reality at most companies today protein binding data will be obtained on many fewer compounds than activity or stability data. Often, fairly large error bars can make initial data only suitable for binning compounds into low, medium, and high binding categories, rather than rank ordering them.

The most practical and increasingly popular way to bring protein binding data into the drug discovery equation is by running in vitro activity assays in the presence of added serum or serum proteins where, of course, protein binding will result in much higher $IC_{50}$ values. The resulting "shift" correlates with the extent of binding: compounds with less binding will have less of a shift. Figure 9.30 illustrates the concept for three different clinical anti-*hepatitis C virus* (*HCV*) candidates in replicon assays.[131] Two of the compounds, VX-950 and BILN2061 are HCV NS3-4A protease inhibitors, while the other, NLM811, an analog of the cyclic peptide cyclosporine (Figure 7.5), is a cyclophilin inhibitor that acts through the virus's NS5B polymerase.[132] Effects of protein binding on a compound, of course, won't necessarily have anything to do with its MOA. It operates by different rules as we'll see, but observing its effect in an in vitro assay can provide some insights into its SPR and sometimes tip one off as to what to expect in vivo.

The same sort of *serum shift* method can be applied to many (but not all) cellular, enzyme, or receptor binding assays and a very nice method for calculating, or at least estimating, $K_D$ values for protein binding from as little as one additional protein-added data point has been published by a group at GlaxoSmithKline.[135] Comparing $K_D$ values rather than percentage of protein bound, whenever possible, is probably a good idea in general. To most of us, 99% and 99.9% protein binding look pretty similar, but the corresponding $K_D$ values, 6.1 µM and 0.60 µM, somehow psychologically make the difference more noticeable.[136]

Finally, computational methods for predicting plasma protein binding have appeared in recent years,[137] and such software is now commercially available.[138]

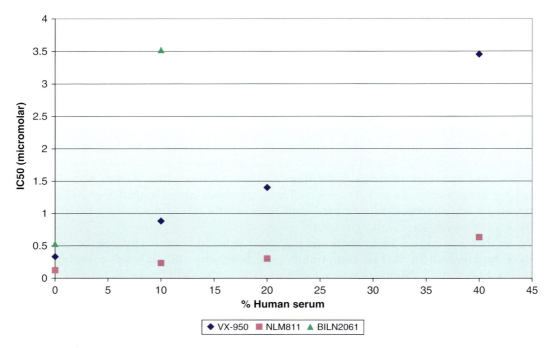

**Figure 9.30** ► Effects of added human serum on HCV inhibitory $IC_{50}$ values for VX-950,[133] NIM811,[134] and BILN2061[134] in replicon assays. Values for BILN2061 not shown are 7.13 µM with 20% human serum and 28.88 µM with 40% human serum.

### 9.3.3 Minimizing Plasma Protein Binding

The property that relates best to protein binding is, once again, lipophilicity, but that relationship isn't always a close one. A recent study involving hundreds of compounds at GlaxoSmithKline examined the relationship between calculated Log$P$ values and binding to rat plasma proteins.[139] It found that overall the correlation coefficient was only 0.24 for the full data set. The correlation was better among acidic ($r^2 = 0.38$) and zwitterionic ($r^2 = 0.53$) compounds, however. More importantly, within a given series of analogs the correlation could be better still. For 8 such analogous series examined, $r^2$ values ranged from 0.41 to 0.84. Figure 9.31 shows the corresponding data for the best and the worst of these, respectively.

When making analogs, then, lowering Log$P$ is often a good way to decrease protein binding and "drive drug action." As we've seen, though, although this can also be good for other properties like solubility, it can sometimes have exactly the opposite effect on the property most valued by medicinal chemists, potency. Figure 9.32 presents a real-world example from the literature. A slight reduction in Log$P$, caused by adding in a methoxy group nearly doubled the free fraction for the resulting analog. A much larger effect on Log$P$ and protein binding was obtained by *replacing* a phenyl ring with a methoxy group instead, but the consequent loss in potency was unacceptable, and instead the other compound was felt to represent "an excellent balance of potency, solubility, and % free."[140]

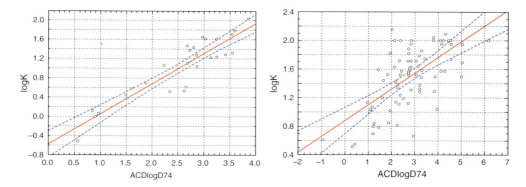

**Figure 9.31** ▶ Relationships between plasma protein binding and $LogD_{7.4}$, as predicted by the program ACD/LogD (Advanced Chemistry Development, Toronto, Ontario) for two different series of analogs. Regression lines and 95% confidence limits are shown. (Reprinted with permission from *J. Med. Chem.* **2007**, *50*, 101–112, copyright 2007, American Chemical Society).

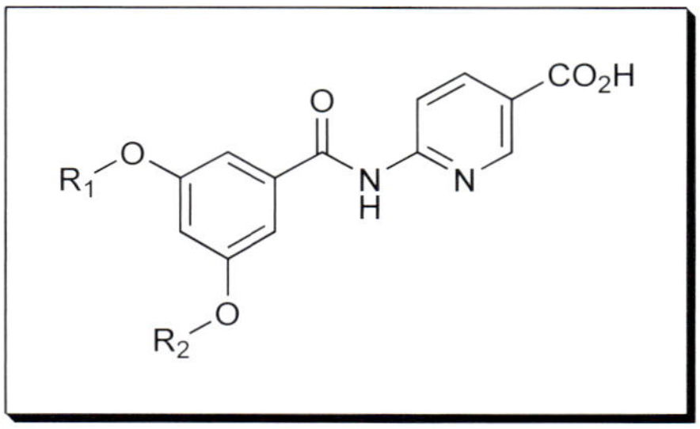

| R¹ | R² | EC$_{50}$ | cLogP | % Free (rat) |
|---|---|---|---|---|
|  | 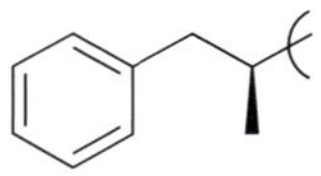 | 20 nM | 5.5 | 0.23 |
| 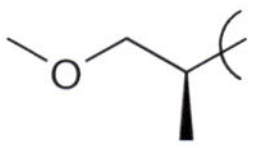 | 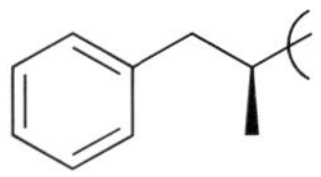 | 30 nM | 4.9 | 0.40 |
|  | 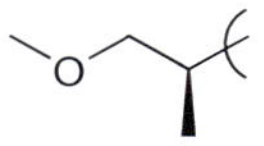 | 610 nM | 3.3 | 5.34 |

**Figure 9.32** ▶ Structures, EC$_{50}$ values for the activation of recombinant human enzyme, calculated LogP values, and unbound percentages for a series of glucokinase activators.[140]

Adding in more hydrophilic substituents or replacing existing hydrophobic ones with them can help reduce plasma protein binding. Which substituents, then, should one use? Carboxylic acids, for example, can lower hydrophobicity quite a bit, but the net effect may well be an *increase* in protein binding due to serum albumin's preference for anionic compounds. In fact, amines are frequently added in and this seems to be a good strategy, probably due to the effects on both charge and hydrophilicity (Box 9.4).

A fascinating study by Leach and co-workers compared a large set of matched pairs of compounds, where the only difference between the two compounds was that in one an aryl hydrogen was replaced by a substituent group.[141] They examined the resulting effects on measured binding to rat plasma proteins as well as solubility and AUC. Of the substituents examined (halogens, $CF_3$, Me, OMe, CN, OH, and $SO_2Me$), the methanesulfonyl group was far and away the most effective, reducing binding affinity by half a log unit. Interestingly, at least a bit of other data in the literature suggests that incorporating a methanesulfonyl or methylsulfoxide group into an alkyl chain may reduce protein binding as well.[142] Other findings were that among the halogens, fluoro substituents caused the least increase in protein binding, and that substitution in the *ortho* position, which forces the conformation away from coplanarity, seemed to cause less of an increase in protein binding than *meta* or *para* substitution, and of course also made for better solubility. N-Methylation of secondary amides seemed to reduce binding interactions and increase solubility, probably by reducing the compound's ability to form H-bonding interactions with plasma proteins.

A second way of reducing plasma protein binding makes use of structure-based drug design (SBDD). In a project designed to produce antagonists of the Bcl-2 family of proteins for anticancer use, researchers at Abbott obtained a lead compound for which in vitro potency

---

**Box 9.4   Key Points About Plasma Protein Binding**

▶ Plasma protein binding reduces the amount of free drug available to the target

▶ It can also, however, increase a compound's in vivo half-life

▶ Many marketed drugs display high plasma protein binding

▶ Many acidic compounds bind to albumin, many basic ones to AAG

▶ Species differences in plasma protein binding are common

▶ "Serum shift" assays are popular ways of looking at protein binding during lead optimization

▶ Plasma protein binding can be minimized by

     Removing or reducing acidity

     Adding in amines

     Lowering lipophilicity (see text)

     Given enough information, using SBDD to disfavor binding

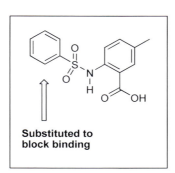
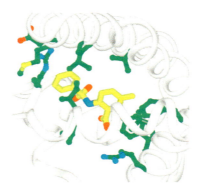

**Figure 9.33 ▶** Structure of a MetAP2 inhibitor (left) and its binding to serum albumin as determined by NMR spectroscopy (right). In a series of analogs, the position *ortho* to the sulfur atom was substituted with amine-containing side chains to hinder binding to the hydrophobic pocket. (Reprinted with permission from *J. Med. Chem.* **2006**, *49*, 3832–3849, copyright 2006, American Chemical Society).

was almost abolished in the presence of 10% human serum. By running the experiments in the presence of the plasma proteins HSA and AAG, the former was found to be responsible for most of the binding, and the majority of it also occurred in the presence of recombinant domain III of HSA. With the knowledge that this was where the troublesome interaction was occurring, an NMR-derived structure of the complex was obtained for a close analog of the lead. Parts of the molecule, where substituents shouldn't be tolerated were identified and used to guide analoging efforts. Despite the infamous flexibility of albumin binding, eventually these efforts resulted in a compound with both improved potency and much less of a shift.[143]

In another example also from Abbott, HSA binding affinity was similarly designed out in a series of methionine aminopeptidase-2 (MetAP2) inhibitors. Here, the NMR-based structure shown in Figure 9.33 was obtained. Some specific amine-containing substituents were introduced into the part of the molecule that bound in a tight, hydrophobic pocket of albumin. The resulting loss in binding affinity, accomplished without loss of target potency, cut the observed shifts in enzyme inhibition in the presence of HSA from about 400-fold to about 10-fold, and improved cellular potencies considerably.[144]

## 9.4  P-Glycoprotein Interactions

### 9.4.1  Structure and Function

*P-glycoprotein* (*P-gp*), the product of the *multidrug resistance 1* (*MDR1*) gene and the member of the *ATP-binding cassette* (*ABC*) family, is a 1280-amino acid glycosylated transmembrane protein that functions as an energy-dependent efflux pump.[145] Initial interests focused on its association with *multidrug resistance* (*MDR*), where cancer cells lose their responsiveness to a wide variety of chemotherapeutic drugs.[146] One—but not the only—thing that happens in MDR is that overexpressed P-gp actively pumps compounds out of cells in an ATP-driven process, thus limiting the intracellular concentrations of structurally quite different drugs like vinblastine, doxorubicin (Figure 1.22), and paclitaxel, allowing tumor cells to survive.

Normally, though, the protein plays a less sinister role in vivo. Its tissue distribution and cellular localization give away its function as a kind of "molecular bouncer" that keeps drugs from getting into cells and being distributed into tissues, as well as helping to eliminate them from the body if they do get in. Figure 9.34 shows that some notable places where P-gp can be found include the apical membrane of enterocytes in the gut (limiting oral absorption), the apical surface of proximal tubule cells in kidney (excreting into urine), the canalicular surface of hepatocytes (excreting into bile), the apical surface of brain capillary endothelial cells (limiting access to the brain), and in a placental location where it can limit fetal exposure to drugs. Clearly, its normal role is a protective one, although it may not be critical to survival in the *absence* of xenobiotics. P-gp knockout mice survive and appear to be normal, displaying no

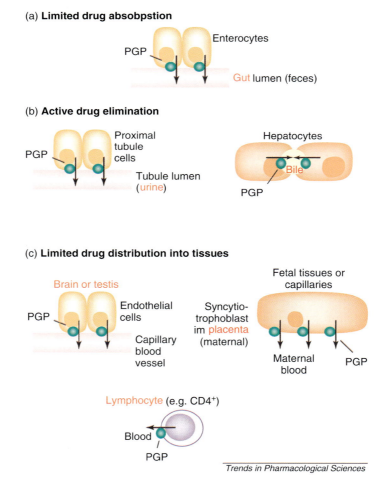

**Figure 9.34 ▶** Expression and function of P-gp in different tissues. (Reprinted from Fromm, M.F. Importance of P-glycoprotein at the blood–tissue barriers, *Trends Pharmacol. Sci.* **2004**, *25*, 423–429, copyright 2004, Elsevier).

overt phenotype except for an unusual sensitivity to xenobiotics, where exposure to normal doses can be toxic.[147]

How exactly does a P-gp work? It's a large and complicated protein, for which no crystal structure yet exists, and in many ways it's very different from most receptors. It exists as a single chain made up of two homologous regions, each of which consists of six transmembrane domains and a cytosolic nucleotide (ATP) binding domain, connected by a flexible linker. Its broad substrate specificity can be explained by attributing to it either multiple (between two and four) binding sites, or, alternatively, a single large, flexible binding site sometimes capable of binding more than one molecule at a time. The latter view has become more widespread recently.[148] Figure 9.35 shows some proposed models of how P-gp might export drugs.

The simplest imaginable way the protein might work is by opening up a pore in the membrane through which a cytosolic drug could be transported, as shown in Figure 9.35(a). Although one still sees this picture presented occasionally, evidence has accumulated that substrates for P-gp transport don't start their journey in the cytosol, but can instead be exported before ever reaching it by interacting with the protein in the membrane bilayer. This "preemptive" action, which might also keep P-gp from exporting normal cytosolic components, seems to begin in the inner, cytosol-facing part of the membrane bilayer and may involve a "gate" between specific transmembrane domains.[149] In the *flippase* model[150] (Figure 9.35(b)), the drug is "flipped" from the inner to the outer layer of the bilayer and might passively diffuse out of the cell from there. In the currently most popular one, the *hydrophobic vacuum cleaner* model[151] (Figure 9.35(c)), compound association with the protein takes place in the inner layer, "concentrating them into the bag of the vacuum cleaner," leading to direct export out of the cell through the central pore.[152]

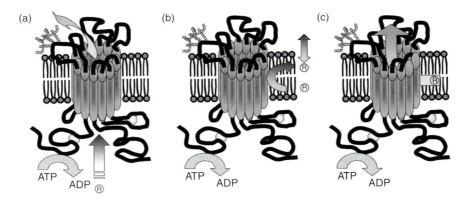

**Figure 9.35** ▶ Three proposed models of P-gp drug efflux. (a) The "pore" model, in which cytosolic drug is transported through a protein channel, (b) the "flippase" model, where drug associated with the inner leaflet of the membrane bilayer is "flipped" into the outer leaflet where it might passively diffuse out of the cell, and, (c) the "hydrophobic vacuum cleaner" model, where drug associated with the inner leaflet of the bilayer is exported out through the protein. (Reprinted with permission from Varma, M.V.S., et al. P-glycoprotein inhibitors and their screening: A perspective from bioavailability enhancement. *Pharmacol. Res.* **2003**, *48*, 347–359, copyright 2003, Elsevier).

All of this might seem very theoretical, but it has some practical implications. If a compound's interactions with the protein occur within the membrane bilayer instead of within an aqueous phase, its ability to partition into the bilayer can be important, and this might involve factors like amphiphilicity. In fact, compounds bearing a charged amine and a lipophilic (especially a planar, aromatic) part often seem to be good P-gp substrates. Also, in such a hydrophobic environment, electrostatic interactions with the protein might be more important than usual. This will come up again later in the context of SAR.

### 9.4.2   Types of P-gp Interactions

A compound can interact with P-gp in several different ways, all of which are of potential interest to the researcher. It might be a substrate that's recognized and transported out of a cell by P-gp, causing it to reach lower than desired concentrations in cells and tissues. This is most likely to be a concern in (1) preventing brain penetration for drugs that need to get into the CNS, and, (2) limiting the absorption of oral drugs. Figure 9.36 shows how much of an effect P-gp has on the plasma and brain concentrations of the HIV protease inhibitor and P-gp substrate nelfinavir (Figure 2.7). Plasma levels of the drug essentially doubled when P-gp efflux was blocked by adding the selective inhibitor LY-335979. Even more startling was its result on brain concentrations, which increased by about 35-fold.[153]

Many experiments to date have shown that P-gp efflux has a bigger effect on brain penetration than it does on oral absorption. It's been argued that this probably reflects relative compound concentrations. In the intestines, P-gp is present in enterocytes and works to limit drug absorption, perhaps, in concert with the metabolizing enzyme CYP3A4, which has similarly broad substrate specificity.[154] But concentrations of oral drugs there can be very

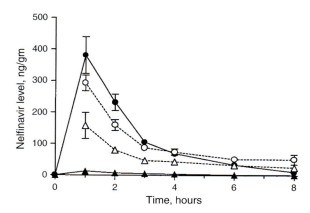

**Figure 9.36 ▶**   Concentrations of [$^{14}$C]nelfinavir (5 mpk) in mdr1a wild-type mice given 25 mpk of the selective P-gp inhibitor LY-335979 [plasma (○), brain (●)] or vehicle [plasma (△), brain (▲)]. (Reprinted with permission from Choo, E.F., et al. Pharmacological inhibition of P-glycoprotein transport enhances the distribution of HIV-1 protease inhibitors into brain and testes. *Drug Metab. Dispos.* 2000, *28*, 655–660, copyright 2000, American Society for Pharmacology and Experimental Therapeutics) (www.dmd.aspetjournals.org/).

high compared to those in the capillary vessels in the brain, where P-gp mans the blood–brain barrier. Being an active transporter, it can only handle limited concentrations of the substrate before it's saturated. When that happens soluble, cell permeable drugs might swamp out its effects by simple passive diffusion, and this is probably what's going on at the high compound concentrations present in intestine but not brain.

A second type of interaction with P-gp is a real concern too. A compound might block the effects of P-gp, thus decreasing its efflux of another drug (as in Figure 9.36), increasing that drug's uptake, and causing a DDI. This would make it a *P-gp inhibitor*, also sometimes called a *modulator*[155] or *antagonist*. Such compounds, intentionally designed for the purpose of increasing concentrations of co-administered cancer drugs in tumor cells are known as *chemosensitizers* or *multidrug resistance reversers* (*MDRR*s). These have been of interest for a long time, and although the *first generation* agents didn't work well in the clinic, *second* and *third generation* compounds are still being examined.[156]

Due to the interest in chemosensitizers over the years, SARs have sometimes been analyzed for P-gp *inhibitors*,[157] but these won't necessarily apply to P-gp *substrates*. Not all inhibitors are also substrates—some might, for example, be acting at the nucleotide binding site in the cytosol or modifying membrane fluidity—and not all substrates seem to inhibit,[158] although some do both. Add to this the fact that some assays, as we'll see, can't really distinguish between the two and you've got a very complex and potentially confusing situation.

A couple of things are clear enough, though. P-gp transport of substrates will be a concern in all medchem projects where oral bioavailability is an issue and especially for cancer projects and those directed at drugs (antipsychotics, HIV drugs, etc.) intended to cross the blood–brain barrier. And although P-gp inhibition may be desirable for some specific applications, in general one tries to avoid it since co-administering such a drug can unintentionally increase plasma levels and toxicity of others. In fact, looking at such potential interactions becomes more important the closer a project gets to the clinic, and FDA submissions will require such information.

DDIs can also occur through *induction* of P-gp, where transcriptional activation could increase P-gp expression levels, thereby lowering plasma levels of co-administered drugs.[159] Notably, St. John's Wort, a popular herbal remedy, is particularly good at inducing P-gp. Keep in mind too that P-gp substrates, inhibitors, and inducers aren't necessarily exclusive groups: a single compound can be a member of all three!

### 9.4.3  Measuring P-gp Interactions

In projects directed at finding new cytotoxic cancer drugs, antiproliferative $IC_{50}$ values obtained in resistant cells, either sublines developed to be resistant to known cancer drugs or *MDR1*-transfected cells, can be compared with those of the corresponding sensitive (parent line or non-transfected) cells.[160] The ratio of resistant-to-sensitive $IC_{50}$ values for the matched pair is known as the *resistance index* and may provide some indication of what to expect in vivo. Alternately, measurements can be made on a single resistant cell line in the presence and absence of a known P-gp inhibitor or antibody to get the same kind of information. Cancer projects directed at either cytotoxic agents or chemosensitizers need to factor in the possible roles of other efflux pumps that might also be overexpressed like the *multidrug resistance-associated protein* (*MRP*) and the *breast cancer resistance protein* (*BCRP*). P-gp can be important, but it probably isn't the only thing involved.

As we've seen, cell monolayer assays like Caco-2 can provide a direct measurement of compound transport. A compound's permeability from the basolateral side to the apical side (representing efflux back into the intestinal lumen) compared to its permeability as determined in the other (influx) direction gives a ratio that if greater than 1 indicates active efflux. Usually, a ratio greater than 1.5 or 2 is considered significant. Caco-2 cells express P-gp, so that ratio probably relates to how well P-gp transports the compound out of cells. However, other transporters like MRP2 and BCRP are also expressed, and the expression levels of all of them can vary from batch to batch, which introduces some uncertainty. Follow-up experiments done in the presence of selective inhibitors can help reduce this.[161] Or different types of cells might be used in monolayer experiments to measure P-gp transport, like MDR1-MDCK cells, dog kidney cells transfected with the *MDR1* gene, or transfected LLC-PK1 cells, from porcine kidney.

Regardless of the cell type, a saturable active transport process is involved, so observed results can depend on compound's concentration. In addition, compounds with very high passive cell permeability can be largely unaffected by P-gp efflux in this assay, with the transporter vainly trying to bale out a sinking ship with a thimble. Although this may reflect the physiological reality, it means the assay won't always agree with others. In this case, the others may show that the compound is a P-gp substrate but the cell transport assay, with the effects of P-gp efflux having been drowned out, won't. The term "nontransported substrates" has been used to describe such compounds.[162]

This being a cell monolayer assay, all the usual problems and limitations like "cacophilicity," the possible opening up of tight junctions, the need for sensitive detection techniques like LC/MS/MS, etc., will apply. Properly carried out and followed up, these kinds of transport experiments are indeed the "gold standard" for determining whether a compound is a substrate for P-gp, but they're time- and labor-intensive and throughput is correspondingly low. Even at big pharma companies with large ADME departments drug discovery programs don't use them for primary screening.

Since efflux by P-gp is powered by ATP, which is bound and then hydrolyzed to ADP and phosphate ion, measuring stimulation of the release of phosphate upon exposure of membrane-bound P-gp to a compound can provide a readout of its transport activity in vitro. This is conveniently carried out using membrane preparations from insect cell lines transfected with the *MDR1* gene.[163] Phosphate ion concentrations are measured colorimetrically and background phosphate ion is subtracted out by running the experiment in parallel with sodium orthovanadate, a transition state mimic of the γ-phosphate of ATP that gums up the works selectively for P-gp. This kind of assay is relatively straightforward and amenable to medium-to-high throughput. The concentration of the compound used can make a difference in results, with the correlation being biphasic, not linear. Additionally, both false positives and false negatives in this assay have been reported, and can't always be rationalized.[164] Still, the ATPase assay is a lot more practical to use at the earlier stages of a drug discovery program than the transport assays described above.

Even more popular for this purpose, though, are the fluorescence-based cellular assays. Figure 9.37 shows the most often seen of these, which use calcein-AM.[165] This compound bears four *acyloxymethyl* esters (the "AM") as well as two acetates per molecule and *isn't*

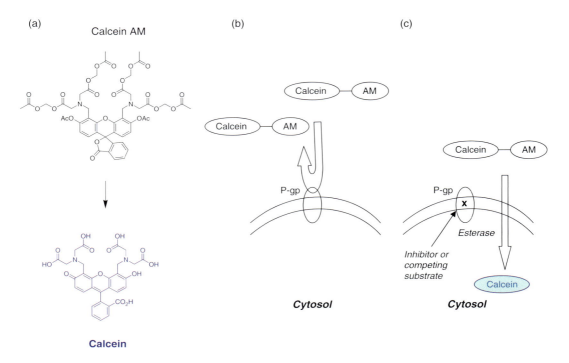

(a) Calcein AM

Calcein

**Figure 9.37** ▶  (A) Structure of calcein-AM, which does not fluoresce, while its fully de-esterified product, calcein, does. (B) P-gp efflux normally prevents calcein-AM from entering cells. All of the esters collectively make up the "AM" unit in this figure. (C) If P-gp function is blocked either by an inhibitor or by a competing substrate, calcein-AM can enter cells, where esterases convert it to the fluorescent calcein.

fluorescent. P-gp is very good at keeping this hydrophobic compound out of cells through active transport, but if this process is interfered with by a P-gp inhibitor or competing substrate some of it will cross the cell membrane and enter the cytosol, where esterases quickly strip off all of the ester groups, exposing calcein itself, which *is* fluorescent.

This assay is relatively simple and well suited to HTS, but it can't really distinguish between a P-gp inhibitor and a substrate, and compounds that are esterase inhibitors may give false negatives. A study at Hoffmann-La Roche compared results of this assay for cells transfected with human *MDR1* and the equivalent proteins in mouse, called *mdr1a* and *mdr1b*, and found some important differences between them, a reminder once again that results obtained for one species won't necessarily apply to another.[164]

This important study also examined the best way to apply the kinds of assays discussed above to early drug discovery. The authors conclude that "indirect fluorescence indicator assays (rhodamine or calcein-AM) should be used as the primary screen followed by an ATPase or transcellular transport assay to distinguish between substrates and inhibitors," and ultimately followed up with in vivo experiments for development compounds. As previously noted, mice homozygous for either *mdr1a*, *mdr1b*, or both are available for such studies.[166]

### 9.4.4    Reducing P-gp Interactions

"The interaction of substrates/inhibitors with P-gp is a complex process that is poorly understood, and an area of great contention."[167] Part of the complexity and much of the contention stem from the two-part process involved: membrane association precedes interaction with the protein, which then takes places in a non-aqueous environment. A single, discrete binding site offering a "lock" that only a very specific "key" will fit is lacking. Instead, either several different binding sites or a single, very large one capable of binding more than one molecule at a time allows a goodly percentage of all possible small molecules to bind there *somehow*. What's a medchemist to do?

Whether or not P-gp substrate affinity correlates with LogP is controversial.[168] Some researchers see such correlations, while others see only their absence. As mentioned before, amphiphilic molecules comprising a positively charged part[169] and a lipophilic part are often good substrates, and it was originally believed that having an amine was a requirement for P-gp binding. Subsequent consideration of some of the molecules P-gp can transport, like steroids, changed this view.

What factors *do* correlate with P-gp substrate affinity? "The physiochemical descriptors that highly correlate to the P-gp affinity are size and shape (MW, R, and V) and the hydrogen-bonding capabilities of the molecule (HB)."[168] So, although reducing lipophilicity per se might not help, making molecules smaller, which in many cases amounts to the same thing, can. One analysis of 84 compounds from National Cancer Institute (NCI) that interact with P-gp found that although the average molecular weight of all compounds in the NCI database was 325 Da, substrates for P-gp in the examined set had an average MW of 636 Da, inhibitors averaged 558 Da, and compounds doing both averaged 739 Da.[170] Large molecules are clearly more likely to interact with P-gp.

Those investigating the requirements for the binding of substrates to P-gp agree that hydrogen bonding and especially HBAs are important to substrate recognition. In one such analysis, a researcher at the University of Basel developed an empirical model that's often used in predicting whether or not a compound is likely to be a P-gp substrate. She defined a pair of HBAs spaced about 2.5 Å apart as *type I units* and either two or three HBAs with the outer groups separated by about 4.6 Å as *type II units* (Figure 9.38). "All molecules that contain at least one type I or one type II unit are predicted to be P-glycoprotein substrates," and "the binding of P-glycoprotein increases with the strength and the number of electron donor or hydrogen bonding acceptor groups forming the type I and type II units."[171] Type I units formed by negatively charged groups like carboxylic acids don't seem to count, however. Transport by P-gp was found to require at least two type I units or one type I and one type II unit.[172] Interestingly, although binding at a different receptor must be involved, molecules containing a type II unit were predicted to be P-gp inducers.

Patterns of HBAs are probably dictated by the corresponding spacing of HBDs in a transmembrane domain of P-gp, and the importance of such interactions is amplified by the low dielectric, non-aqueous environment there. The model doesn't address directionality, which is clearly important in H-bond formation,[173] and different models of substrate recognition have since been proposed,[174] but overall, properly positioned HBAs do seem to be critical to recognition. One example of the effect removing one such HBA might have

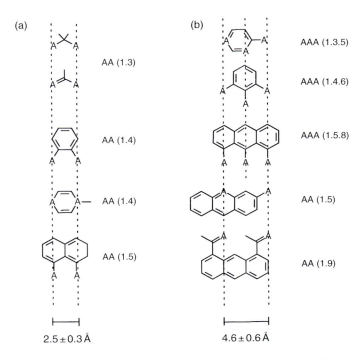

**Figure 9.38** ▶　Two types of HBA patterns are observed in P-gp substrates. (A) Electron pair donors spaced $2.5 \pm 0.3$ Å apart are called type I units. (B) Patterns formed by either three electron pair donors with the largest separation being $4.6 \pm 0.6$ Å or two electron pair donors with the same distance apart are known as type II units. Numbers in parenthesis indicate first and $n$th atoms with free electron pairs. (Reprinted with permission from Seelig, A., A general pattern for substrate recognition by P-glycoprotein. *Eur. J. Biochem.* **1998**, *251*, 252–261, copyrights 1998, Wiley-Blackwell).

on P-gp efflux is shown in Figure 9.39. Merck researchers replaced an amide unit with the trifluoroethylamine isostere (previously mentioned in the section on bioisosteres), and observed a large decrease in the P-gp efflux measured in a cell transport assay. Sadly, though, an even larger decrease in target potency went along with it.[175]

As noted before, P-gp seems to disfavor anionic substrates, so adding in a carboxylic acid can sometimes reduce affinity (Box 9.5). Unfortunately, this is usually anathema to blood–brain barrier penetration, the ultimate property one's often seeking to optimize. But, there is at least one example of this strategy working. The anticancer drug paclitaxel (taxol) isn't orally bioavailable, and doesn't cross the blood–brain barrier even when given intravenously, although co-administration of P-gp inhibitors helps with both. A combinatorial library of taxol derivatives turned up an interesting analog, where the C-10 alcohol had reacted with succinic anhydride to produce the C-10 ester monoacid ($ROCOCH_2CH_2CO_2H$), called Tx-67.[176] In both a brain microvessel endothelial cell monolayer transport experiment and an in situ rat brain perfusion experiment, Tx-67 demonstrated considerably improved permeation relative to taxol. The ester linkage, however, may constitute a hydrolytic liability, and a recent publication shows that replacing the ester mono acid with a carbamate

---

**Box 9.5   Key Points About P-gp Interactions**

▶ P-gp *inhibition* can result in drug–drug interactions that

  Are desirable for multidrug resistance reversal MDRR agents

  Need to be avoided instead for most projects

▶ P-gp efflux of *substrates* can

  Prevent brain penetration of drugs

  Limit oral absorption

  Limit exposure of MDR tumor cells to drugs

▶ P-gp "vacuums" a broad range of substrates out of the membrane bilayer

▶ Assays for interactions with P-gp include:

  Proliferation assays on matched sensitive and resistant cell lines

  Transport assays in monolayers of cells expressing P-gp

  A colorimetric assay for P-gp's ATPase activity

  Cellular assays using fluorescent or pro-fluorescent P-gp substrates

▶ To reduce the effects of P-gp efflux on a compound

  *Improve its passive transmembrane permeability*

  Reduce molecular weight and size

  Remove, replace, or move H-bond acceptors

  Try introducing an anionic group if feasible

---

monoacid ($ROCONHCH_2CO_2H$) maintains improved permeation in MDR-MDCK cells while improving plasma stability.[177]

Probably, the best advice about overcoming a P-gp efflux problem is not to try to reduce it at all! A recent and comprehensive review of published efforts to design out affinity for P-gp efflux, a paper that anyone seriously concerned with this problem is advised to read, concludes that "the most practical approach to de-emphasize the limiting effects of P-gp on a particular scaffold is to increase passive diffusion. *Efflux pumping efficiency may be overcome when passive diffusion is fast enough.*"[178] Suggested steps to increase passive diffusion include halogenation of aryl rings, removal of solvated groups which include both HBDs and HBAs, introducing conformational constraints, and promoting intramolecular H-bonds. Increase passive transcellular transport and you do an end-run around P-gp efflux.

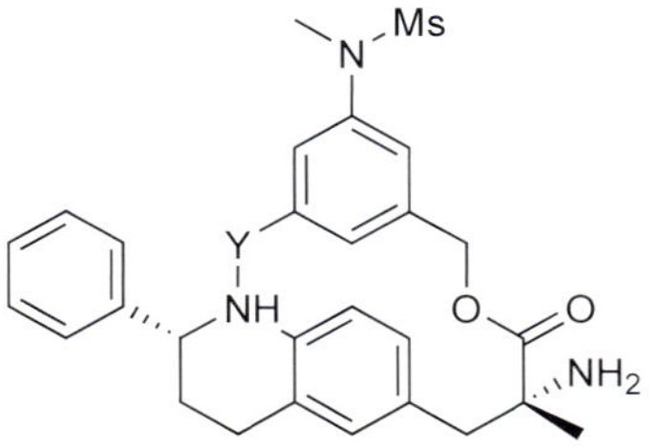

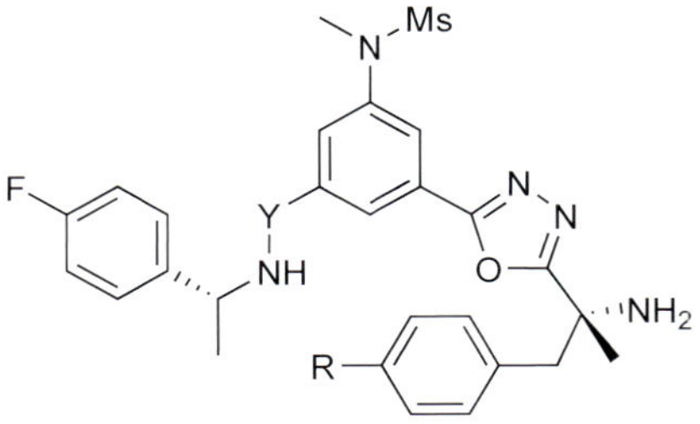

| | Y | BACE-1 $IC_{50}$ (nM)[7] | hu P-gp B-A/A-B[6] |
|---|---|---|---|
| **1a** | CO | 2 | 16 |
| **2a** | $CHCF_3$ | 3700 | 1.3 |

| | Y | R | BACE-1 $IC_{50}$ (nM)[7] | hu P-gp B-A/A-B[6] |
|---|---|---|---|---|
| **1c** | CO | H | 9 | >50 |
| **1d** | $CHCF_3$ | F | 15,000 | 3.2 |

**Figure 9.39** ▶  Effects of replacing an amide with a trifluoroethylamine isostere on target activity and P-gp efflux as determined in a cellular transport assay. References 6 and 7 are from the original publication. (Reprinted with permission from Moore, K.P., et al. Strategies toward improving the brain penetration of macrocyclic tertiary carbinamine BACE-1 inhibitors. *Bioorg. Med. Chem. Lett.* **2007** *17*, 5831–5835, copyright 2007, Elsevier).

# 9.5  Bioavailability

## 9.5.1  Introduction

To recap what was said earlier, in a strict sense *bioavailability* (*F*) refers to both the rate and the extent to which a drug is absorbed and becomes available at the site of action, but in practice it normally refers to the fraction of the administered dose of drug that makes it into systemic circulation. Oral bioavailability, then, is the ratio of how much of an administered drug gets into circulation when dosed orally (*peroral*, or *p.o.*) compared to the amount found in blood upon i.v. dosing. By definition, for i.v. dosing *F* is always 1. Amounts are estimated by determining the *AUC* of a plasma concentration versus time graph over the course of hours, as shown in Figure 9.40.

It's usually impractical to give the same dose to an animal i.v. as p.o., so a correction factor to reflect the different doses is used, and the equation for percent oral bioavailability becomes $\%F = [(AUC)_{oral}/(AUC)_{iv}] \times [D_{iv}/D_{oral}] \times 100$, where *D* is the dose. Bioavailability is an extremely useful parameter, reflecting how much of an orally dosed drug at least "gets on board." As such, it becomes a major driver of compound optimization in most drug discovery projects. Note that PK graphs like the one depicted in Figure 9.40 give other useful information as well, like maximum concentration ($C_{max}$). These and other derivable parameters like clearance (CL), compound half-life ($t_{1/2}$), and volume of distribution ($V_d$) are also important, but won't be discussed here, nor will other types of bioavailability like transdermal or subcutaneous bioavailability. The same basic principles apply to these too, but the barriers

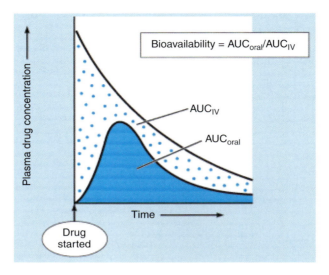

**Figure 9.40** ▶  How plasma concentration versus time curves for i.v. and p.o. dosings are used to determine oral bioavailability. (Reprinted with permission from Brenner, G.M., Stevens, C.W. *Pharmacology*, 2nd Edition (Saunders Elsevier), copyright 2006, Elsevier).

to be crossed are different. Most projects aim for oral drugs, so oral bioavailability will be the only type considered here.

Pharmacokinetic studies used to be carried out on a single compound at a time, but in order to improve throughput the concept of *cassette* or *N-in-1* dosing was introduced.[179] In this case, *N* refers to the number of co-administered compounds, often about five including a standard. Figure 9.41 shows data from a 5-in-1 study in dog and clearly points out the

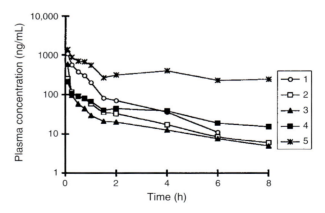

**Figure 9.41** ▶  Concentration versus time profiles for five $\alpha_{1a}$ antagonists after i.v. co-administration (5-in-1) in a dog. (Reprinted with permission from *J. Med. Chem.* **1997**, *40*, 827–829, copyright 1997, American Chemical Society).

advantage of using such studies in comparing compounds. This study, where four of the five compounds turned out to provide values close to those obtained in single compound studies, was done with i.v. dosing, but cassette dosing can also be done with other modes of administration such as p.o. dosing. Both, of course, are needed to determine %*F* for a compound, but i.v. studies may be done first to decide which compounds to put into oral bioavailability studies.

Problems can crop up if one or more of the compounds administered interfere with protein binding, P-gp export, or CYP metabolism, changing the profiles of the rest. Some companies don't believe in running N-in-1 studies for these reasons. To help reduce the risk of such effects, and also the risk of cumulative toxicity, doses of each compound are kept proportionately lower than those used in single agent studies, which mean that very sensitive detection is required. *Atmospheric pressure ionization (API)* LC/MS/MS is often used.[180] Compounds preferably should have different molecular weights, but compounds with identical MWs can be used as long as they separate well on HPLC. An alternative way of improving throughput that avoids some of these problems involves dosing each compound individually, but pooling samples for analysis afterward. This, however, only saves instrument run time, not animals, dosings or blood draws.

### 9.5.2  Understanding and Overcoming Poor Oral Bioavailability

Some of the factors affecting just how much of an oral dose of a drug makes it into the bloodstream are beyond the researcher's ability to control. Variations in transit time in the GI tract, adsorption onto food, PGx effects, and a host of other things can make oral bioavailability variable and hard to predict. Even carefully controlled PK studies done with multiple animals per time point can have substantial error bars, in large part reflecting animal-to-animal variability.

Fortunately, however, many of the important factors impacting bioavailability do lie within the medicinal chemist's power to control. Finding that a compound or series has poor oral bioavailability doesn't mean that it has to be dropped. If the reason behind the low %*F* can be uncovered, there's a good chance that the problem can be fixed. Figure 9.42 illustrates the "usual suspects" in poor oral bioavailability. Each of these is discussed in more detail in the following sections.

#### 9.5.2.1  Solubility

Compounds need to dissolve in order to be absorbed. Whenever a low %*F* is obtained for a compound that the chemist knows is poorly soluble or the formulations scientist couldn't get into solution and had to administer as a suspension, it doesn't take a paid consultant to tell you that poor solubility is the likely culprit. Solubility, as we've seen, depends on crystalline form, and amorphous solids, often the form obtained the first time a compound is synthesized, are likely to have much greater kinetic solubility than crystalline forms of the same compound isolated later. This makes for the "second batch effect" mentioned previously, where a later batch turns out to be much less soluble than the first and calls for a new, less concentrated solution or perhaps even suspension formulation, predictably resulting in much lower apparent bioavailability. This also serves to illustrate the fact that *bioavailability is not*

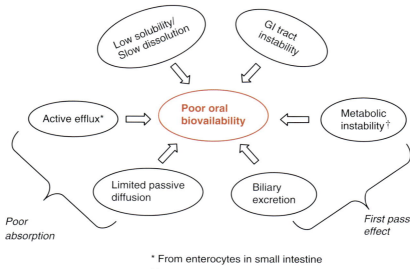

**Figure 9.42** ▶   Factors contributing to poor oral bioavailability.

*a fixed property of a compound like molecular weight, but will vary with formulation, species, etc.* Even among crystalline forms, different polymorphs of the same compound can have different bioavailabilities due to different solubilities and dissolution times. The latter is a factor for real-world drugs, since they're usually administered as solids and have only a limited time (generally 3 or 4 h) in the small intestine to be absorbed. Slow dissolution often goes along with low solubility and depends in part on surface area/particle size, one reason that test compounds are often micronized for suspension dosing studies.

Note that for ionizable compounds solubility can vary considerably with pH and that in traversing the GI tract a compound will be exposed to pH values ranging from about 2 to about 8. This can cause soluble formulations to precipitate in vivo and potentially reduce absorption, for example when a soluble salt formulation of a carboxylic acid meets HCl in the stomach. Formulation designed to inhibit compound precipitation after dosing is a subject of interest for real-world drugs.[181]

In optimizing for oral bioavailability during LO, formulations can take one of two different approaches. Either an optimal soluble formulation will be sought for each compound or series (more work for formulations people!) or a standard suspension formulation will be used for everything. The former gives the best chance of achieving high %$F$, but the latter is closer to the real-world situation where drugs enter the stomach as solids and bioavailability can be limited by their solubilities and dissolution rates in the GI tract. Using a single formulation for all compounds also makes it easier to compare values for different compounds. But since bioavailabilities are likely to be much higher for solution formulations, if a project is pushing to get something into animals for early POC studies, solution formulation may be a better way to go. Prevailing corporate *mores* and project goals are likely to determine which of these two methods gets used in any given case.

Assays which can help determine whether inadequate solubility is a problem for particular compounds include simple kinetic solubility assays run at two different pH values, perhaps 2 and 7.4. Methods of improving compound solubility have been presented in the previous chapter. One word of advice for chemists: avoid suggesting to the formulations scientist that she "just dissolve it in DMSO." This wonderful solvent, with plenty of toxicity and efficacy of its own in many models, is almost always unacceptable for formulations at concentrations over a couple of percent, although that doesn't prevent it from being tried, usually out of desperation.

### 9.5.2.2  GI Tract Stability

It goes without saying that the acidic pH of the stomach represents a challenge to drug stability as well as solubility. Incredibly, the blockbuster proton pump inhibitor omeprazole (see the *Periodic Table of Drugs*, Appendix), has a half-life of less than 5 min at pH 1.[182] Without enteric coating or co-administration of grams of buffering sodium bicarbonate to prevent this rapid degradation its bioavailability would probably approach zero. But acid isn't the only thing drugs in the GI tract have to contend with. The stomach contains pepsin and the intestinal lumen "contains gram quantities of peptidases secreted from the pancreas, as well as cellular peptidases from the mucosal cells, which are constantly sloughed off from the villi."[183] These enzymes, of course, largely serve a digestive function, but can't always tell the difference between an intended drug and food. Peptides, peptidomimetics, and perhaps sometimes amides or esters in general might be sliced, diced, and served up, never making it to their intended targets intact.[184]

One can get an idea of a compound's GI tract stability by monitoring in vitro stability at low and moderate pH values as well as in the presence of *simulated gastric fluid* (*SGF*), and *simulated intestinal fluid* (*SIF*), which can also be used for solubility measurements if compounds prove stable. Ways to avoid such hydrolytic instability were addressed in the "Chemical and Plasma Stability" section of the previous chapter.

### 9.5.2.3  Metabolic Instability

Metabolism in the gut wall or liver, the metabolic component of the first pass effect, is a common cause of poor oral bioavailability. For a discovery stage project the relative contributions of metabolism at each of these sites are unlikely to be determined, although sometimes double cannulation experiments in rats are done to see whether hepatic extraction (liver metabolism and biliary excretion) limits systemic exposure. Here, compounds' concentrations in the portal vein (going *into* liver) are compared with concentrations in systemic circulation (*after* passing through the liver).[185] If much less comes out than goes in, the liver is obviously playing a big role. In general, metabolism in the gut wall plays a lesser role, but one that may not be well-modeled by simply comparing stabilities using intestinal versus hepatic microsomes. This is partly because, as mentioned previously, efflux by P-gp and the resulting cycling may give gut wall CYP3A4 there more than one chance to metabolize a compound, enhancing its effectiveness.[186]

In dealing with low oral bioavailability for a compound, where %*F* isn't limited by solubility or GI tract stability, *it's critical to determine whether absorption or metabolic stability is the bigger problem since each of these will require different kinds of fixes.* Properly diagnosing the problem usually involves putting together in vitro ADME data with in vivo PK data, keeping a shrewd eye on SPR. Sometimes the problem can be obvious even before the compound goes into PK studies—compounds having short half-lives upon exposure to microsomes or poor permeability as determined by PAMPA or Caco-2 assays will almost certainly be problematic. Some functional groups will always raise red flags, like phenols for glucuronidation (metabolism). High LogP compounds are more likely to be metabolized. Comparing bioavailability data for a series of analogs with ADME measures of cell permeability and metabolism can uncover trends. For important compounds, double cannulation experiments can show whether hepatic extraction (usually metabolism) is likely to be involved. So there's no single assay the chemist can submit a compound to that will tell her what the problem is. Instead, there are many from which, perhaps collectively, it can be inferred.

### 9.5.2.4  Biliary Excretion

In the liver a compound or its metabolites can be actively transported out of hepatocytes and into the bile canaliculi by efflux pumps like P-gp and Mrp2. Bile then enters the intestine and if the transported compound isn't further processed, say by gut microbiota, and/or reabsorbed (remember enterohepatic circulation?), it ends up being excreted in feces, as, of course, does drug that never was absorbed in the first place. This other, sometimes neglected part of the first pass effect, which, after all, consists of both metabolism *and* excretion, is sometimes called *Phase III metabolism*. Phase II metabolites like glucuronides and glutathione conjugates are especially good substrates for export, as are large, amphipathic compounds in general. The most notable thing about biliary excretion from the point of view of a drug discovery chemist is that a fairly clear molecular weight "threshold" applies and that this varies greatly by species. In rat, a species particularly good at this, compounds having a molecular weight greater than about 325 Da can be subject to appreciable biliary excretion, while this number is about 440 Da in guinea pig and about 475 Da in rabbit.[187] The molecular weight made famous by Lipinski, 500, magically seems to apply in man.[188]

One implication of this is that poor bioavailability in rodents due to extensive biliary excretion might not translate into other species, notably man, where the MW threshold is higher. Keep in mind though that conjugation of a compound or its Phase I metabolite to glucuronic acid adds 194 Da, so even fairly small molecules can be subject to biliary excretion if they can be conjugated first. The best ways to get around a biliary excretion problem in rodents, then, can sometimes be to test the compound in a different species, assuming efficacy models can be done in it. Removing anionic charges, making molecules smaller, and blocking potential conjugation can potentially help too.

In recent years, in vitro models to predict biliary excretion have been introduced.[189] Prior to that, most models involved animals whose bile ducts were cannulated or ligated. Note that *renal excretion*, not otherwise discussed herein, although not part of the first pass effect, is a very important clearance mechanism for small, polar molecules, in many ways acting as a counterpart to biliary excretion.

## 9.5.2.5  Absorption Problems

Like solubility, absorption into the body from the intestinal lumen is a prerequisite for oral bioavailability. So important is this particular combination of properties that a *Biopharmaceutics Classification System* (*BCS*) has been introduced that sorts clinical drugs into four groups:

▶ Class I – High permeability, high solubility

▶ Class II – High permeability, low solubility

▶ Class III – Low permeability, high solubility

▶ Class IV – Low permeability, low solubility[190]

Compounds in Class I, though constituting the minority of clinical candidates, are expected to have the highest oral bioavailability. Even though discovery stage products are far from the clinic, this is what they aim to produce.

Absorption was addressed earlier in this chapter and the difference between low permeability due to poor passive diffusion and permeability problems caused by P-gp efflux has already been pointed out. Having made that distinction it was mentioned that in practice they really aren't unconnected: poor passive diffusion can magnify the problem caused by P-gp efflux, while improving passive diffusion (or swamping out the active transporter with high local drug concentrations in the small intestine) can render its effect a lot less relevant. Both methods designed to reduce P-gp transport (lower MW, etc.) and those intended to beef up passive transport (increasing lipophilicity, prodrugs, etc.) have been discussed along with the assays normally used to assess cell permeability (e.g. PAMPA, Caco-2). After all that, there isn't much that remains to be said here.

One thing worth mentioning is that a simple and relatively intuitive method for estimating oral bioavailability based on in vitro values for cell permeability (Caco-2) and metabolic stability (microsome or S9) was published a few years ago.[191] The model, among other things, assumes that solubility and GI stability aren't limiting, and that hepatic metabolism and not, say, renal or biliary excretion is the major fate of the compound in vivo. Figure 9.43 illustrates what one might expect in this situation. The panel at upper left shows absorption, which *increases* plasma levels of the compound with time, as modeled by Caco-2. By contrast, concentrations are *reduced* by metabolism, modeled by in vitro stability data, as shown in the upper right. Combining the two curves and integrating the AUC provide an estimate of bioavailability at lower left that approximates the compound's real %*F* (lower right).

Using this model, in vitro ADME data can be plugged in to determine whether a compound should have high, medium, or low oral bioavailability. Microsome data from different species can provide an approximation of what to expect there. Although a model this simple can't be expected to give perfect results, if the assumptions it's based on hold most of the time the binning should be correct and the relative amounts of improvement needed in each property should be highlighted. Perhaps most valuable of all, it forces the chemist to remember that these two properties, with different and often opposing SPRs, need to be optimized in tandem, not always an easy thing to do.

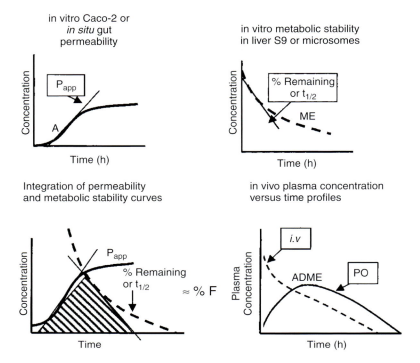

**Figure 9.43 ▶** A method for estimating oral bioavailability based on in vitro permeability ($P_{app}$) and metabolic stability data. (Reprinted with permission from *J. Med. Chem.* **2002**, *45*, 304–311) copyright 2002, American Chemical Society).

## 9.5.3 Things to Keep in Mind

When a compound is found to have poor bioavailability more than one factor can be involved. Oral bioavailability is the *product* of the fraction absorbed from the GI tract, the fraction which escapes first pass metabolism in the gut wall, and the fraction escaping hepatic metabolism and elimination, so that $F = F_a \times F_g \times F_h$. It's unusual for any of these fractions to approach one, so weaknesses in absorption, metabolism, and excretion multiply and the effect on bioavailability is cumulative.

"A prevalent mistake made by many drug discovery scientists is to try to optimize the oral bioavailability (%*F*) of lead drug candidates in animal models. While it is common for the pharmacology of lead drug candidates to be optimized in animal models, these same animal models have not been reliable predictors of oral bioavailability in humans presumably due to anatomy and/or physiology reasons."[192] Although passive absorption doesn't seem to vary too much with species—recall that dog (MDCK) and porcine kidney cells are used in monolayer permeability assays—differences in metabolism and excretion can be profound. Bioavailability will vary accordingly and compounds that work well in mouse or rat may be very different in man, so that plotting animal versus human bioavailability for a set of compounds can sometimes produce data points so randomly scattered that a line might

**TABLE 9.4 ▶ Properties and Processes Which Can Contribute to Poor Oral Bioavailability, Along with Some Relevant Assays and Suggestions for Improvements**

| Property | Assays | Possible solutions |
|---|---|---|
| Low solubility or slow dissolution | Kinetic solubility (acidic, neutral) | Reduce lipophilicity, make salts |
| Low GI tract stability | Stability in buffer (acidic, neutral), SGF, SIF | Replace unstable groups |
| Low metabolic stability | Microsome (hepatic, intestinal) or hepatocyte stability screening for species of interest | Reduce lipophilicity, block metabolic sites, possibly use active metabolite |
| Poor absorption due to low passive diffusion | PAMPA, Caco-2 (A → B) | Remove charges, HBAs and HBDs, try *increasing* lipophilicity, use prodrugs |
| Poor absorption due to active efflux (P-gp, etc.) | Cellular fluorescence assays (e.g. calcein-AM), Caco-2 (B → A/A → B) | Overwhelm by improving passive diffusion, remove or alter HBAs, lower MW |
| Extensive biliary excretion | B-Clear® (see text) or in vivo assays | Reduce MW, remove anionic charge, block conjugation |

equally well (or poorly) be drawn anywhere and in any direction. So, tethering ongoing optimization studies meant for animal models to human microsome stability data by a method like the one shown in Figure 9.43 is always a good idea.

The emphasis on good oral bioavailability that permeates most projects during LO is not without good reason, but it's important to realize that bioavailability shouldn't be the only thing considered in prioritizing compounds. Bioavailability is just one factor in getting effective concentrations of free drugs to their intended sites of action. Potency is another. A poorly bioavailable but potent compound might have a better chance of working in vivo than a less potent compound with a higher %F. Yet again, high potency can save. Protein binding too can make a difference. A less bioavailable compound that's also less protein bound can sometimes be a better choice than a more bioavailable one with a smaller unbound fraction. Pharmacokinetics and the compound selection process being complex, looking solely at bioavailability is too much of an oversimplification.

Table 9.4 summarizes the properties, assays, and possible solutions to bioavailability problems that have been discussed. For those interested in learning more, an outstanding review[193] and book[194] are highly recommended.

# ☐ Notes

1. Sinko, P.J., et al. Differentiation of gut and hepatic first pass metabolism and secretion of saquinavir in ported rabbits. *J. Pharmacol. Exp. Ther.* **2004**, *310*, 359–366.

2. The portal blood definition seems intuitively more appealing but makes it impossible to draw a clean line between absorption and metabolism. My special thanks to Dr. Walt Woltosz (Simulations Plus) and Dr. Gordon Amidon (U. Michigan) for particularly helpful discussions on this subject.

3. A second route which allows compounds to be transported from enterocytes without going through the liver, *intestinal lymphatic transport*, is not discussed here, but those interested should consult Porter, C.J.H., et al.

420 Chapter 9 **ADME and PK Properties**

Lipids and lipid-based formulations: Optimizing the oral delivery of lipophilic drugs. *Nat. Rev. Drug Discovery* **2007**, *6*, 231–248.

4. See www.accessdata.fda.gov/scripts/cdrh/cfdocs/cfcfr/CFRSearch.cfm?fr = 320.1.

5. Dr. Rang provides the best insight into the term "bioavailability" when he says "It is a convenient term for making bland generalizations, but the concept creaks badly if attempts are made to use it with quantitative precision, or even to define it." See Rang, H.P., Dale, M.M. *Pharmacology* (Edinburg, Churchill Livingstone, 1991).

6. Daugherty, A.L., Mrsny, R.J. Transcellular uptake mechanisms of the intestinal epithelial barrier. Part one *Pharm. Sci. Technol. Today* **1999**, *2*, 144–151. Note, however, that bulk pH is not the only pH involved: "microclimates" of different pH values exist around the apical surface of the cells and can be a factor as well.

7. Another possible absorption mode, transcytosis, is believed to be important for relatively few small molecules, like vitamin $B_{12}$ (which isn't all that small) and will not be considered here. See Arturrson, P., et al. Caco-2 monolayers in experimental and theoretical predictions of drug transport. *Adv. Drug Delivery Rev.* **1996**, *22*, 67–84.

8. For a closer look at this, see Pandit, N.K. "Transport Across Biological Barriers" Chapter 8 in *Introduction to the Pharmaceutical Sciences*, (Baltimore, Lippincott Williams & Williams, 2007).

9. Hu, M., Amidon, G.L. Passive and carrier-mediated intestinal absorption components of captopril. *J. Pharm. Sci.* **1988**, *77*, 1007–1011.

10. Zhou, S.Y., et al. Regulation of paracellular absorption of cimetidine and 5-Aminosalicylate in rat intestine. *Pharm. Res.* **1999**, *16*, 1781–1785.

11. Adson, A., et al. Passive diffusion of weak organic electrolytes across Caco-2 cell monolayers: Uncoupling the contributions of hydrodynamic, transcellular, and paracellular barriers. *J. Pharm. Sci.* **2006**, *84*, 1197–1204.

12. See for example, Ouyang, H., et al. Structure-activity relationship for enhancement of paracellular permeability across Caco-2 cell monolayers by 3-Alkylamido-2-alkoxypropylphosphocholines. *J. Med. Chem.* **2002**, *45*, 2857–2866, or the Alba Therapeutics website, www.albatherapeutics.com/Portals/0/pdf/Barrier%20Function%20Technology%204-4-07.pdf.

13. Lipinski, C.A., et al. Experimental and computational approaches to estimate solubility and permeability in drug discovery and development settings. *Adv. Drug Delivery Rev.* **1997**, *23*, 3–25.

14. For a concise review of ROF and its utility in drug discovery, see Lipinski, C.A. Lead- and drug-like compounds: The rule-of-five revolution. *Drug Discovery Today Technol.* **2004**, *1*, 337–341.

15. Kubinyi, H. Drug research: Myths, hype and reality. *Nat. Rev. Drug Discovery* **2003**, *2*, 665–668.

16. Calkins, K. Drug-likeness rules. *BioCentury* January 28, **2002**, p. A1.

17. Veber, D.F., et al. Molecular properties that influence the oral bioavailability of drug candidates. *J. Med. Chem.* **2002**, *45*, 2615–2623.

18. Palm, K., et al. Polar molecular surface properties predict the intestinal absorption of drugs in humans. *Pharm. Res.* **1997**, *14*, 568–571.

19. Clark, D.E. Rapid calculation of polar molecular surface area and its application to the prediction of transport phenomena. 1. Prediction of intestinal absorption. *J. Pharm. Sci.* **1999**, *88*, 807–814.

20. For a comprehensive review, see Van de Waterbeemd, H. "In Silico Models to Predict Oral Absorption", Section 5.28 in *Drug Bioavailability. Estimation of Solubility, Permeability, Absorption and Bioavailability* van de Waterbeemd, H., Lennernäs, H., Artursson, P. Eds. (Weinheim, Wiley-VCH, 2003).

21. For a review, see Clark, D.E., Pickett, S.D. Computational methods for the prediction of 'drug-likeness.' *Drug Discovery Today* **2000**, *5*, 49–58.

22. Ajay, et al. Can we learn to distinguish between "Drug-like" and "Nondrug-like" Molecules? *J. Med. Chem.* **1998**, *41*, 3314–3324.

23. Sadowski, J., Kubinyi, H. A scoring scheme for discriminating between drugs and nondrugs. *J. Med. Chem.* **1998**, *41*, 3325–3329.

24. Cruciani, G., et al. VolSurf: A new tool for the pharmacokinetic optimization of lead compounds. *Eur. J. Pharm. Sci.* **2000**, *11* Suppl. 2, S29–S39.

25. Balon, K., et al. Drug liposome partitioning as a tool for the prediction of human passive intestinal absorption. *Pharm. Res.* **1999**, *16*, 882–888.

26. Pidgeon, C., et al. IAM chromatography: An *in Vitro* screen for predicting drug membrane permeability. *J. Med. Chem.* **1995**, *38*, 590–594.

27. Yang, C.Y., et al. Immobilized artificial membranes—screens for drug membrane interactions. *Adv. Drug Delivery Rev.* **1996**, *23*, 229–256.

28. Kansy, M., et al. Physicochemical high throughput screening: Parallel artificial membrane permeation assay in the description of passive absorption process. *J. Med. Chem.* **1998**, *41*, 1007–1010.

29. For more information on the technical side of PAMPA, see www.millipore.com/publications.nsf/dda0cb48c91c0fb6852567430063b5d6/147bb2231f221e8985256d0b0042433c/$FILE/AN1729EN00.pdf.

30. See, for example, www.pion-inc.com/PAMPA_evolution.htm.

31. Palm, K., et al. Effect of molecular charge on intestinal epithelial drug transport: pH-Dependent transport of cationic drugs. *J. Pharmacol. Exp. Ther.* **1999**, *291*, 435–443.

32. Zhu, C., et al. A comparative study of artificial membrane permeability assay for high throughput profiling of drug absorption potential. *Eur. J. Med. Chem.* **2002**, *37*, 399–407.

33. See, for example, Figure 2 in Veber, Reference 17.

34. Kansy, M., et al. Advances in screening for membrane permeability: High-resolution PAMPA for medicinal chemists. *Drug Discovery Today Technol.* **2004**, *1*, 349–355.

35. Ruell, J.A., et al. PAMPA—a drug absorption *in vitro* model. 12. Cosolvent method for permeability assays of amiodarone, itraconazole, tamoxifen, terfenadine, and other very insoluble molecules. *Chem. Pharm. Bull.* **2004**, *52*, 561–565.

36. Avdeef, A. *Absorption and Drug Development* (New York, Wiley-Interscience, 2003) pp. 128–246.

37. Di, L., et al. High throughput artificial membrane permeability assay for blood-brain barrier. *Eur. J. Med. Chem.* **2003**, *38*, 223–232.

38. Wohnsland, F., Faller, B. High-throughput permeability pH profile and high-throughput Alkane/Water log P with artificial membranes. *J. Med. Chem.* **2001**, *44*, 923–930.

39. See www.fda.gov/cder/OPS/BCS_guidance.htm.

40. Benet, L.Z., Cummins, C.L. The drug efflux-metabolism alliance: Biochemical aspects. *Adv. Drug Delivery Rev.* **2001**, *50*, Suppl. 1, S3–S11.

41. See Lennernäs, H., Lundgren, E. Intestinal and blood-brain drug transport: Beyond involvement of a single transport function. *Drug Discovery Today Technol.* **2004**, *1*, 417–422 and references therein.

42. See, for example, Raeissi, S.D., et al. Interplay between CYP3A-mediated metabolism and polarized efflux of terfenadine and its metabolites in intestinal epithelial Caco-2 (TC7) cell monolayers. *Pharm. Res.* **1999**, *16*, 625–632.

43. Engman, H.A., et al. CYP3A4, CYP3A5, and MDR1 in human small and large intestinal cell lines suitable for drug transport studies. *J. Pharm. Sci.* **2001**, *90*, 1736–1751.

44. Arturrson, P., Magnusson, C. Epithelial transport of drugs in cell culture. II. Effect of extracellular calcium concentration on the paracellular transport of drugs of different lipophilicities across monolayers of intestinal epithelial (Caco-2) cells. *J. Pharm. Sci.* **1990**, *79*, 595–600.

45. A fairly recent publication estimated the cost of Caco-2 screening to be $19 per compound. See Balimane, P.V., Chong, S. Cell culture-based models for intestinal permeability: A critique. *Drug Discovery Today* **2005**, *10*, 335–343.

46. See Artursson, P. Caco-2 monolayers in experimental and theoretical predictions of drug transport. *Adv. Drug Delivery Rev.* **1996**, *22*, 67–84.

47. See www.bdbiosciences.com/external_files/dl/doc/mkt_lit/salesSheets/F04B093.pdf.

48. Braun, A., et al. Cell cultures as tools in biopharmacy. *Eur. J. Pharm. Sci.* **2000**, *11*, Suppl. 2, S51–S60.

49. Matsson, P., et al. Exploring the role of different drug transport routes in permeability screening. *J. Med. Chem.* **2005**, *48*, 604–613.

50. Franke, H., et al. Primary cultures of brain microvessel endothelial cells: A valid and flexible model to study drug transport through the blood—brain barrier in vitro. *Brain. Res. Protoc.* **2000**, *5*, 248–256.

51. Gumbleton, M., Audus, K.L. Progress and limitations in the use of *in vitro* cell cultures to serve as a permeability screen for the blood-brain barrier. *J. Pharm. Sci.* **2001**, *90*, 1681–1698.

52. The complementarity between PAMPA and Caco-2 measurements and a similar staging scheme for their use in drug discovery is advocated in Kerns, E.H., et al. Combined application of parallel artificial membrane permeability assay and Caco-2 permeability assays in drug discovery. *J. Pharm. Sci.* **2004**, *93*, 1440–1453.

53. Gotoh, Y., et al. The Advantages of the Ussing chamber in drug absorption studies. *J. Biomol. Screen.* **2005**, *10*, 517–523.

54. Sable-Amplis, R., Abadie, D. Permanent cannulation of the hepatic portal vein in rats. *J. Appl. Physiol.* **1975**, *38*, 358–359.

55. See Petri, N., Lennernäs, H. "*In vivo* permeability studies in the gastrointestinal tract of humans", Chapter 7 in *Drug Bioavailability. Estimation of Solubility, Permeability, Absorption and Bioavailability* van de Waterbeemd, H., Lennernäs, H., Artursson, P. Eds. (Weinheim, Wiley-VCH, 2003).

56. Quan, M.L., et al. Discovery of 1-(3'-Aminobenzisoxazol-5'-yl)-3-trifluoromethyl-*N*-[2-fluoro-4-[(2'-dimethylaminomethyl)imidazol-1-yl]phenyl-1*H*-pyrazole-5-carboxamide Hydrochloride (Razaxaban), a highly potent, selective, and orally bioavailable factor Xa inhibitor. *J. Med. Chem.* **2005**, *48*, 1729–1744.

57. Rezai, T., et al. Testing the conformational hypothesis of passive membrane permeability using synthetic cyclic peptide diastereomers. *J. Am. Chem. Soc.* **2006**, *128*, 2510–2511.

58. For a comprehensive guide to prodrugs, see *Prodrugs: Challenges and Rewards*, Stella, V.J., Borchardt, R.T., Hageman, M.J., Oliyai, R., Maag, H., Tilley, J., Eds. (New York, Springer, 2007).

59. For more about this and other properties desirable in an ester prodrug, see Beaumont, K., et al. Design of ester prodrugs to enhance oral absorption of poorly permeable compounds: Challenges to the discovery scientist. *Curr. Drug Metab.* **2003**, *4*, 461–485.

60. Gustafsson, D., et al. The direct thrombin inhibitor melagatran and its oral prodrug H 376/95: Intestinal absorption properties, biochemical and pharmacodynamic effects. *Thromb. Res.* **2001**, *101*, 171–181.

61. Clement, B., Lopian, K. Characterization of in vitro biotransformation of new, orally active, direct thrombin inhibitor ximelagatran, an amidoxime and ester prodrug. *Drug Metab. Dispos.* **2003**, *31*, 645–651.

62. See Boudes, P.F. The challenges of new drug benefits and risis analysis: Lessons from the ximelagatran FDA cardiovascular advisory committee. *Contemp. Clin. Trials* **2006**, *27*, 432–440.

63. Beauchamp, L.M., et al. Amino acid ester prodrugs of aciclovir. *Antivir. Chem. Chemother.* **1992**, *3*, 157–164.

64. Landowski, C.P., et al. Gene expression in the human intestine and correlation with oral valacyclovir pharmacokinetic parameters. *J. Pharmacol. Exp. Ther.* **2003**, *306*, 778–786.

65. de Mirand, P., et al. "Mechanisms of the enhanced oral bioavailability of acyclovir with the prodrug valacyclovir HCl (VALTREX™), *Abstr. Intersci. Conf. Antimicrob. Agents Chemother. Intersci. Conf.* **1994**, October. 4–7; 83. Abstract available at http://gateway.nlm.nih.gov/MeetingAbstracts/102213721.html.

66. Friedman, D.I., Amidon, G.L. Passive and carrier-mediated intestinal absorption components of two angiotensin converting enzyme (ACE) inhibitor prodrugs in rats: Enalapril and fosinopril. *Pharm. Res.* **1989**, *6*, 1043–1047.

67. Steffansen, B., et al. Intestinal solute carriers: An overview of trends and strategies for improving oral drug absorption. *Eur. J. Pharm. Sci.* **2004**, *21*, 3–16.

68. Moore, V.A., et al. A rapid screening system to determine drug affinities for the intestinal dipeptide transporter 1: System characterization. *Int. J. Pharm.* **2000**, *210*, 15–27.

69. Andrew P. "Biotransformation of Xenobiotics" Chapter 6 in *Casarett & Doull's Toxicology*, 5th Edition, Klaassen, C.D., Ed. (New York, McGraw-Hill, 1996).

70. Coleman, M.D. *Human Drug Metabolism* (Chichester, John Wiley, 2005).

71. This happened, for example, with some COX-2 inhibitors. See Penning, T.D., et al. Synthesis and biological evaluation of the 1,5-Diarylpyrazole class of cyclooxygenase-2 inhibitors: Identification of the 4-[5-(4-Methylphenyl)-3-(trifluoromethyl)-1*H*-pyrazol-1-yl]benzenesulfonamide (SC-58635, Celecoxib). *J. Med. Chem.* **1997**, *40*, 1347–1365.

72. Clarke, S.E., Jones, B.C. "Human Cytochromes P450 and Their role in Metabolism-Based Drug-Drug Interactions", in *Drug-Drug Interactions*, Rodrigues, A.D., Ed. (New York, Marcel Dekker, 2002).

73. Coon, M.J. Cytochrome P450: Nature's most versatile biological catalyst. *Annu. Rev. Pharmacol. Toxicol.* **2005**, *45*, 1–25.

74. Werck-Reichart, D., Feyereisen, R. Cytochrome P450: A success story. *Genome Biol.* **2000**, *1*, 3003.1–3003.9.

75. For a detailed treatise on CYP structure and mechanism, see Denisov, I.G., et al. Structure and chemistry of cytochrome P450. *Chem. Rev.* **2005**, *105*, 2253–2277. For some unusual reactions CYPs are capable of catalyzing, see Isen, E.M., Guengerich, F.P. Complex reactions catalyzed by Cytochrome P450 enzymes. *Biochim. Biophys. Res. Commun.* **2007**, *1770*, 314–329. For more information about CYPs in general, see *Cytochrome P450: Structure, Mechanism, and Biochemistry*, 3rd Edition, Ortiz de Montellano, P.R., Ed. (New York, Springer, 2004).

76. CYPs are named using the following convention: The first number after "CYP" designates a family which shares at least 40% homology, the letter after that designates a sub-family (at least 60% homology), and the final number refers to an individual enzyme.

77. Guroff, G., et al. Hydroxylation-induced migration: The NIH shift *Science* **1967**, *157*, 1524–2530.

78. (a) Jones, J.P. Computational models for cytochrome P450: A predictive electronic model for aromatic oxidation and hydrogen atom abstraction. *Drug Metab. Dispos.* **2002**, *30*, 7–12. (b) Singh, S.B., et al. A model for predicting likely sites of CYP3A4-mediated metabolism on drug-like molecules. *J. Med. Chem.* **2003**, *46*, 1330–1336.

79. See Coleman, Reference 71, pp. 46–48.

80. For a comparison of these two families of metabolic enzymes, see Cashman, J.R. Some distinctions between flavin-containing and cytochrome P450 monooxygenases. *Biochem. Biophys. Res. Commun.* **2005**, *338*, 599–624.

81. A transformation for which the chemist often needs $BBr_3$ or molten pyridinium hydrochloride!

82. King, C.D., et al. Expression of UDP-glucuronosyltransferases (UGTs) 2B7 and 1A6 in the human brain and identification of 5-hydroxytryptamine as a substrate. *Arch. Biochem. Biophys.* **1999**, *365*, 156–162.

83. See Papageorgiou, C. Enterohepatic recirculation: A powerful incentive for drug discovery in the inosine monophosphate dehydrogenase field. *Mini Rev. Med. Chem.* **2001**, *1*, 71–77.

84. See Roberts, M.S., et al. Enterohepatic circulation: Physiological, pharmacokinetic and clinical implications. *Clin. Pharmacokinet.* **2002**, *41*, 751–790.

85. Source: Humphreys, A. "Med Ad News 200 Best-selling prescription medicines – sales", *Med Ad News*, July 2007, pp. 18–24.

86. Information on metabolism was derived from approved FDA labels which can be found through www.accessdata.fda.gov/scripts/cder/drugsatfda/index.cfm as well as the following:

Lins, R.L., et al., Pharmacokinetics of atorvastatin and its metabolites after single and multiple dosing in hypercholesterolaemic haemodialysis patients. *Nephrol. Dial. Transplant.* **2003**, *18*, 967–976.

Pearce, R.E., et al. Biotransformation of fluticasone: In vitro characterization. *Drug Metab. Dispos.* **2006**, *34*, 1035–1040.

Manchee, G.R., et al. The aliphatic oxidation of salmeterol to alpha-hydroxysalmeterol in human liver microsomes is catalyzed by CYP3A. *Drug Metab. Dispos.* **1996**, *24*, 555–559.

Tang, M., et al. Antiplatelet agents aspirin and clopidogrel are hydrolyzed by distinct carboxylesterases, and clopidogrel is transesterificated in the presence of ethyl alcohol. *J. Pharmacol. Exp. Ther.* **2006**, *319*, 1467–1476.

Roche, V.F. The chemically elegant proton pump inhibitors. *Am. J. Pharm Educ.* **2006**, *70*, 1–11.

Linnet, K. Glucuronidation of olanzapine by cDNA-expressed human UDP-glucuronosyltransferases and human liver microsomes. *Hum. Psychopharmacol. Clin. Exp.* **2002**, *17*, 233–238.

Nakashima, A., et al. Identification of cytochrome P450 forms involved in the 4-hydroxylation of valsartan, a potent and specific angiotensin II receptor antagonist, in human liver microsomes. *Xenobiotica* **2005**, *35*, 589–602.

Chiba, M., et al. Hepatic microsomal metabolism of montelukast, A potent leukotriene D4 receptor antagonist, in humans. *Drug Metab. Dispos.* **1997**, *25*, 1022–1031.

Balani, S.K. Metabolic profiles of montelukast sodium (Singulair), A potent cysteinyl leukotriene₁ receptor antagonist, in human plasma and bile. *Drug Metab. Dispos.* **1997**, *25*, 1282–1287.

87. Di, L., et al. Optimization of a higher throughput microsomal stability screening assay for profiling drug discovery candidates. *J. Biomol. Screen.* **2003**, *8*, 453–462.

88. Gombar, V.K., et al. Role of ADME characteristics in drug discovery and their in silico evaluation: In silico screening of chemicals for their metabolic stability. *Curr. Top. Med. Chem.* **2003**, *3*, 1205–1225.

89. See www.bdbiosciences.com.

90. See Rodrigues, A.D. Integrated cytochrome P450 reaction phenotyping. *Biochem. Pharmacol.* **1999**, *57*, 465–480.

91. See Di, L., Kerns, E.H. High throughput screening of metabolic stability in drug discovery. *Am. Drug Discovery* **2007**, *2*, 28–32.

92. For more about this, see Obach, R.S. Prediction of human clearance of twenty-nine drugs from hepatic microsomal intrinsic clearance data: An examination of in vitro half-life approach and nonspecific binding to microsomes. *Drug Metab. Dispos.* **1999**, *27*, 1350–1359, and references therein.

93. Di, L., et al. High throughput microsomal stability assay for insoluble compounds. *Int. J. Pharm.* **2006**, *317*, 54–60.

94. For more on the merits and demerits of microsomes, cytosol, S9, and other metabolizing systems in predicting drug metabolism, see Brandon, E.F.A., et al, An update on in vitro test methods in human hepatic drug biotransformation research: Pros and cons. *Toxicol. Appl. Pharmacol.* **2003**, *189*, 231–246.

95. Jouin, D., et al. Cryopreserved human hepatocytes in suspension are a convenient high throughput tool for the prediction of metabolic clearance. *Eur. J. Pharm. Biopharm.* **2006**, *63*, 347–355.

96. For more about IVIVE, see Baranczewski, P., et al. Introduction to *in vitro* estimation of metabolic stability and drug interactions of new chemical entities in drug discovery and development. *Pharmacol. Rep.* **2006**, *58*, 453–472.

97. Plant, N. Strategies for using *in vitro* screens in drug metabolism. *Drug Discovery Today* **2004**, *9*, 328–336.

98. See Bjornsson, T.D., et al. The conduct of in vitro and in vivo drug-drug interaction studies: A pharmaceutical research and manufacturers of America (PhRMA) perspective. *Drug Metab. Dispos.* **2003**, *31*, 815–832.

99. For more about FDA requirements for the safety testing of metabolites, see the draft guidance at www.fda. gov/Cder/guidance/6366dft.pdf. Also see Baillie, T.A., et al. Drug metabolites in safety testing. *Toxicol. Appl. Pharmacol.* **2002**, *182*, 188–196, and Hastings, K.L., et al. Drug metabolites in safety testing. *Toxicol. Appl. Pharmacol.* **2003**, *190*, 91–92.

100. See Cruciani, G., et al. MetaSite: Understanding metabolism in human cytochromes from the perspective of the chemist. *J. Med. Chem.* **2005**, *48*, 6970–6979. See also the Molecular Discovery webite at www.moldiscovery.com/soft_metasite.php.

101. Ahlström, M.M., et al. CYP2C9 Structure-metabolism relationships: Optimizing the metabolic stability of COX-2 inhibitors. *J. Med. Chem.* **2007**, *50*, 4444–4452.

102. Thompson, T.N. Optimization of metabolic stability as a goal of modern drug design. *Med. Res. Rev.* **2001**, *21*, 412–449.

103. For more about the use of fluorine to improve metabolic properties of drugs, see Park, B.K., et al. Metabolism of fluorine-containing drugs. *Annu. Rev. Pharmacol. Toxicol.* **2001**, *41*, 443–470. Also of interest is Müller, K., et al. Fluorine in pharmaceuticals: Looking beyond intuition. *Science* **2007**, *317*, 1881–1886.

104. Tandon, M., et al. The design and preparation of metabolically protected new arylpiperazine 5-HT$_{1A}$ ligands. *Bioorg. Med. Chem. Lett.* **2004**, *14*, 1709–1712.

105. This is an interesting reversal of the trend one normally sees, where human metabolism is slower than mouse.

106. Takeda, Y., et al. New highly active taxoids from 9β-Dihydrobaccatin-9,10-acetals. Part 4. *Bioorg. Med. Chem.* **2003**, *11*, 4431–4447.

107. Wu, W.-L., et al. Dopamine D$_1$/D$_5$ receptor antagonists with improved pharmacokinetics: Design, synthesis, and biological evaluation of phenol bioisosteric analogues of benzazepine D$_1$/D$_5$ antagonists. *J. Med. Chem.* **2005**, *48*, 680–693.

108. Noe, M.C., et al. Discovery of 3,3-dimethyl-5-hydroxypipecolic hydroxamate-based inhibitors of aggrecanase and MMP-13. *Bioorg. Med. Chem. Lett.* **2005**, *15*, 2808–2811.

109. Imamura, S., et al. Discovery of a piperidine-4-caboxamide CCR5 antagonist (TAK-220) with highly potent anti-HIV-1 activity. *J. Med. Chem.* **2006**, *49*, 2784–2793.

110. Hwang, S.H., et al. Orally bioavailable potent soluble epoxide hydrolase inhibitors. *J. Med. Chem.* **2007**, *50*, 3825–3840.

111. For an excellent real-world example of balancing optimization of metabolic stability with cell permeability and CYP inhibition, showing just how much effort may be required, see Powell, N.A., et al. Rational design of 6-(2,4-diaminopyrimidinyl)-1,4-benzoxazin-3-ones as small molecule renin inhibitors. *Bioorg. Med. Chem.* **2007**, *15*, 5912–5949.

112. Palani, A., et al. Synthesis, SAR, and biological evaluation of oximino-piperidino-piperidine amides. 1. Orally bioavailable CCR5 receptor antagonists with potent anti-HIV activity. *J. Med. Chem.* **2002**, *45*, 3143–3160.

113. Strizki, J.M., et al. Discovery and characterization of vicriviroc (SCH 417690), a CCR5 antagonist with potent activity against human immunodeficiency virus Type 1. *Antimicrob. Agents Chemother.* **2005**, *49*, 4911–4919.

114. There can be minor differences in the extent of protein binding in serum and plasma, but in general these aren't large and, although they shouldn't be, the two terms are often used interchangeably. See Ohsima, T., et al. Variations in protein binding of drugs in plasma and serum. *Clin. Chem.* **1989**, *35*, 1722–1725.

115. This is a very close to "Free fraction drives the action", a quote from Dr. Robert Ings.

116. For more about plasma protein binding and the free drug principle, see Trainor, G.L. The importance of plasma protein binding in drug discovery. *Expert Opin. Drug Discovery* **2007**, *2*, 51–64, and Trainor, G.L. "Plasma Protein Binding and the Free Drug Principle: Recent Developments and Applications", Chapter 31 in *Annual Reports in Medicinal Chemistry, Volume 42*, Macor, J.E., Ed. (San Diego, Academic Press, 2007).

117. Rang, H.P., Dale, M.M. *Pharmacology* 2nd Edition (Edinburgh, Churchill Livinstone, 1991) p. 104.

118. Kalvass, J.C., Maurer, T.S. Influence of nonspecific brain and plasma binding on CNS exposure: Implications for rational drug discovery. *Biopharm. Drug Dispos.* **2002**, *23*, 327–338.

119. Summerfield, S.G., et al. Improving the *in vitro* prediction of in vivo central nervous system penetration: Integrating permeability, P-glycoprotein efflux, and free fractions in blood and brain. *J. Pharmacol. Exp. Ther.* **2006**, *316*, 1282–1290.

120. He, X.M., Carter, D.C. Atomic structure and chemistry of human serum albumin. *Nature* **1992**, *358*, 209–215.

121. Fitos, I., et al. Selective binding of imatinib to the genetic variants of human $\alpha_1$-acid glycoprotein. *Biochim. Biophys. Acta* **2006**, *1760*, 1704–1712.

122. Benet, L.Z., Hoener, B. Changes in plasma binding have little clinical relevance. *Clin. Pharmacol. Ther.* **2002**, *71*, 115–121.

123. Fuse, E., et al. Unpredicted clinical pharmacology of UCN-01 caused by specific binding to human $\alpha_1$-acid glycoprotein. *Cancer Res.* **1998**, *58*, 3248–3253.

124. Ito, T., et al. Marked strain differences in the pharmacokinetics of an $\alpha_4\beta_1$ integrin antagonist, 4-[1-[3-Chloro-4-[N-(2-methylphenyl)-ureido]phenylacetyl]-(4S)-fluoro-(2S)-pyrrolidine-2-yl]-methoxybenzoic acid (DOI-4582), in sprague-dawley rats are associated with albumin genetic polymorphisms. *J. Pharmacol. Exp. Ther.* **2007**, *320*, 124–132.

125. Edmondson, S.D., et al. (2S,3S)-3-Amino-4-(3,3-difluoropyrrolidin-1-yl)-N,N-dimethyl-4-oxo-2-(4-[1,2,4]triazolo[1,5-a]-pyridin-6-ylphenyl)butanamide: A selective $\alpha$-amino amide dipeptidyl peptidase IV inhibitor for the treatment of Type 2 diabetes. *J. Med. Chem.* **2006**, *49*, 3614–3627.

126. Wan, H., Rehngren, M. High-throughput screening of protein binding by equilibrium dialysis combined with liquid chromatography and mass spectrometry. *J. Chromatogr. A* **2006**, *1102*, 125–134.

127. Zhang, J., Musson, D.G. Investigation of high-throughput ultrafiltration for the determination of an unbound compound in human plasma using liquid chromatography and tandem mass spectrometry with electrospray ionization. *J. Chromatogr. B* **2006**, *843*, 47–56.

128. Fung, E.N., et al. Semi-automatic high-throughput determination of plasma protein binding using a 96-well plate filtrate assembly and fast liquid chromatography-tandem mass spectrometry. *J. Chromatogr. B* **2003**, *795*, 187–194.

129. See, for example Buchholz, L., et al. Evaluation of the human serum albumin column as a discovery screening tool for plasma protein binding. *Eur. J. Pharm. Sci.* **2002**, *15*, 209–215.

130. See Rich, R.L., et al. High-resolution and high-throughput protocols for measuring drug/human serum albumin interactions using BIACORE. *Anal. Biochem.* **2001**, *296*, 197–207. Also see http://web.bf.uni-lj.si/bi/sprcenter/Application%20Note/ApplicationNote30.pdf.

131. The replicon assay results for VX-950 was run by a group at Vertex Pharmaceuticals while the other two reported results are from a Novartis group as per the references shown in the figure. Since there were slight differences in assay conditions between the two the different results may not be 100% comparable.

132. See Watashi, K., et al. Cyclophilin B is a functional regulator of hepatitis C virus RNA polymerase. *Mol. Cell* **2005**, *19*, 111–122.

133. Lin, K., et al. VX-950, a novel hepatitis C virus (HCV) NS3-4A protease inhibitor, Exhibits potent antiviral activities in HCV replicon cells *Antimicrob. Agents. Chemother.* **2006**, *50*, 1813–1822.

134. Ma, S., et al. NIM811, a Cyclophilin inhibitor, exhibits potent *in vitro* activity against hepatitis C virus alone or in combination with alpha interferon. *Antimicrob. Agents Chemother.* **2006**, *50*, 2976–2982.

135. Rusnak, D.W., et al. A simple method for predicting serum protein binding of compounds from $IC_{50}$ shift analysis for in vitro assays. *Bioorg. Med. Chem. Lett.* **2004**, *14*, 2309–2312.

136. Kratochwil, N.A., et al. Predicting plasma protein binding of drugs: A new approach. *Biochem. Pharmacol.* **2002**, *64*, 1355–1374.

137. See, for example, Kratochwil, ibid. and Gleeson, M.P. Plasma protein binding affinity and its relationship to molecular structure: An in-silico analysis. *J. Med. Chem.* **2007**, *50*, 101–112.

138. See, for example ADMET Predictor® from Simulations Plus: www.simulations-plus.com/products/predictor/model_plasma_binding.html.

139. Gleeson, M.P. Plasma protein binding affinity and its relationship to molecular structure: An in-silico analysis. *J. Med. Chem.* **2007**, *50*, 101–112.

140. McKerrecher, D., et al. Design of a potent, soluble glucokinase activator with excellent in vivo efficacy. *Bioorg. Med. Chem. Lett.* **2006**, *16*, 2705–2709.

141. Leach, A.G., et al. Matched molecular pairs as a guide in the optimization of pharmaceutical properties: A Study of aqueous solubility, plasma protein binding, and oral exposure. *J. Med. Chem.* **2006**, *49*, 6672–6682. Chemists grappling with protein binding problems are urged to read this paper.

142. Birch, A.M. Development of potent, orally active 1-substituted-3,4-dihydro-2-quinolone glycogen phosphorylase inhibitors. *Bioorg. Med. Chem. Lett.* **2007**, *17*, 394–399.

143. Wendt, M.D., et al. Discovery and structure-activity relationship of antagonists of B-Cell lymphoma 2 family proteins with chemopotentiation activity in vitro and in vivo. *J. Med. Chem.* **2006**, *49*, 1165–1181.

144. Sheppard, G.S., et al. Discovery and optimization of anthranilic acid sulfonamides as inhibitors of methionine aminopeptidase-2: A structural basis for the reduction of albumin binding. *J. Med. Chem.* **2006**, *49*, 3832–3849. The authors note that improvements in cellular antiproliferative potency may have involved other factors besides just protein binding.

145. Juliano, R.L., Ling, V. A Surface glycoprotein modulating drug permeability in chinese hamster ovary cell mutants. *Biochim. Biophys. Acta* **1976**, *455*, 152–162.

146. See Tannock, I.F., Goldenburg, G.J. "Drug Resistance and Experimental Chemotherapy", Chapter 17 in *The Basic Science of Oncology*, Tannock, I.F., Hill, R.P., Eds. (New York, McGraw-Hill, 1998).

147. Mouse has two genes thought to be relevant to the human *MDR1* gene, called *mdr1a* and *mdr1b*. Both *mdr1a* *(-/-)* and the double knockout *mdr1a /1b (-/-)* survive and reproduce, but an unplanned experiment was carried out when the former along with other mice were sprayed with a dilute solution of ivermectin to control a mite infestation. At doses normally considered quite safe, some of the *mdr1a (-/-)* mice died due to much higher brain exposure levels to the antiparasitic agent. See Schinkel, A.H., Disruption of the mouse *mdr1a* P-glycoprotein gene leads to a deficiency in the blood-rrain barrier and to increased sensitivity to drugs. *Cell* **1994**, *77*, 491–502.

148. See Higgins, C.F. Multiple molecular mechanisms for multidrug resistance transporters. *Nature* **2007**, *446*, 749–757.

149. Loo, T.W., Clarke, D.M. Do drug substrates enter the common drug-binding pocket of P-glycoprotein through "gates"? *Biochem. Biophys. Res. Commun.* **2005**, *329*, 419–422.

150. Higgins, C.F., Gottesman, M.M. Is the multidrug transporter a flippase? *Trends Biochem. Sci.* **1992**, *17*, 18–21.

151. Raviv, Y., et al. Photosensitized labeling of a functional multidrug transporter in living drug-resistant tumor Cells. *J. Biol. Chem.* **1990**, *265*, 3975–3980.

152. Stein, W.D. Kinetics of the multidrug transporter (P-Glycoprotein) and its reversal. *Physiol. Rev.* **1997**, *77*, 545–590.

153. Choo, E.F., et al. Pharmacological inhibition of P-Glycoprotein transport enhances the distribution of HIV-1 protease inhibitors into brain and testes. *Drug Metab. Dispos.* **2000**, *28*, 655–660.

154. See Benet, L.Z., et al. Intestinal MDR transport proteins and P-450 enzymes as barriers to oral drug delivery. *J. Control Release* **1999**, *62*, 25–31.

155. "Modulator" can also refer to a compound interacting with P-gp in the opposite way, that is, stimulating transport.

156. Szakács, G., et al. Targeting multidrug resistance in cancer. *Nat. Rev. Drug Discovery* **2006**, *5*, 219–234. See also McDevitt, C.A., Callaghan, R. How can we best use structural information on P-gylcoprotein to design inhibitors? *Pharmacol. Ther.* **2007**, *113*, 429–441.

157. For an example, see Wang, R.B., et al. Structure-activity relationship: Analyses of p-glycoprotein substrates and inhibitors. *J. Clin. Pharm. Ther.* **2003**, *28*, 203–228.

158. This effect might not be seen except at high concentration, for example, or a substrate could bind at a different site (or different part of the same site) with complex effects on transport.

159. See Xu, C., et al. Induction of phase I, II, and III drug metabolism/transport by xenobiotics. *Arch. Pharm. Res.* **2005**, *28*, 249–268.

160. See, for example, Andrivon, W., et al. A new anthracycline with potent anti-leukemic activity overcomes P-glycoprotein multidrug resistance. *Leuk. Res.* **1998**, *22*, 719–725.

161. See Balimane, P.V., et al. Current Industrial Practices of Assessing Permeability and P-Glycoprotein Interaction. *AAPS J.* **2006**, *8*, E1–E13.

162. Polli, J.W. Rational use of in vitro P-glycoprotein assays in drug discovery. *J. Pharmacol. Exp. Ther.* **2001**, *299*, 620–628.

163. Sarkadi, B., et al. Expression of the human multidrug resistance cDNA in insect cells generates a high activity drug-stimulated membrane ATPase. *J. Biol. Chem.* **1992**, *267*, 4854–4959.

164. Schwab, D., et al. Comparison of in vitro P-glycoprotein screening assays: Recommendations for their use in drug discovery. *J. Med. Chem.* **2003**, *46*, 1716–1725.

165. Homolya, L., et al. Fluorescent cellular indicators are extruded by the multidrug resistance protein. *J. Biol. Chem.* **1993**, *268*, 21493–21496.

166. Schinkel, A.H., et al. Normal viability and altered pharmacokinetics in mice lacking mdr1-type (drug-transporting) P-glycoproteins. *Proc. Nat. Acad. Sci. U.S.A* **1997**, *94*, 4028–4033.

167. Hennessy, M., Spiers, J.P. A primer on the mechanics of P-glycoprotein the multidrug transporter. *Pharmacol. Res.* **2007**, *55*, 1–15.

168. Cianchetta, G., et al. A pharmacophore hypothesis for P-glycoprotein substrate recognition using GRIND-based 3D-QSAR. *J. Med. Chem.* **2005**, *48*, 2927–2935. See also Stouch, T.R., Gudmundsson, O. Progress in understanding the structure—activity relationships of P-glycoprotein. *Adv. Drug Delivery Rev.* **2002**, *54*, 315–328.

169. P-gp doesn't generally transport negatively charged molecules like carboxylic acids very well.

170. Scala, S., et al. P-glycoprotein substrates and antagonists cluster into two distinct groups. *Mol. Pharmacol.* **1997**, *51*, 1024–1033.

171. Seelig, A. A general pattern for substrate recognition by P-glycoprotein. *Eur. J. Biochem.* **1998**, *251*, 252–261.

172. Seelig, A., Landwojtowicz, E. Structure-activity relationship of P-glycoprotein substrates and modifiers. *Eur. J. Pharm. Sci.* **2000**, *12*, 31–40.

173. Smith, D.A., van de Waterbeemd, H. Pharmacokinetics and metabolism in early drug discovery *Curr. Opin. Chem. Biol.* **1999**, *3*, 373–378.

174. For example, see Ciancetta et al., reference 168.

175. Moore, K.P., et al. Strategies toward improving the brain penetration of macrocyclic tertiary carbinamine BACE-1 inhibitors. *Bioorg. Med. Chem. Lett.* **2007**, *17*, 5831–5835.

176. Rice, A., et al. Chemical modification of paclitaxel (Taxol) reduces P-glycoprotein interactions and increases permeation across the blood—brain barrier in vitro and in situ. *J. Med. Chem.* **2005**, *48*, 832–838.

177. Ballatore, C. Paclitaxel C-10 carbamates: Potential candidates for the treatment of neurodegenerative tauopathies. *Bioorg. Med. Chem. Lett.* **2007**, *17*, 3642–3646.

178. Raub, T.J. P-glycoprotein recognition of substrates and circumvention through rational drug design. *Mol. Pharm.* **2005**, *3*, 3–25.

179. Berman, J., et al. Simultaneous pharmacokinetic screening of a mixture of compounds in the dog using API LC/MS/MS analysis for increased throughput. *J. Med. Chem.* **1997**, *40*, 827–829.

180. Olah, T.V., et al. The simultaneous determination of mixtures of drug candidates by liquid chromatography/ atmospheric pressure chemical ionization mass spectrometry as an *In Vivo* drug screening procedure. *Rapid Commun. Mass Spectrom.* **1997**, *11*, 17–23.

181. Remenar, J.F. Improving oral bioavailability through inhibition of crystallization after dosing. *Am. Pharm. Rev.* **2007**, *10*, 84–89.

182. Mathew, M., et al. Stability of omeprazole solutions at various pH values as determined by high-performance liquid chromatography. *Drug. Dev. Ind. Pharm.* **1995**, *21*, 965–971.

183. Woodley, J.F. Enzymatic barriers for GI peptides and protein delivery. *Crit. Rev. Ther. Drug Carrier Syst.* **1994**, *11*, 61–95.

184. For more about this, see Chan, O.H., Stewart, B.H. Physicochemical and drug-delivery considerations for oral drug bioavailability. *Drug Discovery Today* **1996**, *1*, 461–473.

185. Murakami, T., et al. Separate assessment of intestinal and hepatic first-pass effects using a rat model with double cannulation of the portal and jugular veins. *Drug Metab. Pharmacokinet.* **2003**, *18*, 252–260.

186. For more about this and other aspects of gut wall metabolism, see Beaumont, K. "The Importance of Gut Wall Metabolism in Determining Drug Bioavailability", Chapter 13 in *Drug Bioavailability. Estimation of Solubility, Permeability, Absorption and Bioavailability* van de Waterbeemd, H., Lennernäs, H., Artursson, P., Eds. (Weinheim, Wiley-VCH, 2003).

187. Hirom, P.C., et al. Species variations in the threshold molecular-weight factor for the biliary excretion of organic anions. *Biochem. J.* **1972**, *129*, 1071–1077.

188. Eustace, P.W., et al. Biliary Excretion of Diazepam in Man .*Br. J. Anaesth.* **1975**, *47*, 983–985.

189. For example, see the B-Clear® system at www.qualyst.com/products/bclear.php5. See also Sasaki, M., et al. Prediction of in vivo biliary clearance from the in vitro transcellular transport of organic anions across a double-transfected madin-darby canine kidney II monolayer expressing both rat organic anion transporting polypeptide 4 and multidrug resistance associated protein 2. *Mol. Pharmacol.* **2004**, *66*, 450–459.

190. Amidon, G.L. A theoretical basis for a biopharmaceutic drug classification: The correlation of in vitro drug product dissolution and in vivo bioavailability. *Pharm. Res.* **1995**, *12*, 413–420.

191. Mandagere, A.K., et al. Graphical model for estimating oral bioavailability of drugs in humans and other species from their Caco-2 permeability and in vitro liver enzyme metabolic stability rates. *J. Med. Chem.* **2002**, *45*, 304–311.

192. Caldwell, G.W., et al. The new pre-preclinical paradigm: Compound optimization in early and late phase drug discovery. *Curr. Top. Med. Chem.* **2001**, *1*, 353–366.

193. van de Waterbeemd, H., et al. Property-based design: Optimization of drug absorption and pharmacokinetics. *J. Med. Chem.* **2001**, *44*, 1313–1333.

194. *Drug Bioavailability. Estimation of Solubility, Permeability, Absorption and Bioavailability* van de Waterbeemd, H., Lennernäs, H., Artursson, P., Eds. (Weinheim, Wiley-VCH, 2003).

# Chapter 10

# Toxicity-Related Properties

*"What is there that is not poison? All things are poison, and nothing is without poison: the Dosis alone makes a thing not poison."*

—Paracelsus, *Seven Defenses, the Reply to Certain Calumnations of His Enemies*, C. Lilian Temkin, translator

## 10.1 CYP Inhibition

### 10.1.1 Importance

The safety of a drug involves more than just the direct toxicity of the compound and its metabolites. Potentially dangerous drug–drug interactions (DDIs) can arise when one compound alters the effects of another, usually by increasing or decreasing its in vivo concentration. Available studies have defined clinical DDIs in different ways and arrived at very different numbers in estimating their prevalence and consequences,[1] but no one doubts the seriousness of the consequences or the enormous downside associated with such drugs.

In recent years a number of drugs have been withdrawn from the market due to DDIs, including terfenadine (Figure 6.31), mibefradil (Figure 10.1), and astemizole (Figure 1.18). Compounds involved in DDIs can either be responsible for *causing* the interactions themselves ("perpetrators"[2]) or become the source of observed toxicity ("victims") due to co-administration of another compound. Mibefradil, a calcium channel blocker approved by FDA in 1997 and voluntarily withdrawn from the market by Roche a year later, is an example of the former. It was found to be an irreversible inhibitor of CYP3A4 which, as we've seen in Chapter 1, is important to the clearance of about half of all drugs, the cholesterol-lowering agent simvastatin being one of them. In one patient taking both drugs and suffering from rhabdomyolysis, a potential side effect of the statins, a plasma concentration of simvastatin at least five times higher than normal, was noted 24 h after dosing.[3] Eventually, effects of mibefradil on 26 drugs were seen, and the compound was withdrawn even though the toxicities observed didn't stem from mibefradil itself.

Terfenadine (Figure 1.10), discussed in Chapter 6, and astemizole (Figure 1.18) are two examples of "victims" of DDIs withdrawn due to toxicity (in these cases, *torsade de pointes arrhythmia*, often associated with binding to hERG as discussed later) observed only when another drug inhibited their normal metabolic processes, leading to unusually high parent drug concentrations. As one might expect, drugs with narrow therapeutic windows are most likely to be thus victimized. Importantly, inhibiting the enzymes most often involved in drug metabolism is likely to accentuate potential toxic effects in the largest number of drugs. Recalling that CYPs 1A2, 2C9, 2C19, 2D6, and 3A4 are involved in oxidatively metabolizing

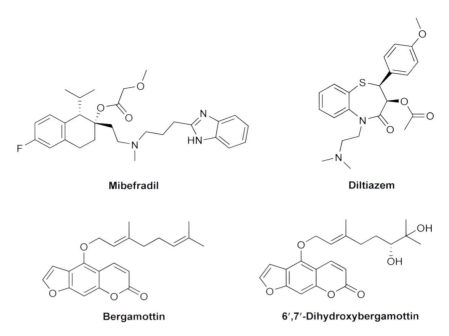

**Figure 10.1** ▶ Some compounds involved in drug–drug interactions.

about 90% of all drugs, it's easy to see that compounds that inhibit CYPs, especially CYP3A4, are the ones most likely to cause DDIs and need to be avoided.

Not only drugs, but any ingested compound can be involved. A classic example is grapefruit juice. Having a glass thereof is an excellent way of selectively inhibiting CYP3A4 in small intestine but not liver, thereby increasing portal vein concentrations ("absorption" by some definitions) of drugs normally metabolized by that enzyme. The natural products bergamottin and 6′,7′-dihydroxybergamottin (Figure 10.1) are believed to be the components most responsible for this. Tobacco smoke, alcohol, dietary components, and street drugs are all potential sources of DDIs as well, and are less likely to be investigated through clinical trials.

Interestingly, CYP inhibition isn't always bad. Inhibition of CYP51, lanosterol 14α-demethylase, is the mechanism of action (MOA) for antifungal drugs like ketoconazole (Figure 10.2). Inhibition of aromatase (CYP19), an enzyme involved in steroidogenesis, is used in the treatment of breast cancer.[4] Neither of these CYP family members, of course, is normally important in drug metabolism. But CYP2C9 is, and its inhibition has been proposed as a potential way of treating ischemic heart disease.[5] And the propensity of CYP3A4 inhibitors to increase drug concentrations has been used for the good, particularly with ritonavir, an HIV protease inhibitor that also inhibits CYP3A4 and is often administered at subtherapeutic doses with other HIV protease inhibitors to boost their in vivo properties. Another example involves the co-administration of the CYP3A4 (and CYP51) inhibitor ketoconazole with cyclosporine which allows the latter, expensive drug to be used at lower doses and/or less often in transplant recipients.[6] As mentioned earlier, at least one company,

**Figure 10.2 ►** The reversible CYP3A4 inhibitor and antifungal drug ketoconazole, the irreversible CYP3A4 inhibitor and proestrogen drug gestodene, and the quasi-irreversible CYP3A4 inhibitor and antibacterial drug erythromycin. Structural features key to their CYP inhibition are shown in blue.

AVI Biopharma, is currently pursuing a clinical candidate, the antisense oligonucleotide AVI-4557, specifically designed to be a CYP3A4 inhibitor.[7] Such cases, however, remain the exceptions to the rule, which is that CYP inhibition is best avoided.

## 10.1.2   Types of CYP Inhibition

CYP inhibitors can be divided into three categories.

1. *Reversible inhibitors* usually compete with substrate for binding at the active site. Ketoconazole (Figure 10.2) is one example of a reversible inhibitor of CYP3A4. Because a rapid equilibrium is involved, the potency observed for such an inhibitor will be independent of incubation time. This isn't true for the other two types of CYP inhibitors, which are therefore known as *time-dependent inhibitors* (*TDI*s).

2. *Irreversible inhibitors* form covalent bonds to the CYP generally after conversion to an electrophilic *reactive metabolite*, thus making them mechanism-based inhibitors. As shown in Figure 10.3, the new bond can be formed at either the heme unit or the CYP

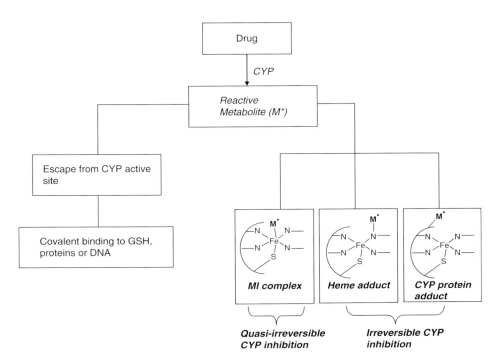

**Figure 10.3 ▶** Formation and possible fates of a reactive metabolite. Once formed at the active site of a CYP, it might dissociate and go on to form a covalent adduct with glutathione (GSH) or nucleophilic groups in cellular components. Alternately, reaction at the active site of the CYP can lead to quasi-irreversible inhibition by complexation with the iron atom of heme or to irreversible CYP inhibition by covalent bond formation either with the heme unit or with a nucleophilic component of the CYP protein.

protein. Either way the CYP is "destroyed" with permanent loss of catalytic activity. Return of CYP activity then depends upon resynthesis of the enzyme. Gestodene (Figure 10.2), a proestrogen used in oral contraceptives, is an example of an irreversible CYP3A4 inhibitor. An excellent review of more than 200 known mechanism-based CYP inhibitors was recently published.[8] Finding that a compound is an irreversible CYP inhibitor will also raise warning flags about what it might also do if some of the reactive metabolite diffuses away from the CYP, taking the path shown on the left of Figure 10.3, which could lead to toxicity.

3. *Quasi-irreversible inhibition* is observed when CYP metabolism produces an intermediate that can form a stable *metabolite–intermediate (MI)* complex. This is another example of mechanism-based inhibition. Erythromycin is one such quasi-irreversible CYP3A4 inhibitor. Upon didemethylation of its tertiary amine group and subsequent oxidation, the resulting nitroso species forms a tight complex with the Fe(II) atom of the CYP's heme unit. Unlike truly irreversible adducts, such complexes can be broken up, say by oxidation with potassium ferricyanide, but under normal physiological conditions this obviously doesn't happen.

---

**Box 10.1    Key General Points About CYP Inhibition**

▶ DDIs have been responsible for clinical toxicity and drug withdrawals.

▶ CYP inhibition can cause DDIs by increasing concentrations of other drugs to toxic levels.

▶ CYP inhibition is sometimes intentionally used to boost levels of other drugs.

▶ CYP inhibition can be reversible, irreversible (via covalent bond formation) or quasi-irreversible (via MI complex formation).

---

Irreversible CYP inhibition would seem to have nothing but downside. The effect could linger long after the inhibitor has been cleared. The adduct formed might conceivably act as a hapten and initiate an autoimmune response.[9] And on a practical level, early-stage primary screens for CYP inhibition, as we'll see, don't involve preincubation of the enzyme with the compound being tested, so that irreversible, time-dependent inhibition might go undetected. All of this, however, only serves to highlight the gap between theory and reality: ritonavir, often used clinically precisely because it inhibits CYPs, thereby boosting concentrations of other drugs (increasing exposure to the co-administered protease inhibitor lopinavir a whopping 77-fold in the combination drug Kaletra),[10] is now known to be an irreversible inhibitor of 3A4 and other CYPs (Box 10.1).[11]

Ritonavir can be used to illustrate yet another point about interactions with CYPs—their complexity. Not only is it a CYP inhibitor, but it's also a CYP substrate *and* a CYP inducer.[12] The groups are not mutually exclusive. Viewing CYPs as "normal" enzymes is also an oversimplification. "It is clear that CYP3A4 does not behave like a traditional lock-and-key enzyme, which has pros (for the ability of the human enzyme to simultaneously handle endogenous and xenobiotics molecules) and cons (for the pharmaceutical industry)."[13] It's long been realized that CYPs don't necessarily follow Michaelis–Menten kinetics,[14–15] and quite a bit of evidence including a recent X-ray crystallographic structure for CYP3A4[16] indicate that more than one substrate might bind these flexible and accommodating enzymes at the same time. "The binding of the first substrate to the active site of the CYP initiates a change in protein conformation that can either promote or inhibit the binding and catalysis of the second substrate."[17] Because of this complexity it may be some time before theory catches up to reality.

## 10.1.3   CYP Inhibition Assays

To measure CYP inhibition, metabolite formation is quantitated in the absence or presence of a potential inhibitor as outlined in Figure 10.4. Metabolite concentration in the presence of inhibitor can be measured at a single concentration of inhibitor or at many to generate a dose–response curve, depending on the tradeoff between accuracy and throughput that's considered acceptable for the assay in question. Data are normally expressed either as percent inhibition at one concentration (e.g. 40% @ 3 μM) or as an $IC_{50}$ value. An estimate of the $IC_{50}$ can be obtained from one-point data as well.[18–19]

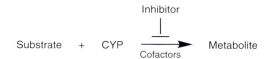

**Figure 10.4** ▶ Measuring CYP inhibition.

The traditional assay for inhibition of a given CYP makes use of a drug substrate that's been found to be selectively metabolized by that enzyme so that microsomes, which contain many of them, can be used. Sometimes substrates would be radiolabeled to aid in metabolite detection. One example is an assay for CYP2D6 inhibition using an $O$-$^{14}$CH$_3$-containing probe drug where demethylation produces labeled formaldehyde that can be detected using SPA beads.[20] Most often, however, isolation and quantitation of the metabolite has been done by LCMS, making this a less-than-high throughput method, particularly when results for 8 or 10 concentrations of inhibitor are analyzed to generate a good curve for every one of the CYP isozymes being examined. This is still considered the gold standard for in vitro measurement of CYP inhibition, and is the type of assay required for IND submission. Its low throughput, however, makes it unsuitable for routine early-stage drug discovery screening.

Combining all of the selective CYP substrates into a "cocktail" which is then treated with microsomes in the presence and absence, respectively, of putative inhibitors can speed things up when data for multiple CYPs are required.[21] Success of the method depends on the substrates being truly selective, not interfering with each other, and getting a good separation of all the metabolites and the test compound by HPLC. Some groups have reported great success with this cassette method[22–23] while others have had some reservations. An alternative consists of incubating each substrate with microsomes separately but then pooling the samples for analysis, thus saving HPLC runtime, although this involves running the samples at higher dilution.[19]

In the *double cocktail* technique a mixture of selective substrates is also used but the microsomes are replaced by a second "cocktail" consisting of a mixture of rCYPs.[24] This has several advantages. Concentrations of each isozyme can be adjusted to achieve the best kinetics, and effects due to compounds sticking to phospholipids are eliminated. The batch-to-batch variability inherent in microsomes is also bypassed. A group at Wyeth has reported the double cocktail method to be "sensitive, selective, and high throughput for use in drug discovery to provide an early alert to potential toxicity with regard to drug-drug interaction, prioritize chemical series, and guide structural modifications to circumvent CYP450 inhibition."[25]

All of these techniques rely on LCMS readouts, so throughput is limited by HPLC runtime. Fast gradients, stepped runs, and various ways of combining or pooling components have, as we've seen, been used to accelerate throughput, but the basic limitation remains. By contrast, assays based on fluorescence readouts are quick and, if not exactly one-size-fits-all, at least one-size-fits-most. Far and away the most common way of looking at CYP inhibition in early-stage drug discovery involves using a coumarin-based substrate that provides a fluorescent product upon CYP-catalyzed $O$-dealkylation at the 7-position. Some commonly used ones are shown in Figure 10.5, and Table 10.1 summarizes these various approaches.

**TABLE 10.1 ▶  Four Types of Assays that Can be Used to Determine CYP Inhibition**

| Assay Type | Substrate | CYP Source | Detection | Throughput |
|---|---|---|---|---|
| **Traditional** | Selective – single drug | Microsomes | LCMS | Low |
| **"Cassette"** | Selective substrate cocktail | Microsomes | LCMS | Medium |
| **"Double cocktail"** | Selective substrate cocktail | rCYP cocktail | LCMS | Medium |
| **HT Fluorescence** | Nonselective – fluorogenic | Single rCYP | Fluorescence | High |

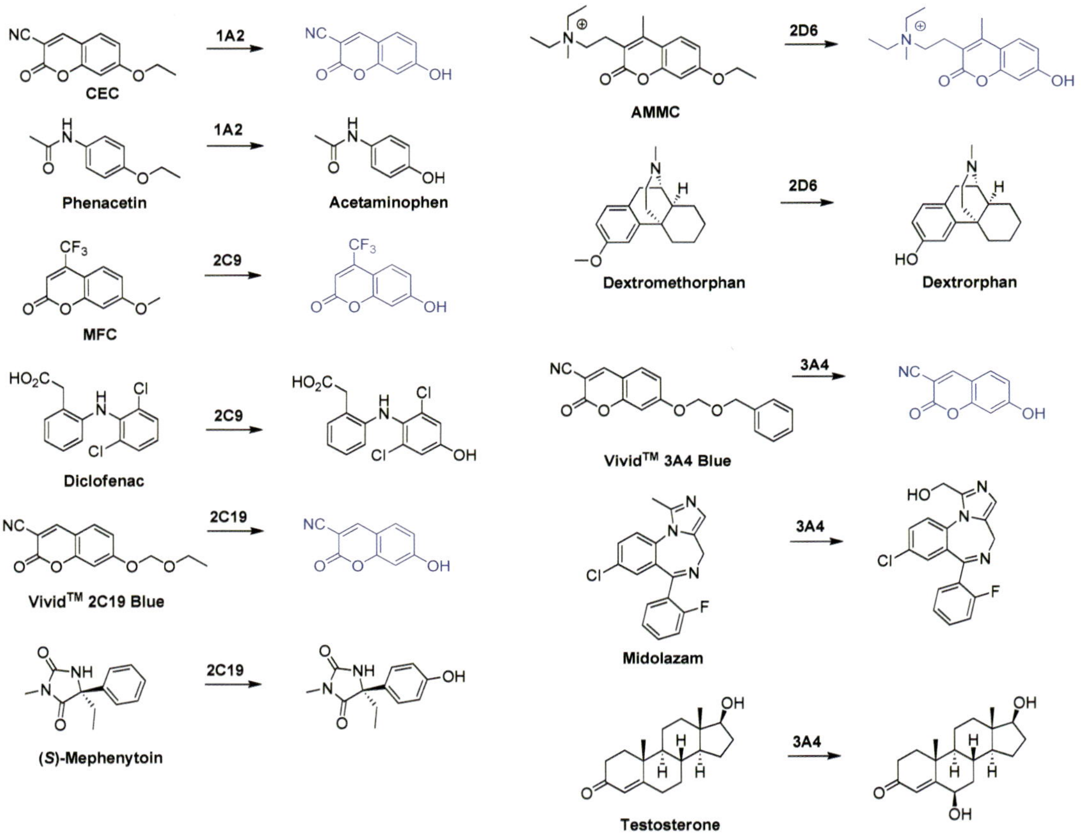

**Figure 10.5 ▶**  Some common substrates and metabolic reactions used in CYP inhibition assays. Products shown in blue are fluorescent. Products in black are normally detected by LCMS.

Fluorescent substrates to date have not proven to be very selective, being cleavable by a number of different CYPs. Because of that, microsomes or cocktails of rCYPs are unsuitable for use in such assays: only one rCYP per assay can be used. Readouts are quick and throughput can be very high, but as in activity screening, test compounds which are either fluorescent themselves or absorb at the emission wavelength can give false negatives and false positives, a major limitation of this technique. Just how well $IC_{50}$ values for CYP inhibition obtained from fluorescence-based assays compare with those obtained from conventional (microsome/LCMS) methods is somewhat controversial,[26–27] but in the early stages of a drug discovery program at least a rank-ordering of compounds based on CYP inhibition data obtained in fluorescence-based HT screens has been the norm for some time now. As compounds progress further into LO, secondary screening using other methods generally follows.

Whether or not a CYP inhibition assay uses a fluorescence readout isn't the only important issue either. For CYP3A4 in particular, the DME (drug metabolizing enzyme) of highest interest in evaluating DDIs, inhibition results are very much dependent on the exact substrate used in the measurement, whether fluorogenic or not. Values obtained for a given compound using the same method can vary more than 200-fold when different 3A4 substrates are used,[28] and can vary at least 10-fold even for the *same substrate* when CYP3A4 gives two different metabolites and data for one or the other are used to calculate the $K_i$.[29] For this reason, two substrates that give the widest possible variation in CYP inhibition values are generally used, such as midazolam and testosterone (Figure 10.5) in the traditional, microsome/LCMS assay. All of this probably stems from the ability of CYP3A4 to bind multiple substrates simultaneously, bringing complex cooperativity effects into play.

For CYP inhibition assays, organic solvent concentrations need to be kept low, usually with DMSO at no more than 0.2%, so once again insoluble compounds will be unsuitable for such assays. When microsomes are used, microsome protein concentration can drastically affect results. At high concentrations extensive binding to phospholipids and rapid metabolism may occur, leading to an underestimate of inhibitory potential.[30] Because of this, microsome protein concentrations need to be kept low. Substrate concentrations can also affect results, and for maximum relevance they should be kept close to concentrations that they might reach in vivo.

Quick primary screens for CYP inhibition can fail to identify time-dependent inhibitors (TDIs) as noted in Box 10.2. These are not at all uncommon. To identify TDIs one sample of the test compound is incubated with microsomes or an rCYP *without* NADPH, while a second one otherwise identical, but containing this necessary CYP co-factor, is prepared. After 30 min the substrate and NADPH (now needed for metabolism to proceed) are added to each, and the amount of metabolite formed is determined for each sample. Time-dependent inhibitors will

---

**Box 10.2    Key Points About CYP Inhibition Assays**

▶ HT CYP assays use fluorogenic substrates and rCYPs; compound fluorescence or absorption at the emission wavelength can give false results.

▶ Secondary assays can use microsomes or rCYP cocktails and selective drug substrates.

▶ Assay results are substrate-dependent, especially for CYP3A4 where multiple substrates are often used to determine inhibition.

▶ Detecting and measuring irreversible and quasi-irreversible CYP inhibition requires as special assay

show significantly greater inhibition for the sample preincubated in the presence of NADPH, which allowed the test compound to be processed to the irreversible inactivator during preincubation, than for the sample lacking it. If both are about the same, a non-time-dependent mechanism is indicated. Keep in mind that some compounds can inhibit CYPs by a combination of both reversible *and* time-dependent processes.

CYP inhibition values can vary from assay to assay, substrate to substrate, and lab to lab, and what's considered acceptable for one indication or at one company might not be acceptable for the next. As a rule of thumb, though, for assays using microsomes and selective drug probes, $IC_{50}$ or $K_i$ values that are submicromolar generate a lot of concern about potential for DDIs, numbers above $10\,\mu M$ are preferred, and single-digit micromolar compounds are somewhere in between. These kinds of data normally come out of secondary screening, but the chemist looking at primary screening data might want to keep these numbers in mind for use later on.

### 10.1.4   Common Structural Features of CYP Inhibitors

Whether or not a compound will inhibit a given CYP isn't always predictable, but there are some common structural features that seem to put compounds at greater risk for CYP inhibition. We've already seen that high lipophilicity is usually associated with an increased risk of metabolism by CYPs—a kind of substrate "cypophilicity"—so it isn't too surprising that the same holds true for CYP inhibitors. Figure 10.6 shows the correlation between $LogD_{7.4}$ and $Log\ IC_{50}$ for CYP3A4 inhibition for a series of chemically diverse compounds.[31] This lipophilicity effect tends to combine with inhibitory affinity that certain chemical groups confer so that, for example, lipophilic compounds containing an imidazole ring tend to be more potent CYP inhibitors than compounds lacking one or the other.

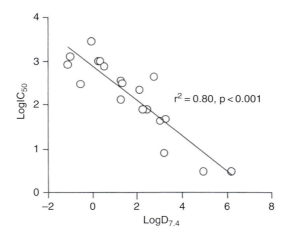

**Figure 10.6 ▶**   Relationship between $LogIC_{50}$ for inhibition of CYP3A4 and $LogD_{7.4}$ for some chemically diverse ligands. (Reproduced with permission from The Thomson Corporation and Riley, R.J. The potential pharmacological and toxicological impact of P450 screening. *Curr. Opin. Drug Discovery Devel.* **2001**, *4*, 45–54. copyright 2001, The Thomson Corporation).

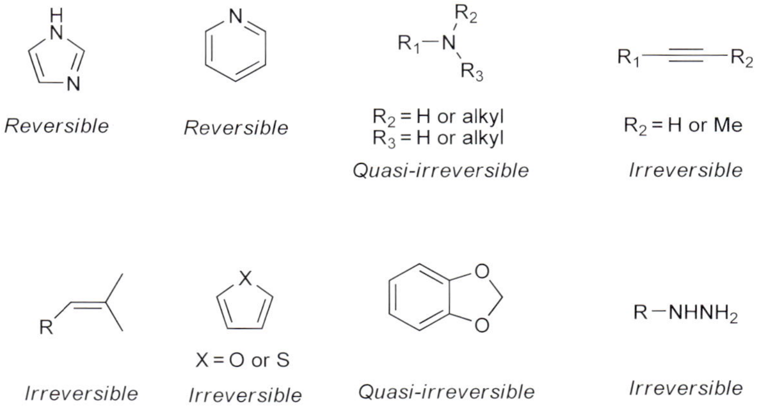

**Figure 10.7** ▶    Some structural features often found in CYP inhibitors and their usual modes of inhibition.

Imidazole and other moieties often found in CYP inhibitors are shown in Figure 10.7. Nitrogen heterocycles like imidazole and pyridine have a tendency to coordinate the heme iron atom, but they do so in a reversible fashion. This, for example, is a key part of the reversible inhibition that ketoconazole (Figure 10.2, LogP 4.0) shows for its target, CYP51, as well as CYP3A4.

Amines can make for quasi-irreversible CYP inhibition. This can happen with primary, secondary, or tertiary amines. Erythromycin (Figure 10.2), for example, is successively metabolized to the secondary amine, primary amine, and nitroso compound, which then forms the MI complex, as shown schematically in Figure 10.8. Many drugs contain amino groups (diltiazem, imipramine, nicardipine, etc.), but this doesn't happen with all of them: azithromycin, another macrolide antibiotic containing a dimethylamino group, doesn't inhibit CYPs much in vitro, nor has venlafaxine, which also contains a dimethylamino group, proven to be problematic in patients. Still, this is a structural feature worth keeping an eye on in terms of potential CYP inhibition.

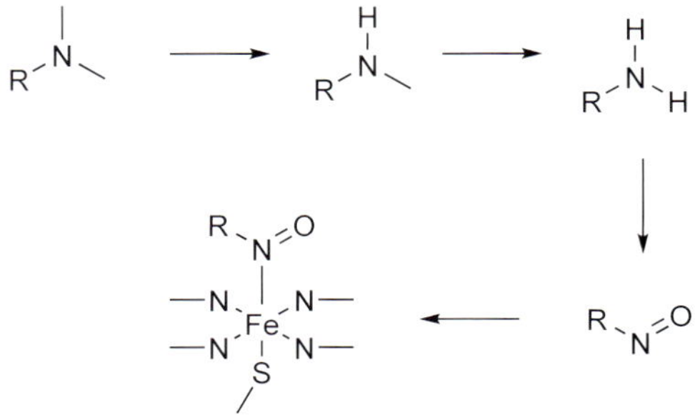

**Figure 10.8** ▶    A common mechanism for quasi-irreversible CYP inhibition. *N*-dealkylation of amines and subsequent oxidation to nitroso compounds can lead to MI complex formation.

Terminal acetylenes are sometimes CYP inhibitors, being activated by oxirene formation and rearrangement to ketenes which can bind to CYPs. This is believed to be the case for gestodene (Figure 10.2), a component of oral contraceptives. In this case, it's more of a problem in theory than in practice, though. The potency of this steroid ensures that probably only subnanomolar concentrations are ever achieved in vivo, far lower than those needed for significant CYP inhibition. Thus, a comparison of plasma levels, a surrogate for the harder-to-measure concentrations in hepatocytes or enterocytes, to $IC_{50}$ values for CYP inhibition is important in deciding how much to worry about DDIs. Once again, high potency can come to the rescue.

Terminal propynes and olefins have also been found to inhibit CYPs via an irreversible mechanism.[32] Furans and thiophenes are frequently found to be irreversible CYP inhibitors. Epoxidation of either system may be involved, and for thiophenes oxidation to the S-oxide can enhance electrophilicity and may be involved in binding to CYPs. Benzodioxoles often display quasi-irreversible CYP inhibition via conversion to the corresponding methylene carbenes which form MI complexes at the CYP heme unit. Compounds containing the hydrazine group, like isoniazid,[33] can also be irreversible CYP inhibitors.

As an example of what to expect with real-world compounds, an analysis of the scaffolds most frequently found in CYP inhibitors from the Sigma Aldrich Library of Pharmacologically Active Compounds (LOPAC™) is shown in Figure 10.9. Looking at this, one can see that, for example, benzodioxoles are 10.9 times more likely to inhibit CYP2C9 than a randomly selected scaffold. Again, one should be aware of the potential of such structural features for CYP inhibition, but keep in mind that this isn't inevitable. Hard data need to be obtained, and even if CYP inhibition is found, it could be at a concentration never reached in vivo, for a compound cleared primarily by a pathway other than hepatic CYP oxidation, relevant to a CYP like 1A2 which is a major metabolic player for relatively few drugs, and/or an effect that structural modifications might mitigate.

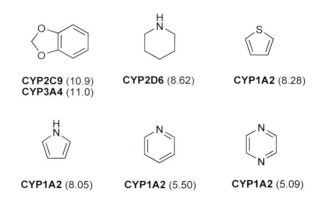

**CYP2C9** (10.9)          **CYP2D6** (8.62)          **CYP1A2** (8.28)
**CYP3A4** (11.0)

**CYP1A2** (8.05)          **CYP1A2** (5.50)          **CYP1A2** (5.09)

**Figure 10.9** ▶  Scaffolds most likely to inhibit assayed CYPs (1A2, 2C9, 2C19, 2D6, 3A4) with odds ratios >5.0 from among the 15 most prevalent scaffolds in the Sigma Aldrich library of pharmacologically active compounds (LOPAC™). The relevant CYPs are shown with odds ratios in parentheses. Information from Kho, R., et al. "Prevalence of Scaffolds in Human Cytochrome P450 Inhibitors Identified using the LOPAC™-1280 Library of Pharmacologically Active Compounds" available at www.sigmaaldrich.com.

## 10.1.5   Ways to Reduce CYP Inhibition

### 10.1.5.1   Reduce Lipophilicity

Since lipophilicity tends to correlate with CYP inhibition as seen in Figure 10.6, the most obvious general way of minimizing CYP inhibition is to lower LogP values. As mentioned in the context of improving metabolic stability, at all but the earliest stages of compound optimization lipophilic groups will probably be there for a reason—usually good potency—so lopping off large chunks of LogP can come at a price. Adding in polarity to lower it can be an option, though, but keep in mind that this might involve installing new groups that could themselves be inhibitory (phenyl → pyridyl substitution, alkylamines that might form MI complexes, etc.) or introducing new problems like carboxylic acids that can't cross the BBB. In practice, it's often not easy to do, but it is worth trying.

### 10.1.5.2   Remove or Replace Offending Structural Features

As we've seen, imidazoles are notorious for their frequent CYP inhibition via coordination with the iron atom of the heme unit. This is especially problematic for compounds targeting histamine receptors, where ligands containing imidazoles are understandably the most likely to be active. The best way around this, of course, is to avoid having an imidazole present in the first place, if at all possible. Examples of this can be seen in the recent literature on $H_3$ receptor antagonists/inverse agonists, of interest for various CNS indications including schizophrenia and ADHD. Such research has been going on since the 1980s, but CYP inhibition is viewed as a major force behind the drive toward non-imidazole-containing compounds in recent years.[34]

Ciproxifan (Figure 10.10), an imidazole-containing $H_3$ receptor antagonist, was found to potentiate the effects of the antipsychotic drug haloperidol in rat, and this was originally

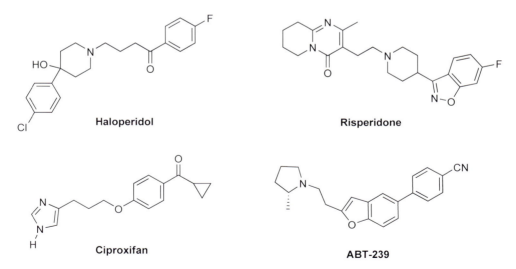

**Haloperidol**

**Risperidone**

**Ciproxifan**

**ABT-239**

**Figure 10.10 ▶**   Top: antipsychotics haloperidol and risperidone. Bottom: H3 receptor antagonists/inverse agonists ciproxifan and ABT-239 (Zhang, M., et al. Lack of cataleptogenic potentiation with non-imidazole $H_3$ receptor antagonists reveals potential drug-drug interactions between imidazole-based $H_3$ receptor antagonists and antipsychotic drugs. *Brain Res.* **2005**, *1045*, 142–149).

thought to "support the existence of direct functional $H_3/D_2$ receptor interactions in striatopallidal neurons of the indirect pathway."[35] A later study, however, showed that no such potentiation could be observed with ABT-239, another $H_3$ receptor antagonist, which lacks an imidazole group and consequent potent CYP inhibition. Furthermore, the well-known CYP3A4 inhibitor ketoconazole was also capable of potentiating the effects of risperidone, another antipsychotic agent. Together this suggested that DDIs caused by CYP inhibition rather than a histamine/dopamine receptor mechanistic linkage may have been involved.[36] If such interactions are to be avoided, compounds like ABT-239 have a major advantage.

Just how much of a difference in CYP inhibition the removal of an imidazole can make is illustrated by the data in Figure 10.11. A group at Bristol-Myers Squibb found that the first compound, containing an $N$-substituted imidazole, had nanomolar activity against the target, IGF-1R, but was even more potent as an inhibitor of CYP1A2 and CYP3A4. The $C$-2 linked 4-methyl analog maintained target potency while reducing CYP inhibition considerably, but replacing the imidazole with an $N$-methylamidine or $N$-methylcarboxamide group had the greatest effect on CYP inhibition. The best balance of target potency versus CYP inhibition was found with imidazolines.[37]

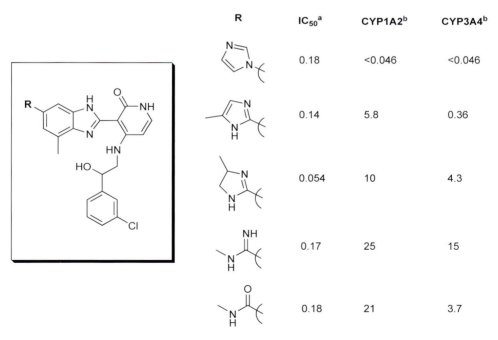

| R | $IC_{50}{}^a$ | $CYP1A2^b$ | $CYP3A4^b$ |
|---|---|---|---|
| | 0.18 | <0.046 | <0.046 |
| | 0.14 | 5.8 | 0.36 |
| | 0.054 | 10 | 4.3 |
| | 0.17 | 25 | 15 |
| | 0.18 | 21 | 3.7 |

**Figure 10.11 ▶** Structures and $IC_{50}$ values for some inhibitors of insulin-like growth factor receptor-1 (IGF-1R) (Velaparthi, U., et al. Imidazole moiety replacements in the 3-(1$H$-benzo[d]imidazol-2-yl)pyridin-2 (1$H$)-one inhibitors of insulin-like growth factor receptor-1 (IGF-1R) to improve cytochrome P450 profile. *Bioorg. Med. Chem. Lett.* **2007**, *17*, 3072–3076.) [a]IGF-1R inhibitory potency (μM). [b]$IC_{50}$ values obtained using Supersomes™ and fluorogenic substrates.

### 10.1.5.3 Sterically Hinder Coordinating Nitrogen Atoms

If the imidazole or other heme-coordinating heterocycle can't be removed, replaced, or reduced, steric effects can often help. This is illustrated by data for some analogs of the HIV protease inhibitor, indinavir.[38] As shown in Figure 10.12, **I**, the *gem*-dimethyl analog of indinavir, inhibits CYP3A4 with an $IC_{50}$ of about 900 nM. Placing a methyl group at the 6-position of the pyridine ring, thereby hindering heme complex formation (**II**), increases this value about 17-fold. The same effect is obtained for **IV** by moving the ring nitrogen of the pyridine *ortho* to the linker. CYP3A4 treated with hindered compounds **II** and **IV** showed UV-visible spectral changes characterized as *Type I*, normally seen for CYP substrates, rather than *Type II*, usually seen for CYP inhibitors and observed for indinavir and **I**.[39] For these and other series compounds, CYP inhibition potencies correlated with microsomal metabolic clearance so that the poorer the CYP inhibition the faster a compound was cleared. This is the other face of the effect warned about in the metabolic stability section, that increasing metabolic stability might mean increasing CYP inhibition.

Steric effects do often work to reduce CYP inhibition, but such modifications should be taken with a grain of NaCl. Aside from the risk of increasing metabolic clearance, there's a possibility that one commonly used modification, placing a 2-methyl group on an imidazole to hinder heme interactions, might reduce reversible CYP inhibition but replace it with mechanism-based inhibition arising through oxidation to an imidazomethide intermediate, an effect that's been seen with CYP1A2 and the 8-methylxanthenes furafylline and cyclohexylline.[40] For these sorts of modifications, time-dependence is worth checking out early.

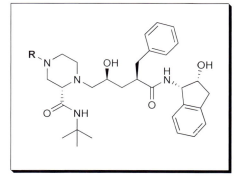

| R | Compound | CYP3A4[a] |
|---|---|---|
| | Indinavir | 0.451 |
| | I | 0.873 |
| | II | 15.1 |
| | IV | 19.3 |

**Figure 10.12 ▶** CYP inhibition by indinavir and analogs. [a]$IC_{50}$ (μM) for the inhibition of CYP3A-catalyzed hydroxylation of testosterone by HLMs (Chiba, M., et al. P450 Interaction with HIV protease inhibitors: Relationship between metabolic stability, inhibitory potency, and P450 binding spectra. *Drug Metab. Dispos.* **2001**, *29*, 1–3.)

Note the contrast, though, in what happens when CYP inhibition is minimized by reducing lipophilicity instead. No reactive metabolites will be involved, and by decreasing affinity for DMEs the approach is unlikely to result in increased oxidative clearance. The downside is that it's probably much harder to do than just adding a methyl group.

### 10.1.5.4  Find a Way to 'Insult' the CYP

Trying to find some structural feature that a CYP won't tolerate can be a challenge too, no doubt because these enzymes were selected for their abilities to metabolize such a wide variety of diverse substrates. Still, they're neither infinitely flexible nor tolerant of every structural change. As with other properties, sometimes the best route to reduced CYP inhibition is via the kind of structural variations and scaffold changes medicinal chemists will instinctively try but for which no scientific-sounding name or profound theoretical basis yet exists—"playing around" with the molecule. Changes at parts of the molecule that don't greatly affect activity can sometimes succeed in reducing CYP inhibition, although it may not always be clear why.

X-ray structures of CYPs and computational models now enable scientists to treat CYPs as target proteins and use SBDD to design binding affinity *out* of a ligand. The prediction of where a substituent on a ligand might "bump" into a target receptor and therefore not bind very well seems to have an above-average success rate for SBDD despite the wildcard of protein flexibility. In some recent work on κ-opioid receptor agonists at Adolor, compounds were docked into a homology model of CYP2D6, inhibition of which had proven to be problematic.[41] In agreement with experimental results for the series of compounds shown in Figure 10.13, the model showed room in the 2D6 binding pocket for *ortho* and *meta*

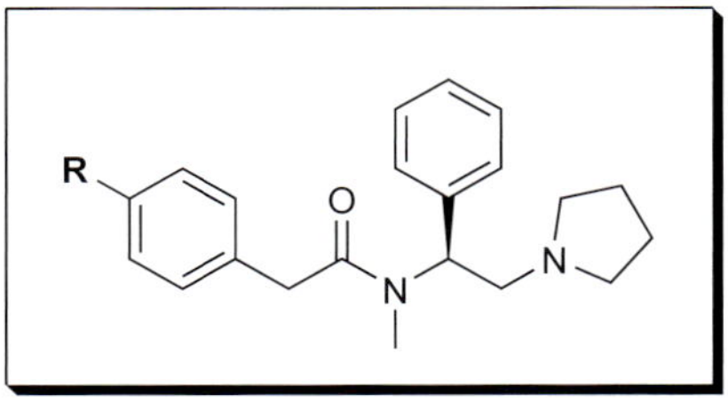

| Compound | R | $K_i (\kappa)^a$ | CYP2D6$^b$ |
|---|---|---|---|
| 2 | H | 5.8 | 46 |
| 5 | -CH$_2$NH$_2$ | 57 | 326 |
| 8 | -CH$_2$NHCOCH$_3$ | 47 | 1108 |
| 12 | -CH$_2$NHSO$_2$CH$_3$ | 2.9 | >10,000 |

**Figure 10.13** ▶  Target and CYP inhibitory potency for some κ-opioid agonists. $^a$Inhibition of binding of [$^3$H]diprenorphine to cloned human κ-opioid receptors expressed in cells, in nM. $^b$IC$_{50}$ values for inhibition of CYP2D6 activity using a fluorogenic substrate with HLMs, in nM (Le Bourdonnec, B., et al. Arylacetamide κ opioid receptor agonists with reduced cytochrome P450 2D6 inhibitory activity. *Bioorg. Med. Chem. Lett.* **2005**, *15*, 2647–2652.)

---

**Box 10.3   Key Points About CYP Inhibitor SAR**

▶ Characteristics of CYP inhibitor structures frequently include

    High lipophilicity

    Imidazoles, pyridines, and other *N*-containing heterocycles

    Alkylamines

    Alkynes and alkenes

    Furans and thiophenes

    Benzodioxoles

▶ CYP inhibition can often be minimized by

    Reducing LogP or LogD values

    Removing or replacing the responsible structural features

    Sterically hindering *N*-atoms coordinating heme iron

    Introducing groups which "bump" at the CYP active site

---

substituents on the left-hand phenyl ring, but little at the *para* position. In fact, a >200-fold decrease in 2D6 inhibition as well as an increase in target potency was observed with the right *para*-substituent (**12**). Box 10.3 provides a summary of this and other aspects of CYP inhibitor SAR.

## 10.2   CYP Induction

CYPs can not only be inhibited, reducing their activities, but, except for CYP2D6 among the major human xenobiotic-metabolizing CYPs, they can also be *induced*, resulting in more protein and a subsequent increase in activity. Although other mechanisms can be involved,[42] CYP induction typically involves binding to ligand-activated transcription factors, primarily AhR (aryl hydrocarbon receptor) for CYP1A2, PXR (pregnane X receptor) for CYP2B6, CYP2C9, and CYP3A4, and CAR (constitutive androstane receptor) for CYP2B6. This is followed by translocation to the nucleus, binding with dimerization partners, binding to DNA promoter regions, and finally activation of gene transcription. The overall process is slow and CYP concentrations can take a few days to rise.

Although undesirable, CYP induction isn't quite as big a problem as CYP inhibition, so there's less of a focus on it at the early stages of drug discovery. This is largely because, as shown in Table 10.2, while a clinical CYP inhibitor might boost levels of co-administered drugs to toxic concentrations by preventing their breakdown, CYP inducers would *increase*

**TABLE 10.2 ►  Possible Effects of CYP Inhibitors and Inducers on Substrate Drugs and Reactive Metabolites.**

| Class | CYP Activity | [Substrate drug] | Possible effect | [Reactive metabolites] | Possible effect |
|-------|--------------|------------------|-----------------|------------------------|-----------------|
| CYP Inhibitor | Decrease | Increase → | Toxicity | Decrease → | Protection against toxicity |
| CYP Inducer | Increase | Decrease → | Loss of efficacy | Increase → | Toxicity |

Safety concerns can theoretically arise from either class, but in most cases toxicity due to increased concentrations of co-administered Drugs is the primary concern in DDIs. For Drugs which both induce and inhibit a given CYP, the net effect can be complex

their metabolism, which runs the risk of (only) causing a loss of efficacy instead. There is an exception, though. Although normally acting to detoxify xenobiotics, CYP metabolism can sometimes give rise to electrophilic *reactive metabolites*, as we've seen. These won't necessarily bind covalently to the CYP that produced it (the route at the right of Figure 10.3), but might instead dissociate from the enzyme and react with something else. That "something" is normally glutathione (GSH), an electrophile quencher found at millimolar concentrations in cells, in a process catalyzed by glutathione *S*-transferases (GSTs). But if local GSH concentrations happen to be depleted or GSTs inhibited, the reactive metabolite can instead form covalent adducts with cellular constituents like proteins or nucleic acids. This can lead to liver toxicity.[43]

The classic example of how a CYP inducer might increase concentrations of reactive metabolites involves induction of CYP2E1 and its effect on acetaminophen metabolism. Acetaminophen, shown in Figure 10.14, contains a phenol group, so it's mainly metabolized by glucuronidation, then excreted. A minor pathway, however, involves oxidation by CYP2E1 to form a quinoneimine, NAPQI. Ethanol is a well-known and commonly self-prescribed CYP2E1 inducer, and its chronic use *might* thereby increase NAPQI production in those also taking this popular OTC analgesic. Clinical data on this effect, however, aren't unequivocal, and effects may depend on the temporary absence of ethanol, which is also a CYP2E1 inhibitor.[44] Still, it underscores the potential for DDIs that might accompany CYP induction.

CYP induction often turns out to be more of a problem for preclinical toxicity studies than something to worry about in the clinic, partly because the high doses used in the former aren't replicated in the latter. At the discovery stage, CYP induction might underlie an apparent loss of bioavailability seen for a compound dosed repeatedly over a number of days in an animal model. But extreme caution must be used in trying to extrapolate CYP induction data obtained in animals to man. Both CYPs and transcription factors differ among species,[45] and in practice, CYP induction has been seen to vary not only by species but also by strain and by sex. For example, the antibiotic rifampicin is an excellent CYP3A inducer in humans, but not in rat. Transgenic mice expressing human CYPs or transcription factors have sometimes proven useful,[46] and transplantation of human hepatocytes

**Figure 10.14 ▶**  Acetaminophen is metabolized by CYP2E1 to the reactive metabolite NAPQI, which can either react with glutathione (GSH) and eventually be excreted or react with cellular components and cause hepatotoxicity.

into immune-compromised mice[47] has produced a chimeric model for human hepatic drug metabolism that mirrored the induction of human CYP3A4 observed for rifabutin, a rifampicin analog.[48] So in vivo models of CYP induction, if somewhat exotic, aren't out of the question.

More convenient and much more relevant to early-stage drug discovery projects, however, are in vitro induction experiments. These are usually done using cryopreserved human hepatocytes. They're incubated with the test compound at several doses for 2–3 days. At the end of that time CYP mRNA levels, protein levels, or activity against selective substrates can be assessed. Increases relative to control indicate enzyme induction. Although activity determination is probably the easiest and most frequently used endpoint, it might not provide the truest measure of CYP induction for compounds like ritonavir that are also CYP inhibitors. Test compounds also need to be very stable to metabolism to survive a days-long incubation with hepatocytes which, after all, specialize in breaking them down.

Reporter gene assays for CYP induction have been developed in hepatic cell lines.[49] These sorts of assays and cassette dosing strategies[50] offer the possibility of higher throughput than the "gold standard" fresh hepatocyte model. To date, however, early-stage drug discovery researchers probably see much less CYP induction data than CYP inhibition data: the former is still typically done on fewer compounds later on in the drug discovery process than the latter.

An interesting new way of using activity-based probes (ABPs) to determine activities of multiple CYPs both in vitro and in vivo, with potential utility in examining both CYP induction and CYP inhibition was recently published by a group at the Scripps Research Institute.[51] This takes advantage of the previously noted ability of terminal alkynes to act as mechanism-based inhibitors of CYPs. As shown in Figure 10.15, after covalently binding the CYP, another alkyne at the other end of the probe allows a copper-catalyzed azide cycloaddition reaction to conjugate the resulting complex to a reporter group. Using both biotin and a fluorescent label in the reporter group allows for easy enrichment via streptavidin-conjugated beads. After SDS-PAGE separation, fluorescence detection, and trypsinization, LCMS can be used to identify the particular CYPs thus labeled. In proof-of-concept work, 2EN-ABP proved capable of detecting induction of CYP1A2 by β-naphthoflavone in treated mice.

Little is known about what structural features make for CYP induction, and reports of CYP induction SAR in the literature are scarce. The PDGF receptor kinase inhibitor RPR101511 (Figure 10.16) is an example of a compound showing lower oral bioavailability on repeated dosing because of CYP induction. Oral dosing studies in rat demonstrated that it had less than half the bioavailability on the fourth day of dosing than it had on the first, and CYPs were found to be upregulated. Other series compounds based on the 2-aminoquinoxaline scaffold, fortunately, didn't induce CYPs, and for these compounds (**2** and **10**), bioavailability didn't decrease significantly with time.[52]

A study of known clinical drugs that are CYP inducers found that "in general structures are diverse but most are lipophilic as defined by a positive calculated LogP value."[53] So, just as for metabolic stability and CYP inhibition, lowering LogP may once again be an effective strategy here. But as the same author points out, perhaps even more important is the clinical dose at which the compound is to be used. Most known clinical CYP inducers tend to be dosed at 500–1000 mg/day and reach total (protein-bound plus unbound) plasma concentrations of 10–100 μM. Troglitazone (Figure 5.3), a PPARγ agonist and anti-diabetic drug

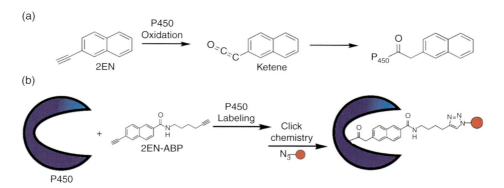

**Figure 10.15** ▶ (a) Mechanism of inhibition of P450 enzymes by 2-Ethynylnaphtalene (2EN) involves oxidation to a ketene followed by covalent labeling. (b) Attachment of a second acetylene that functions as a "click" handle allows cycloaddition to provide a way of attaching the bound complex to a reporter group (red circle). (Reprinted with permission from Wright, A.T., Cravatt, B. F., Chemical proteomic probes for profiling cytochrome P450 activities and drug interactions in vivo. *Chem. Biol.*, **2007**, *14*, 1043–1051, copyright 2007, Elsevier).

| Structure | %F Day 1 | %F Day 4 | P450 Upregulation |
|---|---|---|---|
| <br>**RPR101511** | 49 | 21 | Yes |
| <br>**Compound 2** | 17 | 16 | No |
| <br>**Compound 10** | 82 | 72 | No |

**Figure 10.16 ▶**   Three inhibitors of PDGF receptor tyrosine kinase, their oral bioavailabilities in rat as noted on days 1 and 4 of dosing, and whether or not they induced rat CYPs (He, W., et al. Potent quinoxaline-based inhibitors of PDGF receptor tyrosine kinase activity. Part 2: the synthesis and biological activities of RPR127963, an orally bioavailable inhibitor. *Bioorg. Med. Chem. Lett.* **2003**, *13*, 3097–3100.)

---

**Box 10.4   Key Points About CYP Induction**

▶ CYP induction can also cause DDIs but is of less concern than CYP inhibition.

▶ CYP induction normally involves binding to transcription factors and can vary with species, strain, and sex.

▶ In vitro assays for CYP induction often measure metabolism of selective substrates after a long incubation of the test compound with hepatocytes.

▶ CYP induction SAR is largely unknown, but high lipophilicity and large clinical doses seem to be involved.

---

withdrawn due to liver toxicity, was dosed at 200–600 mg/day and increased CYP3A4 levels 2–3 fold in one study.[54] Rosiglitazone (Figure 5.3), another thiazolidinedione, by contrast is dosed at 2–12 mg/day and doesn't appear to induce CYPs. In a now familiar refrain, target potency—translated into clinical efficacy at low doses—can once again come to the rescue.[53] All of this is summarized in Box 10.4.

## 10.3   Binding to the hERG receptor

### 10.3.1   Introduction

The most interestingly named antitarget the researcher is likely to face is the product of the *human ether-a-go-go related gene (hERG)*. In 1968 a mutagenesis experiment in drosophila produced a mutant that was characterized by rhythmic shaking of the legs with an occasional twitching of the abdomen when the flies were anesthetized with ether.[55] This brought to mind the then-popular dancing at the Whisky A Go-Go nightclub in West Hollywood[56] where the patrons were perhaps semi-anesthetized with a related compound, among other things. The responsible fruit-fly gene, dubbed *ether-a-go-go (eag)*, codes for a voltage-gated potassium ion channel. A human homolog, hERG, is responsible for a protein that, in tetrameric form, comprises a pore through which $K^+$ can flow out of a myocyte in a synchronized step critical to repolarizing the cell and ending the ECG waveform shown in Figure 10.17.

Interfering with the outflow of potassium by clogging up the hERG channel can cause a prolongation of the QT interval. Although prolonging this interval is actually the mechanism of action by which *Class III antiarrhythmic* drugs like amiodarone and dofetilide (both shown in Figure 10.18) exert their effects, the resulting drug-induced *long QT syndrome (LQTS)* is associated with an increased risk of *torsades de pointes (TdP)*, a dangerous type of arrhythmia that can sometimes lead to ventricular fibrillation and *sudden cardiac death (SCD)*. Other risk factors, some known and others not,[57] stand between LQTS and TdP, and years ago drug-induced LQTS didn't set off the alarm bells it does today. But the antihistamine terfenadine was withdrawn from the market for LQTS when co-administered with CYP inhibitors after a rate of less than one reported arrhythmia or other adverse cardiac events for every 4 million doses dispensed.[58]

Today any drug displaying signs of even modest hERG binding at the preclinical stage is subject to increased scrutiny and additional regulatory requirements. Millions of dollars in additional R&D costs are likely to be involved in developing any such compound.[59] It's therefore extremely important to identify and address a potential hERG liability early on in a discovery project. A rule of thumb is that the $IC_{50}$ value for hERG should be at least 30 times greater than the maximum in vivo concentration ($C_{max}$) of free (not protein bound) drug, although a larger safety margin is desirable and in some cases might be required.[60] In early drug discovery, well before human plasma concentrations are known, in vitro hERG $IC_{50}$ values $>10\,\mu M$ are usually

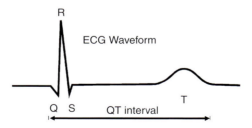

**Figure 10.17** ▶   Idealized ECG waveform. (Reprinted with permission from Vaz, R.J., et al. Human ether-a-go-go related gene (HERG): A chemist's perspective. *Prog. Med. Chem.* **2005**, *43*, 1–18, copyright 2005, Elsevier).

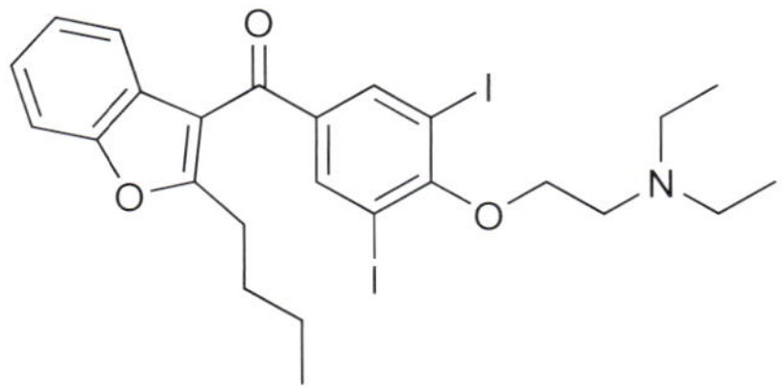

**Figure 10.18** ▶   The Class III antiarrhythmic drugs amiodarone and dofetilide, and MK-499. All of them inhibit hERG.

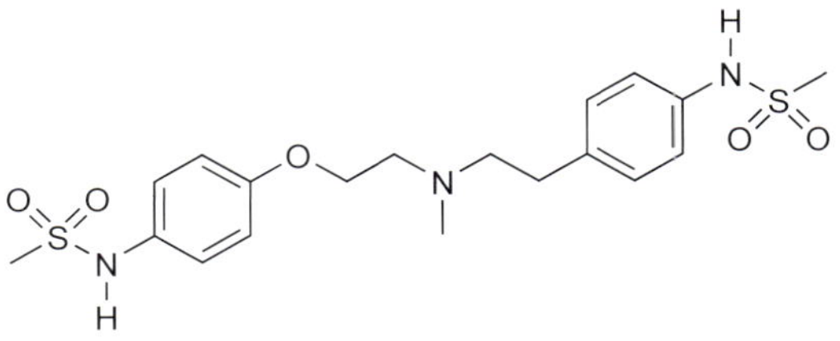

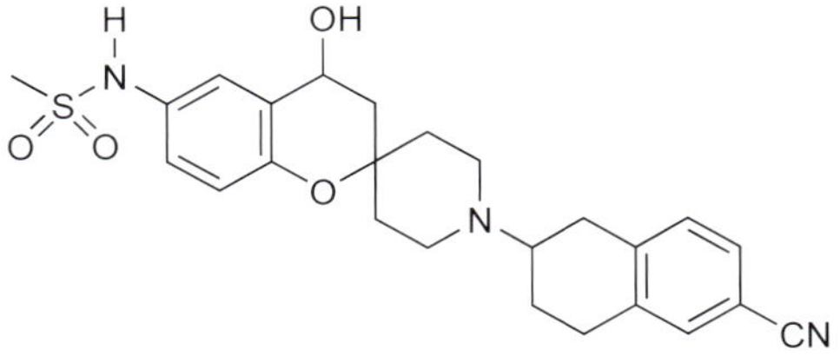

sought, while submicromolar values are considered very worrisome, although these numbers and their limits depend a lot on the methods used to determine them, as we'll see.

Also worrisome is how frequently hERG problems crop up. One estimate is that 50–60% of all discovery compounds may have hERG $IC_{50}$ binding values below 30 μM.[61] Inhibition of the hERG channel in vitro is so commonly detected that the critique has been offered that we may be a bit overzealous in looking for it: "Provided that the concentration is high enough, it seems that almost any compound will inhibit the hERG channel in vitro. One only has to torture nature long enough—in its desperation it will give you an answer."[62] While the physiological significance of hERG binding data obtained at high concentrations may be subject to debate, drug discovery researchers will find that potent hERG binding can certainly result in sudden lead or project death.

## 10.3.2   In Vitro Assays

The "gold standard" for the in vitro measurement of a compound's effects on hERG ion channels is the whole cell patch clamp assay.[63] Here a negative feedback electrical circuit is used to control the voltage potential across a cell when investigating this voltage-gated ion channel. Often *Chinese hamster ovary* (*CHO*) cells transfected with hERG are used. This technique allows the ion flux (current) through open channels to be measured directly. Although it gives direct, functional data, traditional patch clamping involves a lot of skill and manipulation, as in positioning a micropipette to contact a single cell and form a good seal, and is consequently low-throughput and labor-intensive. Largely because of the shortcomings seen with other methods like those discussed below, a lot of effort has gone into finding ways to automate the process and improve its throughput. A major advance has been the *planar patch method* where cells are sucked into place by aspiration through tiny apertures in a plate.[64] Commercial instruments like the Ionworks[TM] HT system and the PatchXpress® 7000A (both from Molecular Devices, Sunnyvale, CA) now allow several thousand data points to be measured in a day. A number of these are needed to test a compound adequately, but at least medium throughput can be obtained.

Flux assays using rubidium are also sometimes used to determine the hERG channel blocking potential of compounds. Here cells are preloaded with RbCl, then washed. Ion channels are activated, usually by exposure to KCl, and the amount of $Rb^+$, potassium's downstairs neighbor in the periodic table, that's effluxed is quantitated with or without preincubation with the test compound. This type of assay tends to be less sensitive and give higher numbers than most others. The assay originally used the nasty radioisotope [86]Rb, a gamma emitter the use of which requires lead shielding, but more recently a method relying on non-radioactive rubidium detected by atomic absorption spectroscopy has been developed.[65] At least one commercial instrument, the ICR 8000 from Aurora Biomed (Vancouver, British Columbia) has automated this process, bringing it up to medium-throughput standards.

A third type of hERG assay commonly found in discovery-stage research involves simple binding competition using a high-affinity radioligand. These assays are often done on human embryonic kidney cells (HEK293) expressing hERG and using [³H] dofetilide, [³H] astemizole or [³⁵S] MK-499 (Figure 10.18), all them being exceptionally good hERG binders. This assay is often criticized as providing only an indirect, not a functional, readout, as conveying no information on whether a given compound is a channel blocker or channel activator, and as being based on the assumption that compounds can bind hERG only at the same site where the radioligand does. Both false positives and false negatives are possible. With all its faults, though, the method is fairly straightforward, inexpensive, and often used for a first look at potential hERG liabilities.

Each of these techniques can give quite different hERG $IC_{50}$ values for the same compound. For example, terfenadine's $IC_{50}$ has been measured at 30 nM in the patch-clamp assay, 4.8 μM in the rubidium flux assay, and 500 nM in a binding assay.[66] Needless to say, numbers from different assays, different cells, or different labs for that matter, can't be directly compared.

### 10.3.3　Models of hERG binding

In the absence of an X-ray crystal structure for the hERG channel, site-directed mutagenesis experiments, homology modeling, and pharmacophore maps have been used to try and establish a structural basis for its broad binding abilities that might allow the *in silico* prediction of whether or not a given compound should be a high-affinity hERG binder. Mutagenesis experiments revealed that two aromatic residues in the S6 transmembrane region, Tyr652 and Phe656 (repeated four times, of course, in this tetrameric pore), are crucial to hERG binding.[67] Examining the sequences of other potassium ion channels shows that hERG is the only one with these residues at those positions. Sequence comparisons also suggest that hERG is unusual in not having a potentially volume-restricting "kink" in one of the channel-lining alpha helices caused by a turn induced by the sequence Pro-Val (or Ile)-Pro. Instead Ile-Phe-Gly is found in hERG, and as a result the volume of the channel may allow for much larger molecules to bind there.

This information, together with homology models built using the related channels KvAP and KcsA, for which X-ray structures are known, has allowed a model to be pieced together which also takes into account hERG's apparent propensity for binding charged amines. Sometimes called the "drain plug model,"[69] this is shown in Figure 10.19(b) .

In this model, binders enter the channel from the intracellular side and line up along the long axis of the channel. A hydrophobic part of the molecule containing an aryl ring interacts with Phe656 via π-stacking; a positively charged amine near Tyr652 is nicely situated for cation-π interactions;[70] and HBAs in another hydrophobic region interact with Ser624 and Thr623 near the selectivity filter part of the channel. Sometimes the central amine isn't present and another aryl ring instead π-stacks with Tyr652, and sometimes only a single hydrophobic region is available for interaction.

A somewhat different model of binding to the hERG channel was recently proposed by scientists at Schrödinger, Inc. and Columbia University.[71] According to this model, a molecule looking into an open hERG pore from the cytosol (Figure 10.20) would see a hollow "crown-shaped" volume lined by pairs of hydrophobic Phe656 and Tyr652 residues and just beyond that a sort of "propeller" consisting of Ser624 and Val625 residues. Unlike the drain plug model, protonated amine nitrogens would be stabilized by electrostatic interactions with Ser624 and nearby backbone atoms, not by cation-π interactions. These amines would also provide a crucial branching point, giving the molecule a U-shaped, not linear, conformation that allows distant hydrophobic groups to be pointed toward multiple Phe and Tyr residues along the "crown" to maximize π-stacking interactions. A given

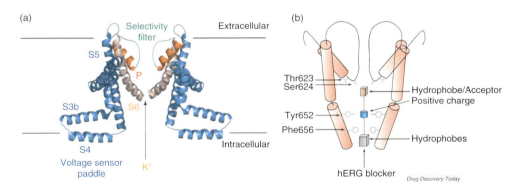

**Figure 10.19** ▶ (a) Important elements of hERG channel topology illustrated by an X-ray structure of the related ion channel KvAP,[68] showing two of the four subunits. (b) The "drain plug" model of the hERG channel, with two of the S6 subunits shown. One of two hydrophobic regions of the binder interacts with the key residue Phe656, while the positive charge from a protonated amine is stabilized by cation-π interaction with Tyr652, and a hydrophobic "tail" of the molecule accepts electrostatic interactions from Thr623 and Ser624 in the selectivity filter. (Reprinted with permission from Aronov, A.M., Predictive *in silico* modeling for hERG channel blockers. *Drug Discov. Today* **2005**, *10*, 149–155, copyright 2005, Elsevier).

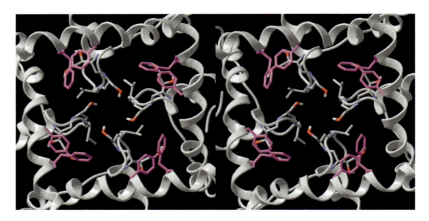

**Figure 10.20** ▶ Stereo ribbon diagram of the open form of a hERG homology model based on an X-ray structure of the KvAP monomer, looking into the pore from the intracellular side toward the selectivity filter. Tyr652 and Phe656 are shown in magenta, lining the pore. Ser624 and Val625 are shown in the center. (Reprinted with permission from Farid, R., et al. New insights about HERG blockade obtained from protein modeling, potential energy mapping, and docking studies. *Bioorg. Med. Chem.* **2006**, *14*, 3160–3173, copyright 2006, Elsevier).

molecule might bind in multiple ways, the hERG channel thus resembling a "low information content, 'host-guest' type binding site" rather than a traditional receptor lock for a specific key.

Rather than dwell on the differences between the "drain plug" and "crown and propeller" models, it's probably most instructive to step back and examine the body of knowledge on which they're based. After all, both attempt to provide a structural basis for why large, mostly flexible molecules that usually contain a basic amine and multiple hydrophobic regions have such a propensity for binding the hERG channel. To date, although useful for understanding hERG interactions, neither these nor any of the many pharmacophore models[72] that have been developed to predict hERG binding can be counted on to accurately predict affinity in a given series. But when it does turn out to be a problem, some ways of attenuating hERG binding are known, and these will now be discussed.

### 10.3.4  Reducing hERG Interactions

Many successful examples of hERG optimization have been summarized in a recent review,[73] and some favorite methods are summarized in Box 10.5. Often quite a bit of trial and error is needed in a given series before hERG affinity can be tuned out and what eventually works may be neither predictable in advance nor easy to rationalize afterward. But a few general strategies that sometimes succeed are given below.

Taking the "floppiness" out of molecules with long linker groups by introducing constraints (e.g. cyclization) sometimes works, but like anything else, it's only useful if target potency doesn't suffer too much. Along these lines, Figure 10.21 shows how macrocyclization of a farnesyltransferase inhibitor reduced hERG binding by more than an order of

> **Box 10.5   Key Points About hERG**
>
> ▶ Blockade of hERG $K^+$ channels can induce long QT syndrome, increasing the risk of torsades de pointes, a dangerous arrhythmia
>
> ▶ Since many compounds bind to the hERG channel, achieving an adequate safety margin can be challenging
>
> ▶ Models show that hERG binding involves $\pi$-stacking and usually interactions with a protonated amine
>
> ▶ Affinity for the hERG channel might be reduced by
>
> Removing, repositioning, or reducing the basicity of amines
>
> Adding in a carboxylic acid or other anionic group
>
> Removing phenyl rings or replacing them with heterocycles
>
> Adding electron withdrawing substituents to aryl rings
>
> Rigidifying or cyclizing the molecule

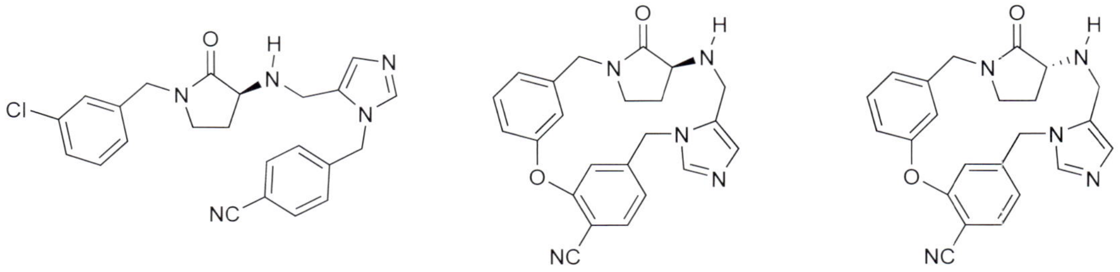

**Compound 1**

**FTase IC$_{50}$ 1.9 nM**

**hERG IP 440 nM**

**Compound 11**

**FTase IC$_{50}$ 3.5 nM**

**hERG IP 4700 nM**

**Compound 12**

**FTase IC$_{50}$ 54 nM**

**hERG IP 11000 nM**

**Figure 10.21 ▶** Three farnesyltransferase inhibitors, their potencies against the desired target, and inflection points for hERG binding determined using a radioligand binding assay (Bell, I.M., et al. 3-aminopyrrolidinone farnesyltransferase inhibitors: Design of macrocyclic compounds with improved pharmacokinetics and excellent cell potency. *J. Med. Chem.* **2002**, *45*, 2388–2409.)

magnitude without very much of a loss in potency. Further optimization of the macrocycle resulted in bioavailable compounds significantly more potent against their target than Compound 1 and with greatly reduced hERG binding.[74] This is easiest to rationalize using the drain plug model, where a long linear binding conformation has been suppressed, but the major conformational changes involved could be a factor in either model. Stereoisomers can differ quite a bit in their hERG binding affinities as well, as seen in the case of Compound 12.

A nice analysis of patch-clamp data for over 1000 discovery compounds at Bristol-Myers Squibb sheds some light on a couple of things that seem well worth trying.[75] Average $IC_{50}$ values were found to be 4.7 µM for basic compounds, 10 µM for neutral compounds, and 95 µM for acidic compounds. The hERG channel, after all, exists to transport potassium *cations*, not neutral or anionic compounds. *Lowering the pKa of an amine involved in hERG binding, thus decreasing the extent of its protonation, can drive down hERG affinity*, regardless of whether cation-π or electrostatic interactions are involved. Amines not critical to target potency, for example an *N*-methylpiperazine group added to a scaffold to increase its solubility, might be removed altogether or banished to another part of the molecule. Alternatively, amine basicity can be lowered by adding an adjacent carbonyl (e.g. turning an *N*-substituted piperidine into a piperidone) or a β-hydroxy or alkoxy group, or perhaps by replacement of a dimethylamine group with a morpholino group. Figure 10.22 illustrates just how much of a difference moving or reducing the pKa of an amine can make to hERG binding.[76]

*The poor affinity for hERG that anionic species have in general is sometimes exploited by introducing a carboxylic acid group into a molecule*, producing a zwitterionic molecule if it has a basic center or an acidic one if it doesn't. Fexofenadine, which, you'll recall, is identical to

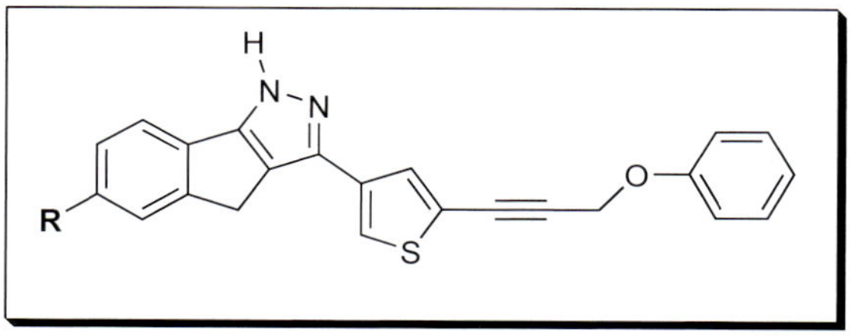

| R | KDR IC$_{50}$ (nM) | hERG IC$_{50}$ (µM)$^*$ |
|---|---|---|
| 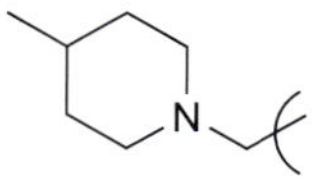 | 69 | 0.08 |
| 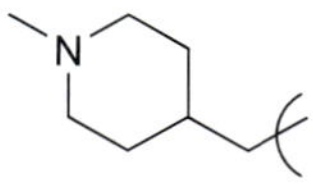 | 11 | 2.1 |
| 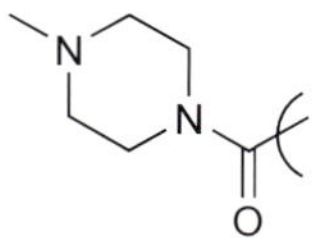 | 44 | 1.72 |

**Figure 10.22** ▶ In vitro KDR kinase inhibitory activity and hERG binding data as determined in a radioligand binding assay for three compounds. (Dinges, J., et al. 1,4-dihydroindeno[1,2-c]pyrazoles with acetylenic side chains as novel and potent multitargeted receptor tyrosine kinase inhibitors with low affinity for the hERG ion channel. *J. Med. Chem.* **2007**, *50*, 2011–2029.)

terfenadine except that it contains a carboxylic acid group in place of one of terfenadine's methyl groups, has a reported hERG $IC_{50}$ value of $23\,\mu M$ versus $56\,nM$ for terfenadine. A recent publication examined the effect of incorporation of a carboxylic acid on hERG binding across a number of different scaffolds, most containing a benzamidine group but some not. The authors concluded that hERG "ligand-binding affinity can be almost universally suppressed by introducing a negatively charged carboxylate group into the ligand, an effect not exhibited by the corresponding neutral ester groups."[77] Furthermore, the effect seemed to be essentially independent of where on the molecule the acid group was placed. Examples of acyl sulfonamides being used to reduce hERG affinity exist as well.[78] Although some might disagree that this method works "universally," it seems a promising strategy to try, provided that effects on other properties like absorption and protein binding don't outweigh the potential beneficial effect on hERG.

High lipophilicity is a feature common to many hERG blockers, and *modifications designed to reduce LogP sometimes succeed in reducing hERG affinity*,[79] particularly in cases where phenyl groups can be removed or replaced with polar heterocycles. In such cases this may, however, not be a kind of general lipophilicity effect but instead a reflection of their reduced propensity for π-stacking, which might also be achieved to a lesser extent by attaching electron withdrawing substituents to an aryl ring. Sometimes modifications that effectively *increase* LogP can actually minimize hERG interactions. Figure 10.23 shows one such case, a reminder that the property-based prediction of hERG binding can be anything but straightforward. In work leading to the CCR5 antagonist and HIV entry inhibitor maraviroc, the lead compound shown at the top in Figure 10.23, although potent, was found to have an unacceptably high hERG affinity. Replacing the benzimidazole with a substituted triazole did much to reduce this, and although switching out the original cyclobutylamide with fluorinated aliphatic amides increased logD values somewhat, hERG inhibition was reduced. "The 4,4-difluorocyclohexyl group is clearly not tolerated within the ion channel due to the steric demands of the cyclohexyl group and also the dipole generated by the difluoro moiety."[80]

## 10.4 Mutagenicity

### 10.4.1 Background

No company wants to introduce a new drug that increases a patient's likelihood of getting cancer, that is, a *carcinogen*. Lengthy and elaborate safety testing is required, much of it in animals models, before any drug can be approved, in order to prevent this from happening. Even long-term toxicity studies in animals aren't totally predictive of what will happen in humans, but they do give the best information currently available. These studies are expensive, resource-and animal-intensive, and represent a commitment to a particular late-stage compound. If a compound drops out at this stage, it's a huge loss for the program and the company. Being able to weed out compounds likely to fail here would be very valuable.

Many, though not all,[81] carcinogenic substances act by inducing genetic mutations. Many, but not all, such *mutagens* interact directly with DNA. A drug itself or, more often, its reactive metabolite like the exo-epoxide formed from aflatoxin B1 (Figure 10.24), might form a covalent bond to DNA, for example by alkylating a nitrogen or oxygen

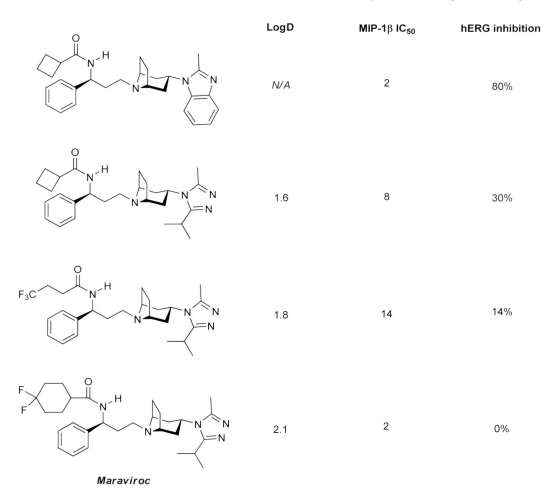

| | LogD | MIP-1β IC$_{50}$ | hERG inhibition |
|---|---|---|---|
| | N/A | 2 | 80% |
| | 1.6 | 8 | 30% |
| | 1.8 | 14 | 14% |
| *Maraviroc* | 2.1 | 2 | 0% |

**Figure 10.23 ▶** LogD, concentration required to inhibit replication of HIV$_{BaL}$ in PM-1 cells by 90%, and percentage hERG inhibition as determined in a radioligand binding assay at 300 nM for the CCR5 antagonist maraviroc and three analogs. (Price, D.A., et al. Overcoming HERG affinity in the discovery of the CCR5 antagonist maraviroc. *Bioorg. Med. Chem. Lett.* **2006**, *16*, 4633–4637.)

atom of guanine.[82] A non-covalent type of interaction often happens too. Relatively flat molecules, typically polycyclic aromatics, can *intercalate*, that is, insert between adjacent base pairs by π-stacking, distorting the structure of the DNA, inhibiting replication or inducing mutations.[83] A mutagen might also cause a covalent interaction with DNA *indirectly* by generating a *reactive oxygen species* (*ROS*) that reacts with it instead.[84] For cytotoxic cancer drugs like doxorubicin (Figure 1.22) and camptothecin (Figure 8.27) binding at DNA, inhibition of replication and the consequent antiproliferative effect on tumor cells is actually the desired mechanism of action, but for most drug discovery projects any interaction with DNA, whether covalent, non-covalent, direct, or indirect, will be shunned.

**Figure 10.24 ▶** Epoxidation of aflatoxin B1, a *promutagen*, by either CYP1A2 or CYP3A4 produces both *endo-* and *exo*-epoxides. Either can be opened up by glutathione in the presence of GSTs, resulting in detoxification and excretion, but the exo-epoxide can also alkylate the $N^7$ position of guanine and is a potent mutagen.

A variety of *in silico* and in vitro methods have been developed with the aim of predicting and measuring mutagenicity. Despite decades of efforts and some notable successes, of all the properties discussed in this book it remains the one least amenable to routine, high-throughput screening. Because of this, even in vitro mutagenicity testing has traditionally been reserved for compounds in preclinical development. Understanding and designing out mutagenicity was hard to do in the absence of much data. The last few years have seen considerable progress, though, and today the discovery stage researcher will likely be exposed to more—although still not high or even medium-throughput—data relating to compound mutagenicity than was formerly the case. Mostly this will come from two sources: the *Ames test* and computational methods.

Back in 1964 the list of ingredients on a potato chip package led geneticist and biochemist Dr. Bruce Ames to ponder the wisdom of introducing new food additives, as well as other compounds, without a way of screening them for mutagenicity first. In the following years he defined and refined a new in vitro mutagenicity test simple enough to become a staple of high school biology labs yet predictive enough to become an important component of government-mandated toxicology testing for new products, from food additives to pharmaceuticals.[85]

A number of different strains of *Salmonella typhimurium* were engineered with mutations in the genes coding for the proteins involved in histidine biosynthesis. These bacteria can therefore only live if histidine is supplied in the growth medium. In addition, their DNA repair capabilities as well as the integrity of their lipopolysaccharide outer coats were intentionally compromised, the latter to better allow test compounds in. Plating these bacteria out on agar containing sparse amounts of histidine causes them to die out after just a few cell divisions, and no colonies will be formed *unless* a mutation takes place to restore their ability to synthesize the amino acid, which is called *reversion*.[86]

If a test compound significantly increases the number of observed colonies relative to a control for spontaneous reversion, which always occurs to some extent, that compound must have caused mutation that resulted in reversion. Each type of strain, of which five are normally used for the full test and two for more limited versions, detects a specific type of mutation: TA98 detects frameshift mutations while TA100 picks up base pair substitutions. Since the mutagenic abilities of compounds often depend on metabolic activation, as it does for aflatoxin B1, the test can be run in the presence of S9 liver fraction from rats whose CYPs have been induced by the polychlorinated biphenyl Aroclor 1254. An increase in reversion only in the presence of S9 shows that the mutagen must be a metabolite.

Running multiple strains, multiple doses, and perhaps duplicates, all in the presence or absence of S9, doing separate tests for cytotoxicity and counting colonies doesn't make this assay a very high-throughput one, and it means that a few hundred milligrams of compound might be required for the "full Ames test." Shortcuts consisting of the use of fewer strains, fewer doses, measuring pH of liquid suspensions in multiwell plates as an endpoint in place of counting colonies on agar in Petri dishes, and other modifications can improve throughput somewhat and reduce compound requirements to only a few milligrams, although the results may not be quite as good. One sees names like "Mini Ames,"[87] "microAmes,"[88] and Ames II$^{TM}$[89] associated with these sorts of modifications.[90]

The Ames test and its Mini-Me versions come with a laudably low rate of false positives compared to other mutagenicity assays. However, they won't necessarily pick up other forms of genotoxicity like *clastogenicity* (the induction of chromosome breakage) or *aneugenicity* (causing the loss of chromosomes), and they use prokaryotic cells, which may not totally reflect what happens with mammalian DNA. For these reasons other assays like the in vitro *micronucleus assay* (*IVMN*)[91] are used in addition to Ames testing for compounds of interest, but these won't be discussed here. It's worth noting too that very recently a new assay said to be suitable for high-throughput screening of compounds for all classes of genotoxicity in human cells called GreenScreen HC has been reported.[92] Only time will tell whether this or any other new assay will become a standard part of discovery research screening.

## 10.4.2 Structural Aspects

It's been recognized for a long time that certain structural features can predispose a compound to mutagenicity. Think "mutagen," and the words "nitroso" and "polyaromatic hydrocarbons (PAHs)" will likely come to mind. In the mid 1980s Dr. John Ashby introduced "alerts" to structural features associated with carcinogens.[93] More recently a nice structural analysis of over 4000 compounds for which Ames test data were available was

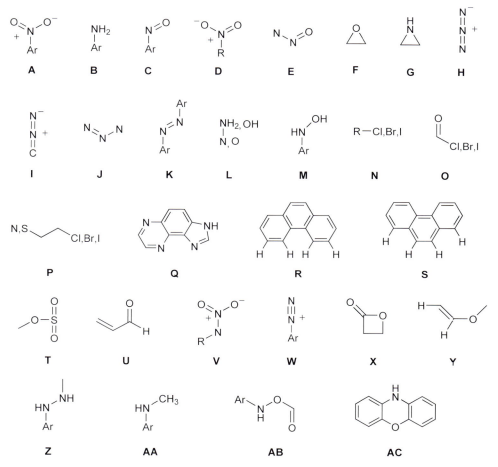

**Figure 10.25 ▶**  Twenty-nine specific toxicophores for mutagenicity as identified by Kazius et al. (Kazius, J., et al. Derivation and validation of toxicophores for mutagenicity prediction. *J. Med. Chem.* **2005**, *48*, 312–320.) (A) Specific aromatic nitro, (B) specific aromatic amine, (C) aromatic nitroso, (D) alkyl nitrite, (E) nitrosamine, (F) epoxide, (G) aziridine, (H) azide, (I) diazo, (J) triazene, (K) aromatic azo, (L) unsubstituted heteroatom-bonded heteroatom, (M) aryl hydroxylamine, (N) alkyl halide, (O) acyl halide, (P) *N*- or *S*-mustard, (Q) polycyclic aromatics, (R) bay-region, (S) K-region, (T) sulphonate-bonded *C*, (U) unsaturated aldehyde, (V) alkyl *N*-nitro, (W) diazonium, (X) β-propiolactone, (Y) unsubstituted α,β unsaturated alkoxy, (Z) 1-aryl-2-monoalkyl hydrazine, (AA) aromatic methylamine, (AB) aryl hydroxylamine ester, and (AC) polycyclic planar system.

carried out at Organon (now part of Schering-Plough).[94] A set of 29 common toxicophores for mutagenicity was identified, and these are shown in Figure 10.25.

Inspection of these toxicophores reveals a number of obvious alkylating and acylating groups along with planar, tricyclic ring systems that seem likely to intercalate with DNA. Many of the aryl rings to which nitrogen in its various oxidation states is attached probably share a common mechanism of mutagenicity as shown in Figure 10.26. Arylamines,

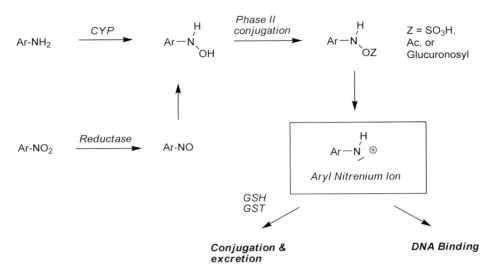

**Figure 10.26 ▶**  A common mechanism by which arenes substituted with nitrogen at various oxidation states can become mutagens. This scheme is highly simplified—other fates might also await the various arenes shown.

nitroarenes, and nitrosoarenes can all potentially end up in an intermediate oxidation state as aryl hydroxylamines. These are known to induce mutagenicity by conjugation with acetate, sulfate, or perhaps the glucuronosyl group followed by decomposition to the aryl nitrenium ion which, if not quenched and excreted, could bind to DNA.

Equally interesting is the fact that arene substitution could sometimes reduce mutagenicity if electron withdrawing groups like trifluoromethyl, sulfonamide, sulfonyl, and sulfonic acid groups were used. Other reports of reductions in mutagenicity caused by substituting N-arenes with bulky alkyl groups either at sterically hindering *ortho* positions or even at quite a distant location have been published.[95] Remember too that as metabolites rather than parent compounds are often involved in mutagenicity, the most effective way of reducing mutagenicity for compounds positive in the Ames test only in the presence of S9 may be to once again reduce the compound's affinity for the activating DME (e.g. reduce lipophilicity). Box 10.6 provides a summary of these methods as well as other key points about mutagenicity.

Another recent study employed pairwise comparisons of compounds tested for mutagenicity and was able to rank different ring systems for their likelihood of being found in mutagens.[96] These investigators were able to produce a useful table suggesting isosteric ring replacements worth trying should mutagenicity become a problem in a project. In addition, they point out that "rings identified in public domain mutagenicity data sets do not reflect the types of scaffolds in drugs and those typically used in medicinal chemistry," findings which "bring into question the utility of predictive models that were derived from public domain data sets."[96]

As previously mentioned, commercial software like DEREK (Deductive Estimation of Risk from Existing Knowledge), MCASE (Multiple Computer Automated Structure Evaluation), and TOPKAT (Toxicity Prediction by Komputer Assisted Technology) are often used

---

**Box 10.6    Key Points About Mutagenicity**

▶ The mutagenicity of a compound *or its metabolites* can contribute to carcinogenesis.

▶ Mutagens can act through

　　Covalent binding to DNA

　　Non-covalent intercalation between adjacent DNA base pairs

　　Generation of other species that interact with DNA

▶ The *Salmonella*/microsome mutagenicity assay (Ames test) is usually used to predict mutagenicity

▶ Structural features frequently found in mutagens include

　　Alkylating or acylating moieties

　　Arenes bonded to nitrogen (arylamines, nitroarenes, etc.)

　　Flat, fused tricyclic ring systems

▶ Mutagenicity can sometimes be reduced by

　　Blocking metabolism to electrophilic intermediates

　　Eliminating known toxicophores

　　Adding electron-withdrawing or bulky substituents to aryl rings

　　Reducing CYP affinity

---

to try and predict genotoxicity.[97] In-house systems may employ such software to provide structure alerts to chemists the same way that ROF violations are automatically pointed out. These programs, of course, can be applied to other types of toxicity too.[98] Although they may add value where project chemists have little familiarity with toxicophores like the ones described above, a recent review concluded that current programs could only correctly identify 43–52% of Ames positive compounds and were particularly likely to miss non-covalent interactions with DNA.[99]

In general, the sparsity of data on mutagenicity owing to the low-throughput nature of testing up until now shows that mutagenicity prediction is still at the early part of its learning curve. Chemists may be wary of over-reliance on a method whose odds of correct prediction at present approximate that of a coin toss. They may also wonder at times whether broadly applied mini Ames testing of discovery stage compounds tells one more about STRs (structure–toxicity relationships) or ITRs (impurity–toxicity relationships). These are issues that will be grappled with in coming years as the best way of incorporating genotoxicity considerations in early-stage drug discovery projects gets worked out. But the need to do so is there, and eventually it will.

# Notes

1. One study attributed up to 2.8% of hospital admissions to DDIs: Jankel, C.A., Fitterman, L.K. Epidemiology of drug-drug interactions as a cause of hospital admissions. *Drug Saf.* **1993**, *9*, 51–59. Another found potential DDIs among 47% of patients admitted to emergency rooms who were either over 50 and taking two or more medications or simply taking three or more drugs: Goldberg, R.M., et al. Drug-drug and drug-disease interactions in the ER: Analysis of a high-risk population. *Am. J. Emerg. Med.* **1996**, *14*, 447–450.

2. Greenblatt, D.J. Absorption interactions to improve bioavailability. Presentation 1279, program and abstracts of the 41st Interscience Conference on Antimicrobial Agents and Chemotherapy, Chicago, Illinois, December 16–19, 2001.

3. Schmassmann-Suhijar, D., et al. Rhabdomyolysis due to interaction of simvastatin and mibefradil *Lancet* **1998**, *351*, 1929–1930.

4. See Bruno, R.D., Njar, V.C.O. Targeting cytochrome P450 enzymes: A new approach in anti-cancer drug development. *Bioorg. Med. Chem.* **2007**, *15*, 5047–5060.

5. Granville, D.J., Gottlieb, R.A. Cytochromes P450 and ischemic heart injury: Potential role for inhibitors in the treatment of myocardial infarction. *Drug Discovery Today Technol.* **2005**, *2*, 123–127.

6. Gerntholtz, T., et al. The use of a cyclosporine-ketoconazole combination: Making renal transplantation affordable in developing countries. *Eur. J. Clin. Pharmacol.* **2004**, *60*, 143–148.

7. See www.avibio.com/devNeugene.html.

8. Fontana, E. Cytochrome P450 enzymes mechanism based inhibitors: Common sub-structures and reactivity. *Curr. Drug Metab* **2005**, *6*, 413–454.

9. Although, once again, current thinking is that more than this might be required to initiate an immune response.

10. Sham, H., et al. ABT-378, a Highly potent inhibitor of the human immunodeficiency virus protease. *Antimicrob. Agents Chemother.* **1998**, *42*, 3218–3224. Note that part of this 'boosting' effect is due to another one of ritonavir's properties, P-gp inhibition.

11. Ritonavir has both a reversible and an irreversible component to its CYP inhibition. See Ernest II, C.S., et al. Mechanism-based inactivation of CYP3A by HIV protease inhibitors. *J. Pharmacol. Exp. Ther.* **2005**, *312*, 583–591.

12. And a P-gp inhibitor to boot!

13. Ekins, S., et al. *In vitro* and pharmacophore insights into CYP3A enzymes. *Trends Pharmacol. Sci.* **2003**, *24*, 161–166.

14. Korzekwa, K.R., et al. Evaluation of atypical cytochrome P450 kinetics with two-substrate models: Evidence that multiple substrates can simultaneously bind to cytochrome P450. *Biochemistry* **1998**, *37*, 4137–4147.

15. See Atkins, W.M. Implications of the allosteric kinetics of cytochrome P450s. *Drug Discovery Today* **2004**, *9*, 478–484.

16. Ekroos, M., Sjögren, T. Structural basis for ligand promiscuity in Cytochrome P450 3A4. *Proc. Nat. Acad. Sci. U.S.A.* **2006**, *103*, 13682–13687.

17. Shou, M. The impact of Cytochrome P450 allosterism on pharmacokinetic and drug-drug interactions. *Drug Discovery Today* **2004**, *9*, 636–637.

18. Gao, F., et al. Optimizing higher throughput methods to assess drug-drug interactions for CYP1A2, CYP2C9, CYP2D6, rCYP2D6, and CYP3A4 in vitro using a single point IC50. *J. Biomol. Screen.* **2002**, *7*, 373–382.

19. Lin, T., et al. *In vitro* assessment of cytochrome P450 inhibition: Strategies for increasing LC/MS-based assay throughput using a one-point $IC_{50}$ method and multiplexing high-performance liquid chromatography. *J. Pharm. Sci.* **2007**, *96*, 2485–2493.

20. Delaporte, E., et al. The potential for CYP2D6 inhibition screening using a novel scintillation proximity assay-based approach. *J. Biomol. Screen.* **2001**, *6*, 225–231.

21. Breimer, D.D., Schellens, J.H.M. A 'cocktail' strategy to assess *in vivo* oxidative drug metabolism in humans. *Trends Pharm. Sci.* **1990**, *11*, 223–225.

22. Turpeinen, M., et al. Multiple P450 substrates in a single run: rapid and comprehensive in vitro interaction assay. *Eur. J. Pharm. Sci.* **2005**, *24*, 123–132.

23. Smith, D., et al. Analytical approaches to determine Cytochrome P450 inhibitory potential of new chemical entities in drug discovery. *J. Chromatogr. B Analyt. Technol. Biomed. Life Sci.* **2007**, *850*, 455–463.

24. Weaver, R., et al. Cytochrome P450 inhibition using recombinant proteins and mass spectrometry/multiple reaction monitoring technology in a cassette incubation. *Drug Metab. Dispos.* **2003**, *31*, 955–966.

25. Di, L., et al. Comparison of cytochrome P450 inhibition assays for drug discovery using human liver microsomes with LC-MS, rhCYP450 isozymes with fluorescence, and double cocktail with LC-MS. *Int. J. Pharm.* **2007**, *335*, 1–11.

26. Cohen, L.E., et al. In vitro drug interactions of cytochrome P450: An evaluation of fluorogenic to conventional substrates. *Drug Metab. Dispos.* **2003**, *31*, 1005–1015.

27. Bapiro, T.E. Application of higher throughput screening (HTS) inhibition assays to evaluate the interaction of antiparasitic drugs with cytochrome P450s. *Drug Metab. Dispos.* **2001**, *29*, 30–35.

28. Obach, S.R., et al. In vitro cytochrome P450 inhibition data and the prediction of drug-drug interactions: Qualitative relationships, quantitative predictions, and the rank-order approach. *Clin. Pharmacol. Ther.* **2005**, *78*, 582–592.

29. Von Moltke, L.L., et al. Midazolam hydroxylation by human liver microsomes in vitro: Inhibition by fluoxetine, norfluoxetine, and by azole antifungal agents. *J. Clin. Pharmacol.* **1996**, *36*, 783–791.

30. See Lin, J.H. Applications and limitations of interspecies scaling and *in vitro* extrapolation in pharmacokinetics. *Drug Metab. Dispos.* **1998**, *26*, 1202–1212, and references therein.

31. Riley, R.J. The potential pharmacological and toxicological impact of P450 screening. *Curr. Opin. Drug Discovery Dev.* **2001**, *4*, 45–54.

32. For a proposed mechanism of inhibition of a terminal olefin to CYPs, see Ortiz de Montellano, P.R., et al. Destruction of cytochrome P-450 by 2-isopropyl-4-penteneamide and methyl 2-isopropyl-4-pentenoate: Mass spectromentric characterization of prosthetic heme adducts and nonparticipation of epoxide metabolites. *Arch. Biochem. Biophys.* **1979**, *197*, 524–533.

33. Wen, X., et al. Isoniazid is a mechanism-based inhibitor of cytochrome $P_{450}$ 1A2, 2A6, 2C19, and 3A4 isoforms in human liver microsomes. *Eur. J. Clin. Pharmacol.* **2002**, *57*, 799–804.

34. See Box 2 in Celanire, S., et al. Histamine $H_3$ receptor antagonists reach out for the clinic. *Drug Discovery Today* **2005**, *10*, 1613–1627.

35. Pillot, C., et al. Ciproxifan, a histamine $H_3$-receptor antagonist/inverse agonist, potentiates neurochemical and behavioral effects of haloperidol in the rat. *J. Neurosci.* **2002**, *22*, 7272–7280.

36. Zhang, M., et al. Lack of cataleptogenic potentiation with non-imidazole $H_3$ receptor antagonists reveals potential drug-drug interactions between imidazole-based $H_3$ receptor antagonists and antipsychotic drugs. *Brain Res.* **2005**, *1045*, 142–149.

37. Velaparthi, U., et al. Imidazole moiety replacements in the 3-(1*H*-benzo[d]imidazole-2-yl)pyridin-2(1*H*)-one inhibitors of insulin-like growth factor receptor-1 (IGF-1R) to improve cytochrome P450 profile. *Bioorg. Med. Chem. Lett.* **2007**, *17*, 3072–3076.

38. Chiba, M., et al. P450 Interaction with HIV protease inhibitors: Relationship between metabolic stability, inhibitory potency, and P450 binding spectra. *Drug Metab. Dispos.* **2001**, *29*, 1–3.

39. See Schenkman, J.B., et al. Substrate interactions with cytochrome P-450. *Pharmacol. Ther.* **1981**, *12*, 43–71.

40. Racha, J.K., et al. Mechanism-based inactivation of human cytochrome P450 1A2 by furafylline: Detection of a 1:1 adduct to protein and evidence for the formation of a novel imidazomethide intermediate. *Biochemistry* **1998**, *37*, 7407–7419.

41. Le Bourdonnec, B., et al. Arylacetamide κ opioid receptor agonists with reduced cytochrome P450 2D6 inhibitory activity. *Bioorg. Med. Chem. Lett.* **2005**, *15*, 2647–2652.

42. Ethanol is believed to induce CYP2E1 by stabilizing the protein, for example.

43. For more about hepatotoxic reactive metabolites, see Park, B.K., et al. The role of metabolic activation in drug-induced hepatotoxicity *Annu. Rev. Pharmacol. Toxicol.* **2005**, *45*, 177–202. For more about their possible roles in hypersensitivity, see Naisbitt, D.J., et al. Reactive metabolites and their role in drug reactions. *Curr. Opin. Allergy Clin. Immunol.* **2001**, *1*, 317–325.

44. See Sikka, R., et al. Bench to bedside: Pharmacogenomics, adverse drug interactions, and the cytochrome P450 system. *Acad. Emerg. Med.* **2005**, *12*, 1227–1235 and references therein.

45. Guengerich, F.P. Comparisons of catalytic selectivity of cytochrome *P*450 subfamily enzymes from different species. *Chem Biol. Interact.* **1997**, *106*, 161–182.

46. Gonzalez, F.J., Yu, A-M. Cytochrome P450 and xenobiotic receptor humanized mice. *Annu. Rev. Pharmacol. Toxicol.* **2006**, *46*, 41–64.

47. Tateno, C., et al. Near completely humanized liver in mice shows human-type metabolic responses to drugs. *Am. J. Pathol.* **2004**, *165*, 901–912.

48. Katoh, M., et al. In vivo induction of cytochrome P450 3A4 by rifabutin in chimeric mice with humanized liver. *Xenobiotica* **2005**, *35*, 863–875.

49. Dickins, M. Induction of cytochrome P450. *Curr. Top. Med. Chem.* **2004**, *4*, 1745–1766.

50. Mohutsky, M.A., et al. The use of a substrate cassette strategy to improve the capacity and throughput of cytochrome P450 induction studies in human hepatocytes. *Drug Metab. Dispos.* **2005**, *33*, 920–923.

51. Wright, A.T., Cravatt, B.F. Chemical proteomic probes for profiling cytochrome P450 activities and drug interactions in vivo. *Chem. Biol.* **2007**, *14*, 1043–1051.

52. He, W., et al. Potent quinoxaline-based inhibitors of PDGF receptor tyrosine kinase activity. Part 2: the synthesis and biological activities of RPR127963, an orally bioavailable inhibitor. *Bioorg. Med. Chem. Lett.* **2003**, *13*, 3097–3100.

53. Smith, D.A. Induction and drug development. *Eur. J. Pharm. Sci.* **2000**, *11*, 185–189.

54. Loi, C.-M., et al. Clinical pharmacokinetics of troglitazone. *Clin. Pharmacokinet.* **1999**, *37*, 91–104.

55. Kaplan, W.D., Trout III, W.E. The behavior of four neurological mutants of drosophila. *Genetics* **1969**, *61*, 399–409.

56. See Swain, C. Open access and medicinal chemistry. *Chem. Cent. J.* **2007**, *1*, 2–3.

57. See Yap, Y.G., Camm, A.J. Drug induced QT prolongation and torsades de pointes. *Heart* **2003**, *89*, 1363–1372.

58. Lindquist, M., Edwards, R. Risk of non-sedating antihistamines. *Lancet* **1997**, *349*, 1322.

59. Fermini, B., Fossa, A.A. The impact of drug-induced QT interval prolongation on drug discovery and development. *Nat. Rev. Drug Discovery* **2003**, *2*, 439–447.

60. Redfern, W.S., et al. Relationships between preclinical cardiac electrophysiology, clinical QT interval prolongation and torsade de pointes for a broad range of drugs: Evidence for a provisional safety margin in drug development. *Cardiovas. Res.* **2003**, *58*, 32–45.

61. Whitebread, S., et al. *In vitro* safety pharmacology profiling: an essential tool for successful drug development. *Drug Discovery Today* **2005**, *10*, 1421–1433. The authors do note, however, that this data set may be biased by the selective submission for testing of compounds and analogs known or anticipated to be hERG binders.

62. Hoffmann, P., Warner, B. Are hERG channel inhibition and QT interval prolongation all there is in drug-induced torsadogenesis? A review of emerging trends. *J. Pharmacol. Toxicol. Methods* **2006**, *53*, 87–105.

63. Wood, C., et al. Patch clamping by the numbers. *Drug Discovery Today* **2004**, *9*, 434–441.

64. See *Patch-Clamp Analysis: Advanced Techniques*, Walz, W., Boulton, A.A., Baker, G.B., Eds. (Totowa, NJ, Humana Press, 2002). See also Behrends, J.C., Fertig, N. Planar Patch Clamping, available at www.nanion.de/pdf/PlanarPatchClamping.pdf.

65. Terstappen, G.C. Functional analysis of native and recombinant ion channels using a high-capacity nonradioactive rubidium efflux assay. *Anal. Biochem.* **1999**, *272*, 149–155.

66. From a Cerep data sheet. See www.cerep.fr/Cerep/Users/pages/Downloads/Documents/Marketing/Pharmacology%20&%20ADME/Application%20notes/cardiactoxicity.pdf.

67. Mitcheson, J.S., et al. A structural basis for drug-induced long QT syndrome. *Proc. Nat. Acad. Sci. U.S.A.* **2000**, *97*, 12329–12333.

68. Jiang, Y., et al. X-ray structure of a voltage-dependent K+ channel. *Nature* **2003**, *423*, 33–41.

69. Pearlstein, R.A., et al. Characterization of HERG potassium channel inhibition using CoMSiA 3D QSAR and homology modeling approaches. *Bioorg. Med. Chem. Lett.* **2003**, *13*, 1829–1835.

70. Cavalli, A., et al. Toward a pharmacophore for drugs inducing the long QT syndrome: Insights from a CoMFA study of HERG $K^+$ channel blockers. *J. Med. Chem.* **2002**, *45*, 3844–3853.

71. Farid, R., et al. New insights about HERG blockade obtained from protein modeling, potential energy mapping, and docking studies. *Bioorg. Med. Chem.* **2006**, *14*, 3160–3173.

72. See, for example, Cavalli, Reference 71; Aronov, A.M. Predictive *in silico* modeling for hERG channel blockers. *Drug Discovery Today* **2005**, *10*, 149–155; Aronov, A.M., Goldman, B.B. A model for identifying HERK $K^+$ channel blockers. *Bioorg. Med. Chem.* **2004**, *12*, 2307–2315; Johnson, S.R., et al. Estimation of hERG inhibition of drug candidates using multivariate property and pharmacophore SAR. *Bioorg. Med. Chem.* **2007**, *15*, 6182–6192.

73. Jamieson, C., et al. Medicinal chemistry of hERG optimizations: Highlights and hang-ups. *J. Med. Chem.* **2006**, *49*, 5029–5046.

74. Bell, I.M., et al. 3-aminopyrrolidinone farnesyltransferase inhibitors: Design of macrocyclic compounds with improved pharmacokinetics and excellent cell potency. *J. Med. Chem.* **2002**, *45*, 2388–2409.

75. Johnson, et al., Note 72.

76. Dinges, J., et al. 1,4-dihydroindeno[1,2-*c*]pyrazoles with acetylenic side chains as novel and potent multi-targeted receptor tyrosine kinase inhibitors with low affinity for the hERG ion channel. *J. Med. Chem.* **2007**, *50*, 2011–2029.

77. Zhu, B.-Y., et al. Inhibitory effect of carboxylic acid group on hERG binding. *Bioorg. Med. Chem. Lett.* **2006**, *16*, 5507–5512.

78. Blum, C.A., et al. Design, synthesis, and biological evaluation of substituted 2-cyclohexyl-4-phenyl-1*H*-imidazoles: Potent and selective neuropeptide Y Y5-receptor antagonists. *J. Med. Chem.* **2004**, 2318–2325.

79. For some specific examples of this, see Jamieson, Note 73.

80. Price, D.A., et al. Overcoming HERG affinity in the discovery of the CCR5 antagonist maraviroc. *Bioorg. Med. Chem. Lett.* **2006**, *16*, 4633–4637.

81. There are also *promoters* like phorbol esters that seem to act through increased cell proliferation.

82. See Singer, B., Kuśmierek, J.T. Chemical mutagenesis. *Annu. Rev. Biochem.* **1982**, *52*, 655–693.

83. Ferguson, L.R., Denny, W.A. Genotoxicity of non-covalent interactions: DNA intercalators. *Mutat. Res.* **2007**, *623*, 14–23.

84. Feig, D.I., et al. Reactive oxygen species in tumorigenesis. *Cancer Res.* **1994**, *54*, 1890s–1894s.

85. Ames, B.N., et al. Methods for detecting carcinogens and mutagens with the salmonella/mammalian-microsome mutagenicity test. *Mutat. Res.* **1975**, *31*, 347–364.

86. For an overview and procedures for running the Ames test, see Mortelmans, K., Zeiger, E. The Ames *Salmonella*/microsome mutagenicity assay. *Mutat. Res.* **2000**, *455*, 29–60.

87. See www.cerep.fr/cerep/users/pages/downloads/Documents/Marketing/Pharmacology%20&%20ADME/Application%20notes/Genetictoxicity.pdf.

88. See www.midwestbioresearch.com/ames.htm.

89. See www.aniara.com/xenometrix.htm.

90. Interestingly, a Google search reveals that both "AMES II" and "Micro-AMES" also refer to Northrop Grumman's Electronic Warfare Threat Environment Simulator for military use. See www.dsd.es.northrop-grumman.com/DSD-Brochures/amherst/Micro_AMES.pdf.

91. See Fenech, M. The *in vitro* micronucleus technique. *Mutat. Res.* **2000**, *455*, 81–95.

92. Hastwell, P.W., et al. High-specificity and high-sensitivity genotoxicity assessment in a human cell line: Validation of the GreenScreen HC *GADD45a-GFP* genotoxicity assay. *Mutat. Res.* **2006**, *607*, 160–175. See also www.gentronix.co.uk/.

93. Ashby, J. Fundamental structural alerts to potential carcinogenicity or noncarcinogenicity. *Environ. Mutagen.* **1985**, *7*, 919–921.

94. Kazius, J., et al. Derivation and validation of toxicophores for mutagenicity prediction. *J. Med. Chem.* **2005**, *48*, 312–320.

95. See Klein, M., et al. From mutagenic to non-mutagenic nitroarenes: Effect of bulky alkyl substituents on the mutagenic activity of 4-nitrobiphenyl in *Salmonella typhimurium*. Part I. Substituents ortho to the nitro group and in 2′-position. *Mutat. Res.* **2000**, *467*, 55–68, Klein, M., et al. From mutagenic to non-mutagenic nitroarenes: Effect of bulky alkyl substituents on the mutagenic activity of 4-nitrobiphenyl in *Salmonella typhimurium*. Part II. Substituents far away from the nitro group *Mutat. Res.* **2000**, *467*, 69–82, Glende, C., et al. Transformation of mutagenic aromatic amines into non-mutagenic species by alkyl substituents. Part I. Alkylation *ortho* to the amino function. *Mutat. Res.* **2001**, *498*, 19–37, and, Glende, C., et al. Transformation of mutagenic aromatic amines into non-mutagenic species by alkyl substituents. Part II. Alkylation far away from the amino function. *Mutat. Res.* **2002**, *515*, 15–38.

96. Kho, R., et al. Ring systems in mutagenicity databases. *J. Med. Chem.* **2005**, *48*, 6671–6678.

97. See Greene, N. Computer systems for the prediction of toxicity: An update. *Adv. Drug Delivery Rev.* **2002**, *54*, 417–431.

98. See Blagg, J. Structure-activity relationships for *In vitro* and *In vivo* toxicity. *Annu. Rep. Med. Chem.* **2006**, *41*, 353–368, which also provides a good overview of other types of toxicophores and toxicity considerations.

99. Snyder, R.D., Smith, M.D. Computational prediction of genotoxicity: room for improvement. *Drug Discovery Today BIOSILICO* **2005**, *10*, 1119–1124.

**Chapter 11**

# A Career in Drug Discovery Research

## 11.1  Hiring: A Good Match

Hiring someone involves an expensive commitment. Each full-time employee (FTE) in the US now costs a company somewhere in the neighborhood of $300,000 per year, including salary, benefits, and overhead. So even the biggest and most profitable companies require a lot of justification internally before a job opening is approved and candidates can be interviewed. Job openings can result either from attrition (someone leaving the company or moving to a different position within the company) or from expansion (adding a new position). If it isn't already explained to him, the savvy job candidate should ask which of the two a particular job opening represents. Either way, the commitment of nearly one-third of a million dollars a year means that the company has specific needs that the person hired will be expected to fulfill and that those needs couldn't be met in a cheaper way like adding them on to the responsibilities of current employees, contracting the work out, or hiring a "temp."

The bottom line is that hiring doesn't just "happen" to keep the HR folks busy. Every position represents a strong internal demand to fill a specific need. As a prospective employee, you will be evaluated depending on how well you fill that need. So it's in the best interest of both employer and job candidate that the right person is chosen for the position. To understand what makes a good match, we need to look at both sides of the coin and examine the needs and expectations of both parties.

### 11.1.1  What Do Employers Want?

#### 11.1.1.1  The Candidate Selection Process

Ultimately, how good a given job candidate might be can only be determined by months or years of experience working at the company, but running through a group of candidates this way obviously isn't possible. At times, for entry-level "research associate" or "research assistant" positions companies will hire "temps" and give them offers later on if they're happy with the temps' performance. This can amount to a sort of trial period, but usually a firmer commitment is required from the start, so making the right decision is important. Short of being able to "try them out," then, selecting the right candidates to bring in involves taking shortcuts: narrowing down the list by degree level, previous experience, publications, and possibly recommendations. Keep in mind, though, that once hired, even regular full-time employees are usually subject to an initial 6-month probationary period during which time it's much easier for the employer to terminate them.

Initial selection of candidates, then, will be based mostly on the information provided by a candidate's resume, possibly supplemented by checking with references. When the candidate list has been sufficiently narrowed down, candidate interviews are used to determine whom to hire.

#### 11.1.1.1.1  The Resume

Employers first need someone with the technical skills required to do the job properly. Although a fundamental requirement, it isn't all that easy to judge, particularly when a position is advertised and a foot-high stack of resumes comes in. Largely as a result of this, degree requirements and sometimes the number of years of previous industry experience are usually used to make the first "cut." Often a starting "research scientist" level job will require a Ph.D. degree and a resume coming in to the HR department won't be considered if the applicant doesn't have this. One could argue that this isn't necessarily fair. Gifted researchers like Russell Marker, whose efforts resulted in the first commercially viable route to steroids (and who invented the octane system for gasoline to boot),[1] and Nobel laureate Gertrude Elion, whose important work on antimetabolites was briefly mentioned in Chapter 2, lacked Ph.D. degrees, except for the honorary ones conferred years later. But at least for job candidates just coming out of school, having a Ph.D. degree means having taken more courses, passed additional exams, and done a number of years of usually grueling, semi-independent laboratory research that candidates without this degree will probably lack. By the same logic, in a tight job market postdoctoral experience might be required as well. Sometimes "scientist-level" job ads will say something like "Ph.D. and 0–3 years experience or MS degree and 6–10 years experience," to acknowledge the fact that one might learn in industry as well as academia—something that should be obvious to this book's readers by now.

*A scientist applying for a job will almost always have a better chance of at least getting her* resume *to the right person by forwarding it to a contact within the company, if known,* rather than, or, even better, in addition to sending it to HR. Resumes sent to HR can be lost or perhaps discarded via sorting by keyword, for example. Web forms can seemingly evaporate into cyberspace. Neither obliges the recipient to even acknowledge that the resume has been received. Sending your resume to someone you know at the company in addition to the other route is just more likely to succeed in getting the information to the hiring manager, along with a good word about your abilities, and ensures that it's at least been delivered. Many companies award cash bonuses to employees who bring in new hires, thereby giving them additional motivation to pass such resumes along. Unfortunately, though, going by this route is tough for new job candidates who need it most but haven't yet built up a network of contacts.

The subject of resume preparation has been done to death by numerous books, websites, and professional services. In fact, most resumes today are very well-written, so only a couple of points need be made here. Resumes of candidates who aren't native English speakers need to be critically examined by people who are before being sent out. Since the standard is set so high these days errors in grammar and spelling, not all of which word processing programs will catch, can stand out and sow the seeds of the idea that the candidate might have a communication problem.

Second, many scientists include along with the resume a summary of research they've done. *Having a well-prepared research summary is a great idea*: after all, you're being considered for a research job. Such a summary can showcase the work you've done beyond that talked about at the interview seminar, and this can sometimes provide a "hook" if the company happens to be interested in one of those areas. But keep it brief. Don't turn it into a dissertation.

### 11.1.1.1.2 Recommendations

Recommendations deserve a few comments. First of all, not everyone knows that it's often considered unethical and a breach of company policy to solicit recommendations from people not listed as references by the candidate. Lots of times, reviewing a resume turns up a university or company where the hiring manager has contacts, and naturally he'll want to check with them and see what they think of that candidate. But the candidate may need to give at least verbal permission first before such a recommendation can be requested. Check with HR: they'll know the rules. In addition, calling anyone at the candidate's current employer for a recommendation, unless explicitly authorized to by the candidate, is a definite no-no for obvious reasons.

Second, giving a recommendation for anyone involves another legal issue as well. In our litigious society it's conceivable that one could be sued for giving an unjustifiably bad *or* an unjustifiably good recommendation. Aside from possible legal repercussions, the reputation of the person giving the recommendation is on the line too if it later appears that he'd told a "whopper." Companies often instruct their own employees to give out neither references nor information on former employees, but to turn such inquiries over to HR, which will probably only provide dates of employments, job titles, and salary levels instead. Having a current supervisor give out a good recommendation for a valued employee now pursuing another job for whatever reason is far from unknown, but these are very much confidential agreements between the two, unsanctioned by official company policy. Legal considerations have also helped make the telephone call, which leaves no written record, the favored format for recommendations. Sometimes, though, formal letters of recommendation are still solicited.

Finally, few candidates will list a reference unlikely to give at least a "very good" recommendation, so the range one sees begins there and runs to "outstanding" (from B to A+). The skill of the interviewer (usually the hiring manager) in asking the right questions of the reference about qualities he or she *hasn't* brought up can make a big difference in the final impression conveyed.

### 11.1.1.1.3 The Interview

When the list of candidates has been narrowed down by degree, experience, and possibly recommendation, a number of them might be interviewed. The interview process might just involve a phone call to set it up and a single half-day or all-day visit, or it could use a more extensive phone interview as an initial prescreen and then one or two onsite-interview visits as the candidate list gets narrowed down. It all depends on the company, the job market, and the particular situation.

Most often, the onsite interview will consist of one-on-one talks with the hiring manager (who's usually the potential supervisor), other members of the group, representatives of other groups and other departments within the company, someone from HR, and perhaps an executive or two.[2] Sometimes group sessions are held, which can be a little intimidating for the candidate. And, hopefully, a barrage of questions won't prevent the proper consumption and digestion of lunch. The kind of questions asked can range from general ones about previous experience and career goals to specific details of reaction mechanisms, for example, for prospective medicinal chemists. Candidates with industrial experience will always be asked why they left their previous company or are considering leaving their current one. Hating your job or having a tyrant for a boss, even if true, isn't a good answer here, so you may want to rephrase it to reflect your receptiveness to new and better opportunities. Finally, the candidate should keep in mind that an interview isn't a one-way street, but instead represents an opportunity to ask questions and gather information. Some of the ones that may be worth asking are listed later on.

An interview seminar, which often proves pivotal in the hiring decision, will probably be requested for those having previous published research, whether academic or industrial. This may not apply in cases where scientists with B.S. degrees interview for entry-level positions or for those with industry experience in more applied fields like formulations or analytical method development where research publications are few and almost everything they've done is proprietary. The candidate should keep in mind that not only will the contents of the seminar matter, but the way it's presented and how questions are handled can be equally important.

Finally, the job candidate owes it to himself to become familiar with the company's products and/or research pipeline before interviewing there. This information is now so readily available on the web that not checking it out can make interviewers wonder whether you're really serious about wanting to work there. Imagine applying for a postdoc with a professor you don't know and haven't bothered to check out! Better still, do an online search of the company's publications and patents. This is best done even before sending your resume, as it can help you decide whether it's the kind of place you'd want to work at. If possible, try and remember a name or two from the publications, or just bring a copy. Some of the authors or inventors might be on your interview schedule. *Candidates familiar with the company's situation and research are just more likely to be taken seriously by interviewers.*

### 11.1.1.2 Selection Criteria

Box 11.1 shows one breakdown of the major qualities that employers would like to see in a drug discovery scientist. Core scientific knowledge, critical thinking, scientific creativity, and the ability to use good practical judgment and solve problems are important contributors to being able to do the job. These qualities, along with all of the others listed, are really what interviewers will be trying to judge while interviewing a candidate and listening to his seminar. But by themselves, qualities which speak of competency won't help, if they can't be properly combined with those of coworkers, and this involves communications and teamwork skills.

Being able to communicate well is critical. According to Dr. Michael Varney, Vice President of Small Molecule Drug Discovery at Genentech, "An enormous amount of whether they [candidates] get hired or not will rest on their ability to communicate effectively during the interview process, which usually includes giving a seminar. I would encourage them to practice extensively before they go out and interview."[3] Communications are easily impeded when rivalries crop up, personalities clash, or hidden agendas become involved, so candidates who seem like they may be prone to these things will not be selected.

---

**Box 11.1    Key Qualities that Employers Look for in Candidates**

▶ Competence

    Scientific knowledge

    Critical thinking

    Creativity

    Problem solving abilities

    Good judgment

▶ Teamwork and communication skills

    Ability to integrate well with project team, management, and reports

    Ability to thrive in a cross-disciplinary environment

    Ability to work under pressure

▶ Motivation

    Ability to get the job done without excessive prodding

---

This book has repeated over and over just how much drug research is a cross-disciplinary effort these days. Although universities have begun to take steps to have this reflected in their curricula,[4] most job candidates just out of school have concentrated on a single discipline. Medicinal chemistry is traditionally populated by those having strong synthetic organic chemistry background. Some big companies used to (and may still) require that any prospective candidate for a scientist-level position have completed the total syntheses of one or more complex natural products. This kind of candidate would certainly be capable of facing the challenges involved in industrial synthetic routes. But without additional capabilities would he or she necessarily be the best person for the position? Consider the bullet points shown in Box 11.2, which represent some desirable qualities for medicinal chemists that have been suggested by industry sources.

The same reference suggests that universities should educate students desiring to become medicinal chemists in more than a dozen fields besides organic chemistry. These include molecular biology, biochemistry, pharmacology, pharmacokinetics, physiology, toxicology, and metabolism. This list is more than a decade old, and these days more subjects, like pharmacogenomics, could well be added. In the real world, no candidate comes in knowing about all the various disciplines involved, so the ability to learn some important aspects of each discipline on the job—for the most part independently and without formal coursework, a kind of "guerilla education"—becomes critical in contributing to the project and the company. The candidate who walks in the door believing that his education is over is in for a surprise!

Although one can't know everything, it's very advantageous to have at least some knowledge of the non-chemistry issues involved in drug discovery, some of which this book has

---

**Box 11.2  Desirable Qualities for Medicinal Chemists as Suggested to IUPAC by Various Industry Representatives.**

▶ Ability to fit into multidisciplinary teams and interface with diverse biological scientists

▶ Ability to search for novel molecules

▶ Knowledge of how to synthesize molecules suitable for focused biological testing

▶ Understanding of reasons for making compounds

▶ Understanding of drug design

▶ Insight in SAR with insufficient data

▶ Knowledge in collateral fields (pathophysiology, cell biology, genetics)

▶ Familiarity with new technologies

From Ganellin, C.R. et al. Educating medicinal chemists. *Annu. Rep. Med. Chem.* **1995**, *30*, 329–338.

---

hopefully conveyed. *Candidates, especially those without previous drug discovery experience, who can demonstrate some basic knowledge about subjects like bioavailability and the drug discovery process have a great advantage over those who can't* because they've proven themselves willing and able to learn some necessary, non-core subjects.

Two other factors that employers normally look for in job candidates are the ability to tolerates stress, which one is liable to get plenty of with modern short deadlines and high expectations, and, especially, self-motivation. The difference between a motivated and an unmotivated employee in any industry at any level is a quantum one. It's actually better to leave a position unfilled than to hire an individual lacking motivation.[5] Candidates most likely to be hired will be the ones who seem to enjoy their work and give the impression that they'll *get the job done*, the ultimate industry accolade.

## 11.1.2  What Should a Candidate Look For?

### 11.1.2.1  The Company

Market conditions (available candidates versus available openings), not to mention talent, experience, and luck, will dictate how much of a choice of potential employers the candidate will really have. Still, it's always a good idea to picture what an ideal situation would be as a benchmark to compare real opportunities with. Industrial research on small molecule therapeutics can take place in big pharma or at smaller companies, which may be called "biotechs" whether or not they also try to develop biologics. The alternative term, "small pharma" might also describe them, as we've seen.

How many job openings will there be at each? Unfortunately, statistics on the number of chemists employed in drug discovery research at both types of organizations are elusive.

A survey of American Chemical Society (ACS) members in 2004 showed that roughly equal number of member chemists working in industry worked for large employers and small employers, but that wasn't broken down by type of industry or job function. Government employment statistics also don't offer the desired breakdown. It's easy to gather one's own statistics on publication output (not job numbers) by going through a few journals and patent searches and tallying things up. A quick search through the most recent issues of *The Journal of Medicinal Chemistry* and *Bioorganic and Medicinal Chemistry Letters* showed that of papers originating in industry (roughly 40% of the total), about 70% came from big pharma.[6] Similarly, in one very informal search, about 62% of recently published US patent applications listing assignees (not all of them do) and involving three different current therapeutic target categories went to big pharma companies, with biotechs or smaller companies being responsible for the rest.[7] Again, though, these don't necessarily translate into job numbers.

It's quite possible that employment at a biotech or small pharma company will have a different "feel" to it compared with a job in big pharma. Pride in entrepreneurship, an emphasis on early-stage projects and ideas, focused but limited resources, and the need to get rapid results to ensure continued funding are frequent attributes of the former. Pride in having marketed drugs, more extensive resources, sales that support the organization, and a more bureaucratic structure are natural facets of the latter. Job stability is unfortunately not an attribute of either these days. Long ago a scientist coming out of school could get an entry-level job with a company, work for 40 years while concentrating only on his field of specialization and moving steadily up the ladder, then retire with a nice pension. That rarely happens anymore. Today's new job candidate can expect to have a number of different employers over the course of a career,[8] perhaps even having to change careers at some point.[9] While somewhat unsettling, this effect isn't restricted to drug discovery scientists—it's a fact of modern economic life.

Like statistics on employment of chemists in drug discovery research, detailed statistics on job tenure and the likelihood of layoffs among drug discovery scientists aren't readily available. Global outsourcing of jobs affects domestic employment in both pharma and biotech. Failure of funding, often due to dropped collaborations, clinical trial failures, or just bad market conditions often cause biotech companies to "let employees go." Even success can be the reason: companies with limited resources have been known to deemphasize "R" and push "D" when good results begin appearing from the clinic. But big pharma is far from immune to layoffs and has gone through a lot of these recently (see Figure 11.1) owing to its failure to meet shareholder expectations and the necessity for cutting costs. "Just a decade ago, the drug industry looked like a safe bet for scientists seeking long-term job security. Now the tables have turned, as fewer companies are delivering the blockbusters their size demands."[10] A number of different sources suggest that the prospective job candidate now owes it to himself not only to do due diligence on the products and research publications of the prospective employer, but even to judge the prospects of its research pipeline and the abilities of its executives. This seems a lot to expect, and a cynic might argue that the candidate who could really do all those things may be squandering his true talent and forgoing greater rewards as a market analyst or venture capitalist. But when it comes to employment the stakes are high, and it's only prudent to check out all available information.

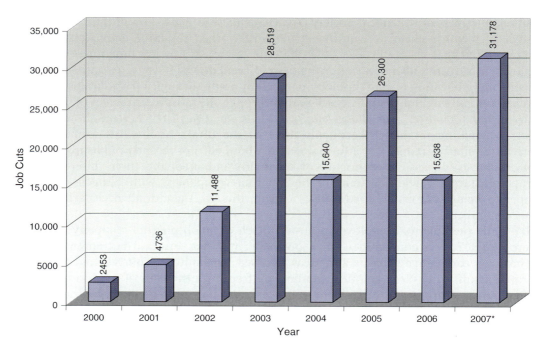

**Figure 11.1 ▶** Job cuts in pharma during 2000–2007. Cuts in total jobs, not just R&D, are shown. *Data through December 7, 2007. Data from Challenger, Gray & Christmas, Inc. reports.

Things may seem particularly dicey when interviewing for a position at a small biotech startup company. Industry consultant Dr. Cynthia Robbins-Roth has some advice in this case. "If the company is still private and you're looking at accepting a job there, you need to know how much money they have in the bank right now and how long is that going to keep them afloat before they have to go and raise more funding. And how much more money are they going to have to raise?"

"It's very easy," she says, "to ask those questions. And they may hem and haw, and you say 'Look, you're asking me to join your organization. I understand that it's a risky business so I need to understand how this works.' The other thing you can ask is what benchmark or milestone is the company going to have to reach to be able to go out and raise that money. If you have to have something in the clinic to be able to raise more money, then you've got to use your own knowledge to see if you think that's probable or possible."

The management factor, she advises, is also worth checking out for small companies. "When you're interviewing, it's very easy to look up the bios of all the people in management. Look up their backgrounds. Is this somebody who's right out of academia and this is his first company? That means something: it means they're practicing on you. If they're scientists and they're running science, maybe that's not a big deal. But if they're in charge of regulatory affairs I think it's a problem. Who's the chief financial officer? Is it somebody who's been at other biotech companies and who's been involved in many other fund raisings? If they have, the probability is pretty good that they know how to do it, they've got a network, and they'll be able to get it done.

"Business development is probably the most crucial job in some of these companies. Is the VP of business development somebody who's done it before or is this his first deal? Is it someone from the science side of the company who's trying to do it just because they needed somebody and, what the heck, this person wanted to do it? Do your homework. And then talk to people at different levels within the company that you're contemplating joining. Ask them how they experience the corporate culture. What's it like to them? What are the kinds of things that are encouraged and discouraged?"

Aside from financing, management, and corporate culture issues another factor candidates may want to consider is, as in real estate, location. Pharma and biotech companies aren't dispersed uniformly throughout the United States, but are heavily concentrated in a just a few areas. Surveys show that the top three biotech locations are currently the San Francisco Bay Area, Boston/Cambridge, and San Diego. Quite a few big pharma companies do research in northern New Jersey, but several also have locations in San Diego and one (Roche Palo Alto), in the San Francisco area. Companies are, of course, located elsewhere too (e.g. Lilly in Indianapolis and Abbott just north of Chicago), and states like Florida and North Carolina are actively pursuing biotech companies all the time; but for the most part California, Massachusetts, and New Jersey are the states where drug discovery jobs are most likely to be found.

Location is important not only from the point of view of personal preference but it can matter years down the road as well. Getting a job at, say, a startup biotech company in Des Moines, Iowa might be fine, but if layoffs follow or you decide you want another job a few years later, there probably won't be many local places for you to apply to. To get a new job, you may well have to move (most likely to New Jersey, California, or Massachusetts). If you're 30 years old and single this might not bother you; it might even be exciting. But if you're 50 and have a family, it's something quite different. So you may want to consider proximity to other potential jobs as a factor in your calculations when hunting for one, particularly if you have a spouse working in the industry too, which doubles the stakes.

Nothing makes a bigger difference to what a job's going to be like than the supervisor involved. It's far easier to work for a good supervisor at a flawed organization than for a flawed supervisor at a good one. There's a lot of communication and guidance (not to mention evaluation) involved in a good reporting relationship, and if mutual respect is lacking, the results can only be disastrous. As mentioned before, "people skills" courses aren't generally requirements for a management position in science regardless of the number of reports. Companies often give supervisors special in-house classes to try and improve their managerial abilities, but these might be too little too late. So managing employees end up being perhaps the most important examples of on-the-job learning. Like any other skill, some are more adept at it than others, and *probably the most important information a job candidate can gather is who his supervisor would be and what he or she would be like to work for.*

This isn't always possible. Sometimes a job opens up before it's been decided which of the several positions would be filled. You, the candidate, might even be asked whether you have a preference among a couple of potential supervisors you've spoken with. Sometimes hiring plans can change, and you might come in working for someone other than the person you expected to report to. But always ask in any case. When interviewing with a potential supervisor, always try to get a feeling of what that person would be like to work for. Is he a good listener? Has he had a lot of supervisory experience before? What's he looking for in the people who report to him? Members of his group can be asked what it's like in their experience to work for him. You won't always get a totally frank answer—group members may fear the repercussions of less-than-positive remarks that might find their way back to the supervisor—but it's a place to start.

One of the hardest things to do in a job interview is to get a sense of the amount of work and the hours the new hire will be expected to put in. It's important to know whether or not graduate school-like hours will apply so that not showing up every Saturday will be seen as shirking responsibility.[11] But it's a very hard question to ask without arousing suspicion on the part of interviewers that the candidate's goal is to slide by with a minimal effort. Rumor has it that a sort of "drive-by approach" can be used instead. Here you just drive by the company, if you can, on a Saturday afternoon or at 8 p.m. on a weeknight and see how many cars there are in the parking lot! Of course, employees can be motivated to spend weekends and evenings at work either because they enjoy their jobs and want to or because they have to, or often some combination of both. You won't be able to tell this by counting cars, but at least you'll have some idea of what to expect. Or so the rumor goes.

### 11.1.2.2 Compensation and Benefits

Salaries in drug discovery research vary widely with such factors as degree level, scientific field experience, size of the company, and geographical region. The best and most current source of information on how all these factors affect the amount of money a chemist can expect to make is the annual American Chemical Society (ACS) salary survey, results of which are published in the *Chemical & Engineering News*. Every candidate should check this out before interviewing. Some pertinent data from the most current one (2006) are summarized in Table 11.1.

Location becomes important in salary too, as some areas like the Pacific coast (e.g. California) on average offer higher paying jobs than others like the "east south central" states (e.g. Alabama). But note that one needs to consider local cost-of-living factors: a 10%

**TABLE 11.1 ▶ Median Annual Salaries of 2006 Chemistry Graduates with Full-Time Employment in Industry and Less than 1 Year of Technical Work Experience Prior to Graduation, and Median Base Salaries for all Chemists with Full-Time Permanent Jobs in Industry as of March 1, 2006**

| Degree | Starting salary | Median base salary |
|--------|-----------------|--------------------|
| B.S.   | 35              | 67                 |
| M.S.   | 52.7            | 82                 |
| Ph.D.  | 75              | 105.5              |

(Both are in thousands of dollars. Data are not restricted to pharma/biotech employment. Data from the ACS employment and salary survey 2006.[12–13])

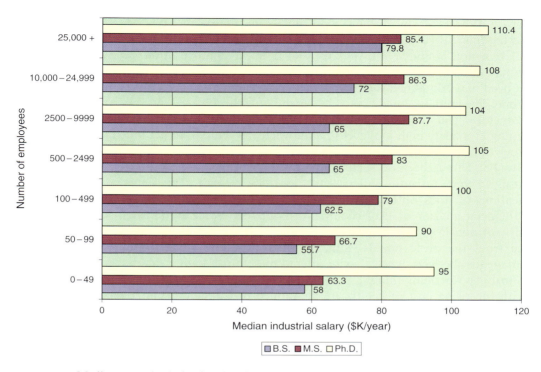

**Figure 11.2** ► Median annual salaries for chemists employed full-time at companies of different sizes as of March 1, 2006. Data (not only for pharma and biotech) is from ACS employment and salary survey 2006. (Heylin, M. Employment and salary survey. *Chem. Eng. News*, September 18, **2007**, pp. 42–51.)

higher offer in California won't mean much if it costs twice as much to live there. Avoid unpleasant surprises. *Candidates considering jobs in locations they're not familiar with may want to check out local housings costs, either rent or purchase prices, before accepting an offer.*

As you can see from Figure 11.2, when it comes to salary, size matters, particularly at the B.S. and M.S. degree levels. Large companies tend to pay more. There are probably many factors behind this, like employee demographics. Bigger companies may be more likely to have employees with more years of experience who therefore draw bigger salaries. But part of it probably is owing to their ability to pay those salaries. Regardless of the reasons behind these or other statistics, the statistics ultimately don't mean much; the candidate will need to evaluate the specific offers he gets. The statistics just add a bit of perspective.

Can a candidate expect to negotiate a higher salary than the one initially offered? Maybe or maybe not. It depends on how much the company really wants the candidate and how much the candidate really wants the job. It works at times, but keep in mind that if you turn down an offer for any reason, the company is not obliged to make you another. Another tricky issue comes up earlier on when the candidate is asked about his/her salary expectations. Name too low a price, and you're liable to get it to your everlasting regret—what you make in years to come depends largely on this starting point. But too high a figure might price you out of a job. Try and avoid answering the question and you'll come across as "evasive." Whatever you decide to do, go to the interview with a prepared strategy for answering this difficult question.

Salary is not, of course, the only form of financial compensation. Many companies offer an annual bonus program, so the candidate should check into this. If one is offered, you might need to work there a year before becoming eligible or perhaps get a partial amount the first time. The amount you do get may depend on your job level, performance appraisal, and goals met by your department and/or the company. Often there's a separate program beginning at a fairly high job level that allows for executives to get much larger amounts.

Stock options are still sometimes offered as well, although they're not as prevalent as they used to be. This is because such grants must, by US law, now be treated as a company expense, whereas up until a few years ago they weren't. The financial possibilities and implications of stock options lie far beyond the scope of this book. You may need to talk to a financial advisor or tax accountant about this. Stock options can be worth a lot of money—or not. Poor decisions about stock options can cost you money, so it's best to seek advice from the professionals. Still, an offer including stock options can definitely be advantageous. Be leery when they're offered in lieu of a higher salary, though. They won't necessarily be worth anything.

With the rising cost of healthcare mentioned in Chapter 1, employer-sponsored health insurance is a very valuable benefit these days in the US. Anyone who's had to pay entirely for his own coverage before will understand this. Companies no longer routinely pay the full amount of the health insurance, so some employee contribution for himself and/or his family may need to come out of the paycheck. Most companies will offer the candidate a choice of plans. Don't assume that because you work in a health-related field your own health insurance will somehow be better or more comprehensive than anyone else's. It just isn't so. Even physicians can end up with less-than-desirable health insurance plans these days.

Company sponsored pension plans, as mentioned before, are a dying breed these days. Should your employer still offer one, consider it a very major benefit and cross your fingers that it lasts. A 401k plan is far more likely to be offered. Here the employee contributes a percentage of his/her salary, some of which may or may not be matched by the employer (find out!), and the whole amount is invested in a special retirement fund. Some very small companies may not offer even this.

Other perks worth weighing are vacation time (often just 10 days per year to begin with in the United States) and tuition reimbursement for college courses, usually restricted to those related to the job or to an MBA degree. Many employers reimburse their employees for professional membership (e.g. ACS) dues, sometimes even paying for a journal subscription too. The company's policy on professional meeting attendance is worth finding out as well. Most will pay for attendance at some selected meetings every year or two, the choice of which is often worked out between the employee and her supervisor, and that may be considered part of one's professional development. Sometimes employees are required to give a talk or present a poster before they're allowed to go, especially if it's an expensive meeting and/or at an exotic locale. The amount of money one's permitted to spend often depends on grade level, as you might guess.

All of these things, from salary to meeting reimbursements, are meant to make for a motivated, productive workforce. A variety of less traditional means of reward are sometimes tried as well. "As an example, most chemists have a genuine interest in science and frequently conduct personal projects to which managers turn a blind eye. Reward for achieving the project goal could, therefore, include time and/or money to pursue the chemist's own research interests; in other words, a form of sabbatical."[14] Being allowed a certain percentage of work time to carry out this kind of "free research" was mentioned back in Chapter 4 as an enlightened policy one can find at some companies. True industrial sabbaticals, although rare, have been known to exist too.[15] Here either long service at a company or perhaps a special award allows the researcher to take 6 months or a year off to recharge one's batteries, write reviews, or learn new skills through courses or a stint with an academic group.

### 11.1.2.3 Some Questions to Ask

▶ Can you tell me whom I'd be working for, and in what group?

▶ Can you tell me anything about the project or the sort of research I might work on?

▶ How many people work in the group and how big is the department?

▶ Is this a new job opening, or a replacement position?

▶ What's the company's current financial standing? (Especially for small ones)

▶ What do you look for in an employee?

▶ What's it like working here?

▶ Does the company encourage its employees to learn on the job and take on new responsibilities?

▶ Is there a lot of work on new therapeutic targets going on at the company?

▶ Who decides on what compounds to make?

▶ Where do new projects originate? Does the company solicit new ideas for projects and compounds?

▶ Do you have consultants? Would I meet with them regularly?

▶ Is there a lot of interaction between chemistry and other departments, like cell biology and pharmacology? Do chemists and biologists get together much outside of project meetings?

▶ How does the project team system work here? Does everyone go to the meetings?

▶ Are employees encouraged to publish their research (after legal approval, of course)?

▶ Are researchers allowed any time to work on their own ideas, even if the ideas are not necessarily project-related?

▶ Can you explain a little about how the evaluation system works here?

▶ If I'm hired, what would my job title be?

▶ What's the next level above that, and what qualities would the company want to see me demonstrate before I could get there?

You'll notice that questions about salary, benefits, and perks have been left out. Most candidates will have no trouble remembering to ask about these, nor will corporate HR people forget to mention them. Keep in mind that interviews are delicate situations. Coming across as being passive and having no questions to ask isn't good, but neither is a perceivable attempt to turn the interview around and avoid answering questions by aggressively badgering the interviewer instead. One wants to make friends here, not enemies. Avoid at all cost giving the impression of being too good for the job, adopting an attitude of superiority—it's the best way of ensuring that you'll get no offer. Try to keep the interview to the level of a friendly conversation between two parties respectfully coming together to decide whether or not they should work together. That's exactly what it is.

## 11.2  Assessing Performance

### 11.2.1  Evaluations

Once hired, how will the new industrial scientist be evaluated? It turns out that evaluating performance in research is notoriously difficult. As with anything else, an extremely bad or an extremely good performance is easy to see. The hard part is evaluating the vast majority of them that fall between the two. In a desperate attempt to arrive at some logical metrics, various measures have at times been proposed that might strike the reader as somewhere between odd and pathetic. The number of trays of dirty glassware removed from each chemist's bench has been used to measure productivity. Of course, as word of this leaked out scores of perfectly clean flasks, especially big ones, found their way into the wash. The number of compounds a chemist registers is, of course, a classic target for metrics. Anyone who's ever synthesized a compound, though, knows that some are orders of magnitude harder to make than others. Legend has it that to compensate for this, one creative manager assumed that, on average, the bigger the molecule, the harder it probably is to make. Therefore, each compound was assigned one point for every 100 Daltons and each chemist's annual point output was tallied up. Needless to say, an incredible number of trityl- and *t*-butyldiphenylsilyl protected intermediates were registered thereafter, in a sort of anti-Lipinski selection process.

The point is that there is no foolproof, totally objective way of measuring research productivity or comparing research contributions. But scientists need to know that real effort will really be rewarded, and consequently some sort of system needs to be in place to ensure that evaluations aren't arbitrary and unmotivating. This is often done by breaking them down into two parts. In one, the skills demonstrated by the researcher (scientific knowledge, problem solving abilities, communicating with other team members, ability to keep good notebook records, etc.) are evaluated over the course of the previous period (usually a year). This is ultimately determined by the line supervisor although input from other sources (his supervisor and/or the project team leader) might be called upon too. In the other part, how well the researcher has met specific goals established at the start of the evaluation period is assessed. Keep in mind that probably no two companies have exactly the same way of doing evaluations, but in many cases this sort of two-part system is what one will find.

Whole volumes could be written about setting goals for drug discovery researchers, and some very different ways of doing it can be found. Goals usually begin at the top and then percolate down through the organization. The company may have a corporate goal of, say, bringing an oral small molecule NCE that's an anti-TNFα agent to the clinic within the next 3 years, which leads to a current project goal of identifying a compound that meets certain criteria, say, efficacy at a specific level in a certain animal model. Responsibility for meeting the project goal will then be divvied up among the researchers working on it. Although the goals upon which the researchers will be evaluated need to be designed with the aim of reaching the project goal and ultimately the corporate goal, if these higher level goals aren't met, everyone down the line can't be given a zero. Research, after all, is research, and no one knows in advance whether it will work out. A way of distinguishing between those who put their heart and soul into trying to achieve the project goal and did not only what they were supposed to do, but even more, and those who put little effort into trying needs to be found.

The way that goals for medicinal chemists will be drawn up depends very much on the company. Some sort of quantitation, say, the number of analogs made by the chemist within a given timeframe, will generally be used. Even this, by the way, is difficult to do. At the beginning of the year it's hard to imagine what sort of compounds one might be making 6 or 10 months down the road, much less how difficult they'll be to make. For this reason, there's usually some flexibility involved so that if the direction of the project changes, the goals can be changed too. Unfortunately, the researcher will really know what to expect in an evaluation only after going through the exercise at least once at his/her company. Non-science goals, like effectively working as part of the team or the ubiquitous "career development" goal, will often be included as well.

An issue that always comes up is where to draw the line between an ambitious goal (often known as a "stretch goal") and an absolutely unachievable one. Even though partial credit is given and effort rewarded, having too many unachievable goals can ultimately lead to despair, whereas having too many unambitious ones leads to complacency. The bottom line is that balance is needed, and in fact the goal-setting, and the overall evaluation process, isn't an easy one. Many managers and HR people consider it one of the most difficult parts of their jobs, consuming vast amounts of time and effort, especially for managers of large groups.

As previously noted, line managers are ultimately responsible for evaluating the people who report to them, but input from other sources, especially project leaders in the matrix management system, is important as well. The line manager's supervisor and supervisors up to the departmental VP are likely to be involved too. This is a result of the need to keep evaluations and the resulting raises fair and to equalize them across the whole department and the whole company. The smaller the company, the less this is needed, of course.

When the evaluations have been approved by the "higher-ups," each employee will probably be assigned an overall performance level. There can be anywhere from three to six of these levels, with names like "superior performance," "above average performance," "good performance," and "needs improvement". The verbiage can vary and one shouldn't spend too much time trying to read between lines to see if, for example, "average" is really the same as "good" or not. Only the definitions of substandard classifications deserve a lot of scrutiny: often something like "needs improvement" means that no raise will be forthcoming, and the employee is at risk of being terminated if things don't change. Again, though, this depends on the company, and hopefully the researcher will never need to find out.

For each of these levels, a range of raises (say 7–10% for "superior performance," 5.5–8% for "above average performance," etc.) will be possible. At smaller companies things might be less formal and less rigidly defined. Larger companies will probably use a normal distribution overall and perhaps by department. Hopefully, though, they'll avoid the statistically unjustifiable folly of taking it down to the level of individual groups so that, for example, if only two people work for a given supervisor, assigning a higher-than-average rating to one means having to give a lower-than-average rating to the other. The actual amount an employee will receive within a given range will probably be determined by the supervisor after approval from his superiors and the HR department.

When the whole process is completed, it's time for the big meeting between employee and supervisor where the evaluation is gone over and the results communicated. Like a candidate interview, this isn't a one-way street. It's important for your boss to hear your views and understand your concerns. Career development discussions are usually part of these meetings too, and either here or, more often, at subsequent meetings goals for the next review period will start being drafted.

On behalf of supervisors everywhere, let me remind researchers that their bosses aren't given free rein to award ratings and raises that they think are right, but have to work within a corporate system that puts limits on the raises and often the ratings that they can give out. Generally they won't even remind you of this for fear of "passing the buck." Good companies try to ensure that employees aren't unpleasantly surprised when they get their evaluations. If potential problems in someone's performance arise, it's wrong to wait till the end of the year to bring them up. Instead, these issues need to be addressed right away. Most companies will have a formal mechanism for doing this earlier at either quarterly or midyear meetings, and supervisors will be able to bring the issues up informally as they happen. Supervisors are usually required to attend special in-house courses or meetings to ensure that performance appraisals are done properly.

If you think your performance appraisal is unfair, most companies will suggest you talk to someone in the HR department. It's rare for evaluations to be changed, but it's important that your point of view should at least be heard.

## 11.2.2   Promotions

All but the smallest companies will have established titles and *grade levels* for the jobs available in research, along with a list of qualifications, job descriptions, and probably salary ranges for each. The list will usually be available to the employees, although salary ranges are usually not revealed. The number of levels and descriptions of each can vary a lot. It's hard to put into words what makes a good researcher, and sometimes comparing job descriptions between adjacent levels can leave one not quite sure what the differences are. These can be subtle: at one level a researcher might be "*capable* of carrying out independent research," at the next one he might be "*expected* to carry out independent research," and above that he might "*routinely*" do so. Increasing independence, by the way, is one of the general hallmarks of career progression and often a requirement for promotion. Most supervisors are "tickled pink" when those reporting to them don't need to be told what to do, as long as they stay on track, get the supervisor's buy in for their plans, and communicate the results. On the other hand, though, no one should ever be afraid to ask a supervisor for advice or ask about specific compounds and priorities when needed.

Some companies have more job levels than others for researchers, and, not surprisingly, where there are a lot of close job levels, promotions to higher ones tend to come quicker, all other things being equal. Keep in mind that job titles don't have universal meanings. A "Research Scientist" at one company might have more responsibilities and a higher salary than an "Associate Director" at another. Job levels also aren't immutable; sometimes they're reorganized and redefined or new ones are added. Promotion isn't an entitlement or something like a raise or bonus that occurs at regular intervals. How long it takes to get one depends on the researcher, supervisor, and organization, and might range from one to many years. Raises usually accompany promotions, but most of the time these aren't guaranteed nor are they necessarily large: At many organizations they depend on where one stands in the existing salary range. If you're near the top, promotion to the next level might involve only a few percent raise, whereas if you're below the mean at your existing job level, you might get a much bigger raise that brings you into the new salary range. Again, there are lots of different ways companies can do it.

Above a certain level, often several levels above that of a starting Ph.D. scientist, one usually finds a branching point for separate "tracks" in science and management. This is the famous *dual ladder* system. At this point, researchers can either continue a career focused on science or, alternatively, become more involved in management and organizational responsibilities. Empirical observations over many years suggest that the latter ladder likely leads to more rapid advancement and generally to higher salaries, but it's not for everyone. In any case this isn't an issue the new researcher will need to face for a long time.

At some companies, employees are considered for promotion only once a year, perhaps at evaluation time. At others the interval can be shorter, or the process can occur at any time. It's something that never happens automatically; it has to be driven by someone, almost always the supervisor. Typically, when she sees that the researcher has been working "at the next level" for some time, she can set the mechanism in motion by requesting and filling out forms and confidentially soliciting letters of recommendation from other people within the company. This will involve other supervisors, some of them probably from different departments (another reason to sharpen you're interdisciplinary skills). The information will be put together and evaluated by a committee convened to consider the most recent list of proposed promotions. A big factor in the decision is what kind of rating the researcher has gotten in recent appraisals. It's hard to justify promoting someone who hasn't gotten above-average evaluations recently.

If the decision is to deny the promotion, you'll probably never find out about it. The whole process is usually hush-hush. If, instead, it's approved, it will be celebration time. Beware of the first year after promotion, though—great things will be expected of you now that the company has recognized your potential!

## 11.3 The Long Haul: Perspectives

*"A career takes more than talent. It takes character."*

Kitty Carlisle Hart[16]

## 11.3.1   Job and Industry Evolution

The Greek philosopher Heraclitus observed that one can never step into the same river twice, and to a large extent the same holds true for new drug research. Plunge into the industrial drug discovery at two different time points, and you'll see two different industries. Its ongoing evolution is a result of selection pressures exerted by advancing scientific knowledge and a changing regulatory and business environment. This being the case, over the course of a career the one thing the researcher is guaranteed to see is change.

Consider the items listed in Table 11.2. All of these subjects have been discussed earlier on, but collectively they make for very different environments. The researcher at home in one of them wouldn't necessarily be pleased with or prepared for the other. Although the evolution of some things, like scientific knowledge, will be universally welcomed, others, like the increasing use of global outsourcing, might not be. Even the state-of-the-art in one's core field will change, and this can require new learning. Table 11.3 lists some of the factors that make the synthetic part of a medicinal chemist's job very different from what it used to be.[17]

What the scientist entering the field today will be facing in another 30 years is anyone's guess. To get there he'll likely have to weather many a storm—things like mergers and acquisitions, possible relocations, poor job markets at times, and the ever-present new, cutting edge technology that, it's feared, will make his job obsolete. A lot will depend on his resourcefulness, tenacity, dedication, and—let's admit it—just plain good luck or lack thereof. It's a safe bet, though, that he'll need flexibility to adapt to a new environment along with the willingness to take on new challenges. As new techniques and approaches ride the waves of their expectation curves and get added to the list of the many things a good medicinal chemist must know, more "guerilla education" is in order. Stop learning—ever—and the industry will move on past you.

**TABLE 11.2 ▶   Some Differences Between Modern and "classic" Drug Discovery**

| Drug Discovery Circa 1977 | Drug Discovery Circa 2007 |
|---|---|
| Some corporate sensitivity to investor pressure | Extreme corporate sensitivity to investor pressure |
| Product safety important | Product safety absolutely critical |
| Readiness to stay with project until goal is achieved | Readiness to kill projects if problems develop |
| Mechanisms-of-action (MOAs) frequently unknown | MOAs usually known |
| Physiology-driven projects using tissue or whole animal testing as primary screen | Target-driven projects using animal testing only on selected compounds later |
| All work done in-house | Tasks outsourced whenever economically viable |
| Project champion stays with project from conception to drug approval | No project champion or one gone by development |
| Project team can be 1 chemist + 1 pharmacologist | Large project teams representing many disciplines including legal and marketing |
| Communications easy and informal | Communicaitons formalized via regular project team meetings |

**TABLE 11.3 ►   Changes in Synthetic Chemistry Over the Years**

| Synthetic Chemistry Circa 1977 | Synthetic Chemistry Circa 2007 |
|---|---|
| Manual literature searching of CA cumulative indices, dusty Beilstein volumes, etc. | Easy online searching of databases |
| Fewer commercially available starting materials, good reagents, and synthetic methods | Many more commercial starting materials, reagents, and methods |
| Gram quantities of compounds synthesized for testing | Milligram quantities of compounds synthesized for screening |
| A few compounds synthesized per group per week | Many compounds synthesized per group per week |
| Compound ID via UV, IR, 60–90 MHz $^1$H NMR | Compound ID via high field $^1$H, $^{13}$C, $^{19}$F, $^{31}$P, or 2D NMR, APCI or electrospray MS |
| Purification by column chromatography, recrystallization | Purification by automated flash chromatography, preparative HPLC |
| Purity determination by elemental analysis, mp | Purity determination by HPLC |

## 11.3.2   The Evolution of a Research Career

Summarizing what's been spoken about at length in previous chapters, Box 11.3 lists some of the roles a medicinal chemist can expect to play. When put this way, some of the terms sound a bit strange, and, of course, no one can be expected to wear all of these hats at the same time; they too change as projects and careers progress.

Knowing how to make molecules lies at the heart of the medicinal chemist's skill set. Without this knowledge, all the interdisciplinarity and communication skills in the world won't make for a good scientist. Not surprisingly, the vast majority of those entering the field were trained as synthetic organic chemists. Especially for the ones with graduate degrees, this

---

**Box 11.3   Some Roles of the Medicinal Chemist in New Drug Discovery**

► Inventor of new molecules

► Maker of new compounds for testing

► Chooser of screening compounds

► Hit assessor and prioritizer

► Scale-up service provider

► SAR and SPR development specialist

► Multiparametric compound optimizer

► Project contributor, leader, originator

► Mentor to those new to the project and/or the field

education was understandably oriented toward the development of new synthesis methods and routes to complex compounds. Coup would have been counted by the publication of articles in journals like *The Journal of the American Chemical Society*.

The transition from academia to an industrial drug discovery involves a major reorientation. A better drug, the ultimate object of your efforts, is obviously not the same as an exciting new chemistry paper. It's also not "a new molecule which injected into mice produces a paper"[18] for the biologists among us. A career progression pattern frequently seen in the industry among medicinal chemists is to start out focusing on synthesis methods and routes, and then slowly expand into broader issues. There's usually a sort of natural progression of one's scientific interest over the years. This might be represented by the questions and areas the medicinal chemist sequentially focuses on as her experience grows:

▶ What's a clever way to make these molecules? [synthesis]

▶ How do I explain observed trends in potency and selectivity? [SAR, biochemistry]

▶ What might be better molecules to make? [SBDD, etc.]

▶ Why don't they work in cells? [in vitro ADME]

▶ Why don't they work in animals? [PK/PD, pharmacology, toxicology]

▶ Why am I making these anyway? [disease pathology, medical markets]

▶ What would be a better target for the indication and how would I approach it? [new project proposal, all of the above issues]

This won't hold true for everyone. Some scientists, for example, are perfectly content to deal only with synthesis issues, as we've seen, although many of them might eventually end up in process chemistry where their expertise is valued most. The interests of most medicinal chemists, though, are likely to progress through something like this sequence at varying speeds if they stay in the field long enough. Note that career interests tend to develop in the opposite way that a company plans for a project, which begins with a medical need, proceeds through defined animal models to be run, and ultimately works its way down to the chemist making molecules at the bench (or hopefully, these days, in a fume hood).

Career development, then, usually involves the slow acquisition of a bigger picture of how new drugs are discovered and developed, a consideration of many more pieces of the puzzle. This isn't to say that the researcher will become an expert in each of the many disciplines involved. Reading a book about cardiology won't make you a cardiologist. Instead, an appreciation of the need for real experts in other disciplines, along with a realization of one's own limits comes with that broader view. Experts are usually available on project teams and getting to know a few basic facts about their disciplines allows for a common language to be spoken. The right questions can then be asked and the answers perhaps understood. Nor does curiosity about other disciplines imply that the researcher will put less emphasis on compound synthesis. A greater appreciation of just how important synthetic chemistry is to the whole process comes with a better understanding of all the hurdles a compound needs to clear.

It isn't always easy for a scientist to broaden out his expertise and explore new fields. Not only might he not want to, but in some cases he might not be encouraged to nor even, perhaps, allowed to. Particularly at large organizations, contact with researchers in other departments might be limited to infrequent meetings or seminars, and wandering around trying to talk to biologists instead of staying at the hood and making compounds might be discouraged. This kind of pigeonholing would be particularly unfortunate considering the vast amount of diverse expertise that probably exists at such companies, making it the kind of situation where the shoemaker's son goes barefoot.

On the other hand, learning outside one's core area of expertise is often expected, maybe even required, at small companies. But here there can be a different problem: there may not be any in-house experts to talk to. And regardless of company size, different supervisors can actively support this to varying degrees in yet another example of why it's important to have a good one. So those of us fortunate enough to have been given the opportunity to learn more should realize that it *is* an opportunity, not an obligation or entitlement, and thank our lucky stars. Ultimately it's good for everyone involved, the company not least.

### 11.3.3  Frustration

There is one career factor for researchers that gets little press. This is the frustration factor. The current "productivity gap," the fact that in recent years the number of approved NMEs has gone down and the costs to come up with one have gone up, apparently despite everyone's best efforts, has already been noted. This has been the focus of much concern in the industry, not to mention among stockholders. It's created a lot of gloom in pharma and led to the suspicion that "a collective depression might be bothering the pharma industry. Perhaps we are starting to lose faith in drug discovery, thus making R&D productivity decline a self-fulfilling prophecy."[19] But what's the effect of all this on researchers themselves?

Even in the best of times research is in many ways a strange and unlikely, almost masochistic occupation. The researcher spends his whole career fighting odds that are stacked against him, the gambler watching his pile of chips get dangerously low as he places his next bet, the one that he's sure will win big and vindicate them all. Both mathematical analysis and common sense dictate that since he's almost certain to lose he'd be better off not trying in the first place. Many things can go wrong in many ways in a project. The odds of them all going "right" are fairly remote, regardless of how one tries to bias them with foresight, technologies, etc.

Earlier on an estimate, and perhaps a slightly optimistic one at that, was presented that gave an average new project a 1 in 57 chance of ever leading to a new approved drug. As we've also seen, research chemistry might be involved in a project for between 2 and 5 years, assuming the project isn't terminated early. It seems a good estimate, in keeping with empirical observation over the years, that on average a research medicinal chemist will get a new project every 2 years or so. Thus, in a 40 year career he or she might expect to work on perhaps 20 projects. The odds, then, will be two out of three that, despite her best efforts, the researcher will go through her entire career and retire without ever having worked on a project that resulted in a new drug. Working on a lot of novel targets aiming for truly innovative drugs will make the odds considerably higher. Notably, high-level managers who supervise many projects increase their odds of being able to claim victory by association. They just get more tickets in this particular lottery. And these numbers only relate to *being associated with* an ultimately successful project. Keeping in mind that projects at the lead-optimization stage often employ 10 or more medicinal chemists, the chances that our retiring career chemist will have ever conceived of and/or first synthesized an NCE herself are very low, probably just a few percent.

Good researchers feel a certain ownership of the projects they work on. This is especially true for medicinal chemists, who produce the compounds upon which everything ultimately depends. In a weird sort of way, these can be almost like their children. When projects fail, *which they normally do*, it can be particularly tough on them and they may tend to view it as a sort of personal failure. Although a few papers or patents might be salvaged, after several years' commitment to a project, researchers reflecting on its termination might cynically wonder whether it would have made any real difference in the grand scheme of things if they'd spent that time doing nothing but twiddling their thumbs instead, a particularly depressing thought. After all, the construction worker can point to buildings and say that he helped put them up and even a salesman can claim a role in the profits his company has made. These things are visible and measurable. But at the end of the day, short of a drug, what can a drug discovery researcher really claim as his own contribution to his business or to the world?

Repeated disappointments eventually exact a toll on most researchers.[20] At least in one anecdotal case the resulting effect was noticed by outsiders, management trainers, who found that "drug discoverers were surprisingly sceptical about their jobs." They "seemed to think nothing would ever come out and nobody could do anything about it, whereas the trainers expected drug hunting to be one of the most rewarding jobs in the world."[19] Clearly, ways of coping with the odds and dispelling this kind of gloom are needed.

Stubborn persistence and endurance are prerequisites. One needs to keep knocking at doors that aren't opened, and sometimes even break them down. A focus on doing the best science that one can combined with an understanding that many of the factors involved are beyond the scientist's control can help as well. The best cure for disappointment in the wake of a scientific idea that fails is to have so many new and exciting ones to try next that there's no time to mourn the passing of the old one.[21] There's one more thing that can cut through the gloom and keep the researcher in the drug discovery business for the long haul too.

### 11.3.4   Hope

*"There really aren't any problems that are more complicated, or more satisfying from a human perspective, than producing novel medicines."*[22]

Dr Mark Murcko, Chief Technology Officer, Vertex Pharmaceuticals

Some people end up in scientific research by long design, others find themselves there almost by chance. Those in drug discovery research might be there for a whole variety of reasons. Curiosity about nature and the desire to expand our understanding of it is one. Most of us have a genuine love for science that makes drug discovery research a particularly attractive career. This being the real world, the desire to make a decent living and support ourselves and our families via a profession we're suited for will be involved as well. As we've seen above, salaries for drug discovery researchers are generally reasonable, but—my apologies to those who may be disappointed—unlikely to make you rich. Over many years, though, I've noticed another factor that's usually involved in choosing and sticking with a job in drug discovery research. I wouldn't say it's the primary reason we're willing to commit to such a potentially frustrating occupation, but without some degree of it we're unlikely to stay in it for long.

Most of us have seen what diseases can do. We may have a friend or relative who's been diagnosed with cancer, disabled by heart disease, or crippled by rheumatoid arthritis. We might know someone with diabetes who needs to forgo holiday treats, measure blood sugar levels, and inject insulin every day or someone who suffers from a psychological disorder that makes for lifelong difficulties. We may have even been forced to deal with serious illness ourselves. One quality that biotech and pharma researchers tend to have is empathy, and the thought that something we can do in our jobs might end up helping real patients goes a long way toward getting through the frustration and the many dry spells.

Although we might personally not end up being the one who made a new drug, collectively our efforts are working. New drugs play a significant role in increasing life expectancy, which in the United States has risen by a full 7 years in the last 34 years.[23] In some cases, the results of new drugs being available has been even more direct and dramatic. Many of us will remember the days when being diagnosed with HIV was essentially a death sentence. The first approved AIDS drug, retrovir (AZT), did much to change this. The availability of subsequent drugs has also played an important role: in the United States, death rates from HIV fell from 16.2 to 4.5 per 100,000 between 1995 and 2004.[24]

Cancer treatment continues to improve as new and better drugs are introduced. Figure 11.3 shows improvements in 5-year survival rates for the four most common types of cancer between

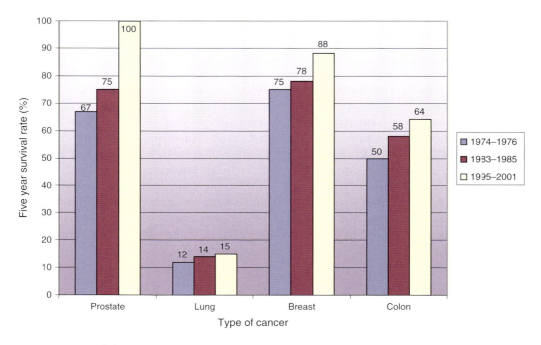

**Figure 11.3** ▶ Trends in 5-year relative survival rates (adjusted for normal life expectancy) for the four most common types of cancer. Trends between the first and last time period are statistically significant in all cases. (Source: *Cancer Facts & Figures 2006*, American Cancer Society.)

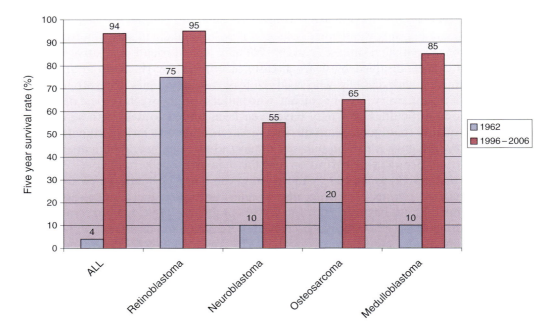

**Figure 11.4** ▶   Trends in survival rates of five or more years for patients with five types of childhood cancer. (Data from St. Jude Children's Research Hospital, ALSAC, 2006.)

1974 and 2001. Patients, of course, are not statistics, but individuals, and a lot depends on what stage the particular cancer was diagnosed in, underscoring the importance of early detection. The current state-of-the-art is far from perfect, but the ongoing positive trends in survival are unmistakable. Figure 11.4 shows that progress in some childhood cancers has been particularly pronounced in recent decades, and here again the availability of newer drugs plays a prime role.

Other signs of hope in large part owing to the availability of better drugs, are all around. Patients now live for years and even decades after organ transplants, something formerly not possible because of rejection. Death rates due to heart disease are now less than half of what they were in 1950, and the death rate due to stroke is down by almost three-quarters in that same time period.[25] Drugs like beta blockers, statins, and tPA are major contributors to these improvements.

Deaths are prevented, and formerly fatal diseases become treatable conditions instead. Not only reduced mortality, but quality of life improvements and economic benefits to society go along with new drugs. In the end, being part of the worldwide team of research scientists who face down the odds every day in their quest for new medicines is a privilege that no other occupation can offer.

## Notes

1. Marker's advisor warned him that if he didn't complete his p-chem requirement and instead left without a Ph.D. (which he did) he would end up as no more than a urine analyst!

2. The highest level executive, say a VP or head of research, will often be listed near the end of the interview schedule. Beware of last-minute cancellation of this part of the interview. It could be that something really did

come up to prevent him from seeing you, but it's also possible he somehow decided that he really doesn't want to.

3. Quoted in Arnette, R. Chemical connections. *Science* **2006**, *312*, 1067–1068.

4. One example is the new graduate program in chemical biology at McGill University in Montreal. See Silvius, J. Strength in diversity: A cross-disciplinary approach to graduate training in chemical biology. *Nat. Biotechnol.* **2007**, *25*, 255–258.

5. Hence the double *entendre* implied by the recommendation "Believe me, *nobody* could do a better job than he!"

6. These data were obtained by manually tallying up the affiliations of first corresponding authors of articles from *J. Med. Chem.* 50, No. 24 and *Bioorg. Med. Chem. Lett.* 17, No. 24. Obviously, results should be taken with a grain of NaCl, the survey not being extensive.

7. The error bars here are particularly big, though, and seem to vary a lot with the target. The most recent 50–100 U.S. patent applications listing the words "kinase," "cathepsin," and "vanilloid" in their titles as of December 5, 2007 were tallied. No attempt was made to separate out small-molecule patents from other types or to identify patents resulting from collaborations between big pharma and biotech. The publication of better representative numbers from a more detailed look at the issue would be welcomed.

8. As a guess based on previous observations and various surveys, a research employee at a drug discovery company in America will probably only stay there for, on average, 3–4 years, which can eventually make for some lengthy *resumés*!

9. Those scientists contemplating a career change, or unsure about going into discovery research to begin with, should read *Alternative Careers in Science*, Robbins-Roth, C. Ed. (San Diego, CA, Academic Press, 2005).

10. Smaglik, P. Naturejobs. *Nature* **2007**, *445*, 677.

11. These kinds of graduate-student or postdoc hours are unusual, if not rare, in industry, though. Many companies find that the safety issues (and potential liability) involved in having an employee working alone in lab outweighs any advantages the additional work would provide.

12. Heylin, M. Class of 2006 salaries and jobs. *Chem. Eng. News*, December 3, 2007, pp. 73–77.

13. Heylin, M. Employment and salary survey. *Chem. Eng. News*, September 18, 2007, pp. 42–51.

14. McDonald, S.J.F., Smith, P. W. Lead optimization in 12 months? True confessions of a chemistry team. *Drug Discov. Today* **2001**, *6*, 947–953.

15. Wilkinson, S.L. On sabbatical: A refreshing pause. *Chem. Eng. News*, April 30, 2001, pp. 22–25.

16. Quoted in Berger, M. Kitty Carlisle Hart, Actress and Arts Advocate, Dies at 96. *New York Times*, April 18, 2007.

17. An outstanding review of industry changes and their effects on the job of a medicinal chemist can be found in Lombardino, J.G., Lowe III, J.A. The role of the medicinal chemist in drug discovery—then and now. *Nat. Rev. Drug Discov.* **2004**, *3*, 853–862.

18. Janssen, P.A. Drug research. *Rev. Med. Brux*. **1980**, *1*, 643–645.

19. Uitdehaag, J.C.M. Where is the optimism? Warrior teachings to regain the drug discovery spirit. *Drug Discov. Today* **2006**, *12*, 105–107.

20. And that even before going home and reading about the latest unsafe prescription drug in the papers or talking to well-meaning acquaintances who just can't understand why drugs should cost so much!

21. My thanks to Dr Todd Somers for introducing me to this philosophy.

22. Dr Mark Murcko, quoted in Clough, J. Mark Murcko discusses enabling technologies in drug discovery. *Drug Discov. Today Biosilico* **2004**, *2*, 51–54.

23. Life expectancy at birth for someone born in 1970 was 70.8 years while for someone born in 2004 life expectancy was estimated to be 77.8 years. Source: *Health, United States, 2007*, National Center for Health Statistics.

24. Source: *Health, United States, 2007*, National Center for Health Statistics.

25. The death rate for heart disease was 586.8 in 1950 and 217.0 in 2004. For stroke, the corresponding numbers are 180.7 and 50.0. Source: *Health, United States, 2007*, National Center for Health Statistics.

# Best-Selling Small Molecule Single Agent Prescription Drugs
## "The Periodic Table of Drugs 2006"

Mendeleev set out to arrange the known elements by their chemical properties as an aid to teaching chemistry students and ended up with something much more profound. Unlike the elements, prescription drugs and their sales are anything but constant and immutable, and no sublime truths should be sought in their juxtaposition. But grouping popular prescription medications by therapy and target can make similarities (especially pharmacophores) and differences readily apparent, give the reader a feeling for what kinds of structural features are acceptable, and correlate therapeutic areas, targets, and drugs at a glance.

To this end, 2006 sales figures for top-selling prescription drugs as listed in the "Med Ad News 200 World's Best-Selling Medicines"[1] were sorted into classes, therapeutic uses, and drug targets. Biologics, large molecules, and combination drugs (more than one API) were

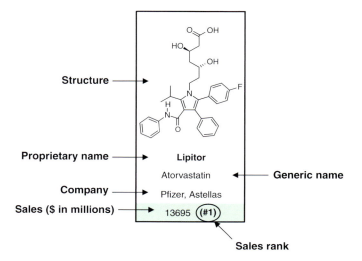

**Figure 12.1** ▶ Key to data fields presented in 'The Periodic Table of Drugs'.

*The Periodic Table of Drugs 2006[a]*

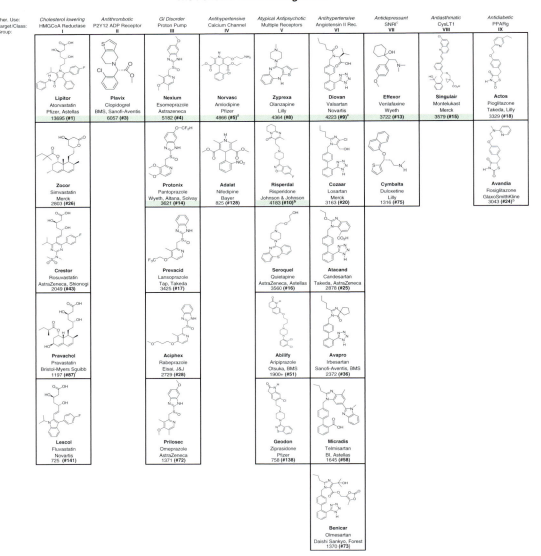

**Figure 12.2** ▶  The best-selling small molecule single agent prescription drugs of 2006.

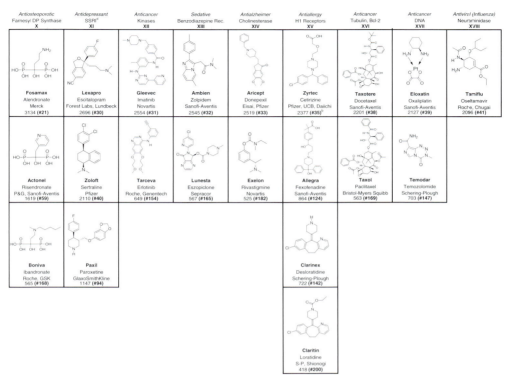

[a]Updates available at www.realworlddrugdiscovery.com

[b]Sales figures include combinations with other APIs.

[c]Serotonin Norepinephrine Reuptake Inhibitor

[d]Selective Serotonin Reuptake Inhibitor

[e]Sales of Mevalotin (Daiichi Sankyo, $799M) not included

[f]Sales of Amlodin (Sumitomo, $506M) not included

Note: Salts have been omitted from structures.

Source: Names, companies and sales data are from Humphreys, A. "MedAd News 200 World's Best-Selling Medicines", *MedAd News* Vol. 26, No. 7, July 2007, pp. 14–40

Abbreviations: "BI" = Boehringer-Ingelheim, "BMS" = Bristol-Myers Squibb, "GSK" = GlaxoSmithKline, "J&J" = Johnson & Johnson, "P&G" = Procter & Gamble, "S-P" = Schering-Plough

**Figure 12.2** ▶ *(Continued)*

eliminated, which explains why neither the second best-selling drug, Advair, nor the sixth, Remicade, for example, is included. The remaining drugs with the highest 2006 sales figures were arranged into columns so that atorvastatin, the #1 bestseller, is at the top of Group I and below it are listed other top selling statins (simvastatin, rosuvastatin, etc.) in order of sales. Still other statins, notably the now-generic lovastatin, were not included because they didn't make the top 200 sales list. Group II consists of the next best-selling single agent, small molecule prescription drug, clopidogrel, the only drug in its class to make the list. Remaining columns are then headed up by the next best-selling drugs from left to right, out to an arbitrary total of 18 columns. Since this number is enough for the real periodic table, it would be tough to justify using more!

Columns are headed up by therapeutic use (e.g. cholesterol lowering for the statins) and "target or class" (HMG-CoA reductase for the statins). Each drug entry shows structure, proprietary name,[2] non-proprietary name, the companies selling the drug, sales figures for 2006 in millions of $US, and sales rank in the top 200 list. A green tab in the sales figure indicates that the drug is one of the top 10 bestsellers. Collectively, the 50 medicines shown accounted for over $126 billion in sales during 2006. Note that counterions and salts (e.g. calcium for atorvastatin) have been omitted in the structures.

As the fruits of drug discovery efforts reach market and gain momentum while drugs lose market share, "go generic," or perhaps are withdrawn, the constituent elements of this table will change. Updates will be posted to this book's website, www.realworlddrugdiscovery.com.

## Notes

1. Humphreys, A. MedAd News 200 World's best-selling medicines. *MedAd. News 26*, No. 7, July 2007, pp. 14–40.

2. At least, the most familiar proprietary name in the U.S.

# Index